Underbalanced Drilling

Gulf Drilling Series

Casing and Liners for Drilling and Completion

Managed Pressure Drilling

Underbalanced Drilling: Limits and Extremes

Underbalanced Drilling

Limits and Extremes

Editors

Bill Rehm
Arash Haghshenas
Amir Paknejad
Abdullah Al-Yami
Jim Hughes
Jerome Schubert

Houston, TX

Underbalanced Drilling: Limits and Extremes

Gulf Publishing Company
2 Greenway Plaza, Suite 1020
Houston, TX 77046

ISBN: 978-1-933762-050
10 9 8 7 6 5 4 3 2 1

Library of Congress Cataloging-in-Publication Data

Underbalanced drilling : limits and extremes / Bill Rehm, consultant.

p. cm. -- (Gulf drilling series)

Includes bibliographical references and index.

ISBN 1-933762-05-5 (alk. paper)

1. Underbalanced drilling (Petroleum engineering) 2. Boring. I. Rehm, Bill, 1929-

TN871.33.U525 2012

622'.3381–dc23

2012005311

Production services and design by TIPS Technical Publishing, Inc.

We dedicate this book to Dr. Leon "Doc" Robinson

The IADC-TPC committee and this series exists because of his passion for transmitting our present technology and years of experience, his ability to encourage and inspire the many authors to go forward with their efforts and his unwavering dedication to make sure the next generation of drillers are left with a bright legacy to follow.

This book was prepared under the auspices of the IADC Technical Publications Committee, but has not been reviewed or endorsed by the IADC Board of Directors. While the committee strives to include the most accurate and correct information, IADC cannot and does not warranty the material contained herein. The reviewers of this book do not represent the IADC Underbalanced Operations & Managed Pressure Drilling Committee, and the committee has not reviewed this book.

The mission of the IADC Technical Publications Committee is to publish a comprehensive, practical, and readily understandable series of peer-reviewed books on the petroleum drilling industry known as the *Gulf Drilling Series* in order to educate and guide industry personnel at all levels.

Contents

Preface xix

Where Have We Been and Where are We Going? xix
Don Hannegan, P.E., Weatherford International Ltd.

1.1 UBD's Contribution to Hydraulic Flow Modeling xx
1.2 UBD's Contribution to the Development of Rotating Control Devices xxi
1.3 From Underbalanced Drilling to Pressurized Mud Cap Drilling xxi
1.4 UBD's Contribution to MPD xxii
1.5 Combinations of Drilling Methods: Conventional Mud Drilling, MPD, UBD xxii
1.6 Going Forward—"Drill-to-the-Limit" (DTTL) Concepts xxiii
1.7 Today and Tomorrow xxvi

Underbalanced Horizontal Drilling: Could it be the Ultimate Completion Technique? xxvii
Jim Hughes, SunStone Technologies LLC

1.8 Introduction xxvii
1.9 Gaseated Fluid and Flow Drilling xxvii
1.10 Underbalanced + Horizontal Drilling = UBHD xxviii
1.11 Natural Fractures xxix
1.12 UBHD Well Construction xxx
1.13 The Drilling Fluid Decision xxxi
1.14 Trip and Complete without Killing the Well xxxii
1.15 Achieving Cost Control Equality xxxiii
1.16 Combination Drilling and Completion Rigs xxxiv

1.17 New UBHD Technology xxxiv
1.18 Conclusion xxxvii
1.19 References xxxvii

Biographies xli

1 Introduction 1
Bill Rehm, Drilling Consultant

1.1 Book Description 1
1.2 Other Options 2
1.3 Introduction to Underbalance Drilling 2
1.4 Drilling Fluid Density 2
1.5 Advantages to UBD 3
1.6 Challenges with UBD 4
1.7 IADC Definitions 5
1.8 General Underbalanced Drilling Equipment 6
1.9 What is Not in This Book 7
1.10 References 8

Section 2
Techniques Common to Underbalanced Drilling 11
Bill Rehm, Drilling Consultant

1.11 Introduction 11
1.12 Well Control in Underbalanced Drilling 11
1.13 Stripping 22
1.14 Pipe Light 27

Section 3
Lessons in Underbalanced Drilling 28
Abdullah Al-Yami, Texas A&M University

1.15 Negative Field Case 29
1.16 Williston Basin 36
1.17 Introduction 36
1.18 Challenges 37
1.19 Final Comment 38

2 Flow Drilling: Underbalance Drilling with Liquid Single-Phase Systems 39
Bill Rehm, Drilling Consultant
Arash Haghshenas, Boots & Coots

2.1 Introduction to Single-Phase Underbalance Systems 39

2.2 Advantages to Drilling Underbalanced with Single-Phase 42
2.3 Increased Drill Rate 43
2.4 Challenges and Limits to Flow Drilling 45
2.5 Flow Drilling: Drilling Underbalanced with a Single-Phase Fluid 48
2.6 Connections 52
2.7 Trips 53
2.8 Solutions and a Short Summary 55
2.9 Questions 56
2.10 References 57

Section 2
Underbalanced Drilling
Experience in the Ghawar Field 59
Mohammad Muqeem, Saudi Aramco

2.11 Introduction 59
2.12 Background 59
2.13 Planning Phase 60
2.14 Initial Wells 61
2.15 Documentation 64
2.16 Sour Gas Provisions 64
2.17 Subsequent Wells 64
2.18 Conventional versus UB Comparisons 65
2.19 Case History of Initial Challenges 66
2.20 Early Experience with (Down-Hole) Isolation Valves 69
2.21 Operational Improvements 70
2.22 Lessons Learned 71
2.23 Important Questions about the Chapter 72
2.24 References 72

Section 3
Friction Controlled Drilling, A Novel Approach to Drilling HPHT Wells Underbalanced 73
Robert L. "Bob" Cuthbertson, P.E., SunTerra Oil and Gas LP

2.25 Introductions and Background 73
2.26 Friction-Controlled Drilling 73
2.27 The Concentric Casing String Theory 74
2.28 Concentric String Operations 75
2.29 Modeling the Operation 75
2.30 Drilling Operations 76
2.31 Built in Kill String 76
2.32 Constant Circulation 76
2.33 Conclusions 77

2.34 Comment on Extreme Temperature 78

Section 4
Rheology of Single Phase Fluids 79
Arash Haghshenas, Boots & Coots

2.35 Flow Patterns 79
2.36 Reynolds Number 79
2.37 Viscosity 81
2.38 Types of Fluids 82
2.39 Pressure Loss Across the Nozzles 89
2.40 API Recommendations 90
2.41 Wellbore Pressure and Temperature Correction 90
2.42 Example 1 94
2.43 Cutting Transport 105
2.44 References 108

3 Gaseated Fluids (Gas-Liquid Mixtures) 109
Bill Rehm, Drilling Consultant
Arash Haghshenas, Boots & Coots

3.1 Introduction to Gaseated Fluids 109
3.2 Advantages and Concerns of Gaseated Systems 113
3.3 Challenges with Operating Gaseated Systems 115
3.4 Flowing Hydrostatic Pressure Prediction 118
3.5 Operations-Basic Gaseated Fluids 120
3.6 General Limits of Gas and Fluid Volumes 123
3.7 Solids Control Equipment 126
3.8 Methods of Gas Injection 126
3.9 Well Kicks (Gas, Oil, or Water Flows) 132
3.10 Operational Concerns and Challenges 134
3.11 Questions 138
3.12 Answers 139
3.13 References 142

Section 2
Using Concentric Casing with
Gaseated Systems, Principles and Examples 145
Paco Vieira, Weatherford Services, U.S., LP

3.14 UBD—Concentric Casing Gas Injection 145
3.15 First Applications 146
3.16 Options to Mitigate the Pressure Fluctuations 146
3.17 Middle East and North Africa Experience 153
3.18 References 156

Section 3
Field Cases for Gaseated UBD Systems 158
Abdullah Al-Yami, Texas A&M University

3.19 Underbalanced Drilling Long-Term Performance 158
3.20 Production Enhancement-Brunei 159
3.21 Fractured Carbonates 163
3.22 Comparison of Conventional and Underbalanced Drilling 165
3.23 UBD North Sea 167
3.24 Kuwait Fractured Dolomite 168
3.25 UBD Mexico Fractured and Depleted Formations 169
3.26 Thailand—Down-Hole Deployment Valve 170
3.27 Underbalanced Experience in Libya 170
3.28 Massive Lost Circulation in Libya 171
3.29 References 172

Section 4
Two-Phase Flow Modeling 174
Arash Haghshenas, Boots & Coots

3.30 History of Two-Phase Flow Modeling 174
3.31 Gaseated Flow 176
3.32 Hydraulic Modeling 180
3.33 Questions and Answers 195
3.34 References 195

4 Foam Drilling 197
Bill Rehm, Drilling Consultant
Amir Paknejad, Add Energy, LLC

4.1 Introduction to Foam Drilling and Workover 197
4.2 History of Foam Systems 201
4.3 Advantages of Foam Systems 204
4.4 Challenges and Technical Limits with Foam Systems 208
4.5 One Pass Systems or Disposable Foam 209
4.6 Recycle Foam 210
4.7 Basic Design of Foam Systems 211
4.8 Water, Gas, and Chemical Agents 222
4.9 Foaming Agents and Foam Extenders 222
4.10 Trips and Connections 225
4.11 Questions 229
4.12 Answers 230
4.13 References 232

Section 2
Field Foam Properties 234
Reuben Graham, Weatherford International

4.14 Introduction 234
4.15 Quality (GVF) and Foam Stability 234
4.16 Carrying Capacity and Settling 235
4.17 Water and Oil Dilution 236
4.18 Modeling and Field Results 236
4.19 Two-Phase Foam 237
4.20 Testing 239
4.21 Circulating Time Tests 239
4.22 Operational Considerations 241
4.23 Chemicals 241
4.24 Pumping Surfactant and Chemicals 242
4.25 Jet Subs 243
4.26 Hammers and Motors 243

Section 3
Oil-Based Foam Drilling Fluid 245
Olusegun M. Falana, Weatherford International

4.27 Introduction 245
4.28 Development of Oil-Based Foam Drilling Fluid 246
4.29 OleoFoam HT System 247
4.30 Features 251
4.31 Challenges 251
4.32 Conclusion 253
4.33 References 254

Section 4
Foam Rheology 255
Amir Paknejad, Add Energy, LLC

4.34 Introduction 255
4.35 Mathematical Concepts 255
4.36 Foam Quality 264
4.37 Foam Specific Weight 266
4.38 Foam Velocity 267
4.39 Foam Friction Factor 267
4.40 Cuttings Removal Phenomena 270
4.41 Background and History 271
4.42 Steady State Foam Flow 273
4.43 Importance of Surface Back-Pressure 285
4.44 Cuttings Transport in Horizontal Wells 286
4.45 Pressure Drop across Bit Nozzles 292
4.46 References 293

5 Air and Gas Drilling (Drilling Dry and with Mist) 297
Bill Rehm, Drilling Consultant
Arash Haghshenas, Boots & Coots
Abdullah Al-Yami, Texas A&M University

5.1 Introduction 297
5.2 Definitions 298
5.3 Rotary and Hammer Drilling 301
5.4 Advantages of Gas Drilling 305
5.5 Limits, Extremes and Challenges to Gas Drilling 307
5.6 Special Rig Equipment for Gas Drilling 317
5.7 Gas Drilling Volume Requirements 322
5.8 Gas Drilling Operations 323
5.9 Mist Drilling Operations 328
5.10 Conclusion 330
5.11 Questions 330
5.12 Answers 331
5.13 References 332

Section 2
Limits to Water Volumes in Mist Drilling 335
Arash Haghshenas, Boots & Coots

5.14 Introduction 335
5.15 Dry Gas Injection Rate 335
5.16 Standpipe Pressure with Different Water Injection Rates 336
5.17 Mist Drilling Requirements 336
5.18 References 348

6 Snubbing and Underbalanced Drilling 349
Mike Ponville, Boots & Coots

6.1 Introduction 349
6.2 Basic Snubbing 350
6.3 Snubbing Units 353
6.4 Well Control 355
6.5 Auxiliary Equipment 357
6.6 Snubbing Operations 359
6.7 Wireline Procedures 364
6.8 General Stripping Procedures 366
6.9 Pipe Handling 368
6.10 Acknowledgments 369

7 Mud Cap Drilling in Fractured Formations 371
Dennis Moore, Signa Engineering

7.1 Introduction to Mud Cap Drilling 371
7.2 Background to Mud Cap Drilling 371
7.3 Mud Cap—Geology and Drilling 376
7.4 Constant Bottom-Hole Pressure 379
7.5 Horizontal Wells 381
7.6 Decision Tree for Drilling Fractured Formations 381
7.7 Stabilizing Conditions with Mud Cap Drilling 381
7.8 Floating Mud Cap Drilling—Depleted Formations 386
7.9 Water Sensitive Formations Exposed 390
7.10 Mud Caps versus Gas Assist UBD 390
7.11 Mud Cap and Hole Cleaning 391
7.12 High Bottom-Hole Temperature 391
7.13 Down-Hole Isolation Valves 392
7.14 Concentric Annuli 392
7.15 Constant Surface Circulation Approach 393
7.16 Different Pressure Regimes 393
7.17 No RCD Available 394
7.18 Deepwater and Floating Rigs 394
7.19 Casing, Cementing and Zonal Isolation 395
7.20 Conclusions 396
7.21 References 396

8 Underbalanced Liner Drilling 399
Robert Sanford

8.1 Introduction 399
8.2 Well Candidate Selection and Design Considerations 400
8.3 Advantages of UBLD 404
8.4 Limits and Challenges with UBLD 407
8.5 Well Control Considerations 408
8.6 Drilling Fluid Considerations 408
8.7 Special Equipment 409
8.8 Future Trends 410
8.9 References 410

9 Coiled Tubing and Underbalanced Drilling 415
Earl Dietrich, Blade Energy Partners

9.1 Introduction 415
9.2 Preplanning 415
9.3 Coiled Tubing Equipment 419

9.4 Operation Comments 427
9.5 Problems and Challenges 428
9.6 Ag-itator and Tractor Systems 432
9.7 Case Histories 432
9.8 References 437

10 Gases Used in Underbalanced Drilling 441
Bill Rehm, Drilling Consultant
Abdullah Al-Yami, Texas A&M University

10.1 Introduction 441
10.2 Air as the Underbalanced Drilling Gas 443
10.3 Natural Gas as the Underbalanced Drilling Gas 446
10.4 Membrane Nitrogen as an Underbalanced Drilling Gas 451
10.5 Cryogenic Nitrogen 453
10.6 Carbon Dioxide (CO_2) as a Drilling Gas 456
10.7 Questions 459
10.8 Answers 459
10.9 References 460

11 Equipment and Equipment Integration 463
Bill Rehm, Drilling Consultant

11.1 Introduction 463
11.2 Planning and Supervision 464

Section 2
Halliburton - GeoBalance
Underbalanced Drilling Services 466
Isabel C. Poletzky, GeoBalance Services, Sperry Drilling

11.3 Project Management Approach 466
11.4 Equipment Requirements 474
11.5 Real Time Reservoir Evaluation (RTRE) 485
11.6 Data Acquisition System 487
11.7 UBD Field Case 488
11.8 Conclusions 491
11.9 References 491

Section 3
Schlumberger's Approach to Underbalanced Drilling, Engineering, Equipment and Services 493
Mike Tangedahl and the M-I SWACO Pressure Control Technical Experts

11.10 M-I SWACO, a Schlumberger Company—Pressure Control 493
11.11 Identifying the UBD Candidate Well 493
11.12 Planning Phases 495
11.13 Project Planning 495
11.14 Well Engineering 496
11.15 Typical UBD Equipment 497
11.16 Pressure Fluids Management System (PFMS) 502
11.17 Membrane Nitrogen Generation Units 503
11.18 Total Gas Containment (TOGA) System 505
11.19 Low-Pressure and High-Pressure Well Types 506
11.20 Final Word about UBD Planning and Equipment 508
11.21 Questions 509
11.22 References 509

Section 4
SunTerra Integrated Underbalanced Non-damaging Reservoir System 510
Dale Cunningham, SunTerra Oil and Gas LP

11.23 Introduction to the SunTerra System 510
11.24 Four Phase Separation System 511
11.25 Choke and Manifold System 511
11.26 Gas, Fluids Measurement, and Geological Samples 513
11.27 Specialty Drilling Fluids 514
11.28 Solids Control System and Transportation of Drilling Fluid 515
11.29 Compression and Nitrogen Systems 516
11.30 Engineering 517
11.31 General Operational Comments 519
11.32 Conclusion 520

Section 5
Weatherford Approach to Underbalanced Operations 522
Brian Grayson, Secure Drilling Services

11.33 Introduction 522
11.34 Phase One—Suitability 524
11.35 Phase Two—In Depth Analysis 525
11.36 Supervision and Monitoring of the Process 527
11.37 Physical Resources to Deal with the Process 528
11.38 Special Mechanical Wellbore Approaches 530
11.39 Special Equipment 530
11.40 Data Acquisition While Drilling 534

11.41 Final Report 535
11.42 Conclusion 536
11.43 References 536

12 Flaring 537
Olavo Cunha Leite, Flare Industries LLC

12.1 Editor's General Comment 537
12.2 Introduction 537
12.3 Safety 538
12.4 Types of Flares 540
12.5 Types of Flared Gases 540
12.6 Smokeless Flaring 541
12.7 Limits and Cautions 542
12.8 Combustion Principles 543
12.9 Flare Header Design 549
12.10 Elevated Flare Components 550
12.11 Ground Flares 561
12.12 Pulsation 564
12.13 Flare Combustion Noise and Spectrum 565
12.14 References 565

Section 2
Flare Systems 566
Olavo Cunha Leite, Flare Industries LLC

12.15 Mathematical Expressions 566
12.16 Hot Spot Temperatures 569
12.17 Grade Level Concentration of Vented Gas 570

13 Corrosion in Drillpipe and Casing 577
Bill Rehm, Drilling Consultant
Abdullah Al-Yami, Texas A&M University
Katherine Dimataris, Lamberti USA

13.1 Introduction 577
13.2 How Corrosion Occurs 578
13.3 Identifying the Corrosion Types 582
13.4 Corrosion Testing 591
13.5 Measuring Corrosion 595
13.6 General Corrosion Prevention and Treatment 598
13.7 Make Up Water Problems and Solutions 609
13.8 Formation Water Quick Solutions 611
13.9 General Recommendations to Minimize Corrosions in UBD Water Based Systems 612
13.10 Questions 613

13.11 Harder Questions 614
13.12 Answers 614
13.13 Answers to Harder Questions 615
13.14 References 616

Index 617

Preface

Where Have We Been and Where are We Going?

Don Hannegan, P.E., Weatherford International Ltd.

The purpose of this book is to discuss the limits and extremes of underbalanced drilling (UBD) technology and its enabling tools, a formidable task by any measure.

The world's first commercially successful hydrocarbons well was drilled underbalanced, with what today would be considered akin to a cable tool rig. It is likely that the world's last well will be drilled underbalanced due to the grossly depleted nature of future reservoirs. In fact, the world's last hydrocarbons well will likely not be for conventional oil or gas because those reserves will have been exhausted. It will likely be drilling for commercial quantities of the world's last abundant resource of hydrocarbons energy, methane hydrates. Given that methane hydrates disassociate in accordance with Boyles Law, it is reasonable to suspect that those drilling programs will dictate the use of UBD concepts and key enabling equipment. This defines the limits of UBD application, and warrants no additional discussion.

The extremes of UBD, on the other hand, causes one to reflect upon the extremities of the technology which have positively impacted the upstream industry and the greater promise it holds for the future. UBD's trademark benefits, which range from drilling into grossly depleted formations without damaging the prospects productivity to increasing recoverable reserves by enabling the drilling of otherwise un-drillable prospects, remain core values of the technology. However, its extremities are also reflected in the manner in which thc technology has facilitated and/or complimented the development of other drilling methods needed to expand exploration and production. UBD has also fostered the development of a suite of technologies that have proven to be invaluable in the development of new tools, and has contributed immensely towards safer drilling practices to the benefit of the upstream industry as a whole.

UBD has been a technology incubator of benefit to the industry as a whole. Drilling methods that do not invite the well to flow in the process of being drilled have benefited from lessons learned from UBD's safe and effective practices. First and foremost, UBD challenged conventional wisdom by encouraging drilling decision-makers to rethink the way they view the hydraulics of drilling. In the process they have become more receptive to alternatives to the conventional wisdom that has been primarily used since the industry learned to drill with weighted mud systems over a century ago.

Therefore, this preface will focus upon the extremes or extensions of UBD's root concepts and the enabling equipment that have had a significantly positive impact on the broader cross-section of the upstream industry, particularly as it relates to drilling complex wells, both onshore and offshore, safely and effectively.

1.1 UBD's Contribution to Hydraulic Flow Modeling

The practice of UBD often involves drilling with two-phase fluids and drilling where surface equipment requirements must handle multiphase annulus returns. Compressible fluids within the drillstring and annulus necessitate a profound development in hydraulic flow modeling technology; to design the UBD fluid, estimate optimum circulating rates, establish surface equipment specifications for pressure containment capability and flow rates, and serve as an invaluable data resource for pre-drill planning, HazId/HazOp processes, etc. Hydraulic flow modeling capabilities initiated by UBD requirements have now been extended to be beneficial to any drilling operation where there may be compressible fluids in the wellbore. For example, early kick detection is enabled by software with UBD roots. Sophisticated "candidate selection" methods of determining whether Air/Mist/Foam drilling, UBD, or managed pressure drilling (MPD) is the best drilling method for a prospect or a zone have been enabled by the teachings of UBD hydraulic flow modeling.

1.2 UBD's Contribution to the Development of Rotating Control Devices

Air/Mist/Foam Drilling, Underbalanced Drilling, and Managed Pressure Drilling require drilling with a closed-loop fluids system. A key enabler is a rotating control device (RCD) of required pressure containment capability and preferably of a design requiring minimum modifications to the existing rigs mud returns system.

Although low-pressure capable RCD's (500 psi, or less) were available to the drilling industry decades earlier, it was not until 1989 that demand for higher pressure capable designs began to surface. Attempting to drill horizontally with conventional methods into the inclined fractures of abnormally pressured Texas Austin Chalk presented a well control concern. Loss circulation occurred, followed almost immediately by a severe influx of reservoir fluids into the wellbore, a "hard kick". The development of the world's first 1,000 psi capable RCD enabled the well to flow safely while drilling proceeded.

The practice of underbalanced drilling played a key role in fostering widespread usage of the RCD's on conventional mud drilling programs. Most importantly, UBD requirements of the tool precipitated the development of high-pressure designs with redundant annular sealing elements.

A widely perceived value of drilling with a RCD on conventional drilling programs was proven to be a statistically valid premise in a study entitled "Recent Trends in RCD Usage and the Incidence of Blowouts" conducted by the University of Texas. The study tested for a statistical relationship between blowouts and Rotating Control Device (RCD) usage on conventional mud drilling programs. It concluded "We find consistent statistical evidence, across a variety of regression models; the use of RCD's decreases the incidence of blowouts."

Today, RCD designs are readily available that are suitable for the practice of all drilling methods that benefit from closed and pressurizable circulating fluids systems, whether the drilling program is on land, shallow water, deep water, or ultra-deep water.

1.3 From Underbalanced Drilling to Pressurized Mud Cap Drilling

Pressurized mud cap drilling (PMCD) is a crossover between underbalanced drilling and managed pressure drilling. The bottom of the hole up to the lost zone is very much an underbalanced operation, and from the lost zone to the top of the hole it is a managed pressure operation. The pressure is managed at the surface but is underbalanced at the bottom of the hole.

Much of the Pressurized Mud Cap Drilling activity in the Asia Pacific region falls into the contingency category. On average, operators incur severe or total loss scenarios on only one out of five wells. Operators are increasingly implementing contingency plans as they balance out the cost of stand-by rates on the equipment against a potential loss of the well. At least one major operator has established an internal practice that suggests if a drilling program manager does not prepare to practice PMCD in a region known to have severe loss-circulation issues, then a formal HazId/HazOp process should be conducted to prove it is better not to invest in a PMCD contingency.

1.4 UBD's Contribution to MPD

UBD root concepts and enabling equipment have played a key role in prompting the development of another drilling method—one that many in the upstream industry today believe will be applied on 40% of all offshore drilling programs within the next 5 years—managed pressure drilling (MPD). MPD is a technology transference from UBD and would not have achieved the broad industry acceptance it has demonstrated without being strongly rooted in the basic concepts of UBD. Although MPD does not invite the well to flow in the process of being drilled, the method requires some of the same equipment: drill-string non-return valves, fit-for-purpose RCD, and a dedicated choke manifold. MPD also requires the same degree of pre-planning, hydraulic flow modeling, HazId/HazOp processes, crew training, and inter-active drilling program implementation that are characteristic of safe and effective UBD.

1.5 Combinations of Drilling Methods: Conventional Mud Drilling, MPD, UBD

As drilling expenditure uncertainties and low energy prices drive operators toward stronger contingency plans that assure a successful drilling program, MPD has gained favor on conventional-wisdom fluids and well construction programs, onshore and offshore. Such a MPD "Contingency Plan" allows operators to react more efficiently and safely when dealing with unexpected downhole pressure environments.

MPD as a contingency to conventional drilling programs sets the tone for an UBD contingency when practicing MPD. MPD does not invite the well to flow, so what would happen if there is a 20% chance if encountering a zone that, for index reasons alone, begs to be allowed to flow while it is being drilled? All of the required surface equipment is in place, with the exception of a means of dealing with the produced hydrocarbons. A 20% chance of encountering a zone that will benefit by drilled underbalanced, e.g., having a UBD contingency plan to a MPD drilling program that includes all the prerequisites for safe implementation, is likely to be a good business case for well productivity reasons on many of the world's remaining prospects.

In actuality, this is not uncommon on MPD applications with tight gas in hard rock, where drilling progresses while de-energizing the stringers of tight gas. One is drilling ahead with the well flowing, but few accept that they are actually practicing UBD: "We didn't invite the well to flow, as in UBD, instead we chose not shutting in and circulating out for "drill-ability" and NPT reasons."

In some regulatory jurisdictions and wise practice MPD as a contingency to a conventional program and UBD as a contingency to a MPD program, both require the pre-planning, training, and regulatory approval process as if the contingency program were the primary drilling program.

1.6 Going Forward—"Drill-to-the-Limit" (DTTL) Concepts

When UBD was introduced to drilling decision-makers, it was a new way of looking at the hydraulics of drilling a well, and where the ultimate prize was increased well productivity associated with minimal damage to pay-zone porosity. The production of formation fluids to surface while drilling is typically necessary to achieve that objective.

MPD focuses upon the drill-ability of difficult prospects by addressing drilling-related challenges with more precise management of the pressure profile in the wellbore, enhancing control of the well and by reducing non-productive drilling time. Production of formation fluids was discouraged, typically by the application of varying amounts of surface backpressure when the rigs mud pumps are off. It too was a new way of looking at the hydraulics of drilling a well.

Both require drilling with a closed and pressure-able circulating fluids system enabled by drillstring non-return valves, RCD, and a dedicated drilling choke. And, both challenged the paradigm paralysis of drilling with an open-to-atmosphere mud system that's been the Albatross around the neck of conventional circulating fluids systems since Spindletop, Beaumont, Texas in 1901.

To grasp the concept of DTTL methods, one must understand that UBD and MPD also encouraged drilling decision-makers to view the circulating fluids system as one may a pressure vessel. In the case of UBD, pressure vessel mentality was mostly applied to surface equipment and their operations. In the case of MPD, pressure vessel mentality has mostly been directed towards a means of maintaining the equivalent weight of the mud (EMW) in the hole at the time within formation pressure and fracture pressure margins, e.g., "drilling window."

For example, the constant bottom-hole pressure (CBHP) variation of MPD typically employs a lighter mud, perhaps slightly hydrostatically underbalanced. Relatively modest amounts of surface backpressure is applied only during jointed pipe connections to counter the loss of circulating annular friction pressure, enabling drilling ahead in narrow or relatively unknown drilling windows.

DTTL draws upon some root concepts of the CBHP variation of MPD and UBD, but approaches the drilling decision-making process quiet differently by considering the totality of the fluids and well construction program. Where MPD may be seen as the root concept, UBD tools and methodology is the primary enabler.

Every pressure containing component, the circulating fluids system, and the fluid itself is evaluated with the idea of drilling with the least expensive fluid, simplifying the casing program, and getting reservoir access with a deeper and larger open hole. Obviously, this is beyond UBD's Holy Grail of increasing the productivity index of the completed well and that of CBHP MPD's to keep the equivalent mud weight EMW within the drilling window.

DTTL is:

- Like MPD and conventional drilling:
  - EMW must remain within the drilling window for drilling to progress.
  - Influx is not invited.
- Like UBD:
  - A drilling fluid that imparts a predetermined degree of hydrostatic underbalance in the zone of interest.
  - Some amount of surface backpressure is required when drilling ahead, and more when the mud pumps are off.
  - High-pressure capable RCD's are required on some applications.
  - Typically most applicable to hard rock or otherwise competent open holes.

- A downhole deployment valve used for trips out of the hole and perhaps for completion to avoid need for pipe-light snubbing unit.

- Unlike either of the above:
  - A drilling fluid that may be hydrostatically underbalanced, perhaps grossly, from the transition zone to total depth (TD).
  - Higher circulating rates to assure cuttings removal with lighter and less viscous mud.
  - There is an equal emphasis upon simplifying:

    1. Fluids program with an eye upon or most readily available.

    2. Casing program.

    3. Getting to TD with larger & deeper open holes for completion.

  - Ultra-high-pressure capable RCD's may be required on some applications, perhaps of differential pressure sharing design (i.e., pressure cascading between multiple annular seal elements), temperature shielding of lower element and other embodiments that increase pressure containment capability and temperature tolerance of the tool.
  - Dual down-hole valves (DDVs) may be required for redundancy when tripping out.

DTTL concepts look to make the most of these relationships when selecting the optimum performance mud and designing casing set points:

- Rate of penetration (ROP) typically increases with decreases in mud weight.
- The less dense and viscosity the drilling fluid, the lower the circulating annular friction pressure (AFP).
- The lower the AFP, the less the EMW fluctuations between pumps on versus pumps on that must be compensated for with CBHP MPD.
- The greater the permissible circulating rate.
- The deeper the casing set point.

- The faster the circulating rate, the better the cuttings transport ratio.

DTTL methods address the fact that a growing percentage of the world's remaining drilling programs are facing lower zonal and reservoir pressures. Many of today's prospects and more in the future could benefit by designing the fluids and well construction programs around DTTL criteria:

1. The minimum mud weight required for wellbore stability, compatibility with formations encountered, and cuttings carrying capacity.
2. The weakest component of the circulating fluids system in respect to pressure containment.

Drill-to-the-limit practices require more attention to mud pumps, standpipe assembly, non return valves (NRVs), casing, casing shoe LOT, FIT, wellbore fracture gradient, ballooning scenarios, RCD, and choke manifold.

1.7 Today and Tomorrow

The conventional wisdom of overbalanced mud densities with open circulation systems has served their purpose and is still a useful standard in many drilling operations. Increasingly, drilling prospects are becoming more difficult and costly to deal with those problems. Different ideas and advanced concepts have to be employed. Underbalanced Drilling, Managed Pressure Drilling, and Drill-To-The- Limit are a small but important part of this new approach.

This book, *Underbalanced Drilling: Limits and Extremes*, is a description of where we have been and where we might be going. It deals with some of the suite of options available to drilling decision-makers confronted with prospects that are growing increasingly more difficult to drill safely and effectively.

Underbalanced Horizontal Drilling: Could it be the Ultimate Completion Technique?

Jim Hughes, SunStone Technologies LLC

1.8 Introduction

Horizontal underbalanced drilling can create a completion technique that delivers more productivity because the reservoir's permeability that has been connected to the horizontal wellbore has not been damaged. There is the potential to eliminate the cost of fracturing, packers, and wellsite surface footprint costs, and yet still have a better well.

For horizontal underbalanced drilling to reach this future as a completion option, three issues regarding its application in the reservoir need to be addressed:

- Proper well construction techniques
- The integration of equipment and services needed to drill horizontally underbalanced
- The development of new technology and equipment to refine the process

The application of horizontal underbalanced drilling is so broad and complex that, for the sake of simplicity and clarity, this discussion is limited to the use of gaseated drilling fluids and flow drilling techniques as part of the completion process to improve productivity in competent reservoirs.

1.9 Gaseated Fluid and Flow Drilling

Underbalanced conditions exist in a wellbore when the hydrostatic pressure exerted by a column of fluid is less than the formation pressure. This underbalanced condition is often achieved by the injection of a gas into the drilling or return fluid to create a gaseated fluid, thereby reducing its density as discussed in Chapter 3. The process typically requires gas compression and surface pressure control equipment. This intentional and controlled method of using compressed gas to lower hydrostatic pressure by creating a lightweight fluid is what differentiates gaseated underbalanced drilling from "flow drilling," which is another type of underbalanced drilling.

Flow drilling relies on reservoir conditions and not on compression equipment to create the underbalanced state. Flow drilling, which is discussed at length in Chapter 2, is a drilling technique that developed in the US Austin Chalk because the reservoir is characterized by lost circulation. Losing circulation lowers the fluid column in a well, and as a result, the hydrostatic pressure is lowered causing a well to become underbalanced. This was usually an unplanned but anticipated event, so controlling a series of kicks while drilling was what generally defined "flow drilling." Drilling in over-pressured formations is another condition that can be taken advantage of to create flow drilling conditions. Flow drilling in over-pressured formations uses techniques and procedures similar to those described in "Managed Pressure Drilling" operations.

1.10 Underbalanced + Horizontal Drilling = UBHD

Drilling underbalanced is one of the key procedures used in a reservoir to prevent formation damage[1]. The reduction or elimination of formation damage has proven to be an effective component in the effort to improve productivity by reducing skin damage. In the past, eliminating skin damage in a reservoir with high permeability and porosity was not a priority with many companies. After all, if a reservoir with 600 millidarcies of permeability had 50 percent of its pore throat system plugged with fines from an overbalanced drilling operation, theoretically there would still be 300 millidarcies of permeability. This is probably the reason for the old saying, "You can't hurt a good reservoir."

Unfortunately, the same overbalanced mud system applied to a reservoir with only 10 millidarcies of permeability usually reduces the permeability in the near wellbore region to zero, hence the need to use a technology such as hydraulic fracturing to reconnect the reservoir to the wellbore. Low permeability reservoirs benefit the most from fracturing because they are extremely susceptible to damage from overbalanced drilling. They are generally drilled the same way medium to high permeability reservoirs are drilled. In other words, they are drilled with little regard for the damage being done to the reservoir because conventional completion practices have typically restored productivity to an acceptable level by reconnecting the wellbore to the reservoir.

Horizontal drilling is another technology that has come a long a way in the past 25 years. It can be utilized both as an exploration and a completion tool. The primary benefit of including horizontal drilling as part of the completion process is that the wellbore can be steered at a bearing that is perpendicular to the primary stress direction, thereby connecting natural fracture permeability to the wellbore. It also increases the wellbore exposure to the reservoir, which increases the drainage area. An obvious benefit of a larger drainage area is a reduction in the number of wells needed to develop a field.

The application of horizontal drilling is important in the exploitation of reservoirs that have limited primary permeability. Horizontal drilling has the potential to overcome this condition by intersecting secondary permeability that is derived from natural fractures. As a result of being able to connect fractures to a wellbore, horizontal drilling technology has turned what were assumed to be non-productive reservoirs into economic successes. Fractured reservoirs drilled vertically will usually have a single highly elliptical drainage pattern due to permeability anisotropy, whereas horizontal wells drilled in the appropriate direction in the same reservoir will connect multiple elliptical drainage patterns to the wellbore, resulting in greater production. Horizontal drilling combined with underbalanced drilling creates a completion technique that delivers more productivity because the reservoir's permeability that has been connected to the horizontal wellbore has not been damaged.

1.11 Natural Fractures

The initial propagation of a natural fracture is normal to the bedding plane[2]. Thus, fractures are near vertical in beds that are flat, which accounts for the need to drill horizontally to connect secondary permeability to the wellbore. Fractures are usually described by their aperture as being either macro or micro.

It is generally understood that macro fractures are ones that can be detected with the naked eye (>40μ), and micro fractures are undetectable by a person with 20/20 vision (<40μ). Fracture permeability can be calculated from the following formula:

$$k_f = 8.444 \times 10^6 \, w_f^2$$

where w_f is the width of the fracture in microns[3].

1.11.1 Micro Fractures

It is important to remember that a micro fracture with an aperture of 25μ (the size of a white blood cell) can exceed 50 darcies of permeability. The short length of a micro fracture is the characteristic that reduces its effective permeability. Fortunately, micro fracturing density can be very high; thus, the distance from fracture to fracture is very short. Studies have shown that there can be as many as 80 micro fractures per one inch of rock[4].

Other issues with micro fractures are that they are easily plugged from fines when drilled overbalanced, and can become blocked by water in a water wet reservoir after being drilled with a water-based fluid or after being hydraulically fractured. This is especially true in a reservoir that is under-saturated with respect to water. Micro fractures have high capillary pressures and do not clean up well, thereby reducing the effective permeability.

1.11.2 Natural Completion

Because of natural fractures, the collective use of a non-damaging drilling technique such as underbalanced drilling and formation-compatible drilling fluids combined with horizontal drilling (UBHD) has the potential to become the ultimate completion technique. This technique will maximize productivity from many oil and gas reservoirs and has the potential to improve the recovery efficiency two to three fold for a given period of time[5].

An example of a "perfect" natural completion that utilized drilling underbalanced with no fluid contamination are the early wells that were drilled in California (USA) with cable tools, the original underbalanced technique. These prolific wells were technically high angle wells because a horizontal well by definition is a wellbore that is drilled parallel to the bedding plane. In California, many reservoirs can have dips that exceed 70°. Thus, vertical wells drilled with cable tools were able to dramatically increase their chances of hitting fractures because they were drilled at a more or less 70° angle when measured from the bedding plane.

1.12 UBHD Well Construction

The number one challenge when using horizontal underbalanced drilling as a completion technique is staying underbalanced or at balance 100% of the time[6]. This is critical because it only takes a few minutes to damage a wellbore from overbalanced conditions. The use of a proper fluid system can help mitigate some of the damage problems and is particularly helpful during periods of reservoir/wellbore pressure balance. The challenge includes staying underbalanced or at balance even when the time comes to get off the well.

Constructing a well using the concentric casing technique is particularly suited for UBHD because it is simple to employ, and it ensures that the underbalanced conditions are maintained at all times[7]. With concentric casing, a dual annulus drilling system is created, and the drilling fluid is not gaseated in the drill pipe[8]. Instead, the return fluid is gaseated downhole through ports that connect the inner annulus to the outer annulus. Compressed gas is pumped down the outer annulus. Two distinct advantage of this technique are the ability to have better control over the degree of underbalance by locating the communication ports at an optimal depth downhole and gas injection can continue without interruption while making a connection.

1.13 The Drilling Fluid Decision

If drilling an underbalanced wellbore into a reservoir to reduce formation damage, then careful attention needs to be paid to the selection of the drilling medium in order to protect permeability. The basic choices for creating a gaseated fluid include gases such as air, nitrogen, or natural gas and liquids such as oil or water.

1.13.1 Liquid Phase Considerations

The proper selection of the liquid phase is especially important in tight sandstones. Tight rocks are noted for having high capillary forces. The potential to imbibe the liquid phase can be stronger than the underbalanced condition if the inappropriate fluid is selected. For example, sandstones that are "water wet" should be drilled with oil because a non-wetting fluid cannot be drawn into a "water wet" formation[9]. If water is used and spontaneous imbibition[10] occurs, the permeability can be reduced through a well-documented formation damage mechanism known as "phase trapping."

Because of their normally high clay content, sandstones can experience reduced permeability when a water-based fluid makes contact with reactive clay minerals[11] such as smectite. Many operators see the greatest potential for underbalanced horizontal drilling (UBHD) in sandstones because it is able to significantly reduce permeability damage due to swelling clays, as well as having a potential for greater fracture intensity[12] than limestones.

1.13.2 Solids Control Equipment

Another often ignored decision of an underbalanced operation is the selection of solids control equipment for the liquid phase of a gaseated fluid in a closed loop operation. High solids content in the liquid phase adds hydrostatic weight to the system. Of course solids control equipment is not needed if a blooie line is used and goes straight to a pit. In this type of operation, a "one pass" fluid can be used, which means the fluid that is returned from the well never goes back down the well.

1.14 Trip and Complete without Killing the Well

An important operational issue to address when planning an underbalanced horizontal well is how to rig down from a live well without killing the well with a heavy drilling fluid. A discussion of three practical options follows.

1.14.1 Snubbing

Snubbing can be used to secure a pressurized well bore by using special equipment and a specially-trained crew. This operation is reasonably safe, but when a regular snubbing unit is used, it can be expensive in both time and equipment. Rig-mounted snubbing cylinders have been successfully used in Canada to avoid nonproductive time (NPT) from rigging up and to limit the cost of a specialized crew. Snubbing is extensively discussed in Chapter 6 of this book.

1.14.2 Downhole Casing Valve

A retrievable downhole casing valve can be installed when a concentric casing string is run into the well. This allows the operator to shut the well in while tripping out of (or into) the hole and when rigging down. This is a good tripping and liner running solution because it keeps all reservoir pressure below the valve and does not limit the length of the bottom-hole assembly (BHA).

1.14.3 Drill-In Liner

A third option is the use of an expendable drillstring (drill-in liner) and bottom-hole assembly (BHA) to drill the lateral. (Underbalanced liner drilling is discussed in Chapter 8.) This option becomes available when the target formation is thick and steering is not required to stay in zone. The expendable components include tubing with premium connections for the drillstring, a non-return valve (NRV) installed near the bit, properly spaced stabilizers to create a packed-hole assembly to hold the angle, and a drill bit. The procedure for planting the string involves pre-planning. Surface equipment requires a tubing head that is installed below the blow-out preventer (BOP) stack to land a tubing mandrel. When the time comes to shut the well in, the mandrel is screwed onto the tubing and is then lubricated through the BOP stack, using the annular preventer and the rotating control device. There must be enough space between these two pieces of surface control equipment to accommodate the length of the mandrel. Once the mandrel has passed through the stack, it can be landed in the tubing head to secure the well for rigging down. The tubing can be perforated later. This is also a good technique for testing reservoir stability without risking expensive BHAs. However, landing the mandrel must occur before the tubing becomes completely immobile due to wellbore collapse.

1.15 Achieving Cost Control Equality

If UBHD is to become an acceptable completion option, cost control equality must be achieved. This means that the cost of utilizing UBHD as a completion practice must become as predictable as currently available conventional completion techniques. If the estimated cost to use a technology is unreliable, its use will always be limited. The unfortunate history of UBHD is that it generally exceeds the AFE. Three basic operational changes must occur to achieve cost control equality:

- A single service company provides and manages the primary pieces of equipment needed to carry out the process on one field ticket.
- The number of personnel needed to provide the technology is reduced through cross-training.
- Conventional equipment is automated and tailored for the UBHD operation.

Currently, performing the UBHD operation can necessitate up to a dozen different companies requiring 30–40 people. The list of services and equipment needed for the operation can include:

- A top drive drilling rig
- A rotating pressure control head and BOP stack
- Four phase separators
- A data acquisition service
- Compression equipment
- Nitrogen membrane units

- Directional drilling services and equipment
- Downhole tool rentals such as a deployment valve and multi-lateral junction equipment
- Solids control equipment
- Gyro orientation service
- Wireline unit
- Snubbing services

Coordinating the many service providers can become a costly logistical nightmare for an operator who is attempting to orchestrate the arrival of several companies to avoid unnecessary standby time. Having multiple companies on location can also cause high cost overruns when one company's piece of equipment is late or in need of repair which may cause the operation to be shut down. The other service providers who are on standby will continue to charge the operation because their equipment is not the problem. This situation is the primary cause for cost overruns. Having one company responsible for the entire operation can eliminate many of the logistical issues that can plague an operation.

1.16 Combination Drilling and Completion Rigs

Multi-lateral UBHD becomes a real challenge when the drilling rig is not capable of efficiently handling all of the required UBHD completion procedures. A good example of a company's effort to improve the performance of the drilling rig for UBHD operations was Engineering Drilling Machinery (now owned by TTS Sense) of Norway. They developed a rig that used a rack and pinion system to replace conventional draw works, blocks, etc. This means they have eliminated the need for conventional snubbing equipment because the rack and pinion drive allows the rig itself to perform subbing operations. They even automated the rig so that only one person was needed to trip pipe.

1.17 New UBHD Technology

Underbalanced drilling has evolved from conventional air drilling, a drilling technique that primarily targets non-reservoir rocks in order to increase penetration rates and eliminate lost circulation. This drilling method has been used for over fifty years, and is just lately seeing new improvements.

Over the past twenty years other new technology has been developed because more operators want to drill the reservoir underbalanced to improve productivity. New equipment was needed to create inert downhole environments, to control high surface pressures while drilling, and to shut-in wells downhole to eliminate killing the well. This need resulted in the development of items such as nitrogen membrane units, high pressure rotating control devices and downhole deployment valves.

The tools and ideas that follow are in the process of being implemented. Whatever improvements are made, time and equipment costs are critical elements for drilling in the continental US, both in non-conventional reservoirs and in the re-development of assumed to be depleted oil fields.

1.17.1 Artificial Flow Drilling

A technology that could have a significant impact on the future of UBHD employs artificial lift while drilling to lower the hydrostatic column in a well to create the underbalanced condition[13]. The conventional UB method utilizes compression equipment to inject gas into the stand pipe or return annulus of a concentric wellbore. The objective is to lower the density of the wellbore fluid, thereby reducing the hydrostatic pressure at the bottom of the hole to a level that is lower than the pressure contained within the formation being drilled.

Other methods have been proposed which could be described as "artificial flow drilling," where the underbalanced condition is caused by a lowering of the fluid column in a well. These other methods include:

- The combination of a dual casing string with a jet pump (aka concentric jet pump) may be a solution[14]. Jet pumping technology has been used as a means of artificial lift for more than 40 years[15]. The jet pump is simple in design because there are no moving parts. It requires a power fluid (instead of compressed gas), which can be pumped down the outer annulus between the concentric string and the production casing. The most significant advantage of the concentric jet pump is that it eliminates the cost of the nitrogen and compression equipment that are normally used to induce underbalanced conditions. The only requirement is a second triplex pump at the surface to pump clean power fluid (also known as fluid under pressure) to energize the jet pump.

- A currently available "Equivalent Circulating Density (ECD) Tool" is a mud-driven turbine, and while it reduces bottom-hole pressure in the range of Equivalent Circulating Values, it does not provide as much reduction in bottom-hole pressure as does a gaseated system.
- Schlumberger has reported using gas lift valves to aid in equalizing lift, but this is not the exact system that is needed.

1.17.2 Smart Drill Pipe

A distinct advantage of having a tool joint that plugs together in specific orientations is that it provides a simple means to directly connect one or more wires[16,17]. This type of connection allows for a continuous wire that can be installed to provide more reliable and improved data. Wires can now be installed under an internal drillpipe coating or in the wall of a thermoplastic drillpipe liner. This allows real time information to be gathered while drilling using a gaseated system without induction coils, repeaters, and battery packs.

This type of connection also allows simple stoking practice to orient the downhole tools without using a gyro[18,19] because an imaginary line can be maintained in the drillstring from surface to TD.

1.17.3 Short Radius Rotary Steerable Drilling Tool

Another way to reduce the cost of UBHD is to shorten the radius of curvature and thereby reduce the time it takes to drill from vertical to horizontal by using a short radius rotary-steerable BHA[20] that can achieve build rates up to 75°/100 ft (25°/10 m) and then drill horizontally for long distances.

1.17.4 Sub-Surface Casing Valve

Tripping pipe while a well is flowing can be dangerous and expensive, especially if snubbing equipment is required. A tool that has proven to be a better option than snubbing[21] is a sub-surface safety valve (downhole casing valve[22]) placed at depth. One new version of the valve can be actuated by a casing jack[23] that lifts and lowers a concentric string of casing. This method to shut in a well downhole has proven to reduce cost and improve safety in an underbalanced operation. An improved casing valve needs to be simple in operation, inexpensive, and reliable; and it should not require special crews.

1.17.5 Rotating Control Device (RCD)

In the UBHD world there is a new low cost, higher pressure-rated rotating control device (RCD) being developed. The diverter is short in height, about one meter tall (3 ft) and will have a working rotating pressure near 5,000 psi (3,500 kPa). It will not require an external cooling system[24] because the element and bearings will be pressure balanced. Reducing the pressure differential across the element will eliminate the heat issue and thus do away with the need for a separate hydraulic unit to cool the head, thereby dramatically reducing the day rate cost.

1.18 Conclusion

Formation damage has been proven to significantly inhibit the recovery of hydrocarbons. This could explain why the average recovery factor for oil in the US is only 5 to 15 percent[25]. Today operators are in a unique position to recover a significantly larger percentage of oil and gas from existing fields by employing non-damaging horizontal drilling technology in the reservoir as a completion method. We should thank Howard Hughes, Sr. for this opportunity because of his invention, the tri-cone bit. While the use of the tri-cone bit brought faster penetration rates than cable tool drilling, it also caused formation damage by the overbalanced mud system that is used when drilling with a roller cone bit.

Thus, the challenge before our industry today is to properly apply the UBHD completion technique in assumed to be depleted oilfields and recover another 15 percent or more of the oil in place without any exploration risk[26,27].

1.19 References

[1]Bennion, D.B. and Thomas, F.B. "Underbalanced Drilling of Horizontal Wells: Does It Really Eliminate Formation Damage?" SPE 27352 presented at the SPE Formation Damage Control Conference, Lafayette, LA, USA, February 7–10, 1994.

[2]Nelson, R.A. *Geologic Analysis of Naturally Fractured Reservoirs*, Gulf Professional Publishing, Boston, MA, USA, 2001.

[3]Tiab, D. and Donaldson, E.C. *Peptrophysics: The Theory and Practice of Measuring Reservoir Rock and Fluid Transport Properties*, Gulf Professional Publishing, Boston, MA, USA, 2004, pp. 429.

[4] Laubach, S.E. "Practical Approaches to Identifying Sealed and Open Fractures," *AAPG Bulletin*, 87, No. 4, April 2003, pp. 561–579.

[5] Cade, R., Jennings, J. and Vickers, J. "Producers Monetize Assets with UBD," *Hart's E&P*, January 2003.

[6,9] Bennion, D.B., Thomas, F. B., Bennion, D.W., and Bietz, R.F. "Underbalanced Drilling, Praises and Perils," SPE 35242 presented at the SPE Permian Basin Oil & Gas Recovery Conference, Midland, TX, USA, March 27–29, 1996.

[7] Saponja, J. "Challenges with Jointed Pipe Underbalanced Operations," SPE 37066 presented at the SPE International Conference on Horizontal Well Technology, Calgary, Alberta, Canada, November 18–20, 1996.

[8] Ralmalho, J. "Changing the Look and Feel of UBD Requires Industry to Break Out of Conventional Thinking," *Drilling Contractor*, July/August 2007, pp. 62–67.

[10] Hoffman, M. "Damaging Relative Permeability by Drilling, Completion and Production Operations," *The Mountain Geologist*, 45, No. 4, October 8, 2008, pp. 99–105.

[11] Civan, F. *Reservoir Formation Damage*, Gulf Professional Publishing, 2000, Chapter 2, pp. 10-48.

[12] Aguilera, R. *Naturally Fractured Reservoirs*, 2nd Ed., Pennwell Publishing Company, Tulsa, OK, USA, 1995, p. 9.

[13] Hughes, J.W. 2005, *Downhole Drilling Assembly with Independent Jet Pump*, US Patent # 6,877,571.

[14] Suryanarayana, P. V., Hasan, ABM. K. and Hughes, W.J. "Technical Feasibilty and Applicability of a Concentric Jet Pump in Underbalanced Drilling," SPE 91595 presented at the IADC/SPE Underbalanced Technology Conference, Houston, TX, USA, October 11–12, 2004.

[15] Figueroa, J., Hibbeler, J., Duque, L. and Perdomo, L. "Skin Damage Removal Using Coiled-Tubing Vacuum: A Case Study in Venezuela's Orinoco Belt," SPE 69532-MS presented at the SPE Latin America and Caribbean Petroleum Engineering Conference, Buenos Aires, Argentina, March 25–28, 2001.

[16] Hughes, J.W. 2003, *Tubing Containing Electrical Wiring Insert*, US Patent #6,666,274.

[17] Hughes, J.W. 2007, *Rod and Tubing Joint of Multiple Orientations Containing Electrical Wiring*, US Patent #7,226,090.

[18] Hughes, J.W. 1999, *Method and Apparatus for Aligning Drill Pipe and Tubing*, US Patent #5,950,744.

[19] http://www.hunting-intl.com/well-completion/connection-technology/seal-lock-ht-s-timed

[20] Hughes, J.W. 2008, *Method and Apparatus for Drilling Curved Boreholes*, US Patent #7,373,995.

[21] Ross, E. "Giving Snubbing the Snub," *New Technology Magazine*, January/February 2003.

[22] Hughes, J.W. 2006, *Flapper Valve and Actuator*, US Patent #7,537,062.

[23] Hughes, J.W. 2004, *Concentric Casing Jack*, US Patent #6,745,842.

[24] Hughes, J.W. 2008, *Rotating Pressure Control Head*, US Patent #7,380,590.

[25]http://en.wikipedia.org/wiki/Extraction_of_petroleum

[26]Shirley, K. "Find Draws Illinois Basin Attention," *AAPG Explorer Magazine*, July 2002, pp. 10, 12, 17.

[27]Haselton, T.M., Kirvelis, R., Minijos Nafta, Pia, G. and Fuller, T. "Wells Yield Direct OBD, UBD Comparison," *Drilling Contractor*, March/April 2002, pp. 22–24.

Biographies

Abdullah S. Al-Yami

Abdullah S. Al-Yami is a petroleum engineer with the Drilling Technology Team, Exploration and Petroleum Engineering Center-Advanced Research Center (EXPEC ARC), Saudi Aramco. Currently he is a PhD Candidate of petroleum engineering at Texas A&M University. He has several granted and pending patents, papers and journals in various topics in the drilling and completion area. Al-Yami previously worked as a lab scientist, tool pusher, drilling engineer and workover engineer which enabled him to gain on-hand and practical experience to handle research related to drilling and completion. He has a BS degree in chemistry from Florida Institute of Technology and a MS degree in petroleum engineering form King Fahd University of Petroleum and Minerals. He is a technical editor for the SPE Drilling & Completion Journal and is a member of the American Association of Drilling Engineers.

Dale Cunningham

Dale Cunningham has over 25 years in the international oil and gas business, most of which has been focused on UBHD product development, project management, operations and sales. Mr. Cunningham is currently the President of SunTerra Well Services, an underbalanced and managed pressure oilfield services company. Dale also served as President of Valesco Energy Ventures, a specialized UBHD equipment and engineering service company and lead projects in Europe and the USA. Dale has successfully managed many UBHD projects globally including what was the largest UBHD contract for Shell in the Middle East (Oman) as well as projects in Europe, Venezuela, USA, Canada and North Africa.

Bob Cuthbertson

Mr. Robert L. "Bob" Cuthbertson has over 40 years of petroleum experience, covering reservoir engineering, design and construction of surface and gas plant facilities and pipelines. His background includes field and office oversight of workovers, completions, drilling, production and operations. Mr. Cuthbertson started his career with Exxon Corporation on the King Ranch, then with Quintana Petroleum Corporation and subsequently his own independent oil company. His operating experience includes South Texas, the Gulf Coast, California, Wyoming, the Appalachian Basin, West Texas, New Mexico, Oklahoma and internationally in Colombia, Venezuela, Mexico, Oman, the North Sea, Indonesia and China.

Mr. Cuthbertson has extensive international and domestic experience in underbalanced drilling and has worked for Weatherford International as their Global UBD Technology Manager, Global Manager for Unconventional Gas and Global Manager of Managed Pressure Drilling. Mr. Cuthbertson most recently served as President of Seven Energy, a Weatherford E&P subsidiary devoted to the establishment of reserves through the use of Weatherford's own proprietary UBHD services. He is currently serving as COO for SunTerra Oil & Gas LP and was a co-founder of that company. Mr. Cuthbertson is an honor graduate of the University of Texas at Austin with a BS degree in petroleum engineering.

Earl Dietrich

Earl Dietrich has 18 years of experience in the oil industry. With more than 12 years experience in advanced drilling techniques including through tubing rotary drilling (TTRD), coiled tubing intervention, coiled tubing drilling (CTD), underbalanced drilling (UBD), managed pressure drilling (MPD), low head drilling, flow drilling, casing drilling, drill in liner systems, expandable tubulars, HTHP wells, geothermal, critical sour wells. His career has taken him worldwide, planning, overseeing, and implementing operations in arctic, jungle, desert, onshore, and offshore locations.

He works for Blade Energy Partners, a critical well engineering company based in Houston, Texas. He has a diploma in petroleum engineering technology from the Southern Alberta Institute of Technology (SAIT) in Calgary, Alberta.

Katherine A. Dimataris

Katherine Dimataris holds a BS in chemistry from Southwestern University in Georgetown, Texas, culminating in her Senior Capstone Research: "Effects of Nanoparticles on Rheology of Drilling Fluids."

Her oilfield career started in 2005 with multiple summer internships in the R&D laboratories of an oilfield service company. Currently, she is a Research and Development Chemist at Lamberti USA designing new chemical products for the oilfield.

Olusegun Matthew Falana

Olusegun Falana is a Research and Development Manager with Weatherford International R&D Center in Elmendorf, Texas, USA. He obtained a BSc in chemistry from Ondo State University, Nigeria. He earned a MSc in chemistry and a PhD in organic chemistry, both from Brandeis University in Massachusetts, USA. He is a MBA holder from Northeastern State University Oklahoma and a member of Delta mu Delta. Segun has more than 20 years experience in general chemical synthesis including polymers. He is specialized in innovation and improvement of chemical products and processes as well as chemical applications. He has spent the last decade in the oilfield industry championing development of chemical additives for conventional and non-conventional drilling fluids, completion fluids, production and cement. He formulates aqueous and non-aqueous foam-based drilling fluids for hostile environments coupled with similar conventional mud systems for drilling operations. Segun has four issued US patents, nine published US applications and authored or co-authored at least 20 publications in peer reviewed journals.

Segun is a member of Society of Petroleum Engineers, American Chemical Society, American Association for Advancement of Science, and the American Oil Chemists Society.

Reuben Graham

Reuben has a BS Degree in chemical engineering. Presently, he is employed by Weatherford International as a Global Technical Specialist. His duties consist of on-site engineering, teaching underbalanced drilling practices to Weatherford employees, and developing software programs for practical field use. He also serves as Project Manager on various underbalanced drilling projects including horizontal foam wells and high pressure underbalanced drilling. Before joining Weatherford, RLG Inc., he worked extensively with the Gas Research Institute and various operators on gas drilling and wellbore stress.

Brian Grayson

Upon graduating from Texas A&M University in 1991 with a BS in aero space engineering, Brian Grayson decided to take a different route and took a position with Weatherford as a R&D engineer. Brian held multiple positions within the R&D group, from a DST equipment design engineer to group product line engineering manager for pressure control systems. After the successful development of new technologies, such as Down-hole Deployment Valve Systems and Rotating Control Devices, he accepted a Global Product Line Management position with Weatherford's Controlled Pressure Drilling Group in 2001. Brian is currently based out of Houston, Texas and holds the position of Global Product Line Manager for Secure Drilling Services for Weatherford International.

Arash Haghshenas

Arash Haghshenas is a well control and hydraulic engineer at Boots & Coots International. Currently, he is involved with dynamic kill and flow modeling for well control and contingency planning. His work interests include UBD, MPD, well control, and well contingency analysis. He taught drilling fluids and well control at Texas A&M University where he earned his PhD degree in petroleum engineering. He holds a MS degree from the University of Louisiana at Lafayette in petroleum engineering where he got involved with well control engineering and wellbore pressure management at the Randy Smith Training Solution. He received his BS degree from Petroleum University of Technology in Iran. Haghshenas is a member of the IADC Technical Publishing Committee, AADE, SPE, Pi Epsilon Tau, and the Petroleum Engineering Honor Society.

Don Hannegan, P.E.

Drilling Hazard Mitigation Technology Development Manager
Secure Drilling Services, Weatherford International Ltd.

Don Hannegan is a Lamar University graduate, Registered Professional Engineer (Texas), recipient of The *World Oil* 2004 Innovative Thinker Award for his role in conceiving and developing offshore Managed Pressure Drilling (MPD) tools and technology, SPE 2006-2007 Distinguished Lecturer, Charter Member of the IADC MPD/UBO Committee and founding officer of the Arkansas SPE Section.

His chapter on MPD is included in SPE's textbook, *Advanced Drilling and Well Technology*, and he is the lead author of a book to be published by the University of Texas Petroleum Extension Service (PETEX) entitled *Drilling Hazard Mitigation Tools and Technology*.

He is recognized as a prolific inventor of enabling equipment and methods to safely and effectively drill with closed and pressurizable circulating fluids systems which are known for reducing the risk of reportable well control incidents, less drilling non-productive time and drilling otherwise technically or economically un-drillable prospects.

W. James (Jim) Hughes

Mr. W. James (Jim) Hughes has thirty years of experience in all phases of the upstream oil and gas business. His first ten years were devoted to drilling and production operations, prospect generation, and acquisitions under the tutelage of Dr. David K. Davies, his first employer and mentor, who taught him extensive completion design practice using formation damage prevention techniques. Over the next ten years, Mr. Hughes developed and utilized short radius, multilateral underbalanced horizontal drilling technology ("UBHD") as a primary completion and recompletion method to improve productivity. After several years of research and development and purchasing his own drilling rig, in 1991, Mr. Hughes, using an air hammer, drilled what is believed to be the first horizontal lateral from a short radius (25 ft) curve. Over the next 3 years, he spent most of his time evaluating reservoirs for the recovery of by-passed reserves, using UBHD technology as a completion technique. During this time, he was in Oman as part of the first independent technical team invited to recommend well construction methods and to evaluate indigenous oil fields for redevelopment, using UBHD technology as a completion technique. Mr. Hughes has devoted most of the last ten years to patenting new technologies related to UBHD, including a new short radius rotary steerable BHA, an artificial lift while drilling process for MPD and UBD operations and smart drillpipe. He currently holds fifteen patents related to UBHD and is a co-editor of the IADC's new book titled *Managed Pressure Drilling.*

Mr. Hughes has a BS degree in geology from the University of Missouri and is one of two founding members of SunStone Technologies, LLC.

Olavo C. Leite

Olavo C. Leite is an expert in process combustion specializing in flaring, incineration, pollution control, and LNG equipment with over 30 years of experience in energy and environmental technology projects. His experience spans a variety of areas in combustion and pollution control systems including: hazardous waste incineration systems, waste fuel burner technology, waste energy and by-product recovery systems, and thermal treatment of wastewater. Mr. Leite has gained recognition as an expert in flare systems. Olavo graduated from the Technical University of Lisbon's five-year graduate program in mechanical engineering. He has authored numerous publications, presentations and seminars. Presently, he is Chief Engineer/Engineering Manager with Flare Industries LLC in Austin, Texas.

Dennis Moore

Dennis Moore attended Texas A&M University, receiving a BS in petroleum engineering. Since graduation, he has worked for major oil companies, independent operators, and consulting engineering companies, serving in a variety of drilling, production, and reservoir engineering positions worldwide. These jobs have provided him with a diversity of both engineering design and wellsite supervision experience on HPHT, horizontal, underbalanced, and managed pressure projects, including drilling with casing and with coiled tubing. He has over 30 years experience in the oilfield, has authored or coauthored several articles on underbalanced and managed pressure drilling and is a registered professional engineer in Texas. Dennis is currently the Vice President, Special Operations and International Business with Signa Engineering based in Houston and can be reached via email at dennisdmoore@yahoo.com or by phone at 281-687-8584.

Muhammad A. Muqeem

Muhammad is a Petroleum Engineering Specialist with Saudi Aramco, one of the world's largest integrated national oil & gas company. Prior to joining Saudi Aramco, he was employed with Northland Energy Corporation (now part of Weatherford International) and Tesco Corporation, both out of Calgary, Canada. His expertise lies in planning, designing and implementation of underbalanced and managed pressure drilling technology in oil and gas wells. He was the project lead while implementing UBD in Saudi Aramco for the first time. He has authored and co-authored numerous SPE papers on multiphase flow modeling, three-phase relative permeability, project management, case histories related to UBD operations. He obtained his BSc and MSc both in chemical engineering from Bangladesh University of engineering & technology as well as PhD in petroleum engineering from University of Alberta in Canada. He is a member of SPE and APEGGA.

Amir S. Paknejad

Dr. Amir S. Paknejad is a petroleum engineer and currently serves as a Technical Services Manager at Add Energy LLC. In 2008, Amir began working for Boots & Coots where his duties included relief-well intervention, multi-phase flow hydraulics, blowout contingency planning, well-control modeling, risk management and developing petroleum related software and procedures. He has been awarded with the graduate study scholarship by the National Iranian Oil Company (NIOC) and the American Association of Drilling Engineers (AADE) for his entire graduate studies. He has also been a member of the IADC Technical Publishing Committee since 2006 and serves as a co-editor on the textbook, *Managed Pressure Drilling.* Amir holds a BSc degree in petroleum engineering from the Petroleum University of Technology in Ahwaz and a MSc and PhD in petroleum engineering from Texas A&M University.

Isabel C. Poletzky

Ms. Poletzky is a Senior Technical Advisor for Halliburton Sperry Drilling. She has earned BS and MS degrees in petroleum engineering from the Universidad Nacional de Colombia and the University of Houston. Ms. Poletzky has 13 years of industry experience including drilling and production engineering, directional and horizontal well planning and design, and she has nine years of experience in underbalanced and managed pressure drilling applications. She also spent two years working as a Drillsite Petroleum Engineer on the Kuparuk Field for ConocoPhillips, Alaska.

Her expertise includes reservoir characterization while drilling, modeling of multi-phase flow, and candidate selection for underbalanced and managed pressure drilling and completion projects. Recent responsibilities have included proposals, well planning and design, training, and coordination of underbalanced and managed pressure projects worldwide. Ms. Poletzky has co-instructed several UBD and MPD courses and has also taught wellbore hydraulics modeling. She has written and presented several papers and served on technical committees for SPE, IADC, and AADE. Ms. Poletzky is a member of SPE and IADC.

Mike Ponville

Boots & Coots, a Halliburton Company

Mr. Ponville's oilfield experience is exclusively in the snubbing industry performing a wide variety of critical well interventions such as blowouts, high pressure washouts, hot tap, gate valve drilling, freeze operations, and all aspects of snubbing operations.

2007–Present: International Operations Manager
2000–2007: Technical Manager (Global)
1988–2000: Hydraulic Well Control, Supervisor/Superintendent
1981–1988: Hydraulic Well Control, Snubbing Operator/Helper

Bill Rehm

Bill Rehm is the principal editor of *Underbalanced Drilling.* Bill is an underbalanced drilling and completions consultant in Houston, Texas. He has a BS in geological engineering and has received the "Legends in Drilling" award from the SPE, entered into the IADC Drilling Fluids Hall of Fame, and received a professional degree in petroleum engineering from Missouri School of Mines. He has, at various times, been a mud engineer, well control supervisor, president of a directional drilling company, R&D manager, and drilling consultant. He has written some 50 papers, which included some of the first publications on well control, underbalanced drilling, and managed pressure drilling. Bill may be reached at rehm@earthlink.net.

Jerome Schubert

Dr. Jerome Schubert, PE, is an Associate Professor in the Harold Vance Department of Petroleum Engineering at Texas A&M University with BS, MEng, and PhD degrees in petroleum engineering from Texas A&M University. He has over 30 years in the petroleum industry with Pennzoil Company, Enron Oil and Gas, the University of Houston-Victoria Petroleum Training Institute and Texas A&M University. He joined Texas A&M as a Lecturer in petroleum engineering in 1994 and was promoted to Assistant Professor in 2004 and to Associate Professor with tenure in 2010. Dr. Schubert's main research areas include deepwater drilling, dual gradient drilling, managed pressure drilling, and well control. Dr. Schubert is a co-editor of the textbook, *Managed Pressure Drilling*, and an author of more than 50 technical papers. He has been a committee member for several Society of Petroleum Engineers (SPE) and International Association of Drilling Contractors (IADC) committees, conferences and events, and a technical editor for *SPE Drilling and Completions*. He serves as Faculty Advisor for the Pi Epsilon Tau and the American Association of Drilling Engineers student section.

Mike Tangedahl

Michael J. Tangedahl, PE, CMfgE is Vice President of ES at M-I SWACO, a Schlumberger company. He has been responsible for project management, engineering, research, and development for underbalanced drilling (UBD) and pressure control products and services worldwide. He has over 30 years of corporate business experience with major oil tool and drilling service companies as director, vice president and chief operating officer in charge of business development, marketing, engineering, operations and manufacturing, both domestic and international.

Tangedahl is a certified registered engineer and is an active member in IADC, SPE, ASM, and the Society for Manufacturing Engineers.

Paco Vieira

Paco Vieira holds a BSc degree in mechanical engineering from the Universidad Metropolitana in Caracas, Venezuela and a MSc degree in petroleum engineering from the Tulsa University in Oklahoma, USA. His oilfield career started in 1995 as a Drilling Engineer for PDVSA (Venezuelan National Oil Company). After 2000, he was part of a team created for the implementation of underbalanced and performance drilling operations. He worked in several studies related to multiphase flow hydraulics. In 2004, he joined Weatherford controlled pressured drilling and testing services in the Middle East and North Africa region. Presently, he is the Regional Engineering Manager for the Secure Drilling Services product line.

CHAPTER 1

Introduction

Bill Rehm, Drilling Consultant

1.1 Book Description

As a matter of convenience and organization, each of the various underbalanced drilling (UBD) techniques is presented as a discrete operation in a single chapter. There are some theoretical and practical limits to the various techniques. The primary part of each chapter will describe the procedure and point to these limits. In some cases, there is a chapter section with a theoretical development of a complex issue. Finally, a number of descriptions or case studies of complex or extreme operations are presented to give the reader guidelines on what might happen in actual operations.

Since the initial plan will often be field modified, the discrete operation can become more complex, and the clear lines between the various techniques often become blurred. The interface between underbalanced operations and managed pressure concepts is often not clear because they may grade into each other. Many long horizontal holes are overbalanced, at balance, and underbalanced because of the effect of Circulating Pressure Loss or Annular Pressure Loss (APL).

This chapter provides a definition of underbalanced drilling (UBD), the advantages and constraints to UBD, and briefly outlines the various underbalanced techniques. Section 2, Lessons in Underbalanced Drilling, includes descriptions of practices that are common to all underbalanced operations. Within these discussions, enough of the basic operating ideas involved in UBD are presented to lead a reader to an understanding of the processes and to point the way for further reading.

Best practices based on present technology are a shifting target. Underbalanced drilling has its place, but other concepts, such as wellbore strengthening, liner and casing drilling, and expandable liners, shift the need or the use of the underbalanced concept to new horizons.

1.2 Other Options

Other technology can provide options to certain drilling problems without using underbalanced drilling techniques:

- New drilling fluids additives can strengthen the wellbore (Wellbore Strengthening).
- New lost circulation materials have improved lost circulation control.
- Drilling fluids that cause less skin damage.
- Expandable casing or liners provide "steel filter cake" for lost circulation or unstable zones.
- Casing and liner drilling.
- Motors increase drill rate (but do even better underbalanced).
- Bit technology has improved drill rate.

However, many new drilling technologies benefit from keeping the wellbore pressure at or below the pore pressure.

1.3 Introduction to Underbalance Drilling

Cable tool drilling, the natural form of underbalanced drilling, gradually lost out to rotary drilling tools because rotary was faster and controlled pressures downhole with drilling fluid density. With the advent of the rotary drill bit and oil booms, the apparent advantage of drilling the reservoir underbalanced was lost to the need for faster and deeper drilling. The old American Petroleum Institute (API) records in 1920 discuss wells drilled underbalanced in Trinidad, Montana, and California. The records note that in 1921, a well in the Panuco District of Mexico was drilled with 600 psi pressure on the well head.

1.4 Drilling Fluid Density

Drilling fluid density, or the density of the mud column, is normally the first barrier against a well kick and often contributes to wellbore stability.

There are three general regimes of drilling fluid density:

- *Normal overbalanced* drilling uses a fluid density that produces about 150 psi overbalance against the bottom hole formation pressure. This may also be also expressed as ½ ppg (.06 kg/L) overbalance. The overbalance is based on experience and prudent drilling practices; however, in some cases it is required by statute. Another requirement for normal overbalance is that the density of the fluid column must be adequate to limit the pressure against the base of the last casing string that is based on a certain kick size (kick tolerance).
- *Balanced* pressure drilling is the province of *Managed Pressure Drilling* where the fluid column, either static or circulating, is balanced against formation pressure with the aid of an impressed surface pressure.
- *Underbalanced* operations, which are the province of this book, are where the fluid column is deliberately kept below the formation pressure (pore pressure). This may include drilling with air or gas, drilling with a light single-phase fluid column, or drilling with a two-phase fluid column that has been made less dense by the addition of a gas.

1.5 Advantages to UBD

The drilling industry today recognizes that while drilling, a wellbore pressure lower than the formation pressure may be advantageous since lower pressure:

- Increases instantaneous drill rate
- Prevents the drilling fluid from entering the reservoir and thus limits skin damage
- Reveals hidden productive formations
- Avoids lost circulation
- Avoids differential sticking
- May allow earlier production
- Provides a condition where reservoir flow measurements may be taken during the drilling operation

One of the driving forces behind many of the underbalanced drilling operations in this decade is the preservation of oil mud. Oil mud, or oil invert emulsions, often start at upwards of USD 150/bbl.

Loss of a thousand bbl of oil mud to lost circulation and the ensuing rig time as well as transportation costs can be an AFE (expenditure) disaster.

1.6 Challenges with UBD

On the other hand, underbalanced wellbore pressures can cause significant challenges, most notably:

1.6.1 Flow of Formation Fluids or Gasses to the Surface

Wellbore fluids to the surface can be fortuitous if the fluids can be sent to the sale line, or problematic if there is no convenient way to dispose of them. This is, of course, speaking of oil, gas, sour gas, salt water, or in some cases, acid salt water. The subject is lightly covered in the appropriate chapters in this book. Disposal is a local problem of regulations and environment and is difficult to briefly cover on a general basis.

1.6.2 Wellbore Instability

The problem of wellbore instability can be a challenge: wellbore instability is one of the main problems that disqualifies or limits the use of UBD. Wellbore instability takes several different forms, some of which include:

- Areas where stress is building or has built due to geologic activity
- Fractured or disturbed zones, especially in high pressure areas found near the junctions of the continental plates
- The younger sediments found in some of the ocean basins where the fracture pressure, pore pressure and stability pressures tend to converge
- Massive shale that has an elevated internal pore pressure (geopressured shales)
- Salt is plastic and will flow into the wellbore when it is penetrated. The rate of flow into the wellbore is a function of the pressure differential, temperature, and the composition of the salt, (primarily how many water molecules are part of the salt structure)

Wellbore instability is not a primary subject of discussion in this book. Wellbore instability, like disposal, is covered in other books in this series.

1.6.3 Borehole Pressure Change Surges Associated with Connections and Trips

Every time the pump is turned on or off there is a change in wellbore pressure. This is normally considered part of Managed Pressure Drilling, and it is the primary factor in dealing with minimal drilling margins between lost circulation and pore pressure or wellbore stability. In the case of UBD the effect is there at all times, but there is also the problem of loss of carrying capacity for wellbore cuttings and cavings. These specific problems are discussed in the appropriate chapters about the various fluids.

As a general practical solution in UBD, constant circulation for connections and trips is the best and simplest solution. Constant circulation generally employs one of the following:

- A dual casing string
- Constant circulating SUBS
- A parasite tubing string
- The Constant Circulating System

1.7 IADC Definitions

The International Association of Drilling Contractors (IADC) Underbalanced Operations and Managed Pressure Drilling Committee defines underbalanced drilling as:

> *A drilling activity employing appropriate equipment and controls where the pressure exerted in the wellbore is intentionally less than the pore pressure in any part of the exposed formations with the intention of bringing formation fluids to the surface.*

The committee further separated gas or air drilling as a separate technique from underbalanced drilling.

The IADC Committee's entire definition for managed pressure drilling and underbalanced drilling is available on the web at *www.iadc.org*. The Underbalanced Operations and Managed Pressure Drilling Committee site also includes hazard charts and training material not covered in this book.

This book uses a more relaxed definition of UBD to include gas drilling and certain lost circulation techniques: "Underbalanced drilling is a procedure where the wellbore pressure is deliberately kept less than the exposed formation pressure." The following chapters are a description of the present technology. In some cases, the chapter on the description of the process is followed by sections outlining a case history or a general problem. Figure 1–1 shows a closed loop UBD circulation system.

1.8 General Underbalanced Drilling Equipment

Definitions and details of general underbalanced drilling equipment are covered in other volumes. The equipment discussed in detail in Chapter 11 is specific to a special problem or detail with underbalanced drilling.

The standard set of special equipment used with UBD operations consists of:

- Rotating control head
- Drill pipe non-return valves (NRV)
- Choke and manifold system
- Separator system
- Flare or flare system
- Surface valves and piping

The system may also include:

- Downhole casing non-return valve
- A method for constant circulating
- Air compressors and boosters
- Nitrogen generators
- Special instrumentation
- Special chemical injection equipment
- Mud treating equipment
- Gas detection and analysis

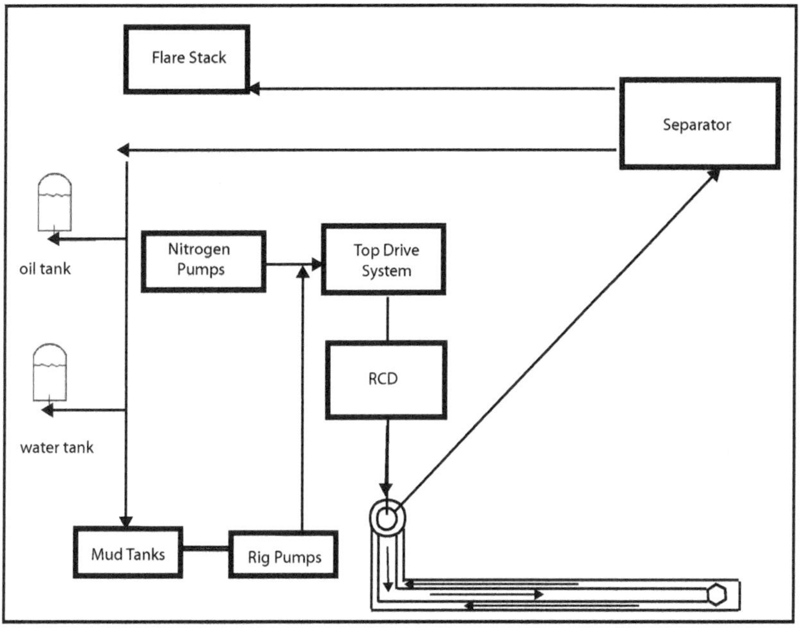

Figure 1–1 *A closed loop UBD circulation system*

1.9 What is Not in This Book

Underbalanced drilling has other issues that are not covered in this book because of their complexity or because they are properly considered in other disciplines.

1.9.1 Dual Gradients

Dual gradient drilling as a deep marine process is not covered in this book. It is discussed in *Managed Pressure Drilling* (Rehm et al., 2008).

However:

- Dual gradient in another form is discussed in Chapter 2 where a dual casing string is used with high mud velocity to provide a dual gradient.
- The use of the dual casing string with gaseated mud is another form of dual gradient, and it is discussed in Ch 3.
- Underbalanced drilling with gas injection is a dual or multi-gradient problem that is discussed in detail in Chapters 3 and 4 of this book.

1.9.2 Casing Drilling and Liner Drilling

Drilling with casing is a new and rapidly evolving technique that requires much more space to properly detail than is available in this book. Casing drilling systems do more than underbalanced drilling, even though they may use the same level of down-hole pressure. Casing drilling and the "smear" effect are an important part of the concept of wellbore strengthening. The limits and extremes of underbalanced liner drilling are briefly discussed in Chapter 8.

1.9.3 Coiled Tubing Drilling

Please refer to Chapter 9 (page 415) for information on this topic.

1.10 References

The following is a sampling of basic references on underbalanced drilling. The following chapters include more detailed reference on the particular subject.

The IADC website for the Underbalanced Operations and Managed Pressure Drilling Committee is the basic reference for nomenclature, hazard identification, and training requirements. IADC also provides a specialized Underbalanced Operations Tour Report form which is available for download from *www.iadc.org*.

Abel, W., Bowden, J. Sr. and Cambell, P. "Fire Fighting and Blowout Control," Private Publication, Wild Well Control, Inc., Spring, Texas, USA, 1994.

Brantly, J.E. *History of Oilwell Drilling,* Gulf Publishing Co, 1971.

Chafin, M., Medley, G., and Rehm, B. "Underbalanced Drilling and Completion Manual," Maurer Engineering, DEA101 Project Manual, 1998.

Davoudi., M, Patel, B., Smith, J., Chirinos, J. "Evaluation of Alternative Initial Responses to Kicks Taken During Managed Pressure Drilling," SPE 128424 presented at the IADC/SPE Drilling Conference, New Orleans, LA, USA February 2–4, 2010.

Ellis, B.J., and Cuthill, J. "Pressure Drilling," *Journal of API,* 1930, pp. 361–378.

Elliott, D. *Quantifying the Input of Overbalance Reduction on the Reservoir,* presented at the SPE Applied Technology Workshop, Phuket, Thailand, 2006.

Garcia, D. and Moreyra, J. "Underbalanced Drilling Technology for the Exploration of Tight Sands in the Neuquen Basin: An Evaluation of Results from 11 Wells," SPE 108335 presented at the IADC/SPE Managed

Pressure Drilling and Underbalanced Drilling Conference, Galveston, TX, USA, March 28–29, 2007.

Gedge, B. *Brownfield Development & Production Optimization*, presented at the IADC/SPE Managed Pressure Drilling and Underbalanced Drilling Conference, Bergen, Norway, 2005.

Graham, R.A., and Culen, M.S. "Methodology for Manipulation of Wellhead Pressure Control for the Purpose of Recovering Gas to Process in Underbalanced Drilling Operations," SPE 91220 presented at the IADC/SPE Underbalanced Technology Conference, Houston, TX, USA, October 11–12, 2004.

Hallman, J.H., Cook, I., Muqeem, M.A., Jarrett, C.M., and Shammari, H.A. "Fluid Customization and Equipment Optimization Enable Safe and Successful Underbalanced Drilling of High H_2S Horizontal Wells in Saudi Arabia," SPE 108332 presented at the IADC/SPE Managed Pressure Drilling and Underbalanced Operations, Galveston, TX, USA, March 28–29, 2007.

Hooshmandkoochi, A., Zaferanieh, M. and Malekzadeh, A. "Optimum Technique Selection for Underbalanced Drilling in Iranian Oil Fields-A Review of Three Major Oil Fields," SPE 108329 presented at the IADC/SPE Managed Pressure Drilling and Underbalanced Operations Conference, Galveston, TX, USA, March 28–29, 2007.

Jin, L. "Quantitative Formation Damage Evaluation Using Dynamic/Static Drill-In Fluid Filtration Test Data," SPE 118659 presented at the SPE/IADC Drilling Conference and Exhibition, Amsterdam, The Netherlands, March 17–19, 2009.

Kozicz, J. *MPD: A Contractor's Viewpoint*, presented at the IADC/SPE Managed Pressure Drilling and Underbalanced Operations Conference, Bergen, Norway, 2005.

Lyons, W.C., Guo, B. and Seidel, F. *Air and Gas Drilling Manual*, McGraw Hill Publishing Co., 2001.

McLennan, J., Carden, R., Curry, D., Stone, C.R., and Wyman, R., *Underbalanced Drilling Manual*, GRI Ref No. 97/0236, Gas Research Institute, Chicago, IL, USA, 1997.

Nas, Steve. "Introduction to Underbalanced Drilling," Weatherford Private Publication Ref: APR-WUBS-WFT-001, 2006.

Nguyen, C., Somerville, J.M., and Smart, B.G.D. "Predicting the Production Capacity During Underbalanced-Drilling Operations in Vietnam," SPE 122266 presented at the IADC/SPE Managed Pressure Drilling and Underbalanced Operations Conference, Galveston, TX, USA, March 28–29, 2009.

Ramalho, J. "Changing the Look and Feel of Underbalanced Drilling," SPE 108358 presented at the IADC/SPE Managed Pressure Drilling and Underbalanced Operations Conference, Galveston, TX, USA, March 28–29, 2007.

Ramalho, J. and Davidson, I.A. "Well Control Aspects of Underbalanced Drilling Operations," SPE 106367 presented at the IADC/SPE Asia Pacific Drilling Technology Conference, Bangkok, Thailand, November 13–15, 2006.

Rehm, B., Schubert, J., Haghshenas, A., Paknejad, S.A., and Hughes, J. *Managed Pressure Drilling,* Gulf Publishing Company, Houston, TX, USA, 2008.

Rehm, B. *Practical Underbalanced Drilling and Workover,* Petroleum Extention of The University of Texas (PETEX), Austin, TX, USA, 2003.

Rommellvert, R. *Moving for UBD to MPD,* presented at the IADC/SPE Managed Pressure Drilling and Underbalanced Operations Conference, Bergen, Norway, 2005.

Salimi, S., Naziri, A., Ghalambor, A. and Tronvoll, J. "Application of UBD Technology to Maximize Recovery from Horizontal Wells in the Naturally Fractured Carbonate Reservoirs," SPE 122275 presented at the IADC/SPE Managed Pressure Drilling and Underbalanced Operations Conference, San Antonio, TX, USA, February 12–13, 2009.

Silva, V., Jr., Shayegi, S., and Nakagawa, E.Y. "System for the Hydraulics Analysis of Underbalanced Drilling Projects in Offshore and Onshore Scenarios," SPE 58972 presented at the SPE International Petroleum Conference and Exhibition, Villahermosa, Mexico, February 1–3, 2000.

Tan, S., Medley, G., Stone, R. *Tripping Operations of Managed Pressure Drilling,* Presented at the IADC/SPE Managed Pressure and Underbalanced Conference, Denver, CO, USA, 2009.

Section 2
Techniques Common to Underbalanced Drilling

Bill Rehm, Drilling Consultant

1.11 Introduction

The following drilling techniques are important to underbalanced drilling because they are part of daily operations for most wells. Well Control, Stripping, Mud Caps, and "Pipe Light" are placed in this section to avoid repetition in the following chapters.

1.12 Well Control in Underbalanced Drilling

The basic concepts of well control and actual operating procedures are a part of underbalanced drilling operations. Since underbalanced drilling tends to bring formation fluids to the surface while there is some annular pressure on the wellhead, it is important to understand how to handle flow from the formation and what is reasonable and safe, and what is pushing the limits of prudent operations.

Before dealing with actual well control procedures, it is worth while to review some of the basic gas laws and how they are related to drilling.

1.12.1 The General Gas Law

While the following paragraph is not totally specific to underbalanced drilling, it is important to understand the effect of a gas bubble in a wellbore. In the case of a well that shows gas cutting, the gas cut may not significantly change the bottom-hole pressure.

When a gas bubble under pressure displaces up the hole, such as with gas cutting, trip, or connection gas, the pressure/volume relationship has to take into effect the reduction in pressure above the bubble of interest. The General Gas Law as a practical field equation can be expressed as:

$$\frac{P_1 V_1}{Z_1 T_1} = \frac{P_2 V_2}{Z_2 T_2} \quad (1.1)$$

where

P = Pressure (Absolute)

V = Volume

T = Absolute Temperature

Z = A dimensionless function of gas composition, pressure, and temperature. The value of Z can be estimated using pseudo reduced chart or correlations.

$_1$ = An initial value

$_2$ = A final value

1.12.2 Strong-White Equation for Gas Cutting

The Strong-White equation calculates static bottom-hole pressure reduction due to gas cut of the drilling fluid when flow drilling or drilling overbalanced. Some rules of thumb with this equation:

- It shows that normal gas cutting does not significantly reduce bottom-hole pressure when the mud is cut 50% by volume with gas.
- The lesson is to be careful about increasing the mud density due to gas cutting while drilling or on a connection.
- The equation is difficult to utilize with large gas injection volumes and does not allow for friction due to circulating, "equivalent circulating density" (ECD).
- With two-phase flow, gaseated, or foam operations, this equation is not suitable because the surface ratio of gas to liquid is not 1:1, but is more in the order of 300:1. The two-phase systems are more strongly effected by circulating friction loss than the normal single-phase fluid.

$$h = \frac{1}{G_P}\left[P_{atm} + \frac{n}{100} \times p_{atm} \times \ln\left(\frac{P_{atm} + p_{atm}\left(1 - \frac{n}{100}\right)}{p_{atm}\left(1 - \frac{n}{100}\right)}\right)\right] \tag{1.2}$$

where

h = Depth, ft

G_p = Hydrostatic pressure gradient in ATM/ft

P_{atm} = Hydrostatic pressure at bottomhole, ATM

p_{atm} = Backpressure at the surface, ATM

$n/100$ = volume fraction of gas in the mud at the surface

If the wellhead is open to atmosphere, the equation reduces to:

$$h\,G_P - P_{atm} = \frac{n}{100-n}\cdot\ln\left(P_{atm}+1\right) \tag{1.3}$$

$(h\,G_P - P_{atm})$ is the amount of pressure reduction at the bottom of the hole caused by a gas cut. Iteration is required to solve the equation for P_{atm}. This equation is derived for low percentage of the gas cut of drilling fluid. This is a good approximation of the reduction in bottom-hole pressure due to gas cutting. At a high percentage of gas volume this model is not stable and provides erroneous results.

Haston (1975) provided a simplistic model of Strong-White Equation in oilfield terms.

$$\Delta P_{atm} = \left(\frac{W_1 - W_2}{W_2}\right)\times 33.81\cdot Log\left(\frac{p_{atm}}{14.7}\right) \tag{1.4}$$

where

ΔP_{atm} = Bottom-hole pressure reduction in psi

w_1 = Weight of uncut mud in pounds per gallon

w_2 = Weight of cut mud (at the surface) in pounds per gallon

p_{atm} = Hydrostatic pressure of mud in atmospheres

Goins and O'Brien published a chart in 1962 illustrating that surface gas cutting of a single-phase drilling fluid with an open annulus had little effect on bottom-hole pressure (see Figure 1–2). This is a particularly important issue with flow drilling in an underbalanced condition.

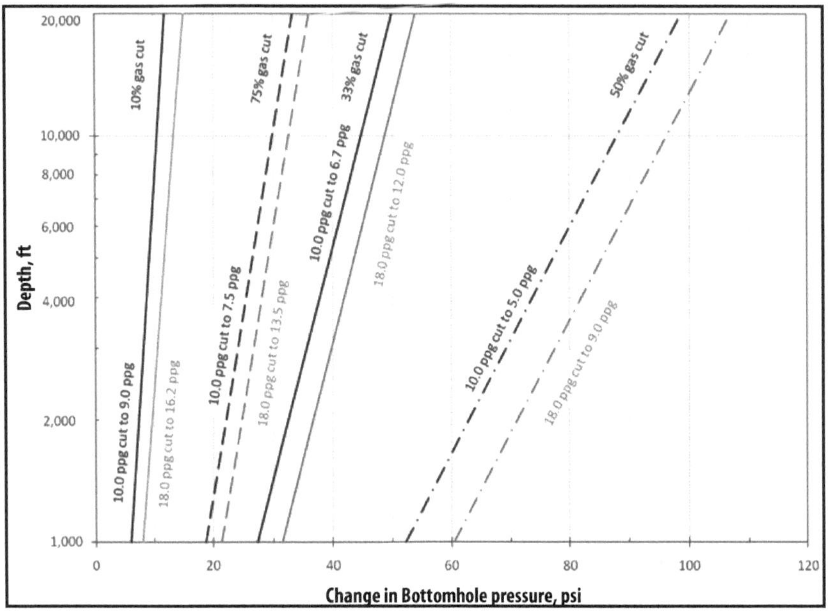

Figure 1–2 *Effect of gas cut at the surface on bottom hole pressure change*

Example 1–1

Depth 14,000 ft
Cut mud 8.0 pounds per gallon
Mud weight 17.5 pounds per gallon

Find:

(1) Bottom-hole pressure reduction.
(2) Equivalent Mud Weight (EMW) reduction.

$\Delta P_{atm} = 33.81 \times (17.5 - 8/8) \times \text{Log}((14{,}000 \times 17.5 \times 0.052)/14.7)$
$= 118$ psi

Average gradient reduction: $118/(14{,}000 \times 0.052) = 0.16$ ppg

where

0.052 = units conversion, psi/ft/pound per gallon, 1 ppg mud exerts .052 psi pressure per foot of depth.

1.12.3 The Effect of Annular Pressure Loss (APL) on Bubble Size

Annular Pressure Loss (APL), when circulating, affects bubble expansion because of increased pressure. The result is that the Strong-White equation and the Goins and O'Brien chart for bottom-hole pressure reduction due to bubble expansion will give an answer that is too great when the hole is being circulated. At this point in the discussion, without going into the complex solutions for APL, surface impressed pressures, and bubble volumes, it is evident that gas mud density measured at the shale shaker has little relation to bottom-hole pressure.

There are no simple solutions for bottom-hole pressure when circulating two-phase flow. Two-phase fluid solutions are complex. The most applicable solutions are found in gaseated mud or foam modeling that combine bubble expansion and flow rate in the bottom-hole pressure calculation. Table 1–1 shows some common two-phase fluid models.

1.12.4 Basic Well Control

Almost all UBD operations (except gas drilling) involve circulating a well as a closed system with constant pump rate and choke control. These UBD techniques tie back to the basic Drillers Method of Well Control. Fortunately, basic well control solutions do not require the complex calculations that are required for planning bottom-hole or wellbore pressures with two-phase flow found in gaseated mud or foam.

1.12.4.1 Underbalanced Drilling with Formation Fluid Flow

When drilling underbalanced with a RCD and choke system, the bottom-hole pressure can be maintained by using a constant pump rate and controlling the standpipe pressure with the choke as in the Driller's Method of Well Control. Changing bottom-hole pressure to respond to too much or too little formation fluid flow is simple by:

- Increasing or decreasing the choke pressure. This is a good immediate response to increased flow.
- Changing the density of the drilling fluid with single-phase flow.
- Changing the liquid to gas ratio with two-phase fluids.
- Changing pump rate—this is not often done, but for an example see Chapter 2, Section 3, Friction Controlled Drilling (page 28).

Table 1–1 Some Common Two-Phase Fluid Models

Company	Name of Model	Design Basis
Petris	DrillNet Hydraulics for UB Drilling	Published Two-phase Models
Shell/Landmark	Flodrill	Mechanistic (steady state) model
Nowsco	Circa	Combination of various correlations
Weatherford	AMFM (foam only)	Chevron foam model
Petrobras	SIDHAM	Unknown
Schlumberger	Sidekick (dynamic)	OLGAS (blowout and well control, HTHP wells)
SPT Group (Neotec)	UBD/MPD Flow Modeling	Basic flow equation modified by field analysis
Wellflo Dynamics	Flow Model (dynamic)	OLGAS blowout model in competition with sidekick
Signa Engineering	HUBS	Mechanistic (steady state) model
Scandpower	Ubits	Dynamic simulator based on Olgas' model.
Pegasus Vertex Inc.	UBDPRO	Aerated mud (Beggs-Brill method) and foam rheology (Bingham plastic, power law, Chevron's model, Reidenbach and Harris model)

In flow drilling (with a single-phase fluid), there is no gas in the drillpipe and the typical well control rule applies: using a constant pump rate, changes in bottom-hole pressure are reflected in the change in drillpipe (circulating) pressure.

With two-phase flow, (gaseated or foam systems) while the direction (±) of the change in bottom-hole and wellbore pressure follows the drillpipe pressure change, the exact amount of change requires a model or some hard math because when there is gas in the drillpipe (such as in gaseated or foam drilling), compression of the gas changes the volume of liquid in the drillpipe.

Precise well control ideas apply to a very specific condition of no lost returns: a minimal amount of gas spread out through the mud column and no gas in the drillpipe. UBD control of the well makes a great deal more sense if the operator is familiar with well control techniques. If there are any questions about the following short description of well control, it is important to go back and review well control principles because UBD is partly a well control process. Figure 1–3 shows a typical well control sheet.

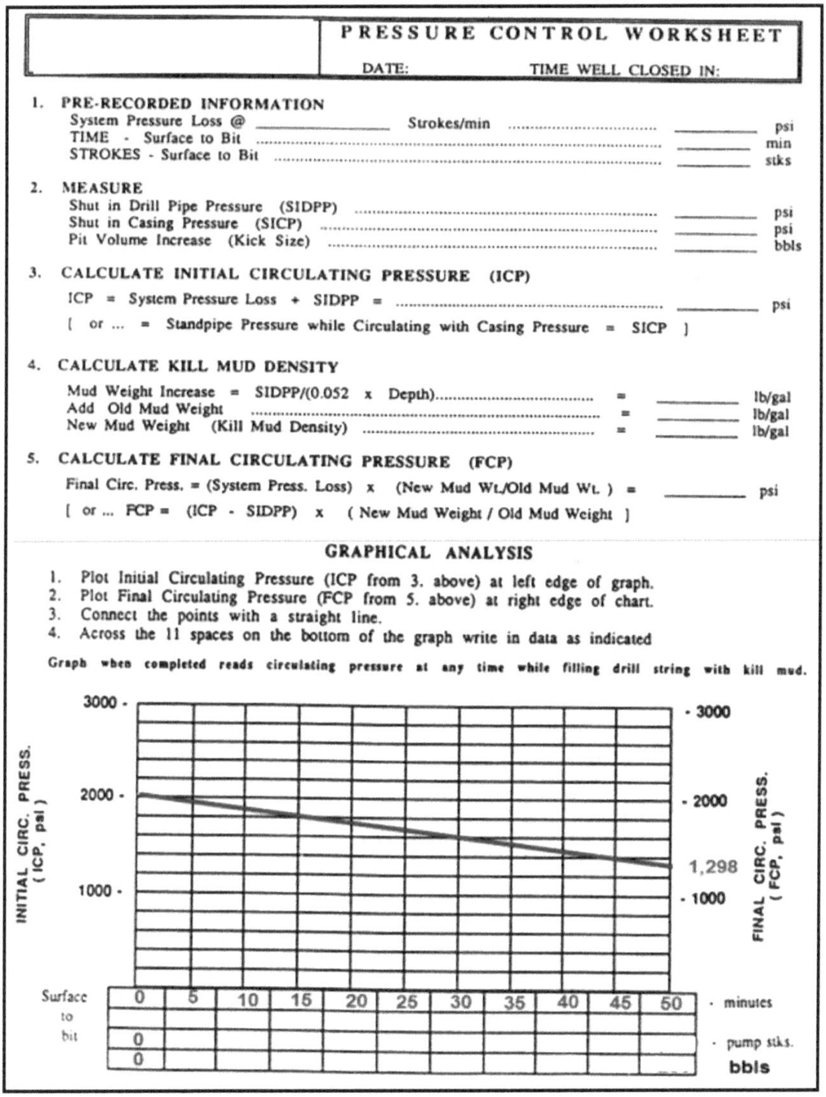

PRESSURE CONTROL WORKSHEET

DATE: TIME WELL CLOSED IN:

1. PRE-RECORDED INFORMATION
System Pressure Loss @ ________ Strokes/min .... ________ psi
TIME - Surface to Bit .... ________ min
STROKES - Surface to Bit .... ________ stks

2. MEASURE
Shut in Drill Pipe Pressure (SIDPP) .... ________ psi
Shut in Casing Pressure (SICP) .... ________ psi
Pit Volume Increase (Kick Size) .... ________ bbls

3. CALCULATE INITIAL CIRCULATING PRESSURE (ICP)
ICP = System Pressure Loss + SIDPP = ... ________ psi
[or ... = Standpipe Pressure while Circulating with Casing Pressure = SICP]

4. CALCULATE KILL MUD DENSITY
Mud Weight Increase = SIDPP/(0.052 x Depth)........ = ________ lb/gal
Add Old Mud Weight .... = ________ lb/gal
New Mud Weight (Kill Mud Density) .... = ________ lb/gal

5. CALCULATE FINAL CIRCULATING PRESSURE (FCP)
Final Circ. Press. = (System Press. Loss) x (New Mud Wt./Old Mud Wt.) = ________ psi
[or ... FCP = (ICP - SIDPP) x (New Mud Weight / Old Mud Weight]

GRAPHICAL ANALYSIS

1. Plot Initial Circulating Pressure (ICP from 3. above) at left edge of graph.
2. Plot Final Circulating Pressure (FCP from 5. above) at right edge of chart.
3. Connect the points with a straight line.
4. Across the 11 spaces on the bottom of the graph write in data as indicated

Graph when completed reads circulating pressure at any time while filling drill string with kill mud.

Figure 1–3 *A typical well control sheet*

1.12.5 The Driller's Method of Well Control

The driller's method of well control is as follows:

1. Shut in the well on a kick.
2. Read the Shut-In Drill Pipe Pressure (SIDP), Shut-In Casing Pressure (SICP), and kick size (pit volume increase).
3. Start circulating by holding the casing pressure constant at the shut in pressure with the choke until the pump rate is up to about one half of the normal drilling rate. This is the planned slow rate circulating pressure.
4. When the pump is up to the planned (slow) rate, hold the initial circulating pressure (ICP) constant on the standpipe using the choke on the annulus. Keep the pump strokes constant.
5. Continue circulating keeping the ICP and pump rate constant.
6. Use the choke to control the ICP.
7. Circulate until the kick is out of the hole.
8. Stop circulating and shut the well in.
9. Calculate as shown in Table 1–2 the:
   - Mud density increase the kill mud weight
   - The time required for the mud to fill the drillpipe (surface to bit time)
10. Mix up the kill mud weight.
11. Start pumping at the constant rate and hold the annulus pressure constant at the shut in pressure until the new heavier mud fills the drillpipe (surface to bit time).
12. Then hold the new drillpipe pressure constant until the well is clean of the kick gas or water. At that point, the Shut-In Drill Pipe Pressure (SIDPP) and Shut-In Casing Pressure (SICP) should both approach zero (several circulations may be required to completely clear the wellbore).

1.12.6 The Wait and Weight Method of Well Control

The wait and weight method is a normal technique for killing a well. The below items (particularly 5, 6, 7 and 8), along with suitable math and graphic values, are sometimes used with UBD.

1. Check the pump pressure at "1/2" the normal drilling rate and record the pressure as Slow Rate Circulating Pressure (SRCP) and the pump rate.
2. When a kick occurs, shut the well in.
3. Record the SIDPP, SICP, and the pit volume increase (Kick Size).
4. Calculate the following (using the well control work sheet):
   - The new mud weight (W2)
   - Initial Circulating Pressure (ICP)
   - Circulating time down the drillpipe (Tdp)
   - Final Circulating Pressure (FCP)
   - The plot and graphical value for drillpipe pressure drop
   - Total kill time (TK)
5. Increase the mud density in the pits enough to kill the kick (W2).
6. Start circulating at the slow rate and control the stand pipe pressure at the ICP.
7. Follow the values for drillpipe pressure drop.
8. Finally, circulate the well clean using the FCP.

1.12.7 Lag-Time Choke to Bottomhole, or Choke to Standpipe

In well control operations, lag time for a choke closure or opening to show a response back to the standpipe gage is estimated to be 1 minute/1,000 ft of total distance. This is a reasonably accurate time lag for well kick operations.

If there is any significant amount of gas in the hole as with UBD operations, the lag time becomes the sum of velocity in a mixed system under various pressures (which is slower), plus compression or de-compression time. In UBD operations, lag time may be up to one hour with foam operations, but it is typically 15 or 20 minutes with gaseated systems.

Table 1–2 Basic Well Control Formulas

$$W_2 = W_1 + \frac{SIDPP}{0.052 \times TVD} \tag{1.5}$$

$$ICP = SIDPP + SRP \tag{1.6}$$

$$FCP = SRP \times \frac{W_2}{W_1} \tag{1.7}$$

where

W_1 = Initial Mud Weight or Density

W_2 = Final Mud Weight or the Mud Weight required to kill the well

$SIDPP$ = Shut In Drillpipe Pressure

TVD = Vertical Depth

ICP = Initial Circulating Pressure

SRP = Slow Rate Circulating Pressure

FCP = Final Circulating Pressure

New circulating pressure at different pump rate can be estimated as:

$$\text{New Pressure} = \text{Old Pressure} \times \left(\frac{\text{New Pump Rate}}{\text{Old Pump Rate}}\right)^2 \tag{1.8}$$

$$\text{Pumping Time (min)} = \frac{\text{Volume (bbls)}}{\text{Pump Factor } \left(\text{bbls}/_{\text{stk}}\right) \times \text{Pump Rate (spm)}} \tag{1.9}$$

$$\text{Surface to Bit Time (min)} = \frac{\text{Drill String Volume (bbls)}}{\text{Pump Factor } \left(\text{bbls}/_{\text{stk}}\right) \times \text{Pump Rate (spm)}} \tag{1.10}$$

$$\text{Length of Influx (ft)} = \frac{\text{Pit Gain (bbls)}}{\text{Annular Capacity } \left(\text{bbl}/_{\text{ft}}\right)} \tag{1.11}$$

Table 1–2 Basic Well Control Formulas (cont'd)

Expected pit gain and surface pressure when the gas is at the surface can be estimated as following:

$$\text{Pit Gain} = 4 \times \sqrt{\frac{\text{Formation Pressure} \times \text{Initial Pit Gain} \times \text{Annular Capacity at the Surface}}{W_2}} \quad (1.12)$$

$$\text{Maximum Surface Pressure} = 0.2 \times \sqrt{\frac{\text{Formation Pressure} \times \text{Initial Pit Gain} \times W_2}{\text{Annular Capacity at the Surface}}} \quad (1.13)$$

Pressure propagation in fluid is analogous to sound velocity in that medium. The time, which is required for the pressure pulse to travel from the choke to a desired target, is called pressure transient lag time. Usually, the desired target is either the bottomhole or the standpipe. Since applying backpressure does not pressurize the well instantaneously, the pressure adjustment is confirmed by the pressure change at the standpipe.

In Chapter 2, Flow Drilling, notice that the change in the *standpipe pressure* is equal to the *change in wellbore pressure as well as in bottom-hole pressure.*

In Chapter 3, Gaseated Fluids (Gas-Liquid Mixtures), and Chapter 4, Foam Drilling, the lag time will always be on the slow end of the scale because of the time it takes to pressurize or de-pressurize the wellbore.

If there is no gas in the drillpipe (using a parasite or dual casing string) the change in drillpipe pressure will show the change at the injection point, not bottom-hole pressure. The pressure in the wellbore will follow the change (±), but the precise value will have to be calculated for any point of interest.

If there is gas in the drillpipe (drillpipe injection), the change in pressure of the drillpipe will be less than the change in bottom-hole pressure because the gas in the drillpipe will compress and make the fluid column in the drillpipe heavier. Shallow wells or wells with high gas ratios in the drillpipe will show the greatest variation from bottom-hole pressure change. Figure 1–4 shows how the gas in the drillpipe changes the stand pipe gauge reaction.

With gaseated fluids (Chapter 3), the use of a jet sub in the drillstring makes the response of the standpipe gauge change in the casing pressure, due to choke movement and the subsequent change in bottom-hole pressure, difficult to guess. Probably the direction change (±) in pressure will be the same, but it must be modeled to find out a numerical result.

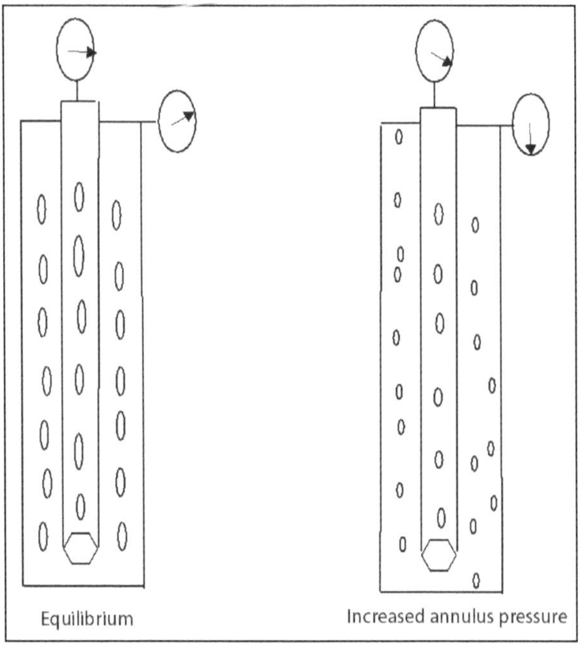

Figure 1–4 *Drill pipe pressure change is less than bottom hole pressure changes because gas in drillpipe is already compressed due to the jets and/or motor making drillpipe column denser*

1.13 Stripping

For the particular points to be made in this book, "stripping" is moving the drillpipe when the wellbore is sealed by the RCD, the annular preventer, or the pipe rams, and there is zero to limited shut in annular pressure. Stripping is common to underbalanced wells where there is a chance the hydrocarbon gas or fluids may come to the surface.

Snubbing is when pressures exist in the wellbore so that it is desirable to tightly control the movement of the drillpipe. In many cases, whether the rig is stripping or snubbing depends upon the equipment involved. In this book, snubbing is carried as a separate chapter (Chapter 6).

Normal stripping operations are carried out either as a safety precaution with zero shut-in annular pressure or with low shut-in annular pressure. While it is typically not prudent, "Low" pressures may be as high as 5,000 psi or more as long as the operating pressure of the surface sealing element is not exceeded and the weight of the drillpipe is more than the upward thrust of the wellbore pressure on the end of the pipe (see the following comments on "Pipe Light").

1.13.1 Wear on the Annular Sealing Element (RCD), or Annular Preventer

Stripping causes wear on the annular sealing element. The amount of wear depends upon the condition or type of pipe, the shoulders of the tool joints, wear pads, pulling or running speed and closing force on the sealing element. The following points are important to minimize wear on the annular sealing element:

- The stripping element should be a compound that is compatible with the drilling fluid and temperature at the wellhead.
- Square shoulders on the tool joints cause excessive packer wear.
- The type of hard banding is critical to packer wear. The hard banding should be smooth to avoid cutting the packer element.
- Closing pressure is important. The most common RCD type has a force fit and is wellbore pressurized, so high annular pressures can be expected to cause extra wear. Rotating Control Devices (RCD) with controllable closing pressure and annular preventers should use the minimum closing pressure to seal the wellbore to avoid excessive wear.
- The pipe should be slowed when pulling or pushing the tool joint through the packer element. In the case of the well pressure energized packers, the "rubber" packer element needs time to distort and allow the tool joint to pass without tearing the packer surface.
- With hydraulically closed and pressured RCDs and annular preventers, the hydraulic oil squeezing the packer needs some limited time to flow back through the relief valve or system to allow the packer to relax enough to pass the tool joint.
- With single element RCDs or annular preventers under pressure, it is possible to have a small release of well fluid from the passage of each tool joint. This is harmless unless H_2S is present.
- The rig needs to be level and centered over the hole to avoid excessive wear with any type of stripping or snubbing equipment (see Figure 1–5).

1.13.2 Mud Displacement when Pulling Pipe

Underbalanced wells make hole fill up easy when tripping, but it is very different from a normal trip fill up. The following items are important when pulling out of the hole (POH):

- If there is no permeability, the bottom-hole pressure could be kept at about the same pressure by adding enough fluid to compensate for drillpipe displacement in the "wet" part of the hole.
- If there is reasonable fluid permeability and/or lost circulation zones, the fluid is going to seek its own level. Nothing in the way of the drillpipe hole fill up will make a difference. This is a place where a mud cap might be used to keep the formation fluid from coming to the surface (see Section 1.13.4 Mud Cap or Chapter 7, Mud Cap Drilling in Fractured Formations).
- If there is a danger of wellbore fluid or gas to the surface, fill up can be done on a normal drilling basis. The ideal solution would be a down-hole casing valve. The next best choice is the mud cap. It is a heavier, thicker column of mud placed at some distance downhole to overbalance the well and keep formation fluid from coming to the surface.

Figure 1–5 *Eccentric wear (Upton, 2009)*

1.13.3 Drill Pipe out of the Hole

When the bottom-hole assembly reaches the surface, the RCD packer must be removed or the annular preventer opened. The pressure gauge may be inaccurate, so a bypass to the annulus should be opened to check that there is no flow or pressure (note also the following discussion on Pipe Light).

If there is no pressure or flow, the blind ram should be closed after the bottom-hole assembly (BHA) is clear and the wellbore should be checked before opening it for insertion of the BHA. Remember that the BHA displaces fluid (and gas) and the simple insertion of the BHA may cause the well to flow before the BHA clears the BOP stack.

Have the proper distance of the annular preventer posted so that it can be closed either above or below any stabilizers. If there is pressure in the wellbore, use a mud cap to balance the pressure. With planning, the annular preventer can be used with a lower pipe ram to "lubricate" the BHA in or out of the hole. This is a procedure that needs to be practiced in advance of any need. The spacing must be able to hold the drill collar or motor, so the distances should be carefully measured and posted where they are convenient to the driller.

1.13.4 Mud Cap

The mud cap is a heavier, thicker mud column that overbalances the well and keeps formation fluid from rising to the surface (see Figure 1–6).

> **Viscosity**—To minimize mixing with the drilling fluid, the viscosity of a mud cap is usually high. In the case of similar muds, the viscosity of the cap should be greater on the order of 10 seconds of funnel viscosity. Too high a viscosity in the mud cap could lead to swabbing.
>
> **Mud Density**—With the practical limit of a mud cap length near 1,000–2,000 ft (300 to 600 m), the mud density in the cap is generally two to three ppg (0.12 to 0.36 gm/cm^2) heavier than the drilling fluid. Greater mud density increases are possible, but not normally practical.

Mud cap length as a practical measurement depends upon the storage capability of the drilling rig to save and isolate the volume of the mud cap.

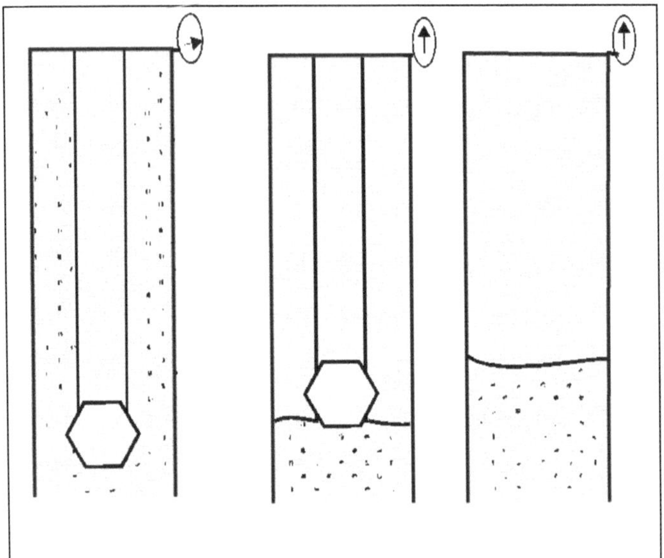

Figure 1–6 *Using a mud cap (high density mud) to kill a well on a trip*

$$\Delta p = L * (m_2 - m_1) * c \quad (1.14)$$

where

Δp = Wellbore pressure increase below the mud cap

L = Length of the mud cap

m_2 = mud cap density

m_1 = wellbore mud density

c = units constant

A constraint to using a mud cap is the limited ability to control the pressure with mud density alone without making a very long column. In aerated or foam columns where the density in the top of the casing down to 1,500 ft (500 m) is very low, a floating mud cap is easy to set up and use. The challenge to mud cap operation in a gaseated or foam system is that the heavy mud cap needs to be circulated out of the annulus as soon as possible when tripping in the hole. Extra pressure on the mud cap can easily exceed the fracture pressure of the formation being drilled, resulting in lost circulation.

Another challenge is viscosity. A thicker mud will significantly inhibit gas or fluid from rising through the mud cap. Thicker muds have a tendency to require a higher pressure to "break circulation," and this may cause lost circulation.

The transfer in and out of the annulus can be sloppy with portions of the mud cap system mixing with the regular mud system. A water-based mud cap can be made with a high viscosity using a drilling mud polymer. Pure oil "mud caps" have a low viscosity, so mud caps for diesel oil or "synthetic" oil systems have been made with invert emulsions or plain water-base mud.

1.14 Pipe Light

"Pipe light" is the term for the condition where the force on the bottom of the drillpipe or tubing is greater than the downward force (weight) of the pipe. This is based on the area of the section of pipe, collar, or tool in the annular preventer or RCD. Pipe heavy is when the weight of the pipe will cause it to fall of its own accord. Pipe light is always possible when stripping or snubbing and there is pressure in the wellbore.

The practical measure of pipe light is, with the annulus closed (with a RCD, annular preventer, or ram) and the pipe pulls with less than 2,000 lbs or 1 metric ton, the light weight indicates that there is an upward force that is trying to force the pipe out of the hole.

Despite the practical approach, the pipe light point should be calculated in advance and a graph posted to alert the driller to the potential of this occurring. In using the equation below, the force required to push the pipe through the preventer of RCD is a hidden margin.

In constant units, the break-over or pipe light point is reached when:

$$W_{ds} - (\pi r^2 * P) \rightarrow 0 \tag{1.15}$$

or

$$(\pi r^2 * P) \rightarrow W_{ds} \tag{1.16}$$

where

W_{ds} = weight of the drillstring

r^2 = Radius of the largest tool joint, drill collar, or tool in the annular preventer or RCD

P = Annular surface pressure

Section 3
Lessons in Underbalanced Drilling

Abdullah Al-Yami, Texas A&M University

Ramalho, J. "Underbalanced Drilling In The Reservoir, An Integrated Technology Approach," SPE 103576 presented at the SPE Russian Oil and Gas Technical Conference and Exhibition, Moscow, Russia, October 3–6, 2006.

The detailed method of modeling formation damage induced by underbalanced drilling was given by Suryanarayana et al.

Shell applied MPD in 450 wells, 380 wells were drilled underbalanced in a sandstone and carbonate reservoir. About 150 were oil wells and the remaining were gas wells.

Shell first drilled underbalanced in offshore operations. The first two wells were drilled underbalanced, but the tripping and completion were performed in overbalanced mode. The underbalanced drilling does not meet its full objectives if it does not maintain the underbalanced condition at all times. In fact, initial production rates from wells that were drilled underbalanced but were killed for trips were similar to wells that were drilled overbalanced. If the underbalanced conditions were maintained, the average initial production rate can be improved by a factor of 4 compared to overbalanced drilled wells.

The increase in production when drilling underbalanced also caused an increase in water production. This problem was solved by utilizing swelling packers and expendables.

The use of oil-based mud is less damaging than water-based mud in horizontal wells. Water-based mud "caused" a loss of production over 3 years; however, wells drilled by oil-based mud showed half the loss over the same period of time. The long term solution was to switch from using water-based mud to oil-based mud.

Kimery, D., and McCaffrey, M. "Underbalanced Drilling in Canada: Tracking the Long-Term Performance of Underbalanced Drilling Projects in Canada," SPE 91593 presented at the IADC/SPE Underbalanced Technology Conference and Exhibition, Houston, Texas, USA, October 11–12, 2004.

Different field cases of wells drilled underbalanced are compared to offset wells drilled conventionally. The comparison was done by using decline analysis and economic techniques.

Field Case Canada 1—Elkton Formation-Harmattan East Field: the formation is dolomitized carbonate with a permeability of 0.1 to 5 mD, porosity of 6 to 12%, initial water saturation of 11 to 30%, gross pay of 8.7 to 32.8 meters, and initial pressure of 12.3 to 21.6 MPa.

The wells were initially drilled vertically and hydraulically fractured. Horizontal wells were done later to maximize the production. However, drilling horizontal wells overbalanced did not show any improvement compared to vertical wells even after stimulation. Therefore, it was decided to use underbalanced drilling to drill horizontal wells, and the first horizontal well was drilled. The initial production rate was higher by 24% than conventionally drilled wells.

Field Case Canada 2—Pekisko Formation-Three Hills Creek Field: the formation is clean limestone. It is coarsely crindoidal and fragmental to fine-grained with a permeability of 0.25 to 5 mD, porosity of 4.5 to 11%, initial water saturation of 20 to 30%, gross pay of 1.7 to 10.3 meters, and initial pressure of 3 to 12 MPa. Horizontal underbalanced wells increased the initial production rate by an average of 238% over conventional wells. The conventional wells were drilled vertically, and stimulated (with hydraulic fracturing or acid fracturing).

Field Case Canada 3—Gething X pool-Kaybob Field: the area is a highly heterogeneous, fluvial-incised vally. The formation has a permeability of 0.07 to 4.2 mD, porosity of 10.5 to 19.7%, initial water saturation of 23 to 47%, gross pay of 1.9 to 11.5 meters, and initial pressure of 11.9 to 15.0 MPa. The horizontal underbalanced wells showed an increase in initial production rate by 254%.

1.15 Negative Field Case

If the reservoir is not suitable for underbalanced drilling and/or formation damage is the result of operational techniques, the following example shows that underbalanced drilling is not a suitable approach.

Field Case Canada 4—Cardium Formation-Ansell Field: the formation is sandstone with a permeability of 0.05 to 1.7 mD, porosity of 9.5 to 13%, initial water saturation of 17 to 37%, gross pay of 5 to 19 meters, and initial pressure of 15.4 to 21.9 MPa. Initial production rate from conventionally drilled wells were higher by 27% because the Cardium formation is not suitable for horizontal drilling because of its low permeability. Fracturing gave better results than horizontal drilling. In this example underbalanced drilling operations were not optimized to prevent formation damage.

Sarssam, M., Peterson, R., Ward, M., Elliott, D., and McMillan, S. "Underbalanced Drilling For Production Enhancement in the Rasau Oil Field, Brunei," SPE 85319 presented at the Middle East Drilling Technology Conference and Exhibition, Abu Dhabi, UAE, October 20–22, 2003.

The following three cases show some of the complexities of trying out a new system. The first attempt often has marginal or poor results.

The Rasau field has a sandstone reservoir (1–100 md) that is formed in a shallow marine environment. This results in a laterally stacked formation sequence. The producing formation is at 4,900–6,500 ft. The oil is light (40° API) with different gas cap sizes. Produced oil and gas were used to save cost on operation and minimize formation damage. Wells were drilled at a high angle through the target reservoir intersecting with many thin layers of reservoir to maximize exposure. However, this leads to great exposure to shale formation.

Underbalanced drilling was initiated in an attempt to minimize formation damage. A skin factor of up to 100 was recorded in wells drilled conventionally. The highest skin factor was recorded for wells that were drilled with unconditioned drilling fluid, and wells that were completed as open hole with wired wrapped screen (WWS)s.

A special tank was required, as well as solid-separation equipment. All elastomers in surface well control, separation system and down-hole motors needed to be resistant against the field oil used. Electromagnetic measurement was used while drilling to provide continuous transmission to surface.

Field Case Rasan 1—The pipe got stuck after initiation of two-phase underbalanced drilling. The pipe was released but later stuck several times. Next, the bottom hole assemble was pulled up and run again to establish underbalanced condition. However, the pipe got stuck one more time before reaching underbalanced conditions. This time, oil and brine circulation and jarring was done to release the stuck pipe. It was not known where the underbalanced conditions or the use of brine caused the hole to collapse which caused pipe-stuck problems. A second whipstock was installed above the first whipstock, and the formation was drilled overbalanced. However, when an attempt was done to reduce the overbalance below 200 psi the hole began to cave and the operation was abandoned. The hole was displaced to water-based mud and drilled conventionally. It is suggested that there was a high degree of depletion and a high percentage of shale in this formation.

Field Case Rasan 2—The target reservoir was not completely depleted from the original hydrostatic pressure, so its risk level is reduced in terms of wellbore stability problems. Seven inch casing was installed and cemented, and the float shoe was drilled. Drilling was switched to underbalanced mode. Field oil was used as the drilling fluid. After drilling to TD and pulling the bit to the shoe, a three rate production test was performed; the well productivity index was 16.8 BOPD/psi, the GOR was 968 scf/B, and there was no water production. A skin factor of zero resulted and the range of flowing bottom-hole pressure was 1,870–1930 psi.

Field Case Rasan 3—This field case was also a horizontal well producer where high percentage of shale was expected. Seven inch casing was installed in the shale just above the sand formation. Drilling changed to underbalanced using field oil and injected gas to achieve 1,770 psi. The reservoir pressure was 1,870 psi.

After drilling to TD and pulling the bit to the shoe, a three rate production test was performed; the well productivity index was 9.9 BOPD/psi, the GOR was 478 scf/B, and there was no water production. A skin factor of zero resulted and the range of flowing bottom hole pressure was 1,800–1,870 psi.

Murphy, D., Al-Busaidi, R., Wind, J., Davidson, I., Mykytiw, C., Kennedy and Arsenault, L. "Applications of Underbalanced-Drilling Reservoir Characterization for Water Shutoff in a Fractured Carbonate Reservoir—A Project Overview," *SPE Drilling and Completion*, September 2006, pp. 153–157.

Reservoir characterization was performed by underbalanced drilling of the formation, and gathering and observing surface and bottom-hole flowing parameters by using pressure while drilling sub.

Tight Shuaiba carbonate can be damaged easily when exposed to overbalanced conditions. To prevent this, a down-hole isolation valve was installed for drilling and for completion to avoid any potential of killing the well when tripping out of the hole. The down-hole isolation valve was installed as deep as possible in the concentric casing; the concentric casing was used instead of drillpipe for gas injection. The use of the down-hole isolation valve also prevented fluid from flowing to the surface when there was no pipe installed.

Completion (packer and tail pipe) was run on a drillpipe while the down-hole isolation valve was closed. Both were set in the liner below the concentric casing polished bore receptacle. The packer with its preset plug isolated the reservoir which allowed the removal of the down-hole isolation valve and concentric casing. Production tubing was then installed in the well by stabbing it into the tail pipe.

Field Problems—The down-hole isolation valve failed due to the damage of the sealing flapper. Therefore, the well was killed before tripping the drillpipe. Killing the well in this way damaged the reservoir. A clear reduction in the production rate was observed.

Enlarged borehole diameter prevented the use of solid expandable tubing (SET). Better practices were developed to solve this problem such as:

- Using bit size of 6 1/8 in. instead of 6 in.
- Using a motor with a reduced bit to bend length in order to reduce the bend required, reducing the amount of flow test which reduced the time in open-hole
- Reducing circulation time before well testing
- Minimizing drawdown

However, no actual reasons were given behind over-gauge hole in underbalanced drilling for this formation.

Drilled solids that gathered behind the concentric casing and while retrieving the concentric casing the solids to settle and accumulated on the plug installed in the tail packer. This incident occurred in three wells out of four. A coil tubing unit was used to remove the solids from the plug in order to remove it successfully. To prevent this from happening again, the packer assembly was raised just above the perforated joint and the plug was retrieved from the first time.

Tellez, C.P., Urbieta, A., Lupo, C., Castellanos, J.M., Ramirez, O., Puerto, G., Bedoya, J., Gabaldon, O., Beltran, J., and Castiblanco, G. "MPD Concentric Nitrogen Injection Used to Drill a Successful Horizontal Well in Fractured and Depleted Mature Reservoir in Mexico South Region," SPE 122982 presented at the Latin America and Caribbean Petroleum Engineering Conference, Colombia, May 31–June 3, 2009.

Concentric casing nitrogen injection was used to drill a high angle well. The use of such a method allowed the use of a conventional mud pulse MWD/LWD tool and kept the bottom-hole circulating pressure as desired. The use of electromagnetic tools is limited in high temperatures.

A temporary tieback was set on top of an intermediate liner, and gas was injected through ports into the tieback micro. In this case the well was drilled conventionally with only fluids pumped inside the drillpipe to minimize the effect of nitrogen on the conventional mud pulse.

In concentric casing nitrogen injection, the slug phenomena might occur due to gas compressibility in a concentric annulus. First nitrogen was injected in the annulus; then, mud and nitrogen were pumped inside the concentric annulus to decrease injection pressure and returned fluid.

Timms, A., Muir, K. and Wuest, C. "Down-hole Deployment Valve-Case History," SPE 93784 presented at the Asia Pacific Oil & Gas Conference and Exhibition, Jakarta, Indonesia, April 5–7, 2005.

When tripping out of the hole during underbalanced operations, the well might be killed which in turn causes formation damage. Another option was to use a subbing unit or trip out while the well was still flowing, but both scenarios imposed safety risks. The use of the valve will eliminate the use of a snubbing unit since the reservoir pressure can be balanced below the valve location. The down-hole valve is run with the casing and is managed using control lines from the surface. This case discusses the use of the down-hole valve in Thailand.

Snubbing takes a longer time than conventional tripping and requires up to 8 people to handle the operation in 24 hours.

A 7 in. casing was set at the top of the reservoir and 6 1/8 in. hole section reservoir was drilled underbalanced and completed with 4 1/2 in. slotted liner. The down-hole valve is a full bore casing shut-off valve run with the casing and can allow down-hole tool entry when in open position. It is a flapper valve providing a 5,000 psi metal to metal seal. It is available in different sizes and works in temperatures up to 300°F.

The down-hole valve was opened by movement of an internal sleeve when operated hydraulically. The fluid was pumped to the valve by ¼ in. tubing hydraulic lines clamped to the casing by cross-coupling clamps (one line for opening and the other one for closing). The down-hole valve could be designed to be retrieved when used in concentric casing or cemented in place for permanent installation. The casing was cemented using an inner string cementing technique to prevent risk of damaging the valve.

The average time saved by using down-hole valve was 1.5 days on bit trip, and 1.5 days on running the perforated liner. In Phu Horm, 4 bit trips and one perforated liner were run which saved a total of 7.5 days.

Safar, H.A., Majdoub A.A., Azhary, S.A., Qutob, H., Chopty, J. and Obeidat, H. "The First Horizontal Underbalanced Drilling Well in Libya: A Case Study," SPE 101073 presented at the IADC/SPE Indian Drilling Technology Conference and Exhibition, Mumbai, India, October 16–18, 2006.

Sabah G55 was the first well drilled horizontally underbalanced in Libya. The reservoir (Beda C) was depleted with an estimated pressure of 1,050 psi. A total of 62 wells have been drilled, and lost circulation was experienced in part of this reservoir due to its good porosity (up to 35%).

One of the 62 wells was drilled horizontally (G53). Severe lost circulation occurred when drilling the top of Beda C with viscous diesel. The drilling fluid was changed to water-based mud, however, the circulation was lost completely and only 580 ft of horizontal section was drilled.

Underbalanced drilling was initiated in G55 to drill from 5,961 ft MD to 7,628 ft MD. The objectives were to minimize formation damage, minimize lost circulation, characterize the reservoir, and improve rate of penetration. The well was drilled conventionally and casing was run and cemented at 5,011 ft MD. Then a 6 in. horizontal well was drilled in the reservoir with two phases (nitrogen and Sabah crude oil) to 7,331 ft MD.

The horizontal well drilled underbalanced resulted in higher production, and lower water cut and gas oil ratio than the horizontal well drilled unconventional.

Safar, H., Azhary, S., Hijazi, A., Qutob, H., Chopty, J. and Pham, C. "Dual Lateral Horizontal Well Drilled Underbalanced in a Depleted Reservoir in Libya-Case Study," SPE 107307 presented at the Middle East Drilling Technology Conference and Exhibition, Cairo, Egypt, October 22–24, 2007.

The major problem with the Facha reservoir was enormous lost circulation and differential sticking, especially in horizontal well sections. Underbalanced drilling was initiated mainly to eliminate lost circulation and minimize formation damage. This is the second well that was drilled underbalanced after the G55. The field had a salt creeping problem which required the use of a higher collapse resistance casing. However, this type of casing was not available and to avoid the salt creeping problem two stringers (9 5/8 in. casing and 7 in. liner) were used to isolate the salt beds.

Concentric casing gas injection was used utilizing 7 in. tieback to surface with perforated interval 38 ft above the top of the liner. Several benefits were realized with concentric casing injection, such as the possibility of using conventional logging while drilling (LWD) since one-phase drilling fluid is pumped inside the drillpipe.

The hole was displaced with crude oil and the concentric casing annulus was displaced with nitrogen. The circulating pressure was stabilized with 200 gpm of oil and 1,300 scfm of nitrogen. The first lateral section was 2,053 ft. The second lateral was 1,517 ft. The underbalanced drilling program resulted in 700–950 BOPD while drilling. A total of 10,918 bbl of oil was produced and sent daily to the Fidda field station.

After completing the underbalanced drilling, the crew was not able to revive the 7 in. tieback due to cuttings accumulation and settling in the 9 5/8 in. × 7 in. annulus below the perforated interval. To avoid this, the perforated interval should be located 5 ft above the

TOL instead of 38 ft. The ROP needs to remain below 50 ft/hr to avoid loading the annulus with cutting and to maintain bottom-hole pressure below 1,900 psi.

COMMENTS ON DRILLING THE BAKKEN/THREE FORKS
THE WILLISTON BASIN, USA
Mike Cannon
Drilling Supervisor, Williston S.D.

1.16 Williston Basin

The Williston Basin covers a vast area in North Dakota, Northern South Dakota, and Eastern Montana in the United States, and extends up into the Canadian Provinces of Saskatchewan and Manitoba. The present active areas are primarily in Western North Dakota and Eastern Montana. The Bakken Shale and the Three Forks below it span the Lower Missippean and Upper Devonian age.

While this discussion is about the Bakken/Three Forks reservoir zone, there are often significant lost circulation, wellbore stability, salt, and hydrogen sulfide problems in the upper part of the hole.

1.17 Introduction

Underbalanced drilling in the Williston Basin is worth a comment because of the number of rigs drilling in the area and the oil potential of the basin. Drilling techniques in the Basin have evolved and show considerable differences between the various areas. The comments below are somewhat generalized.

Drilling in the Bakken and Three Forks in the Williston Basin is a good example of the values of keeping the wellbore pressure just at or below the formation pressure to protect the reservoir formation and increase drill rate. It is also a good example of the use of mud caps on a trip.

The Bakken Shale is unstable, and early efforts required oil mud while drilling in the shale. Present efforts in the Bakken use salt water and require staying just above or below a two foot thick siltstone stringer in the Middle Bakken. There is the normal horizontal drilling problem of turning into and remaining in the target zone. This is complicated by a dip in the basin, local structures, and faulting. Drilling out of zone into the Upper or Lower Bakken Shale for any distance runs the chance of a stuck pipe or difficulty in running casing.

The Three Forks is a dolomitic formation just below the Bakken. The target zone is about ten feet below the Lower Bakken. The Three Forks drills more slowly and the main effort is to stay within the target zone without going into the Lower Bakken.

1.18 Challenges

The primary problems with underbalanced in the Bakken/Three Forks are potential wellbore stability and balancing the Equivalent Circulating Density (ECD) between the heel and toe of the horizontal wells. In the gassier areas, gas influx must be controlled to avoid pressure surges or well kick on trips and connections. In the few wells utilizing invert oil emulsion mud, the oil flow from the reservoir needs to be controlled to avoid excessive oil dilution of the invert emulsion mud.

Rotating heads (RCD) and simple gas busters are used for the low or normal pressured drilling in much of the area. In the geopressured areas in the Center of the Basin near Watford City, RCD, chokes, and gas separators are used to control the gas volumes that show up while drilling, especially after connections and trips.

However, primary wellbore pressure control is simply mud density. This relates to the low effective permeability in the fractured Bakken Shale and Three Forks.

1.18.1 Pressures and Drilling Fluids

Formation pressure varies from a normal gradient to slightly overpressured. Most of the drilling utilizes horizontal wellbores with casing set at horizontal in the Bakken or Three Forks, while the horizontal hole is cased at TD. "Underbalanced" drilling is a matter of minimal under balance to increase drill rate and protect the formations.

At present, saline production water at about 9.5 ppg (SpG 1.4) is a primary drilling fluid, with salt added to increase density to 10 ppg (1.2 SpG) as needed. Trips, mud caps, or drilling fluid "pills" weighted to about 11.2 ppg (SpG 1.35) with Calcium Chloride ($CaCl_2$) are used to control gas migration.

With the low viscosity clear drilling fluid, high viscosity sweeps are used at intervals to clean the hole.

The biggest drilling fluid problem is solids control. The drill cuttings from the Bakken and Three Forks are very fine; constant centrifuging is only partly effective and the salt systems need to be diluted to keep the solids content low.

1.19 Final Comment

While drilling in the Bakken/Three Forks area does not seem to fit the classic description of underbalanced drilling, it is clearly a great example of flow drilling near the balance point is used to increase the drill rate and limit drilling fluid loss to the formation with mud caps used on trips. The mud cap is made (of the drilling salt water) with $CaCl_2$ added for more density.

The Bakken Shale is unstable and close horizontal control is critical. The evolution of drilling procedures and fluids required to keep the hole open for a casing string have been studied by the local operators group as a matter of practical experience and wellbore experimentation.

CHAPTER 2

Flow Drilling: Underbalance Drilling with Liquid Single-Phase Systems

Bill Rehm, Drilling Consultant
Arash Haghshenas, Boots & Coots

This chapter describes the use of single-phase liquid drilling systems in underbalanced drilling (UBD). This is generally described as flow drilling. These fluid systems use a single liquid as the drilling fluid, unlike gas drilling which is a single-phase with gas or two-phase systems which use a liquid and a gas, i.e., foam drilling or gaseated drilling. The single-phase fluid can be water, mud, oil, or an oil/water invert emulsion. This chapter discusses the reasons for, and the general limitations, of using single-phase underbalanced systems.

The preceding paragraph is important to the concept of this book. The various chapters limit the discussion to a single topic or technique. In oil and gas drilling operations, the various ideas, techniques, and equipment are often interchangeable; and what appears in this book as a single subject may be only a small part of a larger, more complex operation using the ideas and equipment from several different procedures. Following the introductory discussion on the basic ideas of flow drilling, there are chapter sections which illustrate the actual field processes and a summary of flow dynamics. At the end of the chapter there are additional references and a set of questions.

2.1 Introduction to Single-Phase Underbalance Systems

The deliberate use of a single-phase liquid system for underbalanced drilling (see Figure 2–1), or "flow drilling," is not a new or an especially high tech approach. While there are records in the API Journals from the 1930's, the largest and most detailed group of older literature describes the use of a salt water system by the Gulf Oil Company

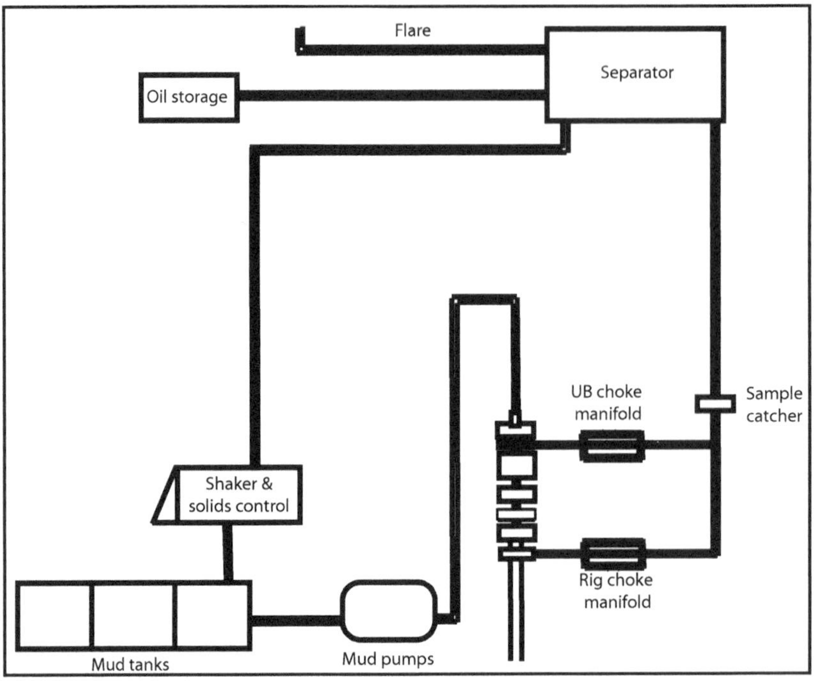

Figure 2–1 *Flow drilling diagram*

in West Texas in the early 1950's. Gulf Oil was faced with drilling a large field that contained several thousand feet of Permian age red beds (fine grained siltstones). The red beds had low permeability but high pore pressure, and required a drilling fluid weight of about 16 ppg (1.92 kg/l) in order to suppress the drilling gas and limit large volumes of trip and connection gas. The mud density restricted the drilling rate to about 3 ft/hr (1m/hr). The operator found that if the red beds were drilled with 9.8 ppg (1.17kg/l) salt water, the drilling rate increased ten fold to 30 ft/hr (10m/hr) and the gas at the surface could be controlled by a simple separator.

The ten fold increase in drilling rate when drilling through 6,000 to 8,000 ft (1,000–2,500 m) of "red beds" provided a significant economic incentive for the perceived risk of drilling underbalanced. The whole point of the operation was to reduce costs by increasing the drill rate. The Gulf Oil experience probably was not the first, but it was unique in that it used a heavy salt water fluid system. Most of today's systems use lighter, fresh water or invert emulsions to minimize the density of the drilling fluid and exert extra surface pressure with a choke.

The Austin Chalk Drilling in Texas and Louisiana is an example of where the formation pressure increases from normally pressured to over-pressured while going from west to east. The mud properties change from normal pressure to underbalanced to balanced. It is an example of a reservoir where both flow drilling and mud cap drilling techniques are applied based on reservoir characteristics.

- Flow drilling requires a reservoir (or formation) that is either over-pressured or has a lost circulation zone, in which case, it requires the underbalanced condition.
- Mud cap condition in flow drilling requires both a reservoir and a lost circulation zone either in or above the reservoir, see Figure 2–2. Similar problems with balancing the down-hole pressure with single-phase fluids occur in the Arab Limestone of the Middle East and in the Southeast Asia area. The actual solution varies with the location.

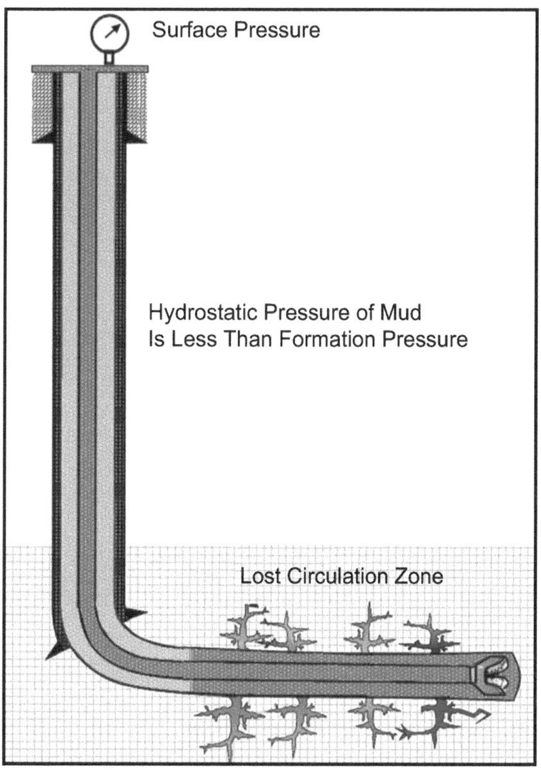

Figure 2–2 *Pressurized mudcap drilling (Rehm et al., 2006)*

2.2 Advantages to Drilling Underbalanced with Single-Phase

Although the Gulf Oil Company in West Texas drilled underbalanced solely for the purposes of increasing drilling rates, there was also a significant reduction in costs from using a simple fluid system. Underbalanced drilling can eliminate differential sticking and lost circulation, in addition to a reduction in reservoir damage.

As noted in the introduction in Chapter 1, a driving force for underbalanced drilling is the use of invert oil emulsion muds, which cost upwards of USD 150/bbl. Lost circulation, with these expensive systems and the concurrent NPT cost, make it very attractive to keep the pressure exerted by the mud column below the pressure that causes lost returns.

Finally, underbalanced is a very broad term. It encompasses systems that are close to balance as well as systems significantly underbalanced. The fluid must be tailored to the situation or problem.

2.2.1 A Simple System

The use of a single fluid that provides an environment for underbalanced drilling simplifies the entire process. Gaseated, or foam, and two-phase systems are more difficult to control and are generally more expensive to use. The single-phase simplifies the rheology. While there is always a potential problem with equivalent circulating density (ECD), changes that occur when moving the pipe up or down or changing the pump speed in a single-phase system is easier to predict and control than dual-phase systems. Underbalanced single-phase systems may phase into managed pressure drilling (MPD) with a minor change in down-hole pressure regime.

2.2.2 Lower Cost

Using single-phase systems are less costly than using two-phase systems since there are no costs for gas or a gas compressor.

2.2.3 Conventional Mud Motors and MWD Units Can Be Used

Since there is no gas in the drillpipe in UBD flow drilling, conventional mud pulse tools can be used such as MWD, LWD, and PWD (Vieira et al., 2007).

2.3 Increased Drill Rate

Drill rate with conventional bits is highly responsive to the differential pressure between the wellbore fluid and the formation pressure. The greatest effects from an underbalanced-overbalanced drilling fluid (where the wellbore pressure is less than the formation pore pressure) are:

1. The rock tends to be more brittle.
2. The brittle chips "explode" from under the bit cutter and are rapidly swept away.

Drilling rate increase, with reduced wellbore pressure, is shown as a general form curve in Figure 2–3.

The increase in drill rate is not entirely lost with PDC bits, but PDC bits do not have the same magnitude of gain in rate as conventional (cone) bits. The scraping action of the PDC is considerably different than the crushing or chipping action of the tricone bit. And while the PDC bit responds to a lower differential pressure with increased drill rate, the rate increase is not as regular or dramatic as with the conventional bit. In some recent cases, PDC bits in shale or shale-like formations have shown a significant drill rate increase with

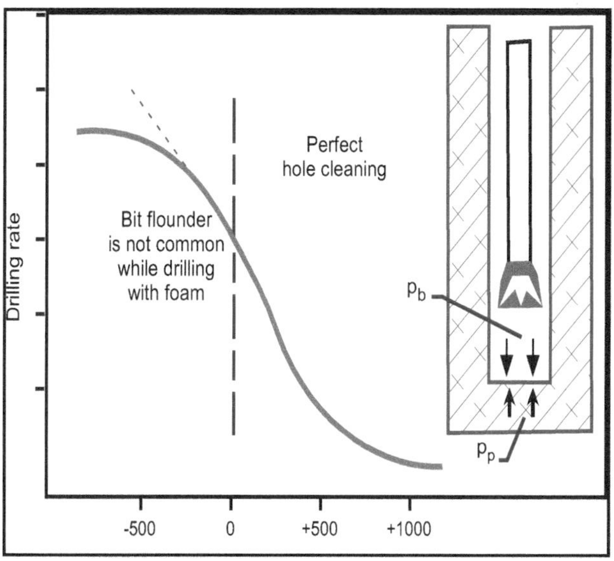

Figure 2–3 *Drilling rate increase with tricone bit*

increased wellbore pressure required for wellbore stability. This increase has been brought forward by Exxon in their "FAST" drilling concept (DuPriest et al., 2011).

2.3.1 Avoid Lost Circulation

The earliest use of underbalanced drilling fluids was to avoid lost circulation. This use is still one of the primary uses of underbalanced drilling. The best way to avoid lost circulation is to keep the ECD values below the pressure required to fracture the formation, or below the pore pressure in the case of high permeability or fractured formations.

2.3.2 Avoid Differential Sticking

Differential sticking requires a stick medium (filter cake) and positive differential pressure from the wellbore to the formation. When underbalanced, drilling fluid does not enter the wellbore to leave a filter cake. Also, the formation pressure is greater than the wellbore pressure, negative differential pressure. Therefore, differential sticking does not occur during underbalanced drilling.

2.3.3 Limited Reservoir Damage

As long as flow is out of the reservoir, skin damage due to wellbore fluids and solids is reduced. A single-phase fluid used to drill into a reservoir may be tailored to cause minimum damage to the reservoir. Single-phase systems can be operated with limited surging and easily maintained below reservoir pressure in vertical wells. Long horizontal wells may pose problems because of the difference in the ECD between the heel and toe of the horizontal section. See the example in Section , Section , for a further discussion of this subject.

2.3.4 A Test of Productivity while in the Drilling Mode

Underbalanced conditions while drilling the reservoir can indicate well potential. In the drilling mode, the degree of underbalanced can be varied (with the choke) to provide an original test of productivity before any cement or long term damage can occur.

Productivity tests require precise measurements of pressure and flow, and generally need special gages and flow measurement tools.

2.3.5 An Indicator of Hidden Productive Zones or Horizontal Productive Intervals

Careful attention to mud logging information, and logging while drilling (LWD) tools, will usually indicate hidden productive zones. These zones may be difficult to identify when overbalanced. Like a productivity test, hidden sweet spots or thin fractures in underbalanced operations are more evident before the well is cased and cemented.

2.3.6 The Potential to Produce Reservoir Fluids while Drilling and Completing

The flow of reservoir fluid to the surface creates potential for gas or oil sales. Hydrocarbon sales, while drilling or completing, are a common procedure where the infrastructure is stable enough to take variable amounts of gas or oil from the ground. Oil or gas sales have proven to be economic drivers for UBD. Documented examples have shown more than fifty percent of the daily drilling costs were recovered from gas sales.

2.4 Challenges and Limits to Flow Drilling

2.4.1 Wellbore Instability

The single greatest limiter to underbalanced operations is the geological constraint of wellbore stability. Wellbore stability is considered to be the lower limitation for underbalanced drilling (Figure 2–4, Kenneth et al., 2009). Wellbore stability takes different forms between the "hard rock" of most land operations and the softer, more recent marine and coastal formations.

In the older, inland formations, wellbore instability is often experienced in massive geopressured shales. The borehole pressure required to promote wellbore stability has to be greater than the shale pore pressure. Areas where formations are intensely fractured because of stress may require wellbore pressures to be higher than pore pressure in order to stabilize the hole. Finally, residual stress from geological activity is common in many of the basins and becomes more evident with increasing depth.

The marine and coastal formations are quite different with wellbore stability in deeper holes responding to stresses that require wellbore pressures above pore pressure. In the shallow and mid-range depth, wellbore instability is often due to poorly cemented sand and

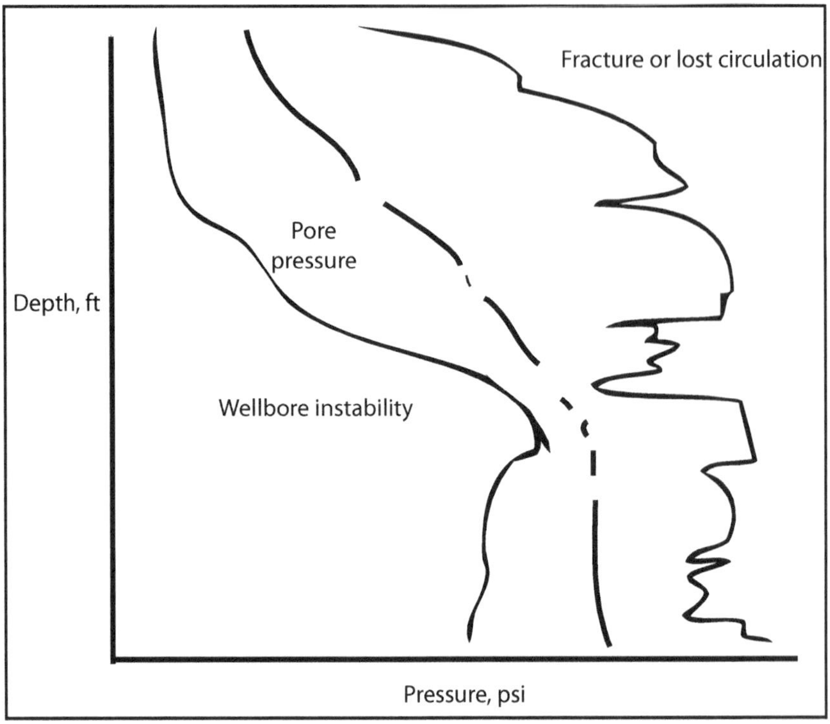

Figure 2–4 *Wellbore instability is a limitation for UBD (Kenneth et al., 2009)*

silt stones, but this is not always the case. There are examples where poorly cemented sands have been successfully drilled underbalanced. There are likewise examples where wellbore stability was a problem at shallow depths.

In horizontal wells, the orientation of the lateral wellbore relative to the primary stress direction can have a significant effect on borehole stability.

2.4.2 Fracture Permeability and High Reservoir Flow Rates

Reservoir flow is a constraint on underbalanced operations because it presents issues of technique, safety, and disposal. Fluid flow rate in and out of fractured formations, occurs at the minimum pressure differential between the wellbore and the formation. Therefore, the issue would be deciding whether to flow the formation fluid to the surface, or push it into the formation. In fractured conditions, flow drilling is often subordinate to mud cap drilling.

A single fractured zone in a vertical well can usually be balanced or controlled by careful selection of mud density. In a horizontal well or high-angle hole, even in the same pressure zone, the ECD increases from the heel to the toe of the hole. Therefore, horizontal wells require either a finer control over surface flow or lost returns, or the use of mud cap drilling and the acceptance of total down-hole loss of the drilling fluid to avoid surface flow.

Several fractured or otherwise high permeability zones that are open in a single vertical or horizontal well could cause lost circulation or a well kick. It is difficult, if not impossible, to maintain a constant return circulation. Mud cap drilling is a solution that depends upon drilling with water that is totally lost to the formation. The constraint on mud cap drilling depends upon either having enough water to drill with total lost returns, or returning a high pressure gas with the drilling fluid to the surface and controlling it with chokes, separators, or friction.

2.4.3 Handling H_2S Gas

One of the limitations to UBD is the presence of H_2S. The gas is common in some Canadian and Saudi Aramco wells, and there is a great deal of formal literature on the subject. H_2S can be neutralized but it causes problems related to urbanization of the area, and the perceived risk. Chapter 13 discusses acid gasses as a problem with corrosion and how to neutralize H_2S. The health and liability problem is not discussed in this book except to note that H_2S is a partial limit to UBD.

2.4.4 Barrier against Uncontrolled Flow

Always be aware that the RCD and choke, not the drilling fluid, have become the primary barriers against uncontrolled flow. Limitations and maintenance to this equipment are practical safety limits to any UBD project.

What is often overlooked in measuring productivity of the different zones is a very close measurement of the depth interval. Limitations in precision pressure gages and flow equipment on the drilling rig, as well as the high cost of time to make underbalanced flow measurements, limits the technique to very few operators.

2.5 Flow Drilling: Drilling Underbalanced with a Single-Phase Fluid

With the exception of using UBD for the sole purpose of increasing drilling rate or to avoid lost circulation, the key to using underbalanced drilling in a reservoir (or as a completion technique) is to maintain underbalanced levels in the reservoir while drilling and completing the well. Even short periods of overbalance will cause damage to the reservoir and may initiate lost circulation. Single-phase fluid systems allow a much tighter control of bottom-hole pressure than does gaseated fluid or foam, however, some items need careful planning.

Careful consideration needs to be given to annular pressure loss (APL) to be sure that the well remains underbalanced. APL is affected by friction losses, internal liquid friction, fluid density, and penetration rate. This is especially critical in two areas:

1. Pressure surges from the upward or downward movement of the bottom-hole assembly (BHA) during connections and trips.

2. Pressure changes from changes in pump rate.

It may not be possible to stay underbalanced to the reservoir at all times. During a trip or down time almost all reservoirs will flow enough for the annulus pressure to become balanced with the reservoir pressure. If the flow is a liquid, the reservoir will come into balance. If the flow contains gas, a closed-in annulus will become overpressured as the gas migrates up through the liquid in the hole. In long horizontal wellbores, the ECD may force the bottom-hole pressure above pore pressure at the toe of the well, Figure 2–5 (Oyeneyin, 2003).

2.5.1 Flow Drilling with Normal Returns

The flow line returns and pit volume should be always monitored. The standpipe gauge should be also observed for any reduction in pressure which indicates presence of gas in the annulus and signifies an approaching kick. When the RCD is opened, the annular preventer or a ram should be closed. Bit floats are required to prevent kicks through the drillpipe.

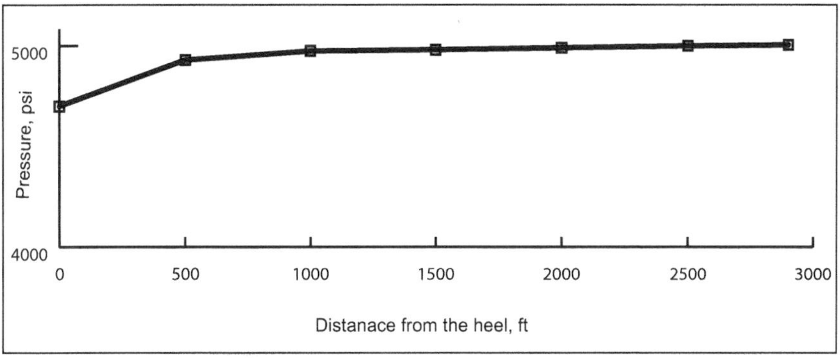

Figure 2–5 *Annular pressure as a function of distance from the heel (Oyeneyin, 2003)*

2.5.2 Flow Drilling with Formation Gas or Fluid Returns

Note from Chapter 1 that the appearance of a gas cut does not necessarily mean a significant reduction in bottom-hole pressure. Drill gas, and even connection gas, normally do not effect bottom-hole pressure enough to justify increasing the wellbore pressure.

- With severe and increasing gas cutting during a connection, shut in the well to buildup pressure in the wellbore and reduce flow into the hole and gas migration.
- When circulating, increase the standpipe pressure with the choke to minimize the influx (the Driller's Method of Well Control, found in Chapter 1, Section 1.12.5, page 18).
- Increase mud density to minimize the flow into the well.

2.5.3 Flow Drilling with No Returns

Flow drilling without returns can be done until a casing point is reached, but this should only be for a limited time. Mud or water can be pumped into the annulus while drilling to flush the cutting into the lost zone. Without pumping mud or water in the annulus the cuttings may pile up and cause stuck pipe.

2.5.4 Flow Drilling with No Returns and with Gas Rising to the Surface

This is when the pressurized mud cap (PMC) technique should be used. Water or mud is pumped down the annulus to flush the gas into the lost zone or to repress the gas in the annulus. The annular pressure is monitored as a measure of gas percolation up the annulus. Pressurized mud cap drilling is discussed in detail in Chapter 7.

2.5.5 Drilling in a Long Horizontal or High Angle Well

Long "horizontal" wellbores develop significant annular pressure loss during circulation, which makes the wellbore pressure at the toe of a well higher than at the heel. The pressure difference is not always a problem since in some cases there is an adequate margin between lost circulation and a well kick or wellbore instability. In other long wellbores, this causes lost circulation at the heel and a well kick at the toe. For initial planning estimates in a worse case scenario, assume a pressure increase of 0.1 psi/ft or 2.25 kPa/m in the horizontal section of the hole. This can be modified or re-calculated later in the planning process.

To reduce the ECD in the planning stages:

- A larger wellbore size can be designated.
- The drillpipe size can be reduced at least near the bottom-hole assembly.
- The fluid viscosity-resistance to flow can be reduced.
- The flow rate can be reduced.

However, all that can be done is to reduce the trend of the pressure difference. The pressure can only be controlled at a single point in the horizontal annulus, whether at the heel, the middle, or the toe.

2.5.6 Circulating Density

The dynamic bottom-hole pressure increases with flow. The annular pressure loss (APL) is the pressure at the bottom of the hole or at any point in the annulus due to fluid flow. There is a ΔAPL due increase/decrease in pressure from increasing/decreasing flow rate or flow directional change during rapid pipe movement. Flow rate changes the well pressure as a result of friction between the fluid, the wellbore, and drillpipe, as well as some increase due to pipe rotation (APL will also increase with drilling rate due to a higher cuttings load).

APL occurs as the result of fluid movement. ΔAPL or APL changes occur with:

- Start up of the pump
- Gel strength in the mud

- Pump rate changes/velocity changes
- Changes in direction of pipe movement
- Pipe rotation

The most common time for lost circulation is when the pipe is running in the hole or when the pump is turned on. Well kicks often occur when the pump is turned off or when the drillpipe is pulled from the hole.

A more detailed look at the mathematics of pressure loss due to friction or APL and the relationships of flow rate and flow rate changes due to pipe movement is explained in further detail in Section , Rheology of Single Phase Fluids.

2.5.7 "Gel Strength" and Inertial Forces

When the pumps are first started, gel strength, or the initial resistance to flow, can cause the well to become overbalanced. Pressure surges due to gel strength after connections or trips can be significant (Figure 2–6). Pressure surges equivalent to an ECD change of ±1 ppg or ±10 kg/m^3 are not uncommon in drilling fluids using bentonite or bentones (a heat modified bentonite) as viscosifying agents. Clear drilling fluids provide minimal gel strength and inertial force. Bottom-hole pressure changes due to pump, pipe movement, gel strength or a constriction in the annulus. All drilling fluids can show an increased APL due to build up of drill solids.

Oil muds (invert emulsions), vary from very low to high gel strengths depending upon their composition and solids content. Some of the older systems that made extensive use of bentones have significant gel strengths with time and temperature. Some of the recent invert emulsion systems show much better flow properties.

"Breaking circulation" is a good test of gel strength and inertial forces in the drilling fluid. Good field practices require a slow start and gradual build up of the pump rate while picking up the drillpipe. A plot of start up pump pressures versus shut down pressure can be made to show a hysteresis that can be attributed to gel strength and inertial forces. The same information can be observed with a bottom-hole pressure gage.

In many hard rock or low permeability drilling operations, wellbore pressure surges below 100 psi (700 kPa) appear to do little damage to the wellbore. There are an equal number of cases on land, and especially offshore, where higher pressure surges in the wellbore create instability and washouts that lead to fill up on trips and tight hole.

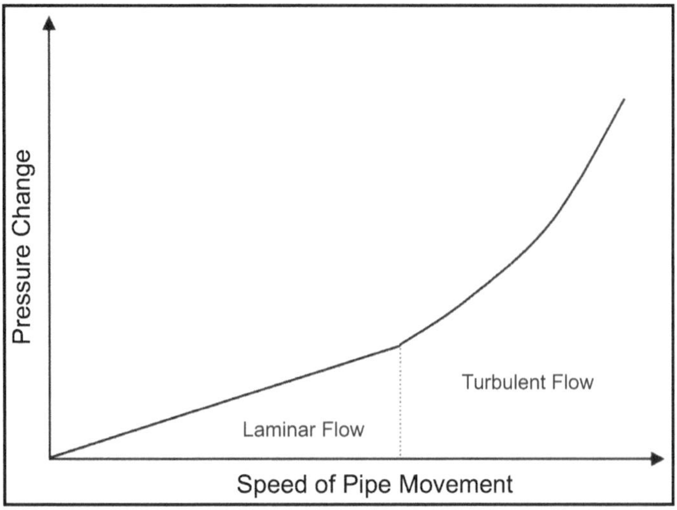

Figure 2–6 *Pressure change in the annulus depends on the speed of pipe movement and fluid flow pattern*

2.6 Connections

In UBD, the drilling fluid is not the primary barrier against a well control incident. The drilling fluid column may, to some degree, inhibit open-hole flow and high surface pressures, but the primary barriers are the rotating control device (RCD) and the choke system. The well may have to be shut in using the RCD and choke system to hold pressure against flow of the formation fluid. However, during a connection, the underbalanced condition actually may revert to a balanced system where the annular surface pressure (from the choke) and mud column pressure balance the formation fluid pressure. On a long connection, some upward migration of gas will occur and surface pressures may rise; as a result, formation fluid may be pushed back into the formation.

When the drillpipe is lifted off bottom on a connection, a swabbing effect occurs (APL) that is dependent upon the rate of pipe movement, the hydraulic diameter of the hole, and the drilling fluid properties. General good practice requires that the pump be kept on until the drillpipe is in the slips. This limits some of the swabbing effect due to pipe movement.

When the pump is turned off, the bottom-hole pressure from the APL is reduced by the friction element. Managed pressure drilling operations compensate for the reduction in bottom-hole pressure by increasing the surface shut in pressure. This is not normally done in underbalanced operations, since any flow from the formation will pressure the wellbore to a balance point.

After the pipe connection, the pump is started, the choke is opened, and the pipe is lowered back to the bottom. If there is significant gas in the annulus, the choke opening rate and the pump start rate are not critical because there is gas compression or expansion to cushion any hydraulic shocks. With limited gas in the annulus, pump start rate and choke opening rate should be brought on slowly to limit the hydraulic shock to the hole.

A better method is to use the simple step-wise transition method proposed by Medley et al., 2008. When shutting the pumps on or off, a table is made of stepwise annular pressure and pump rate values. A plot is made to show how annular pressure should increase as the pump speed decreases. The plot is used to maintain the desired wellbore pressure. For example, to reduce the pump speed, first reduce the choke opening to increase the back pressure, then reduce the pump speed to a value matching the back pressure from the plot, and then repeat this until zero pump speed at the desired maximum back pressure is reached (see Figure 2–7).

2.7 Trips

Short trips for hole evaluation or trips out of the hole should follow proper connection procedures. In low pressure wells, the well can be left open or shut in against the RCD. In a well where fluids can come to the surface, the RCD and well choke may have to be closed as the pump is turned off to leave the well under pressure while the pipe is stripped out of the hole. Pipe should be pulled slowly through the RCD to limit wear and tearing of the packer element.

Fill-up of the hole with a single-phase drilling fluid on an "underbalanced" trip is generally not necessary, or is limited to the displacement of the drillpipe. A well drilled in a highly permeable or fractured formation will quickly come to a pressure balance, so it is not necessary to "fill the hole" since it automatically fills.

In wells, when any gas is present, the surface pressure needs to be monitored during trips to avoid excessive surface pressure buildup. Under most conditions, a pressure buildup at the surface will only push formation fluid back into the reservoir. However, surface pressure may pose a problem when stripping back into the well. A mud cap is often a temporary solution to stripping-in problems.

When tripping out of the hole in very low permeability formations, it may be desirable to keep the fluid level in the hole constant;

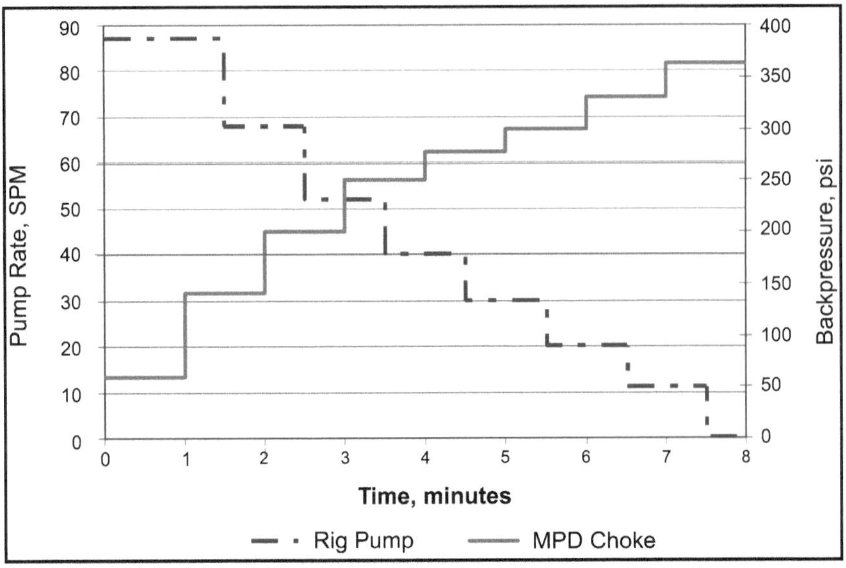

Figure 2–7 *Example back pressure versus pump rate (Medley, 2008)*

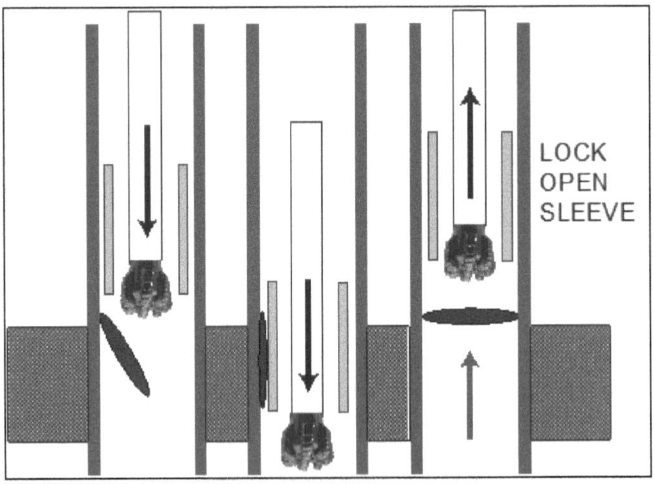

Figure 2–8 *Downhole isolation valve (Ramalho and Davidson, 2006)*

in those cases, fill-up equal to the displacement of the drillpipe is sufficient.

The surface pressure needs to be kept constant by releasing fluid through the choke when tripping back in the hole with the RCD engaged. The drillpipe should be filled at suitable intervals, generally every 10 to 12 stands, but the drillpipe should not be overfilled.

At the end of the trip, or if the hole is to be circulated at the bottom of the casing, the pump should be started slowly to break any gel strength in the mud. The pump start up procedure with a closed-in well is similar to the driller's method well control. Casing pressure can be held constant by the choke until the mud pump gets up to speed. After that, the standpipe pressure should be held constant by adjusting casing pressure with the choke to avoid pressurizing the hole in case there is gas coming up the annulus.

The use of a down-hole casing, or isolation, valve effectively closes off the hole during a trip, improves safety, simplifies procedures, and reduces time for tripping. In a wellbore where gas is capable of flowing to the surface from a high permeability formation, the down-hole casing valve is a significant improvement over mud caps and snubbing (see Figure 2–8) (Ramalho and Davidson, 2006).

2.8 Solutions and a Short Summary

The long list of concerns does not mean that flow drilling is an especially difficult technique. This is born out by the example of extreme flow drilling projects in the following sub-chapters. The previous concerns primarily deal with ECD and the gel strength when the pumps are started after a down period. Here is a list of solutions:

- One solution to the connection and trip problem is the use of a dual casing string (concentric casing). Drilling fluid is passed down the secondary annulus between the primary casing and the secondary casing during connections and trips, which results in a continuous circulating system (Vestavik et al., 2009). Chapter 3, Gaseated Fluids, contains a discussion of the dual-casing system. It is also shown as a working technique in Chapter 2, Section , Friction Controlled Drilling, A Novel Approach to Drilling HPHT Wells Underbalanced.
- Another solution would be the use of Continuous Circulating System (CCS) which allows circulation during the connections. The CCS creates a chamber around the pipe joints and circulates into the drillstring when the pipes are not connected (Rehm et al., 2008). There are also continuous circulating subs which serve the same purpose.
- ECD can be modified by limiting the length of slim hole or increasing the hole diameter to lower the ECD.

- ECD can be modified by changing the mud properties. As a simplistic comment, the lower the drilling fluid viscosity and gel strengths, the lower ECD and ECD changes will be.
- Reservoir analysis while drilling the zones may require more redundancy or more special gauging. If produced gas or oil is to be marketed while drilling, other equipment will be required for the sales line or tank. Equipment systems suitable for underbalanced drilling are discussed at length in Chapter 11.

2.9 Questions

Many of these questions do not have a single answer. The well site often does not have a single answer. Think about the question and what is involved in the real problem.

1. What is a single-phase drilling fluid and what is a dual phase drilling fluid?
2. An 8 ½ in. horizontal gas and condensate well is to be drilled in a fractured and vugular limestone with 9 ppg saline water. The formation pressure at the heel of the well is equivalent to 9.5 ppf. The lateral is planned to be 5,000 ft long. What general problems would be present in any well of this type?
3. How does a down-hole casing valve help make a trip?
4. How would you control the reservoir flow into the wellbore?
5. How would you pull the BHA out of the BOP stack?
6. In your experience, what kind of ECD would there be between the heel and toe of the well?
7. What is the difference between managed pressure drilling and flow drilling?
8. Which one is more affected by ECD changes?
9. Why does drilling underbalanced enhance ROP, and which shows the greatest effect, tricone bits or PDC bits?
10. What are the forms of wellbore instability? Can we use underbalanced drilling if we have wellbore stability problems? Explain.

11. What determines the direction of drilling in horizontal or deviated wells? What happens if we do not follow the recommended direction?

12. What is the trend for ECD in horizontal lateral? How is the effect of this ECD trend minimized when drilling underbalanced?

13. How is APL affected by swabbing and surging effect and how is this effect minimized?

14. Define hydraulic shock and how to minimize it.

2.10 References

Colbert, J.W. and Medley, G. "Light Annular MudCap Drilling—A Well Control Technique for Naturally Fractured Formations," SPE 77352 presented at the SPE Annual Technical Conference and Exhibition, San Antonio, TX, USA, September 29–October 2, 2002.

DuPriest, W. et al. "Borehole Quality Design and Practices to Maximize Drill-Rate Performance," *SPE Drilling and Completion*, June 2011, pp. 303–316.

Hallman, J.H., Cook, I., Muqeem, A.M., Jarrett, C.M., Shammari, H.A. "Fluid Customization and Equipment Optimization Enables Safe and Successful Underbalanced Drilling of High-H2S Horizontal Wells in Saudi Arabia," SPE 108332 presented at the IADC/SPE Managed Pressure Drilling and Underbalanced Operations Conference and Exhibition, Galveston, TX, USA, March 28–19, 2007.

Malloy, K., Stone, C.R., Medley, G.H., Hannegan, D., Coker, O., Reitsma, D., Santos, H., Kinder, J., Eck-Olsen, J., McCaskill, J., May, J., Smith, K. and Sonneman, P. "Managed-Pressure Drilling: What It Is and It Is Not," SPE 122281 presented at the IADC/SPE Managed Pressure Drilling and Underbalanced Operations Conference and Exhibition, San Antonio, TX, USA, February 12–13, 2009.

Medley, G.H., Moore, D. and Nauduri, S. "Simplifying MPD: Lessons Learned," SPE 113689 presented at the SPE/IADC Managed Pressure Drilling and Underbalanced Operation Conference and Exhibition, Abu Dhabi, UAE, January 28–29, 2008.

Muqeem, M.A., Jeffre, A.M., Jarrett, C.M., Khanferi, N.M., Killip, D.R. and Abdul, H.J. "Underbalanced Drilling in Saudi Arabia—Start Up Experience," SPE 102026 presented at the IADC/SPE Asia Pacific Drilling Technology Conference and Exhibition, Bangkok, Thailand, November 13–15, 2006.

Muqeem, M.A., Jarrett, C.M. and Abdul, H.J. "Underbalanced Drilling of Oil Wells in Saudi Arabia: Case History and Lessons Learned," SPE 114258 presented at the IADC/SPE Managed Pressure Drilling and Underbalanced Operations Conference and Exhibition, Abu Dhabi, UAE, January 28–29, 2008.

Oyeneyin, M.B. "Challenges of Underbalanced Drilling for West Africa," SPE 85662 presented at the SPE Nigeria Annual International Conference and Exhibition, Abuja, Nigeria, August 4–6, 2003.

Ramalho, J. and Davidson, I.A. "Well Control Aspects of Underbalanced Drilling Operations," SPE 106367 presented at the IADC/SPE Asia Pacific Drilling Technology Conference and Exhibition, Bangkok, Thailand, November 13–15, 2006.

Rehm, B. *Practical Underbalanced Drilling and Workover*, Petroleum Extension Service, University of Texas, Austin, TX, 2003.

Rehm, B. et al. *Managed Pressure Drilling*, Gulf Publishing, Houston, TX, USA, 2008.

Surewaard, J., De Koning, K., Kool, M., Woodland, D., Roed, H. and Hopmans, P. "Approach to Underbalanced Well Operations in Petroleum Development Oman," SPE 35069 presented at the IADC/SPE Drilling Conference, New Orleans, LA, USA, March 12–15, 1996.

Sutherland, I., Grayson, B. "DDV Reduces Time to Round-trip Drillstring by Three Days, Saving £ 400,000," SPE 92595 presented at the IADC/SPE Drilling Conference, Amsterdam, The Netherlands, February 23–25, 2005.

Vieira, P., Larroque, F., Al-Saleh, A.M., Ismael, H., Qutob, H.H. and Chopty, J.R. "Kuwait Employs Underbalanced Drilling Technology to Improve Drilling Performance while Simultaneously Evaluating the Reservoir," SPE 106672 presented at the Offshore Europe Conference, Aberdeen, Scotland, UK, September 4–7, 2007.

Vestavik, O., Kerr, S. and Stuart, B. "Reelwell Drilling Method," SPE 119491 presented at the IADC/SPE Drilling Conference and Exhibition, Amsterdam, The Netherlands, March 17–19, 2009.

Section 2
Underbalanced Drilling Experience in the Ghawar Field

Mohammad Muqeem, Saudi Aramco

2.11 Introduction

A major campaign was implemented in the Ghawar field to evaluate underbalanced drilling (UBD) in Saudi Arabia. The primary objective of drilling wells underbalanced was to eliminate formation damage, thereby eliminating the need for acid stimulation. Due to the inherent low risk of applying this technology in power water injection wells (PWI), UBD was first introduced to these types of wells. Additionally, reduction in both drill time and cost resulted with the increase in rate of penetration (ROP).

This resulted in improved injection rates, and the routine practice of acid stimulation was eliminated. ROP increased significantly when compared to offset PWI wells drilled conventionally.

This chapter describes Saudi Aramco's start-up experience while introducing underbalanced drilling technology in the Ghawar field. The intention was to prove that a step change in performance could be achieved by the application of this technology. Thus the intent was to optimize well design that will result in reduced unit well cost.

2.12 Background

PWI wells are drilled in the Ghawar field to maintain pressure for optimum production of oil from the Arab-D reservoir. Arab-D formation is a predominantly fractured, oil-bearing carbonate reservoir. Hydrocarbon recovery to date has been through the drilling and completing of overbalanced vertical and deviated wellbores. Conventional operations in some instances have been exacerbated by drilling-related problems (i.e., stuck pipe problems and loss of circulation). Traditionally, PWI wells are drilled with 200 psi overbalanced through the reservoir section. Therefore, invasion of mud filtrate and drilled solids into the fractures of the carbonate formation occurs. Consequently, this type of formation damage requires extensive acid stimulation to bring back the injectivity of these wells. Further, drill-string sticking and lost circulation results in excessive non-productive time (NPT).

Saudi Aramco has identified the minimization of drilling fluid losses into the reservoir, formation damage, and less operational problems as key well objectives. UBD was considered an enabling technology to help Saudi Aramco achieve its objectives.

Based on the reservoir and well characteristics, the UBD concept was a simple one. Using water as the circulating medium resulted in underbalanced conditions across the reservoir section, which enabled utilization of a less complex fluid as the drilling medium.

2.13 Planning Phase

The geology of the area drilled is a carbonate/dolomite zone with varying degrees of porosity. Zone 2A is the most prolific zone with the highest porosity. Zone 2B is a tighter zone and usually exhibits lower porosity values. Zone 3 is the densest zone with the lowest porosity. The wells that were drilled cut a path through all three zones.

As noted in the first paragraph, due to the inherent low risk of applying this technology in power water injection wells (PWI), UBD was first introduced to these types of wells.

2.13.1 UBD Operational Envelope

Figure 2–9 shows a plot of the bottom-hole circulating pressures versus fluid injection rates at varying choke pressures. The operating envelope that was established was based on the current wellbore configuration. The operating envelope is shown within the shaded area in Figure 2–9. The star highlights the optimum combination of both injection rates and choke pressure to achieve the desired bottom-hole pressure.

To manage hole cleaning and inflow, additional constraints must also be addressed within the operating envelope during UBD operations. A maximum drawdown of 200 psi at the bit was designed to minimize the water influx and to account for transients during connections. The 200 psi underbalanced was incorporated into the design to ensure that underbalanced conditions in the well were maintained at all times. The minimum and maximum down-hole motor equivalent flow rates were between 180 gpm and 250 gpm, respectively on these wells. The down-hole conditions were also constrained by the ability of the circulating system to effectively achieve an underbalanced state while providing adequate hole cleaning.

Performing the initial detailed engineering study and establishing clear operating envelopes allowed the PWI wells to be successfully

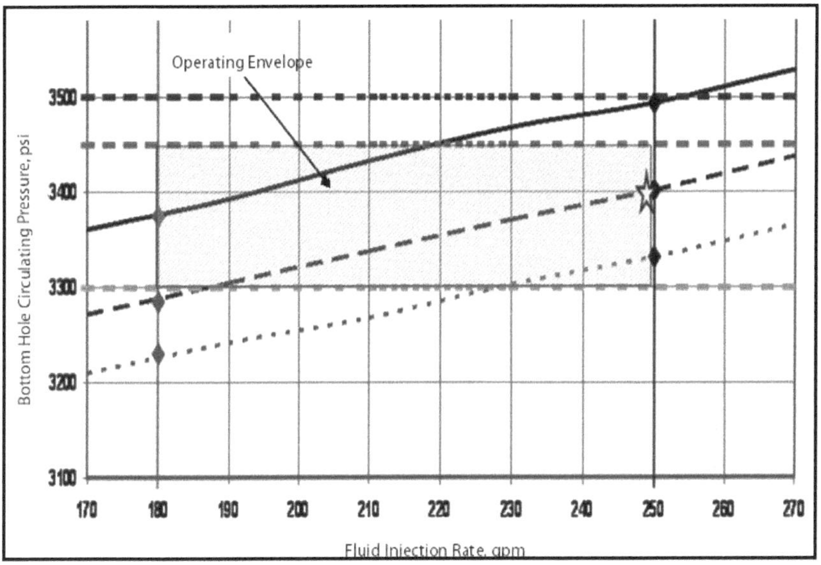

Figure 2–9 *UBD-operating envelope*

drilled underbalanced while maintaining hole cleaning objectives and ensuring that equipment constraints were adequately maintained.

2.13.2 Training

Extensive crew training was conducted by a third party consultant specializing in UBD operations prior to the UBD project start-up. Additionally, key personnel from Saudi Aramco were included in the training process. Both onsite hands-on and classroom training was conducted. Furthermore, all key personnel were certified with IADC approved UBD Well Control Supervisory course.

2.14 Initial Wells

The first three wells drilled in the UBD campaign encountered several problems due to lack of an integrated approach to equipment selection and well design.

The first package of surface control equipment was selectively picked from a number of service providers which resulted in confusion in responsibility and leadership of the project. Steps were eventually taken to have a single service provider supply all the surface equipment for this package.

Dual laterals and very long single laterals were the norm. No changes to well design were made for the initial UBD wells. This resulted in questionable underbalanced conditions in the second of the dual laterals as no attempt was made to isolate the first lateral while drilling the second. In addition, the very long single laterals resulted in fishing jobs due to hole tortuosity and probable poor hole cleaning, as the toe end of the wells became overbalanced due to the high influx at the heel end.

The initial wells drilled underbalanced had a single waste pit system aligned to be downwind of the rig at the prevailing wind direction. Reversal of wind direction commonly occurs in desert locations, and during such times operations had to cease due to blowback of sour gas from the water in the pits. The long laterals also meant that the rig was drilling for much longer and producing much more water to the surface. Coping with this vast amount of water became a logistical and environmental challenge.

2.14.1 Change in Casing Point

In this UBD campaign, only the reservoir section was drilled underbalanced. All other well design remained the same. In the initial stages of the UBD project, the 7 in. liner was set about 2 ft TVD below the Arab-D in Zone 2A reservoir top to cover the anhydrite layer as was the practice in conventional drilling. This created problems in UBD since Zone 2A had a higher pressure than Zone 2B. Most of the time this design resulted in underbalanced conditions in Zone 2A and overbalanced conditions in Zone 2B. Consequently, this culminated in cuttings buildup in 2B that led to hole cleaning, torque- and drag-related problems.

Consequently, UBD candidate wells were later designed to isolate Zone 2A of the Arab-D formation, thereby placing the entire lateral length in Zone-2B, the target reservoir. Thus, the 7 in. liner/casing point was set 2 ft TVD below the Arab-D Zone 2B reservoir top. The reservoir sections were drilled as an underbalanced open hole. This configuration paved the way for optimum UBD operations, resulting in getting the most out of this enabling technology. Figure 2–10 and Figure 2–11 illustrate the step change in well design.

2.14.2 Change in Laterals

Realization of performance improvement (increased injectivity) with UBD resulted in dual-lateral wells to be redesigned as single-laterals. Furthermore, this rationalization of well design also led to shorter

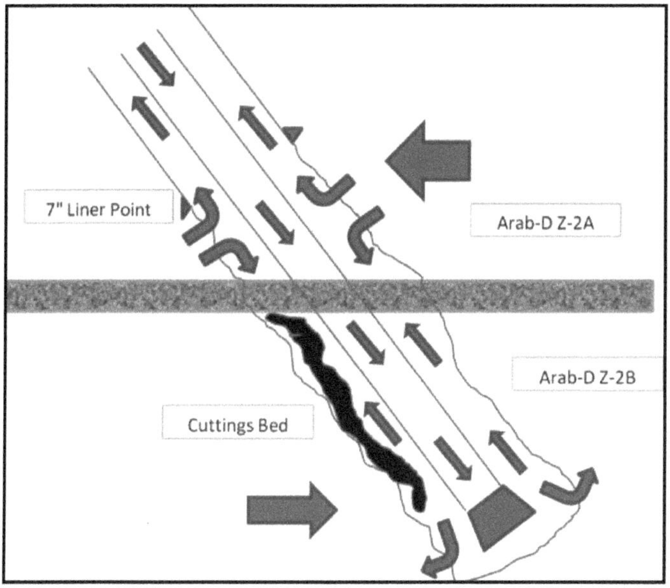

Figure 2–10 *Initial casing setting point*

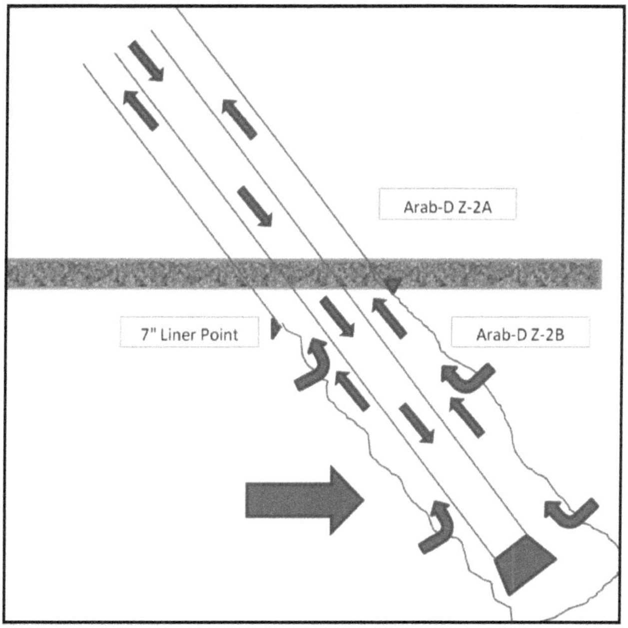

Figure 2–11 *Step change in underbalanced well design*

laterals being drilled. This yielded savings in rig days (from the point of shorter drilling length and increased ROP), and but still achieved or surpassed the projected injectivity requirements of the candidate wells.

2.14.3 Change in Wellpath

Similarly, the requirement for a less tortuous wellpath was adopted and the focus was changed to keep the wellbore as smooth as possible within the bounds of the targets set. Shorter drilling times led to less overall fluid produced to the surface, reducing the size needed for storage pits.

2.15 Documentation

By implementing UBD technology in Saudi Arabia for the first time, the project generated various documents which were additional to the conventional drilling operations. Several operational procedures new to Saudi Aramco were written and additional pieces of equipment were introduced. All these extra documents need to be controlled and the information in them needed to be disseminated thoroughly among the team members directly involved with the UBD operations. As part of this philosophy, a UBD Manual specific to Saudi Aramco UBD Operations was generated. This manual includes the following among many other UBD specific operations, such as UBD Equipment, UBD Operations, H_2S Management, UBD Well Control, and Snubbing Operations.

2.16 Sour Gas Provisions

Most of the wells are sour in varying degrees. H_2S Management plays a key role in ensuring that proper safety measures are in place depending on the specific well types before any UBD operation commences. A guideline has been established in conjunction with a Safety and Environmental Compliance Group that includes an UBD pre-startup checklist and UBD HSE equipment requirements.

2.17 Subsequent Wells

Using the single source approach for the surface package and going through a thorough HSE planning process paid dividends. A second complete UBD package was mobilized and ten wells were subsequently drilled underbalanced with very few major problems, although many other lessons were learned along the way. The very nature of drilling water injectors underbalanced was not common practice worldwide, and the service companies involved had an equal learning curve to optimize the equipment for the purpose.

2.18 Conventional versus UB Comparisons

Comparisons between conventionally drilled offset wells and UBD wells are shown in Figure 2–12. In all cases, there is a marked difference between conventional and underbalanced drilled wells.

2.18.1 Drilling Fluid Costs

The fluid of choice in drilling power water injection (PWI) wells was simply produced formation water with some caustic added to control the pH. This light fluid was used to reduce the wellbore hydrostatic pressure. This resulted in immediate cost savings over conventional mud. The need to perform acid cleanouts after drilling operations was no longer needed as the wells were no longer damaged by conventional mud filtrate and fines invasion.

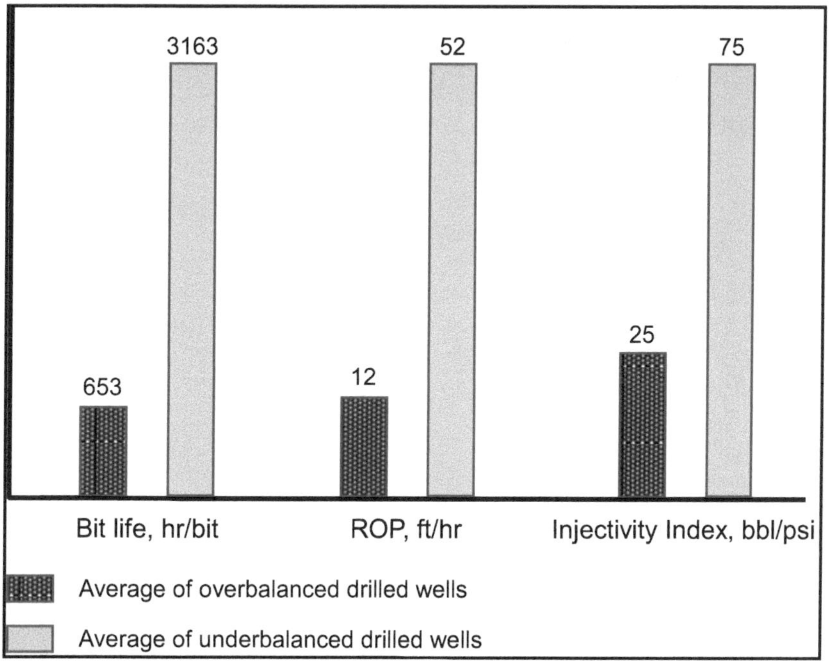

Figure 2–12 *Comparison of overbalanced and underbalanced wells*

2.18.2 Bit Life

It was common practice to use more than two bits to drill the long laterals in conventional drilled wells. The average life achieved for a bit run was of the order of 653 ft. Underbalanced wells are now routinely drilled with a single bit, the average bit run achieves approximately 3,163 ft. This is illustrated in Figure 2–12. Increased bit life also leads to other savings such as reduced tripping time as well as the cost of the bit itself.

2.18.3 Rate of Penetration

Rate of penetration (ROP) can be affected by many variables, including: hydraulics, weight-on-bit, rotational speed, bit wear, lithology and *bottom-hole circulating pressure.* It is this latter factor that is most important when ROP is discussed in relation to underbalanced drilling. The increase in ROP is attributed to the reduction in the confined strength of the rock associated with the lower bottom-hole circulating pressure. The inflow from the reservoir, due to the positive pressure, greatly assists in the removal of drill cuttings at the bit, thereby reducing and/or eliminating the common phenomenon known as chip hold-down. In Saudi Aramco's UBD campaign, the average ROP achieved was more than three times the conventional performance in the nearby offsetting wells. This performance is again illustrated graphically as Figure 2–12. Significant savings of rig time and greater efficiency of the operation was realized due to the three-fold increase in ROP in conjunction with the increased bit life.

2.18.4 Injectivity Index

The higher injectivity in a shorter lateral length is the final piece of the success story. Compared to conventionally drilled and acid-treated wells, the average UBD well realized two-fold increase in the injectivity index, Figure 2–12.

2.19 Case History of Initial Challenges

This well was planned to be drilled as a single horizontal lateral with an open-hole length of approximately 8,300 ft. After drilling 1,050 ft, high standpipe pressure was observed and a decision was made to pull out of the hole and investigate. When the drillstring was recovered no obvious indications were present, so the motor and bit were changed and the string was run back into hole. After a further 2,811 ft was drilled, difficulty in slide drilling was occurring so the drillstring was again tripped to reconfigure the position of the HW drillpipe in the string. A further 1,106 ft of hole was drilled before high standpipe pressure was again encountered. At this stage 4,967 ft of reservoir had been drilled and the time taken was 11 days.

After another trip, during which it was discovered that the standpipe pressure increase had been due to scale buildup inside the BHA components, the motor was changed out and the new assembly run back to bottom. After a further 929 ft of formation had been drilled, sliding problems re-emerged. A decision was made to trip the pipe to reconfigure the string components, but the pipe became stuck while pulling it out. Attempts to free the pipe resulted in the pipe backing off downhole. Further attempts were made to screw the pipe back in, but strong flowback was observed up the drillpipe and the shear rams were deployed to shut the well in. Eight days were subsequently spent fishing the two pipe sections. Drilling then continued and a further 409 ft was drilled before another trip to reconfigure pipe became apparent. The next drilling session achieved a further 256 ft before directional control problems necessitated a trip to change the BHA again. Drilling then resumed and an additional 1,753 ft was drilled before excessively high torque made further drilling impossible and TD was called.

The 7 in. liner could not pass 7,020 ft and was pulled out of the hole, and the hole was reamed. The 7 in. liner was then run again, but could not pass 7,475 ft and was pulled out of the hole; again the hole was reamed. Finally, on the third attempt, the liner was run and cemented. After retrieving the cementing sleeve, it was found packed with cement when inspected at the surface. The down-hole isolation valve failed to seal and it was assumed that residual cement in the lower valve cavity caused the valve not to function as planned.

Total time taken over the reservoir section was 30 days and the injectivity results were similar to a conventionally drilled well. This could have been due to the repeated killing of the well with brine for every trip made. Time and unnecessary expense could have been saved had the well design been optimized earlier on with a shorter lateral length. An injectivity test should have been conducted at the 11 day mark when nearly 5,000 ft of the well had been drilled to ascertain the injectivity of the well.

Despite the setbacks, lessons were learned and improvements made in terms of well planning and operational procedures.

2.19.1 Second Well-Challenges

The 7 in. casing was run to 6,920 ft and the pipe became stuck. Several attempts to free the pipe were not successful. Finally, the liner was backed off above the valve and the well was sidetracked. The section was re-drilled and the decision was made not to run another isolation valve in the section. The large outer diameter of the tool was thought to contribute to the difficulties in getting the casing to bottom.

2.19.2 Third Well-Challenges

As a result of having difficulties getting the tool down in the previous two wells, the hole was drilled and opened simultaneously to 9-7/8 in. The 7 in. casing with the isolation valve was run, and there were no difficulties going into the hole or cementing the casing. Additional measures such as spotting acid to remove any residual cement were performed. Several unsuccessful attempts were made to engage and open the valve. Eventually the valve was milled prior to drilling.

2.19.3 Fourth Well-Challenges

There were no problems encountered with the running of the tool in the 7 in. tie-back string aside from the necessary pre-planning required to enable the isolation valve control lines to be run in circumstances that were not normal for a standard operation. Initial tests showed the tool held underbalanced pressures. After drilling the section underbalanced and pulling the drilling string above the tool, the tool failed to hold pressure for a period of time. The valve was flushed with fluid and finally held pressure. It was suspected that there may have been some debris in the flapper valve allowing the well pressure to bypass the valve.

2.19.4 Case History of a Recent Well

This well was to be drilled as a single horizontal lateral with a target length of approximately 3,235 ft. Drilling the section was achieved in a single bit run in a time of approximately 53 hours, maintaining good underbalance and inflow at all times. The well was only killed once with filtered brine to allow removal of the BHA from the hole, and the injectivity was the highest yet achieved in the UBD drilling campaign.

2.20 Early Experience with (Down-Hole) Isolation Valves

One objective of any well drilled underbalanced is to maintain total underbalanced conditions (i.e., while drilling and completing operations). This is necessary to eliminate the introduction of foreign solids into the zone of interest. Killing the well is one way of tripping drillstring into and stripping out of the well. This defeats the purpose of utilizing underbalanced technology in the first place.

Current methods of achieving and maintaining underbalance conditions primarily involve the use of a snubbing or hydraulic workover unit to pull out of the hole or run in the hole with drilling tools and equipment. This process increases tripping times, adding appreciable operational complexity and expense. In addition, the snubbing unit cannot seal around complex completion assemblies (Grayson, 2004).

One evolving means of maintaining an underbalance is with a down-hole isolation valve, either mechanical or hydraulic. The isolation valve is run as an integral part of the casing program. When it becomes necessary to trip the drillstring, the string is stripped out until the bit is above the valve, at which time the valve is closed and the casing string above the valve is bled off. At this time, the drillstring can be tripped out of the well without the use of a snubbing unit and at conventional tripping speeds, thus reducing rig time requirements and providing improved safety. The drillstring can then be tripped back into the well until the bit is just above the down-hole valve, at which time the valve can be opened and the drillstring run in to continue drilling operations. The down-hole valve can be run and cemented in place with the casing or with a liner hanger and tie-back assembly. Some of the benefits of an isolation valve are highlighted below:

- Isolates the reservoir while tripping during drilling operations
- No snubbing operations required
- Enables the safe deployment of slotted liners or production screens without killing the well
- Faster tripping times
- Reduces costs and increases safety

The objective of testing the two types of tools was to determine if they could be utilized as a downhole mechanical barrier to eliminate the need for a hydraulic workover unit.

2.20.1 Mechanically Operated Isolation Tool

A mechanically operated isolation valve was run into a PWI well to determine if it was suitable and could be classified as a barrier and eliminate the requirement for a hydraulic workover unit. In this case the mechanical isolation tool was run with the 7 in. casing. The tool required no external control lines, utilized a flapper valve, and used metal to metal seals. The tool was run initially with a cementing sleeve installed to protect the flapper valve during the cementation process. This sleeve is then retrieved, allowing the flapper to close and isolating the reservoir in an underbalanced state. Another sleeve is run into the hole with the drilling assembly and is locked into place across the flapper valve, protecting the valve while the drilling tools are tripped below the valve. There was a problem running the bent housing bottomhole assembly through the bore of the tool without damaging the assembly.

2.20.2 Hydraulically-Operated Isolation Tool

A hydraulically-operated isolation tool was run into the fourth well to determine if it was suitable and could be classified as a barrier and eliminate the requirement for a hydraulic workover unit. The hydraulically operated isolation tool was run with a 7 in. tie-back string after running and cementing the 7 in. liner. The tool required two control lines and also utilized a flapper valve. The valve functions by pumping hydraulic fluid down the control line. It operated with minimum difficulties.

2.21 Operational Improvements

- A stepwise approach to UBD operations was used, allowing newly learned lessons to be implemented continuously without the usual management pressures that new technology brings. The first few wells were used to safely develop the technology and the subsequent wells were used to improve it.
- Operational issues that need to be implemented immediately were addressed individually until confidence and competence were achieved.
- Team continuity allowed focus from conception to implementation and continuous improvement, including excellent teamwork between all service providers.

- Conflicts were avoided due to good supervisory management, which in addition to the appropriate education of the UBD process, allowed individuals to understand their roles and responsibilities.
- Close scrutiny and qualification, including auditing by third party specialists of all equipment and drillpipe components, were critical to the project's success.

2.22 Lessons Learned

- UBD technology was proven to be a safe method of drilling in Ghawar field. No major safety or environmental incidents have occurred in the project so far.
- UBD resulted in improved ROP and longer bit life. Reduction in formation damage has resulted in improved injectivity.
- Selection and capacities of the rig, drillpipe, BOPs, rotating control device (RCD), separation system, scrubber system, and flare stack will be critical for future operations to ensure the safety and environmental compliance, as well as the success of this enabling technology.
- Utilization of a single service provider for all the surface UBD equipment resulted in greatly improved operation. A single focal point for the UBD operation on site improved planning, communication, and safety.
- Health and Safety procedures were more rigorously pursued and full HAZOP and HAZID programs were initiated complete with change management procedures. Many items were identified during these processes and the appropriate measures taken to solve the problem or mitigate the risk involved. All parties involved worked towards a process of continual improvement and rigorous documentation of the changes made.
- The ability to work safely in the presence of H_2S gas was verified with the first wells drilled.
    - H_2S monitors and gas detectors were employed at critical areas around the rig, including the discharge lines at the waste pit.
    - A dual-produced water pit system was introduced so that fluid could be diverted to either pit depending on the prevailing wind direction. This only partially solved the problem as the pits had to be built on the same side of the

rig at a ninety degree angle to each other, and there were times when this solution was less than optimal.

- Data acquisition was continually enhanced during the campaign, giving added value to the operation and greater knowledge of the reservoir as it was being drilled.

2.23 Important Questions about the Chapter

What are the prerequisites for the application of underbalanced drilling technology in any given reservoir/field?

What are the key requirements before any UBD operation is implemented in a candidate well?

What sort of training is required for the UBD Team?

Is ongoing analysis of the UBD technology results important? Why?

2.24 References

Grayson, B. "DDV Usage Reduces Time to Round Trip Drillstring, Cut Costs," *World Oil,* May 2004, pp. 25–28.

Hallman, J.H., Cook, I., Muqeem, M.A., Jarrett, C.M., Shammary, H.A. "Fluid Customization and Equipment Optimization Enable Safe and Successful Underbalanced Drilling of High H_2S Horizontal Wells in Saudi Arabia," SPE 108332 presented at the IADC/SPE Managed Pressure Drilling and Underbalanced Operations Conference and Exhibition, Galveston, TX, USA, March 28–29, 2007.

Muqeem, M.A., Jeffri, A.M., Jarrett, C.M., Khanferi, N.M., Killip, D.R., Abdul, H.J. "Underbalanced Drilling in Saudi Arabia—Start Up Experience," SPE 102026 presented at the IADC/SPE Asia Pacific Drilling Technology Conference and Exhibition, Bangkok, Thailand, November 13–15, 2006.

Muqeem, M.A., Jarrett, C.M., Jeffri, A.M., Weisbeck, D., Mallahy, M., Forge, N., Branch, A.J. "The Introduction of Electromagnetic LWD Technology in Saudi Arabia—A Case History and Future Application to Underbalanced Drilling Campaigns," SPE 105471 presented at the IADC/SPE Drilling Conference, Amsterdam, The Netherlands, February 20–22, 2007.

Muqeem, M.A., Jarrett, C.M., Abdul, H. J. "Underbalanced Drilling of Oil Wells in Saudi Arabia—Case History and Lessons Learned," SPE 114258 presented at the IADC/SPE Managed Pressure Drilling and Underbalanced Operations Conference and Exhibition, Abu Dhabi, UAE, January 28–29, 2008.

Section 3
Friction Controlled Drilling, A Novel Approach to Drilling HPHT Wells Underbalanced

Robert L. "Bob" Cuthbertson, P.E.,
SunTerra Oil and Gas LP

2.25 Introductions and Background

Except for sub-normally pressured or depleted reservoirs that require a multi-phase drilling fluid or gas to lower equivalent circulating density (ECD) to any desired point, there are presently only two ways the industry uses to control higher bottom-hole pressure:

- Increase the mud weights to a higher level to increase hydrostatic pressure at depth
- Increased pressure at the surface against the rotating pressure control device

Increasing drilling fluid density requires either going to a high solids weighted system, or (if a solids free fluid is desired for UBD drilling) going to a more expensive and more corrosive salt system such as calcium chloride, calcium bromide or, in the most severe case, zinc bromide to achieve the desired density. Either of these alternatives is less than completely desirable because of cost or operational considerations.

Because of the above limits, most higher-pressured wells are drilled using elevated pressures at the surface against the acrilonitrile elastomeric or polyurethane elements of the RCD. Therefore, the control of high pressure is at the surface where it directly exposes personnel and equipment and it is least wanted.

2.26 Friction-Controlled Drilling

A third relatively unknown method to more safely control this pressure is available. It is called *friction controlled drilling*. The concept is relatively simple to understand: reduce bottom-hole pressure to a safe operating limit before it reaches the surface by intentionally creating pressure drop using friction as the fluid moves from the bit to the surface. Because so few operators know or understand this concept, abnormally high-pressured reservoirs have rarely been drilled using the non-damaging benefits of UBD.

How does one go about creating friction downhole to serve this purpose? The answer is also simple, but it requires advanced planning and design as well as additional equipment to achieve the goal. The friction drilling concept requires two additional components:

1. A concentric casing string to create a second annulus in the well.

2. A dedicated auxiliary pumping system to constantly circulate fluid into this additional annulus.

2.27 The Concentric Casing String Theory

The volume of drilling fluid circulated through the drillstring is generally limited because of the operating window of a drilling motor's performance; it is also limited by surface pressure limits of the mud pumps, standpipe, and hose due to the relatively small diameter of the drillpipe and drill collars, and the friction that they cause on the downward fluid path to the bit. Therefore, increasing standpipe injection volume is generally not an option to raise total fluid coming back to the surface on the return annulus.

By creating a dual annulus with the concentric casing string, another injection path is available: one that does not require its fluid volume to pass through the drillstring or mud motor and thereby is free to pump a wider range of volumes. The ideal friction-controlled drilling design would be one with a large outer annulus to keep surface injection pressure minimized on the concentric casing injection, and a relatively small inner return annulus in order to create additional friction pressure drop as the fluids return to the surface. The drilling fluids used to circulate through the standpipe and into the drillstring can also be used to circulate down a second annulus, thereby simplifying the surface fluid handling system.

Considering that in typical conventional well designs there is no additional fluid injection annulus, any additional volume added to the return stream will create additional friction pressure in the return annulus. The more of this fluid that is added via the auxiliary annulus, the more impact this extra stream of fluid has on the cumulative return friction. Because of the nature of the Reynolds Law and critical flow rate, this additional return volume greatly increases the friction pressure drop along the entire return annulus flow path to the surface.

2.28 Concentric String Operations

How does the theory of concentric string translate into the actual drilling of a UBD high pressure high temperature (HPHT) well? Has such a well ever been attempted in the industry? The answer to the latter question is a resounding yes, but it came about after a number of critical questions had been successfully answered. A case history is presented where a 15,000 psi (103,000 kPa), 15,000 ft (4,600 m) TVD well was directionally drilled using friction-controlled drilling technique of UBD. To statically control such a well, mud weights of nearly 19.0 ppg (2.3 SG) were needed. Drilling in such an environment required approximately an 18.5 ppg fluid and its ECD friction to operate safely. The well was actually drilled using a 16.0 ppg (1.9 SG) mixture of three brines (calcium chloride, calcium bromide, and zinc bromide) to achieve a solids-free drilling fluid. The additional 2.5 to 3.0 ppg (0.3 to 0.36 SG) was made up of intentionally added friction to safely control the well.

2.29 Modeling the Operation

Before such an ambitious project could be attempted, a number of factors had to be considered and safely resolved. A leading industry multi-phase flow modeling program used to forecast expected flow conditions in an underbalanced well design is Neotec's Wellflo 7 program. Before this model could be used with confidence in this case, however, its calculations and predictions had to be verified because it had never been used in extreme conditions, 408 degrees Fahrenheit BHT, >15,000 ft TVD and >18 ppg fluid density, (205°C; 4,600 m; 2.2 SG) The only way to confirm that the Wellflo 7 predicted pressures and flow rates at these conditions was to conduct an actual experiment at these conditions in the well before drilling out of drilling liner into live well conditions.

The operator agreed to conduct an experiment in the well using surface read out (SRO) gauges, special electric drive 15,000 psi triplex injection pumps on the injection annulus, and cryogenic nitrogen to simulate a 20 MMcf/D (390 m^3/d) gas influx downhole. The questions to be determined were:

- Could the ECD at depth be adequately increased by adding a secondary return fluid via the concentric annulus?
- Could this ECD be driven to a pressure that would exceed actual BHP?
- What injection fluid rates would be needed to do this?

These questions seemed intuitive, but the answers needed to be precise because a wrong assumption could lead to catastrophic consequences in terms of well control and its environmental impact on a possible release of deadly H_2S gas present in the reservoir. The cost just to conduct this experiment exceeded $1.4 million, but it had to be run.

The test was run without any complications and the results exceeded predicted calculations in every category, confirming that the model could be used with safety and reliability at those extreme conditions. In fact, the test showed that nearly 5 ppg (0.6SG) of added friction pressure could be created by ramping up annulus injection. It was demonstrated that enough friction could be created simply by increasing injection rates to create an ECD greater than the BHP. In other words, it was possible to completely stop influx into the wellbore and create an overbalanced well condition to fully control the well even at its 15,000 psi pressure level.

2.30 Drilling Operations

The actual drilling operation that followed resulted in a 3,000-ft horizontal lateral being drilled in the harshest of all possible drilling conditions. Not once did the system fail to function exactly as predicted. It was possible to create any level of ECD with extreme accuracy simply by changing the injection GPM up or down.

2.31 Built in Kill String

The wellbore geometry also created an added bonus: there was a built-in "kill string" present in the well at all times such that the concentric annulus could be used to kill the well should it be necessary regardless of the depth position of the drillstring as in the case of tripping.

2.32 Constant Circulation

Furthermore, the typical swings in ECD consisting of an abnormally low pressure at the start of a connection as friction is eliminated without pump flow, followed by a high pressure spike after a connection caused by the pump's restarting now could be manipulated so that the reservoir never "sees" any drop or increase in pressure during connections. That is controlled by the timing of standpipe and annulus injection at each step during a connection.

Before shutting down standpipe injection as in normal connections, increase the annulus injection by a similar volume during the shutdown phase. After making the connection, simply reverse this process by slowly adding standpipe injection while reducing annulus injection by a similar volume. In this manner, all connections using friction-controlled drilling become "invisible" to the reservoir because the exact same total GPM is being pumped during drilling and during connections.

2.33 Conclusions

Because the excessive pressures of a high pressure reservoir are dealt with downhole rather than at the surface, the result is a safer, lower surface pressure environment. The excess pressure is "consumed" by intentionally created friction on the way to the surface. By injecting the desired volumes down the concentric casing, any desired amount of excess pressure can be eliminated by controlled friction assuming that proper well geometries are selected in the design and planning phase of the well, namely casing ID, concentric casing ID and OD, and drillpipe OD. Again, the best design allows ample room between the casing ID and the concentric casing OD to allow a wide range of fluid injection capacity without excessive surface injection pressure due to downward flow friction. At the same time, the optimum design has a very restricted return annulus between the concentric casing ID and the drillpipe OD; in order to create friction as the return fluids, both standpipe injection and annulus injection volumes flow back to the surface. Deep wells have an advantage over shallow wells because the friction path is considerably longer, thus allowing more length to create more friction before reaching the surface.

The planning of a friction-controlled drilling well, whether it is intended as a UBD project, a balanced or "low head" well, or as a conventional overbalanced well requires exact tolerances on the pipe selections as well as detailed flow modeling. The predictions resulting from these flow models can reliably be used now that they have been confirmed at extreme operating conditions thanks to a costly groundbreaking initial experimental project.

The use of friction-controlled drilling now enables operators a means to an effective, safe method to drill HPHT wells in an underbalanced mode to benefit those reservoirs which can be impaired by solids plugging, filtrate invasion, clay swelling and relative permeability alterations just like normally pressured reservoirs. In point of fact, friction-controlled wells are actually safer to drill than conventionally-drilled wells that do not have a permanent kill string or a method of controlling slight to major influxes of reservoir fluids, especially gas, while drilling. The term "connection gas" can be made obsolete using proper techniques during friction controlled drilling because the pressure drops associated with pump shutdown during connections are eliminated by increasing the annulus injection volumes to a like volume.

2.34 Comment on Extreme Temperature

It should be noted that extreme bottom-hole temperature as seen in the case study above creates special considerations aside from well control issues. The use of a weighted multi-brine fluid system at temperatures in excess of 400 degrees Fahrenheit will result in excessive corrosion to all downhole tubular, casing, liners, concentric casing and drillpipe, and collars unless a proactive corrosion chemical program is implemented from the start. Also, extreme depths can create a guidance issue for directional operations because electromagnetic measuring while drilling (EMMWD) requires a cable antennae system that is also extremely sensitive to corrosion at elevated temperatures. Mud pulse MWD systems may have a temperature limit on downhole components. If a mud motor or other downhole electronics are to be used, elastomers and electronics do not fare well at temperatures much higher than 325 to 350 degrees Fahrenheit. In the case well, a metal-to-metal turbine was used rather than a mud motor to eliminate the temperature limits of its rubber components. Although not used in the case well, another alternative to control excessive temperature is the use of a mud chiller to bring bottom-hole circulating temperatures within the limits of the downhole assembly components.

Most abnormally pressured candidate wells will not be as severe as the initial test case. Unless the expected pressures are extremely high (15,000 psi or above), it is doubtful that 16.0 ppg drilling fluids would be needed. If at all possible, a friction controlled design should use a solids-free saturated sodium chloride before needing to resort to the more costly and troublesome calcium, bromide, or zinc brines. With the added friction of up to 5.0 ppg equivalent mud weight that can be created in a well engineered friction design, most applications can use common NaCl without the need for more exotic fluids. Corrosion inhibitor and possible oxygen scavenger chemicals still should be used to prevent potential corrosion even at these reduced danger levels.

Section 4
Rheology of Single Phase Fluids

Arash Haghshenas, Boots & Coots

2.35 Flow Patterns

Rheology is defined as the science of studying the deformation of fluids at different conditions. The behavior of a fluid depends on the flow regime, which is a function of several parameters such as velocity, viscosity, geometry, etc. There are several different flow regimes and transitions that may prevail in the wellbore during the drilling operation. However, for the practical aspect of pressure profile calculations in the wellbore, the flow regimes can be divided into laminar and turbulent flow.

As illustrated in Figure 2–13, in turbulent flow, the fluid moves in eddies and swirling motions. Therefore there is a complex relationship between the flow characteristics and friction factor. In laminar flow, the fluid moves in layers parallel to the boundary of a channel or conduit. The friction factor has an inverse relationship with the velocity of the fluid and it is not affected by the roughness of the boundaries.

2.36 Reynolds Number

Reynolds number is generally used to determine the flow regime. The Reynolds number (N_{Re}) is a dimensionless value that represents the ratio of inertial forces to viscous forces in the fluid. The general form of the Reynolds number is

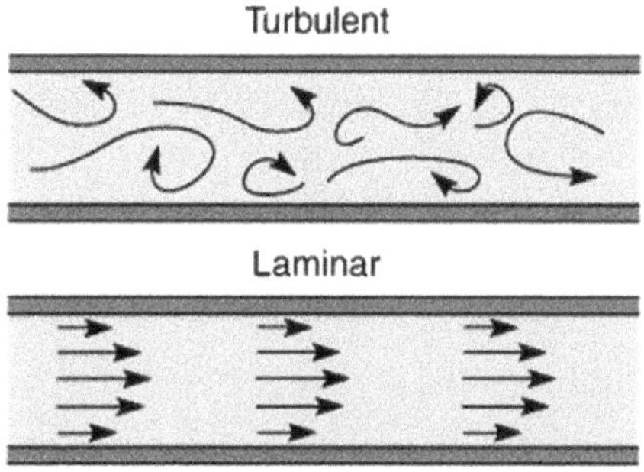

Figure 2–13 *Illustrate flow pattern in fluid for laminar and turbulent flow*

$$N_{Re} = \frac{D \times V \times \rho}{\mu} \tag{2.1}$$

where

D = Diameter of the flow channel, ft or m

V = Average velocity of the fluid in the channel, ft/s or m/s

ρ = Fluid density, lb_m/ft^3 or kg/m^3

μ = Viscosity of the fluid of Newtonian fluid or effective viscosity of non-Newtonian at the flow condition, lbm/ft.s or kg/m.s

Reynolds number in field units is expressed as

$$N_{Re} = \frac{in\left(\frac{ft}{12\ in}\right) \times \frac{ft}{s} \times \frac{lbm}{gal}\left(\frac{7.48\ gal}{ft^3}\right)}{cp\left(\frac{6.72 \times 10^{-4}\ lbm/ft.s}{cp}\right)} \tag{2.2}$$

$$N_{Re} = \frac{928 \times D\ (in) \times V\ (ft/sec) \times \rho\ (lbm/gal)}{\mu(cp)} \tag{2.3}$$

2.36.1 Critical Reynolds Number

A critical Reynolds number is determined as a limit where the laminar flow changes to turbulent flow. If the calculated N_{Re} is greater than the critical Reynolds number N_{Rec}, the flow regime is turbulent; otherwise the flow regime is laminar. For Newtonian fluids N_{Re} of 2,100 is usually used as the critical Reynolds number. However, the value of the critical Reynolds number depends on the fluid behavior and properties. For typical drilling fluids the critical Reynolds number is in a range of 2,100 to 4,000. API recommends using the following equation to determine the critical Reynolds number

$$N_{Rec} = 3470 - 1370\ n \tag{2.4}$$

where n is Power Law fluid index explained in Power Law model section.

2.37 Viscosity

Viscosity is defined as the ratio of shear stress to shear rate. If the shear stress is measured in dynes/cm^2 and the shear rate in sec-1, the viscosity will have the units of poise. Poise is a large unit for measuring the viscosity of common fluid, therefore centipoises which is equal to one hundredth of Poise, g/cm/s, or one millipascal-second (SI unit), are commonly used.

$$\mu = \frac{\tau}{\gamma} \tag{2.5}$$

where

μ = Viscosity of the fluid

τ = Shear stress

γ = Shear rate

Shear stress is measured in unit of force per area and is the force required to sustain a particular rate of the fluid flow. Shear rate is the velocity gradient of the fluid measured across the flow. The unit of shear rate is (1/sec). Shear rate is not constant in circulating fluid. For example, in laminar flow in pipes, the shear rate at the pipe wall is high and reaches a minimum value at the center of the flow. Average prevailing shear rate in the pipe may be used for calculation purposes.

Jimmy Fann was the first to develop the concentric cylinder rheometer currently used by most mud engineers today. Fann designed the gap between the bob and rotating cylinder, and the spring constant that the equipment measures the viscosity of the fluid in Poise at 3 RPM.

To change the dial reading to the unit of dynes/cm^2, the dial reading is multiplied by 5.11. To change the RPM to reciprocal seconds, the RPM is multiplied by 1.703. The slope of the fluid behavior in the given units for shear stress (dynes/cm^2) and shear rate (1/sec) produces the viscosity in unit of poise, dynes.sec/cm^2. The commonly used unit of viscosity is centipoises, and consequently the 5.11 is normally multiplied by 100 to convert the value to centipoises. The ratio of 5.11 divided by 1.703 results in a conversion factor of 300. In other words, when a ratio of readings from a properly calibrated oilfield concentric cylinder viscometer is multiplied by 300, the ratio is the viscosity in centipoise. Obviously, the dial reading at 300 RPM will be the viscosity of the fluid at that shear rate. The dial reading at 600 RPM will be twice the viscosity (in centipoise) at that shear rate.

The ratio of any point on the shear stress-shear rate curve can be used to calculate viscosity, which is referred to as effective viscosity. A statement is then made that this would be the viscosity of a Newtonian fluid. Most frequently, a curve which represents the relationship between shear stress and shear rate for a drilling fluid is made confusing by a line from some point to the origin. Figure 2–14 presents three different fluid flows, Newtonian, Herschel-Buckley, and a shear-thickening fluid. At 300 RPM, all fluids present the same dial reading of 30 (30 cp). Although the dial readings for all samples are similar, each fluid exhibits different rheological behavior. Rheological models try to model the behavior of the fluid for the entire range of shear rate occurred in the wellbore. The obtained viscosity of the fluid by connecting a line from a point on the rheological model to the origin only represents one peculiar point. Therefore, the behavior of the fluid cannot be determined by this method.

2.38 Types of Fluids

Fluids are divided into several categories according to their rheological behaviors as observed in shear stress-shear rate plots. In general, fluids are divided into the two broad categories of Newtonian and non-Newtonian fluids. Water and oil are examples of Newtonian

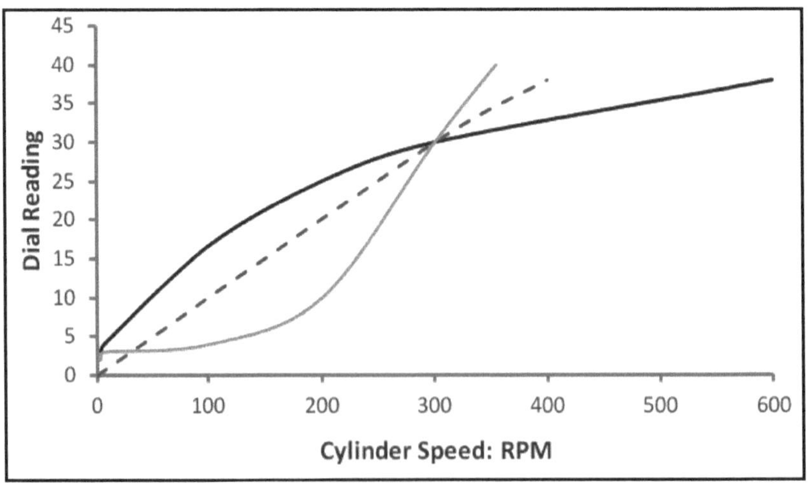

Figure 2–14 *Three different fluid types exhibit similar viscosity at 300 RPM*

fluids. Presence of clays, polymers, and several additives in drilling fluids creates non-Newtonian fluids.

2.38.1 Newtonian Fluids

Newtonian fluids exhibit constant viscosity at different shear rates and constant temperature. As shown in Figure 2–15, the relationship between shear stress and shear rate is a straight line starting passing through the origin. The rheological behavior of Newtonian fluids can be written as

$$\tau = \mu \times \gamma \tag{2.6}$$

where μ is the viscosity of the fluid.

2.38.2 Non-Newtonian Fluid

Non-Newtonian fluid viscosities vary at different shear rates. Drilling fluids are normally shear thinning fluids, which means the viscosity of the drilling fluid decreases with increasing the shear rate. Shear thinning fluid exhibits restively low viscosity in the drillstring, where the shear rate is high, causing less frictional pressure drop. A shear thinning fluid is easier to pump at high shear rates. In the annulus where the velocity of fluid and shear rate is relatively low, the drilling fluid exhibits high viscosity and assists carrying cuttings out of the wellbore.

Another type of non-Newtonian fluids is shear-thickening fluid which the viscosity of the fluid increases as the shear rate increases. Examples of shear-thickening fluids are methyl-methacrylate and corn starch. While measuring the rheological properties of a shear-thickening fluid, it may behave like Polyox and have a large normal stress component that makes it want to climb up the stirrer's shaft instead of forming a vortex. Shear-thickening fluids are not favorable as drilling fluid because they create excessive pressure on the pumps and in the wellbore.

2.38.2.1 Bingham Plastic Model

The Bingham plastic model is the most common rheological model used in the drilling industry. This model is a two parameter model that includes yield stress and plastic viscosity of the fluid. As it is shown in Figure 2–15, the fluid initially resists flowing until the shear stress exceeds a certain value. After the fluid starts to flow there is a linear relationship between shear stress and shear rate.

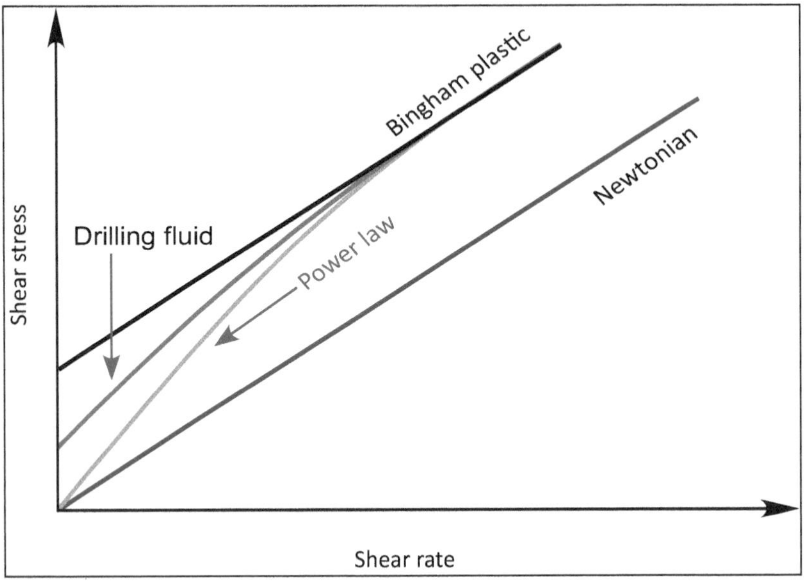

Figure 2–15 *Illustrates rheological behavior of different types of fluids*

$$\tau = \mu_p \gamma + \tau_y \tag{2.7}$$

$$\mu_p = R_{600} - R_{300} \tag{2.8}$$

$$\tau_y = R_{300} - \mu_p = 2R_{300} - R_{600} \tag{2.9}$$

where

μ_p = Plastic viscosity, cp

τ_y = Yield point, $lb_f/100ft^2$

R_{300} = Fann reading at 300 RPM

R_{600} = Fann reading at 600 RPM

If μ_p and τ_y are known for a Bingham plastic fluid, dial readings at 600 and 300 RPM can be determined from Eq. (2.10) and Eq. (2.11).

$$R_{600} = 2\mu_p + \tau_y \tag{2.10}$$

$$R_{300} = 2\mu_p + \tau_y \tag{2.11}$$

Drilling fluids initially resists flowing as shown in Figure 2–15. In the Bingham plastic model, the shear stress should exceed a certain value to break the gelation bonding of the drilling fluid and allow it to flow. This behavior enables drilling fluid to suspend the drilling cuttings and solids within the drilling fluid when the circulation stops. Fluids that exhibit gelling property are called thixotropic. Gelling strength of drilling fluids is time dependant. High gel strength may cause excessive pressure surge when the circulation starts and fractures the formation. The Bingham plastic model became widely used because it is simple and estimates pressure loss in a turbulent condition with accuracy close to the other models.

For drilling fluid treatment purposes, the Bingham plastic model is superior to other models as it indicates the nature of contamination of the drilling fluid and the required treatment. For instance, an increase in plastic viscosity of the fluid indicates solid contamination, while an increase in yield point suggests chemical contamination.

2.38.2.2 Power Law Model

The power law model describes the shear thinning effect of the drilling fluid. This model has two parameters to describe the behavior of the fluid. As shown in Figure 2–15 the shear stress—shear rate relationship of the fluid passes through the origin with a power law shape. Eq. (2.12) describes the behavior of a power law fluid.

$$\tau = K \times \gamma^{n} \tag{2.12}$$

In the power law fluid model fluid starts to move as a shear rate applies to the fluid, which does not explain the thixotropic properties of the drilling fluid. In general, power law fluid underpredicts the behavior of the drilling fluid at low shear rates because the model is forced to pass through the origin of a shear rate-shear stress plot.

In an attempt to improve the accuracy of the power law model (using a VG meter), the laminar flow region (3–100 RPM) and the turbulent region (300–600 RPM) are modeled separately. However, the power law model for the low shear rate section still passes through the origin and does not explain the thixotropic behavior of the drilling fluid. In the drillstring where high shear rate flow prevails, 600 RPM and 300 RPM data are applied to determine the flow parameters.

$$n_t = 3.32 \times Log\left(\frac{R_{600}}{R_{300}}\right) \tag{2.13}$$

$$K_t = \frac{R_{600}}{1022^{n_t}} = \frac{R_{300}}{511^{n_t}} \tag{2.14}$$

In the annulus where low shear rate flow prevails, 100 RPM and 3 RPM data are applied to determine the flow parameters.

$$n_l = 0.657 \times Log\left(\frac{R_{100}}{R_3}\right) \tag{2.15}$$

$$K_l = \frac{R_{100}}{107.3^{n_l}} = \frac{R_3}{5.11^{n_l}} \tag{2.16}$$

where

K= Consistency index, lb.s^n/100 ft^2

n = Flow behavior index

and t and l subscripts indicate turbulent and laminar flow conditions respectively.

2.38.2.2.1 Units of Power Law Parameters

The general form of power law model as given in Eq. (2.12)

$$\tau = K \times \gamma^n$$

If the rheological properties of the fluid are known for two points, then the power law flow parameter, n, can be determined as follows:

$$\begin{cases} \tau_1 = K \cdot \gamma_1{}^n \\ \tau_2 = K \cdot \gamma_2{}^n \end{cases} \rightarrow \frac{\tau_1}{\tau_2} = \left(\frac{\gamma_1}{\gamma_2}\right)^n \rightarrow Log\left(\frac{\tau_1}{\tau_2}\right) = n\ Log\left(\frac{\gamma_1}{\gamma_2}\right)$$

The units of shear stress and shear rate cancel each other, and as a result n is dimensionless. If the rheological properties of a power law fluid at 600 and 300 RPM are known then,

$$\begin{cases} n = \dfrac{Log\left(\dfrac{600\times 1.703}{300\times 1.703}\right)}{Log\left(\dfrac{R_{600}\times 1.0678}{R_{300}\times 1.0678}\right)} \\ n = \dfrac{Log\left(\dfrac{600\times 1.703}{300\times 1.703}\right)}{Log\left(\dfrac{R_{600}\times 511}{R_{300}\times 511}\right)} \end{cases} \rightarrow n = 3.32\ \times Log\left(\dfrac{R_{600}}{R_{300}}\right)$$

In the above equations, if Fann 35 dial readings are multiplied by constant 1.0678, the unit of shear stress is $lb_f/100\ ft^2$. If constant 511 is used, the unit of shear stress is $g/100\ cm/s^2$.

After the value of n is determined, K is calculated as

$$K = \frac{\tau}{\gamma^n} \tag{2.17}$$

For a given data at 600 RPM,

$$K\left[\frac{lb_f\ .\ s^n}{100\ ft^2}\right] = \frac{1.0678\times R_{600}}{(1.703\times 600)^n} = \frac{1.0678\times R_{600}}{1022^n}$$

$$K\left[\frac{g}{100\ cm\ .\ s^n}\right] = \frac{511\times R_{600}}{(1.703\times 600)^n} = \frac{511\times R_{600}}{1022^n}$$

In the theory when power flow exponent, n, is equal to one, the power law model reduces to the Newtonian fluid model and consistency index, K, has the unit of viscosity. If K is expressed in $lb_f.s^n/100\ ft^2$ when n is equal to 1, the unit of K reduces to $lb_f.s/100\ ft^2$. The main advantage of applying the conversion factor of 511 to the fluid consistency equation is to maintain all the units of viscosity in cp. To maintain consistency with API RP 13D, all equations are expressed as mentioned in the recommendations.

2.38.2.3 Herschel-Bulkley (Modified Power Law) Model

The Herschel-Bulkley model is also referred to as the modified power law model, which is a power law model with the addition of yield stress to the model. This model is one of the complex models which has three parameters and defines the behavior of the drilling fluids better than the other models. API RP 13D recommends using this model to predict pressure profile in the wellbore. The behavior of a Herschel-Bulkley fluid is described as

$$\tau = \tau_0 + K \times \gamma^n \tag{2.18}$$

where τ_0 is the initial resistance of fluid to flow.

The Herschel-Bulkley model is a general model that can be reduced to the Bingham and power law model. If the τ_0 is zero, then the Herschel-Bulkley reduces to the power law model. If n is equal to 1, then the Herschel-Bulkley reduces to the Bingham plastic model. The concept of the τ_0 and τ_y are very different. τ_y in the Bingham plastic model is determined at high shear rates (300 to 600 RPM) while τ_0 is determined at low shear rates (3 to 6 RPM) to estimate fluid behavior more accurately.

The numerical method of calculating the three factors of Herschel-Bulkley requires a trial-error method to match the model to all available data. However, the parameters can be approximated as follows,

$$\tau_0 = 2R_3 - R_6 \tag{2.19}$$

$$n = 3.32 \times Log\left(\frac{R_{600} - \tau_0}{R_{300} - \tau_0}\right) \tag{2.20}$$

$$K = \frac{R_{300} - \tau_0}{511^n} \tag{2.21}$$

2.38.2.3.1 Pressure Drop Calculation

The Bingham plastic model and the power law models have been used in the drilling industry to calculate the pressure drop. As shown in Figure 2–14, the Bingham plastic overpredicts the fluid behavior at low shear rates while the power law model underpredicts it. However, regardless of the model, fluid behavior can be modeled with reliable accuracy at very high shear rates.

Before the new API RP 13D release in 2006, API recommended a two part power law model to predict fluid behavior. One part modeled the low shear properties, equal to 3 to 100 RPM that prevails in the annulus, and another part to predict the fluid behavior at high shear rates, 300 to 600 RPM that prevails in the drillstring. For more information, readers are referred to API RP 13D released in 2003.

In 2006 API recommended using the Herschel-Bulkley to predict the fluid behavior and pressure drop calculations more accurately for deep and complex wells. The application of the power law and the Herschel-Bulkley models are described in an example at the end of this section.

2.39 Pressure Loss Across the Nozzles

While passing through the nozzles, the velocity of the drilling fluid increases and pressure decreases. The pressure loss is determined using the kinetic energy of fluid in the upstream and downstream of the nozzles. For simplicity, the velocity of the drilling fluid above the nozzles is ignored to derive an equation for the pressure drop. High shear rate prevails at the nozzles and the length of the nozzles is very short. Then according to the Bernoulli's equation, the pressure drop across the nozzles is solely related to the kinetic energy of the drilling fluid through the nozzles, and pressure loss is not function of the fluid rheology. A simplified equation to calculate the pressure loss across the nozzles is given as

$$\Delta P_{Nozzles} = \frac{156 \times \rho \times Q^2}{\left(D_{N1}{}^2 + + D_{Ni}{}^2\right)^2} \tag{2.22}$$

$$\Delta P_{Nozzles} = \frac{8.33 \times 10^{-5} \times \rho \times Q^2}{C_d{}^2 \times A_t{}^2} \tag{2.23}$$

where

ρ = Fluid density, ppg

Q = Flow rate, gpm

D_{Ni} = Nozzle sizes, 1/32 in.

C_d = Discharge coefficient

A_t = Total area of nozzles, in.^{2}

The value of C_d depends on the type of the bit, nozzels, and flow rate. At very low flow rates, the value of C_d can be as low as 0.8. However, for the current range of flow rates for drilling operation, Robinson (2010) recommends a value of 1.03. The current recommended value in API RP13D is 0.98.

2.40 API Recommendations

In the latest recommended practices major changes were added. In addition to recommending the Herschel-Bulkley, effect of pressure and temperature on the drilling is discussed. It is suggested to divide the wellbore into several sections. New correlations are available to estimate temperature profile in the wellbore. Average temperature in each section can be used to determine rheological properties of the drilling fluid at a given pressure.

2.41 Wellbore Pressure and Temperature Correction

Pressure in the wellbore should be corrected for the compressibility of the drilling fluid. The compressibility of the drilling fluid depends on the compressibility of the different components of the drilling fluid and their volume fraction. If high pressure and high temperature in the well are expected, the compressibility factor of the cuttings and solids in the drilling fluid needs to be included.

Pressure increases the density of the drilling fluid below the rig floor, and temperature reduces the density. Moreover, pressure and temperature change the rheological properties of the drilling fluid. The combination of both pressure and temperature should be considered to determine the bottomhole pressure. If a surface property of the drilling fluid is applied for pressure calculations, bottom-hole pressure calculations might be significantly in error.

The effect of pressure on the density of the non-aqueous drilling fluids is more significant than water. The effect of temperature may reduce the density of the drilling fluid and consequently reduce the bottom-hole pressure of a well drilling in a high-temperature gradient field, such as drilling a geothermal well. On the other hand, in a deep well with low-temperature gradient, the density of the drilling fluid increases in the wellbore and bottomhole pressure will be greater than calculated using the surface data. This effect is very important in offshore drilling where the column of water around the drillstring and riser act as a cooling medium and cools the drilling fluid before it circulates into the well, and cools it off again when the drilling fluid reaches the floor. The cooling effect of the seawater can be observed as higher standpipe and bottom-hole pressure when starting recirculation of the well after a few hours.

Pressure and temperature affect the density of solids. For practical purposes, the effects of pressure and temperature on the density of the drilled cuttings are not significant. Most of the time the correction of fluid phase volumes is considered and the effects of pressure and temperature on the density of solids are ignored.

Drilling cuttings also affect the pressure profile in the wellbore. It is assumed that drilling cuttings are inert solids. Then the presence of cuttings only affects the density of the drilling fluid and their effect on the rheology is ignored. Bottom-hole pressure and equivalent mud weight calculations should be modified for the presence of the drilling cuttings.

Figure 2–16 illustrates the effect of the temperature and pressure on the hydrostatic pressure of a 14.5 ppg non-aqueous fluid with synthetic fluid fraction of 0.60. The hydrostatic pressure calculated using API recommendations for pressure and temperature correction is compared to hydrostatic pressure calculation using surface data. A positive difference indicates that API corrected pressure is greater than conventional calculation. In this figure, effects of pressure and temperature on solids are ignored.

Generally, it is assumed that effect of pressure and temperature cancel each other and the hydrostatic pressure calculation is simplified

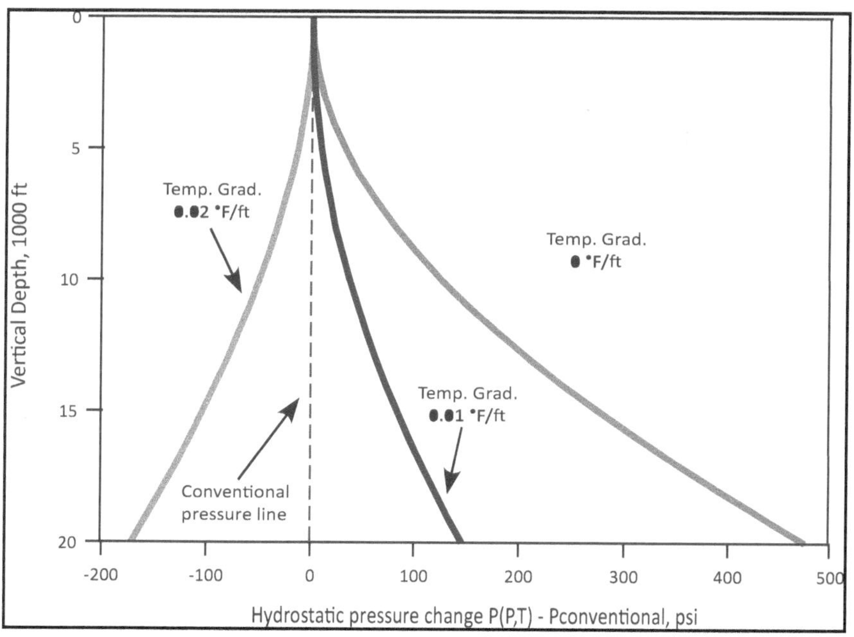

Figure 2–16 *This figure illustrates the difference between pressure and temperature corrected hydrostatic pressure against conventional hydrostatic pressure calculation for a 14.5 ppg drilling fluid with 60% synthetic fluid volume*

using available surface data. The graph with a temperature gradient of zero illustrates the effect of pressure on the density of the drilling fluid. As the pressure increases with depth, the bottom-hole pressure deviates from conventional calculations. For the given non-aqueous drilling fluid, bottom-hole pressure may be different by 475 psi. The temperature gradient in the wellbore counteracts the effect of pressure on the density of the drilling fluid. A temperature gradient of 0.01°F/ft decreases the difference between the API corrected method and conventional calculations. However, the conventional method still underpredicts the bottom-hole pressure.

If the temperature gradient is 0.02°F/ft the bottom-hole pressure will be less than expected pressure calculated conventionally using surface data. For this case, the bottom-hole pressure is 160 psi less than expected. The conventional pressure calculation becomes erroneous as the bottom-hole pressure and temperature increases. In offshore wells, the seawater around the riser and return lines cools the drilling fluid and complicates the situation even more.

Figure 2–17 illustrates the similar situation presented in Figure 2–16 for an offshore situation of 5,000 ft of water depth. The temperature changes linearly from surface to 40°F at the depth of 3,000 ft the temperature remains at 40°F through the depth of 3,000 ft of water to the seafloor. Beyond the seafloor, the temperature gradient varies as illustrated in the Figure 2–16.

Figure 2–17 shows that the cooling effect of water depth increases the density of the non-aqueous drilling fluid, and the conventional hydrostatic pressure calculations underpredict the bottom-hole pressure. As the temperature gradient increases in the formation, the density of the drilling fluid decreases. In areas with a high geothermal gradient, the effect of temperature and pressure on the hydrostatic pressure should be studied, as the bottom-hole pressure may be greater or less than expected from conventional calculations.

Pressure and temperature affect the rheology of the drilling fluid; the relationship is more complex and is related to the type of the fluid and additives present in the fluid. Unfortunately, the effects of fluid types and different additives have not been sufficiently investigated to provide guidelines or models. The effect of pressure on water-based fluids is not significant, while pressure affects the rheological properties of non-aqueous fluid. However, for both types of fluids, a water-based and a non-aqueous fluid, temperature significantly affects the rheological properties of the fluids.

Figure 2–18 illustrates the effect of temperature on the rheological properties of three drilling fluid samples. For sample DF1, temperature reduces the viscosity of the fluid and levels out after a certain

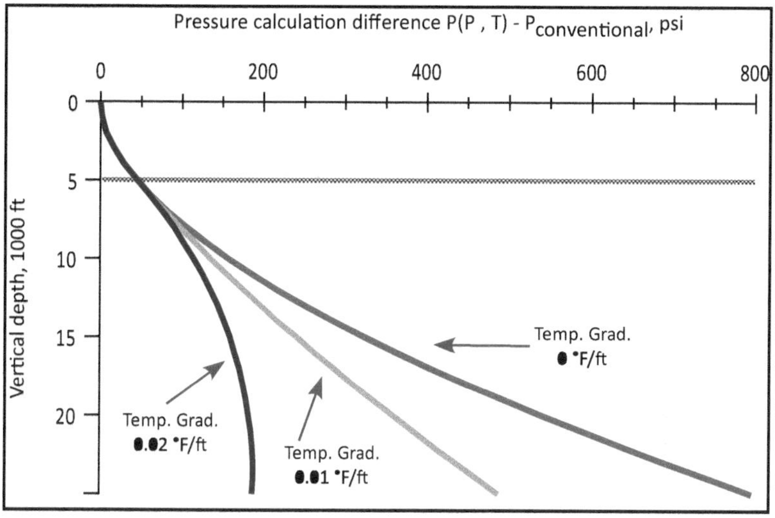

Figure 2–17 *This figure illustrates the difference between pressure and temperature corrected hydrostatic pressure against conventional hydrostatic pressure calculation for a 14.5 ppg drilling fluid with 60% synthetic fluid volume in deepwater with water depth of 5,000 ft*

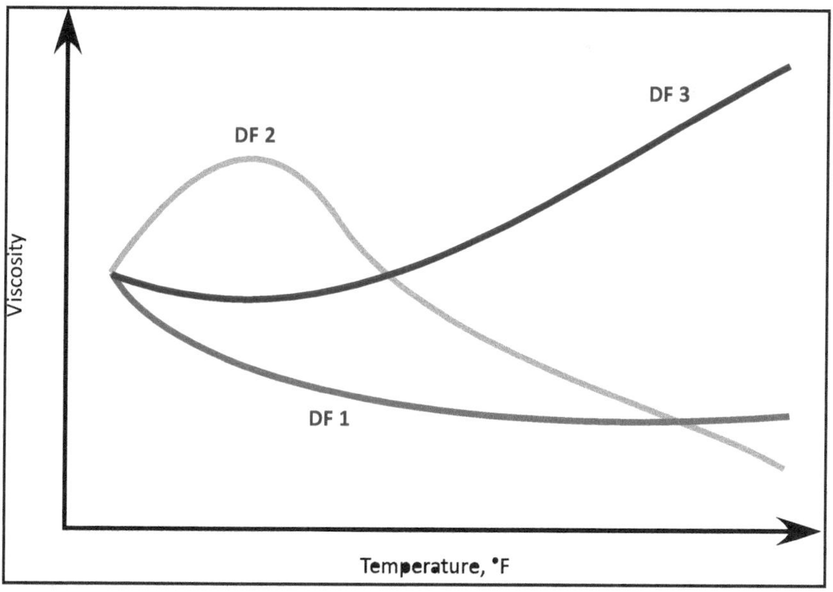

Figure 2–18 *Effect of temperature on the rheological properties of the drilling fluid depends on the type of fluid and additives present in the fluid (Courtesy of Dr. Leon Robinson)*

temperature. This behavior may occur when temperature degrades the properties of additives. The viscosity levels off at a value close to the rheological properties of the base fluid. For sample DF2, increasing the temperature initially increases the viscosity of the fluid. However, after a certain temperature, viscosity starts to decrease. The other observed behavior is constant increase of the viscosity by increase of temperature as shown for sample DF3.

Viscosity hysteresis may occur if the temperature exceeds the chemical stability of the drilling fluid additives. In viscosity hysteresis, the viscosity path while heating the sample is different than the viscosity path while cooling the sample to the starting temperature. If the temperature degrades the chemical bonding of the additives, the observed viscosity while cooling the sample is less than the value measured while heating up. However, if temperature causes additives and base fluid to gum up, the observed viscosity while cooling the sample would be higher than observed viscosity while heating up the sample.

Most rheological models use the properties of the drilling fluid at 120°F. Although this methodology might be accurate enough for shallow wells in the past, more detailed methodologies are required to model pressure profile in the wellbore in deep wells.

Mullen et al. (2001) presented the effects of pressure and temperature on the fluid behavior and determining the pressure in the wellbore using a field cases. Ibeh et al. (2008) illustrated the effect of pressure and temperature on the rheological properties of a drilling fluid using lab data and presented a statistical method to determine plastic viscosity and yield point at different pressures and temperatures. Because the effect of pressure and temperature on the drilling fluid is not well known, it is recommended to test the behavior of the fluid at the expected range of operation to examine the stability of the drilling fluid and obtain enough rheological data for predicting the pressure profile in the wellbore. Statistical models can be applied to predict the properties of the drilling fluid at a given pressure and temperature based on measured values.

2.42 Example 1

A vertical well is planned to be drilled to the depth of 10,000 ft using a 14.5 ppg drilling fluid. The average of the wellbore and casing diameter is 9 ¾ in. and mud is circulated at the rate of 450 gpm. Fann 35 data of the drilling fluid are given in Table 2–1. drillstring, hole-geometry, and nozzle sizes are given in Table 2–2. For further simplicity and application of this example, ignore the effect of pressure and temperature on density and the rheological properties of the drilling

Table 2–1 Fann 35 Data of the Fluid in the Example 1

RPM	600	300	200	100	6	3
Dial Reading	80	50	37	25	6	4

Table 2–2 Geometry of Drillstring, Well, and Nozzle Sizes

Hole size	9 ¾ in.
Drillpipe	
OD	5 in.
ID	4.276 in.
Weight	19.5 lb/ft
Length	9,000 ft
Nozzles	2 × 15/32 + 16/32

fluid. Calculate the standpipe pressure while circulating using the power law and Herschel-Bulckley models.

Standpipe pressure is the sum of the pressure loss in the annulus, pressure drop drill bit nozzles, and pressure loss inside the drillstring. In the following example the pressure drop in each section is calculated and added up at the end to determine the standpipe pressure.

2.42.1 The General Steps of Calculating Pressure Drop in the Drillstring and Annulus

- Determine the rheological properties of the drilling fluid, n and K.
- Calculate the average velocity of the fluid.
- Estimate the apparent viscosity of the fluid at the calculated velocity.
- Calculate the Reynolds number.
- Determine the critical Reynolds number and check if the flow is laminar or turbulent.
- Calculate the friction factor related to the flow pattern.
- Determine the pressure loss gradient and pressure loss in the given section.

2.42.2 Annulus

Determining the rheological properties (using the power law values) in the annulus between the drillpipe and the wellbore

$$n_l = 0.657 \times \log\left(\frac{R_{100}}{R_3}\right) \tag{2.24}$$

$$n_l = 0.657 \times \log\left(\frac{25}{4}\right) = 0.523$$

$$K_l = \frac{R_{100}}{170.3^{n_l}} \tag{2.25}$$

$$K_l = \frac{25}{170.3^{0.523}} = 1.703 \frac{lb.s^n}{100\ ft^2}$$

The velocity of the drilling fluid in the annulus is calculated by Eq. (2.32). In this equation, D_2 is the outside diameter and D_1 is the inside diameter of the annulus in inches.

$$V_a = \frac{0.408 \times Q}{D_2^2 - D_1^2} \tag{2.26}$$

$$V_a = \frac{0.408 \times 450}{9.75^2 - 5^2} = 2.62\ ft/\sec$$

Now the apparent viscosity of the drilling fluid in the annulus can be calculated:

$$\mu_{el} = 511\ K_l \left(\frac{144\ V_l}{D_2 - D_1}\right)^{n_l - 1} \left(\frac{2n_l + 1}{3n_l}\right)^{n_l} \tag{2.27}$$

$$\mu_{el} = 511\ (1.703)\left(\frac{144 \times 2.62}{9.75 - 5}\right)^{0.523-1} \left(\frac{2(0.523)+1}{3(0.523)}\right)^{0.523} = 124\ cp$$

The Reynolds number of the drilling fluid in the annulus is

$$N_{Re_a} = \frac{928(D_2 - D_1)V_a \times \rho}{\mu_{ea}} \tag{2.28}$$

$$N_{Re_a} = \frac{928 \times (9.75 - 5) \times 2.62 \times 14.5}{124} = 1,350$$

The critical Reynolds number for this drilling fluid is estimated using Eq. (2.4)

$$N_{Re\ c} = 3470 - 1370\ (0.523) = 2754$$

The Reynolds number is less than the critical Reynolds number and flow is laminar in the annulus, which is consistent with the laminar equation used for the Reynolds number calculation.

The friction coefficient in the annulus is

$$f_l = \frac{24}{N_{Re_a}} \tag{2.29}$$

$$f_l = \frac{24}{1350} = 0.0178$$

If the flow in the annulus was turbulent, the friction coefficient would be calculated using Eq. (2.30) through Eq. (2.32). In Eq. (2.30) N_{Rea} calculated using turbulent flow parameters, n_t and K_t.

$$f_t = \frac{a}{N_{Re_a}{}^{b}} \tag{2.30}$$

where

$$a = \frac{\log(n_t) + 3.93}{50} \tag{2.31}$$

$$b = \frac{1.75 - \log(n_t)}{7} \tag{2.32}$$

The pressure loss in the annulus is

$$\Delta P = \frac{f_p \times V_p^{\,2} \times \rho}{25.81(D_2 - D_1)} \times L \tag{2.33}$$

$$\Delta P = \frac{0.0178 \times 2.62^2 \times 14.5}{25.81(9.75-5)} \times 10{,}000 = 145 \text{ psi}$$

The total pressure loss in the annulus is 145 psi. The bottom-hole pressure while circulating is the sum of the hydrostatic pressure of the drilling fluid and the pressure loss in the annulus. The hydrostatic pressure of the drilling fluid is 7,540 psi, and the circulating bottom-hole pressure is 7,685 psi.

2.42.3 Drill Bit Nozzles

The pressure loss across the nozzles is

$$\Delta P_{Nozzles} = \frac{156 \times \rho \times Q^2}{\left(D_{N1}^{\,2} + D_{N2}^{\,2} + D_{N3}^{\,2}\right)^2}$$

$$\Delta P_{Nozzles} = \frac{156 \times 14.5 \times 450^2}{\left(16^2 + 15^2 +.15^2\right)^2} = 920 psi$$

2.42.4 Drillstring

In the drillstring the flow parameters are

$$n_p = 3.32 \times \log\left(\frac{R_{600}}{R_{300}}\right) \tag{2.34}$$

$$n_p = 3.32 \times \log\left(\frac{80}{50}\right) = 0.678$$

$$K_p = \frac{R_{600}}{1{,}022^{n_p}}$$

$$K_p = \frac{80}{1{,}022^{0.678}} = 0.729 \frac{lb.s^n}{100\ ft^2}$$

The velocity of the drilling fluid in the drillstring is

$$V_p = \frac{0.408\ Q}{D^2} \tag{2.35}$$

$$V_p = \frac{0.408 \times 450}{4.276^2} = 10.0 ft/\sec$$

The apparent viscosity in the drillstring is

$$\mu_{ep} = 511\ K_p \left(\frac{96V_p}{D}\right)^{n_p - 1} \left(\frac{3n_p + 1}{4n_p}\right)^{n_p} \tag{2.36}$$

$$\mu_{ep} = 511 \times 0.729 \left(\frac{96 \times 10}{4.276}\right)^{0.678-1} \left(\frac{3(0.678)+1}{4(0.678)}\right)^{0.678} = 70\ cp$$

The Reynolds number in the drillpipe is

$$N_{\mathrm{Re}_p} = \frac{928 \times D \times V_p \times \rho}{\mu_{ep}} \tag{2.37}$$

$$N_{\mathrm{Re}_p} = \frac{928 \times 4.276 \times 10 \times 14.5}{70} = 8{,}220$$

The critical Reynolds number from Eq. (2.4) is

$$N_{\mathrm{Re}\ c} = 3470 - 1370\ (0.678) = 2{,}542$$

The flow in the drillpipe is turbulent. The parameters of the friction coefficient in turbulent flow are

$$a = \frac{\log(n_t) + 3.93}{50} \tag{2.38}$$

$$a = \frac{\log(0.678) + 3.93}{50} = 0.07522$$

$$b = \frac{1.75 - \log(n_t)}{7} \tag{2.39}$$

$$b = \frac{1.75 - \log(0.678)}{7} = 0.274$$

$$f_p = \frac{a}{{N_{\mathrm{Re}_p}}^{b}} \tag{2.40}$$

$$f_p = \frac{0.07522}{8223^{0.274}} = 0.00635$$

$$\Delta P_p = \frac{f_p {V_p}^2 \rho}{25.81 D} L \tag{2.41}$$

$$\Delta P_p = \frac{0.00635 \times 10^2 \times 14.5}{25.81 \times 4.276} 10{,}000 = 835\ psi$$

If the flow in the drillstring is laminar, the friction coefficient is calculated as follows

$$f_p = \frac{16}{N_{\mathrm{Re}_p}} \tag{2.42}$$

The total pressure loss in the well is

$$\Delta P_{total} = \Delta P_a + \Delta P_{nozzles} + \Delta P_p \tag{2.43}$$

$$\Delta P_{total} = 145 + 920 + 835 = 1{,}900\ psi$$

2.42.5 Herschel-Bulkley

2.42.5.1 Annulus

The Herschel-Bulkley flow parameters are calculated using Eq. (2.23) through Eq. (2.25).

$$\tau_0 = 2R_3 - R_6$$

$$\tau_0 = 2(4) - 6 = 2 \text{ lb/100 ft}^2$$

$$n = 3.32 \times Log\left(\frac{R_{600} - \tau_0}{R_{300} - \tau_0}\right)$$

$$n = 3.32 \times Log\left(\frac{80 - 2}{50 - 2}\right) = 0.7$$

$$K = \frac{R_{300} - \tau_0}{511^n}$$

$$K = \frac{50 - 2}{511^{0.7}} = 0.61\, lb.s^n / 100\, ft^2$$

The values of n, K, and τ_0 are approximates. Using trial-error methods, the exact values of the parameters can be determined. The exact values are:

$n = 0.706$

$K = 0.6205 \text{ lb.s}^n/100 \text{ ft}^2$

$\tau_0 = 2.6 \text{ lb/100 ft}^2$

The velocity of the drilling fluid in the annulus is calculated using Eq. (2.32) and is 2.62 ft/s.

The Herschel-Bulkley model corrects the shear rate in the annulus for well geometry shear rate. In the annulus the correction factor is simplified to

$$G = 1.5\frac{2n+1}{3n} \tag{2.44}$$

$$G = 1.5\frac{2(0.7)+1}{3(0.7)} = 1.71$$

The shear rate at the wall of drillstring and wellbore, y_w, and associated shear stress, t_w, for Herschel-Bulkley are given

$$\gamma_w = \frac{96 \times G \times V_a}{D_2 - D_1} \tag{2.45}$$

$$\gamma_w = \frac{96 \times 1.71 \times 2.62}{9.75-5} = 90.5 \ {}^{1}\!/_{\text{sec}}$$

$$\tau_w = 1.066 \left(1.5^n \tau_0 + K\gamma_w{}^n\right) \tag{2.46}$$

$$\tau_w = 1.066\left(1.5^{0.7}(2) + 0.61 \times (90.4)^{0.7}\right) = 18\, lb/100\, ft^2$$

The Reynolds number of Herschel-Bulkley fluids in a simplified form is

$$N_{\text{Re G}} = \frac{186 \times \rho \times V_a{}^2}{\tau_w} \tag{2.47}$$

$$N_{\text{Re G}} = \frac{186 \times 14.5 \times 2.62^2}{18} = 1029$$

The critical Reynolds number is determined using Eq. (2.4).

$$N_{\text{Re } c} = 3470 - 1370\ (0.7) = 2511$$

The Reynolds number in the annulus is less than the critical Reynolds number and the flow in the annulus is laminar. The friction coefficients for laminar flow is

$$f = \frac{16}{N_{\text{Re G}}} \tag{2.48}$$

$$f = \frac{16}{1029} = 0.01556$$

If the flow is turbulent in the annulus, the friction coefficient is calculated using Eq. (2.49). The parameters *a* and *b* are calculated using Eq. (2.38) and Eq. (2.39) with the power law model flow parameters.

$$f = \frac{a}{\left(N_{\text{Re }G}\right)^{b}} \tag{2.49}$$

The frictional pressure loss in the annulus is determined as follows

$$\Delta P_a = \frac{\rho \times V_a^{2} \times f}{25.816\left(D_2 - D_1\right)} L \tag{2.50}$$

$$\Delta P_a = \frac{14.5 \times 2.62^2 \times 0.01556}{25.81\left(9.75 - 5\right)} 10{,}000 = 126\ psi$$

2.42.5.2 Drill Bit Nozzles

The pressure loss across the nozzles is not related to the fluid model. Use Eq. (2.26) and Eq. (2.23) determine the pressure drop of 920 psi across nozzles.

2.42.5.3 Drillstring

The flow parameters of the Herschel-Bulkley are the same in the annulus and drillstring and do not change as in the power law model. The average velocity of the drilling fluid in the drillstring as calculated with Eq. (2.35), is 10 ft/sec. The Herschel-Bulkley model corrects the shear rate in the drillstring for geometry and shear rate. The simplified correction factor in the drillstring is

$$G = \frac{3n + 1}{4n} \tag{2.51}$$

$$G = \frac{3(0.7)+1}{4(0.7)} = 1.11$$

The shear rate at the wall of the drillstring, γ_w, and associated shear stress, τ_w, for the Herschel-Bulkley flow model are similar to annular section calculations.

$$\gamma_w = \frac{96 \times G \times V_p}{D_1} \tag{2.52}$$

$$\gamma_w = \frac{96 \times 1.11 \times 10}{4.276} = 249.2 \ {}^{1}\!/_{\text{sec}}$$

$$\tau_w = 1.066 \left(1.33^n \tau_0 + K\gamma_w{}^n\right) \tag{2.53}$$

$$\tau_w = 1.066\left(1.33^{0.7}(2) + 0.61 \times (249.2)^{0.7}\right) = 33.55\, lb/100\, ft^2$$

The Reynolds number of Herschel-Bulkley fluids in a simplified form is

$$N_{\text{Re G}} = \frac{186 \times \rho \times V_p{}^2}{\tau_w} \tag{2.54}$$

$$N_{\text{Re G}} = \frac{186 \times 14.5 \times 10^2}{33.55} = 8{,}039$$

From Eq. (2.4) the critical Reynolds number is 2,542. The Reynolds number in the drillstring is greater than the critical Reynolds number; therefore, the flow in the drillstring is turbulent.

When turbulent flow prevails in the drillstring, power parameters of a and b in turbulent condition are applied to estimate the friction coefficient. From Eq. (2.38) and Eq. (2.39), we have

$$a = \frac{\log(0.7) + 3.93}{50} = 0.0755$$

$$b = \frac{1.75 - \log(0.7)}{7} = 0.274$$

The friction coefficient is estimated using Eq. (2.47)

$$f = \frac{0.07522}{8039^{0.274}} = 0.0064$$

The pressure loss in the drillstring is calculated using Eq. (2.55).

$$\Delta P_p = \frac{\rho \times V_p^{\ 2} \times f}{25.816\ D} L \tag{2.55}$$

$$\Delta P_p = \frac{14.5 \times 10^2 \times 0.0064}{25.816 \times 4.276} 10{,}000 = 841 \text{ psi}$$

The total pressure loss in the system is 1,887 psi.

2.43 Cutting Transport

The main function of the drilling fluids is to clean the hole effectively. An efficient hole cleaning reduces operational problems and reduces drag and torque in the wellbore. Slip velocity of the drilled cuttings can be estimated using various equations. The slip velocity of the drilled cuttings is a function of the density difference between the drilled cuttings and the density of the drilling fluid, viscosity of the drilling fluid, and the size and shape of the drilled cuttings. For efficient wellbore cleaning, the velocity of the drilling fluid in the annulus should be at least twice the slip velocity of the cuttings.

An easier and field proven method is to apply the carrying capacity index (CCI) which is valid for wells with an inclination angle of 35 degrees or less. CCI is the ratio of a constant number over fluid properties and annular velocity of the drilling fluid. If CCI equals 1 or greater the hole cleaning is effective.

$$\text{CCI} = \frac{\text{MW} \times \text{K} \times \text{AV}}{400{,}000} \tag{2.56}$$

where

MW = Density of the drilling fluid, ppg

K = Power law parameter of the fluid rheology, cp

AV= Average velocity of the drilling fluid, ft/min

The K value in this equation is not the same as the K values calculated for the power-law and the Herschel-Bulkley models. The unit of K is in cp and calculated using Eq. (2.57).

$$K = 511^{(1-n)} R_{300} \tag{2.57}$$

As indicated in Eq. (2.56), three parameters of fluid density, the K value (rheological parameter), and the average velocity of the drilling fluid affect cutting transportation to the surface. The density of the drilling depends on the pore pressure regime in the wellbore and should be kept as low as possible to enhance the drilling rate. The average velocity of the drilling fluid is selected based on the size of nozzles for optimizing hydraulic impact force or hydraulic horse power at the bottomhole. Other limitations of average fluid velocity are wellbore washout, bottom-hole assembly, and pump capabilities. Therefore, only rheological properties are available for adjustment of cuttings transport efficiency.

The K factor represents the rheological properties of the drilling fluid. Figure 2–19 illustrates K factor as function of the Bingham plastic model parameters, plastic viscosity, and yield point. Increasing values of yield point for a constant plastic viscosity results in higher values of K factor. However, for a constant yield point, increasing the plastic viscosity reduces the value of K. Usually only small quantity of additive is required to increase the yield point and improve cutting transport efficiency. Values of yield point and plastic viscosity can be back-calculated from Figure 2–19 for field application (see Example 2). More discussion on cutting transport efficiency is available in *Drilling Fluids Processing Handbook, 2004.*

2.43.1 Example 2

Determine a minimum annular velocity required to clean a well if the drilling fluid is a 10.0 ppg fluid with a plastic viscosity of 20 cp and yield point of 15 lbf/100 ft^2. If the average velocity in the wellbore is

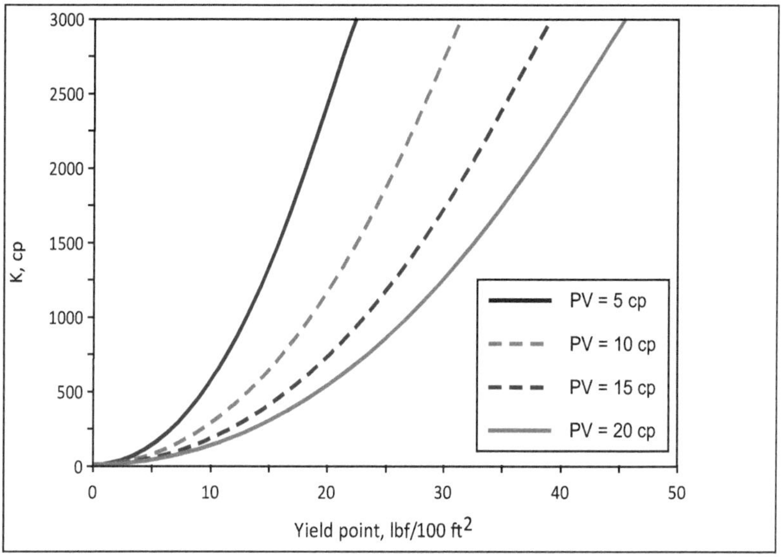

Figure 2–19 K *factor is shown as a function of viscosity and yield point of the drilling fluid (Drilling Fluids Processing Handbook, 2004)*

limited to 90 ft/min, what should be the minimum yield point of the drilling fluid to ensure efficient hole cleaning?

$$R_{600} = 2\mu_p + \tau_y \tag{2.58}$$

$$R_{600} = 2(20) + 15 = 55 \tag{2.59}$$

$$R_{300} = \mu_p + \tau_y \tag{2.60}$$

$$R_{300} = 20 + 15 = 35 \tag{2.61}$$

From the Eq. (2.13) the n parameter of the fluid is

$$n = 3.32 \times \log\left(\frac{55}{35}\right) = 0.652 \tag{2.62}$$

The K factor in the Eq. (2.55) is in cp. The K factor in cp is

$$K = 511^{(1-n)} \times 35 = 307\ \text{cp} \tag{2.63}$$

The minimum velocity for effective hole cleaning occurs when CCI is 1. The Eq. (2.56) is rearranged to find the minimum velocity.

$$AV = \frac{400{,}000}{10 \times 307 \times 1} = 130 \text{ ft/min} \tag{2.64}$$

If the average velocity is limited to 80 ft/min, then the minimum required *K* value of the drilling fluid is

$$K = \frac{400{,}000}{10 \times 80 \times 1} = 500 \text{ cp} \tag{2.65}$$

From Figure 2–19 for plastic viscosity for *K* of 500 cp and plastic viscosity of 20 cp, the yield point of the drilling fluid should be 20 $lb_f/100\ ft^2$ to ensure efficient cutting transport to the surface.

2.44 References

ASME Shale Shaker Committee, *Drilling Fluids Processing Handbook*, Elsevier, 2004.

Ibeh, C., Schubert, J., Teodoriu, C. "Methodology for Testing Drilling Fluids under Extreme HP/HT Conditions," AADE-08-DF-HO-14 presented at the AADE Fluids Technical Conference and Exhibition, Houston, TX, USA, April 8–9, 2008.

Mullen, G., Singamsetty, C., Dye, W., Ledet, D., Rawicki, A., Robichaux, T. and Authement, G. "Planning and Field Validation of Annular Pressure Predictions," AADE-01-NC-HO-08 presented at the AADE National Drilling Conference, Houston, TX, USA, March 27–29, 2001.

Recommended Practice on the Rheology and Hydraulics of Oil-Well Drilling Fluids, 4th Ed., API RP-13D, American Petroleum Institute, Washington, DC, USA, 2003.

Recommended Practice on the Rheology and Hydraulics of Oil-Well Drilling Fluids, 5th Ed, API RP-13D, American Petroleum Institute, Washington, DC, USA, 2006.

Robinson, L. "Drill Bit Nozzle Pressure Loss [Exploration of Finagle Factor Technology]," AADE 10-DF-HO-26 presented at the AADE National Drilling Conference, Houston, TX, USA, April 6–7, 2010.

CHAPTER 3

Gaseated Fluids (Gas-Liquid Mixtures)

Bill Rehm, Drilling Consultant
Arash Haghshenas, Boots & Coots

3.1 Introduction to Gaseated Fluids

The nomenclature of gaseated, aerated, or gas/liquid mixtures is not precise. All of these mean essentially the same thing: the mixture of a gas in a liquid drilling fluid. This discussion will use the term "gaseated."

This chapter discusses the use and theory of gaseated systems, and contains a bibliography as well as a set of questions. Section 2 discusses the special situation of gas injection in a dual-casing system. Section 3 reviews some field examples. Section 4 discusses the rheology of these two-phase systems.

Gaseated fluids, the most versatile of the reduced density drilling fluids, are a simple mixture of a liquid and a gas without any special emulsifier or stabilizing agent. Gaseated systems are forgiving and easy to run. The liquid can be almost any fluid suitable for drilling or workover; the gas can be air, natural gas, nitrogen, or some other gas. The overall characteristics of the system (inhibition, temperature stability, resistance to contamination) will normally be that of the fluid. The gas choice is generally based on the danger of surface fire, down-hole fire, corrosion, gas cost, and/or availability.

The critical issue with gaseated systems is keeping the two-phase system mixed. The gas and liquid separate on a gravity basis and this can create pressure surges in the wellbore. The issue of gas/liquid separation is a major part of this discussion of gaseated systems.

3.1.1 History of Gaseated Drilling

Gaseated fluids, or the IADC formal term of "gas-liquid mixtures," have a long history in drilling, primarily for the mitigation of lost circulation.

The application for using underbalanced drilling was first patented in the United States in 1866. One of the first formal records is a paper read in 1932 to the Imperial Oil Company about using natural gas and bentonite drilling mud to reduce lost circulation in the Atlas Mountains of Persia.

In 1939, in the Sour Lake Field in Southeast Texas, natural gas injected into the mud was used to avoid lost circulation in the depleted parts of the field. A Shaffer Rotating Head was listed in their catalog at this time, so there appears to be other gaseated systems in use.

In the 1960s, aerated systems were used in the Rocky Mountains to increase drilling rate and reduce lost circulation. In the 1970s, aerated systems were used in geothermal drilling, especially in the geysers in California. In the 1980s Canadians were using gaseated systems of natural gas and diesel oil for reservoir protection, and it was also used for lost circulation control in North Africa. During the 1990s, underbalanced drilling was adapted for offshore drilling throughout Europe and the technology moved into the Middle East and Far East shortly thereafter.

3.1.2 Definitions

Gaseated Mud, Aerated Mud, or Gas/Liquid Mixture is a simple mixture of a drilling fluid and a gas.

Surface Ratio of Volumes varies from 1:1 (ft^3 gas at STP: ft^3 fluid), to about 50:1, or in some special cases up to 100:1. The surface ratio of volumes provides a method of measurement for required gas volumes.

Quality is a measurement of the actual gas to liquid volume at any pressure point in the hole. It can be reported as a percent, a decimal, or a whole number. Figure 3–1 shows how hydrostatic pressure changes the ratios and quality at different depths.

Jet Sub is a tool for introducing gas from the drillpipe into the annulus to help eliminate the pressure build-up due to loss of gas in the drilling fluid in the upper section of the annulus.

Concentric String, or Dual Casing String is a method of injecting gas near the bottom of the hole.

Parasite String or **Parasite Tubing String** is a method for injecting gas near the base of the surface casing.

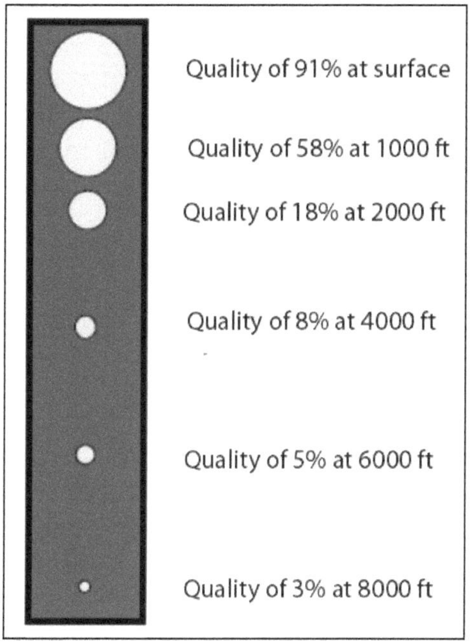

Figure 3–1 *Cross section of the rise of an idealized gas bubble (Medley et al., 1998)*

Constant Circulating Sub provides a method of continuous circulating during a connection.

Dune Effect is when in horizontal holes, the cuttings tend to fall to the bottom of the hole and form "dunes."

60° Zone is when a hole inclines between 50 and 70°, there tends to be an internal flow reversal that deposits cuttings in that area and causes difficulty in pulling the drill collars and bit through the zone.

3.1.3 Method of Reducing Bottom-Hole Pressure

The wellbore hydrostatic pressure is reduced by adding gas into the mud system. The addition of gas decreases bottom-hole pressure by displacing fluid out of the hole (hydrostatic reduction).

In the *hydrostatic regime,* which is at the low end of range of the gas injection volume, the wellbore pressure is very responsive to changes in the gas injection ratio, or impressed surface pressure, and tends to be unstable.

With continued increase in the volume of injected gas, the velocity of the liquid in the upper annulus increases and the flowing friction increases. The system enters the *friction dominated regime,* where the increase in liquid velocity in the upper part of the hole caused by expanding gas creates friction that increases the wellbore pressure at that point. The increased friction pressure keeps the gas from expanding with the result that further increases in gas injection actually tend to increase the bottom-hole pressure. The greater part of the friction effect is from the wetted perimeter—wetted perimeter is the length of the wetted line with the cross section of the conduit—so that this condition is common to 6 in.(152 mm) holes and smaller, and less often observed in larger holes.

In the friction dominated regime, the bottom-hole pressure is marginally responsive to gas injection rate changes, and responds almost as a pure liquid to surface pressure changes, i.e., changes in choke pressure (Nas, 2006) (see Figure 3–2). The high friction pressure (APL) and applied backpressure by choke stabilizes the system and makes it possible to control the natural surging of gaseated systems.

The gas injection point can be in the stand pipe at the surface, downhole from a parasite tubing string, downhole through a ported collar that is run on a concentric string of casing, or through a special dual drillpipe.

3.1.4 Depth Limits

According to the general gas law, gas compresses to half its volume every time the pressure is doubled, limited only by temperature and the gas compressibility factor (z). Near the bottom of a deep well the gas is so compressed that even doubling the injected gas volume does not significantly decrease the volume of fluid in that interval.

Most of wellbore pressure reduction takes place in the top 3,000 ft (1,000 m) of the hole, while bottom-hole pressure reduction depends upon vertical depth. In a shallow well, a gaseated system can reduce the bottom-hole pressure to as little as 25% of the pressure exerted by a full column of water. However, since depth is a critical element, the percent reduction in bottom-hole pressure decreases close to hyperbolically with depth. Generally in deep holes (15,000 ft, 4,500 m) the reduction in bottom-hole pressure is limited to about 75% of that of a full column of water. Beyond this depth, the gas in the fluid column is so compressed that the ensuing deep-hole pressure gradient is close to that of the fluid alone.

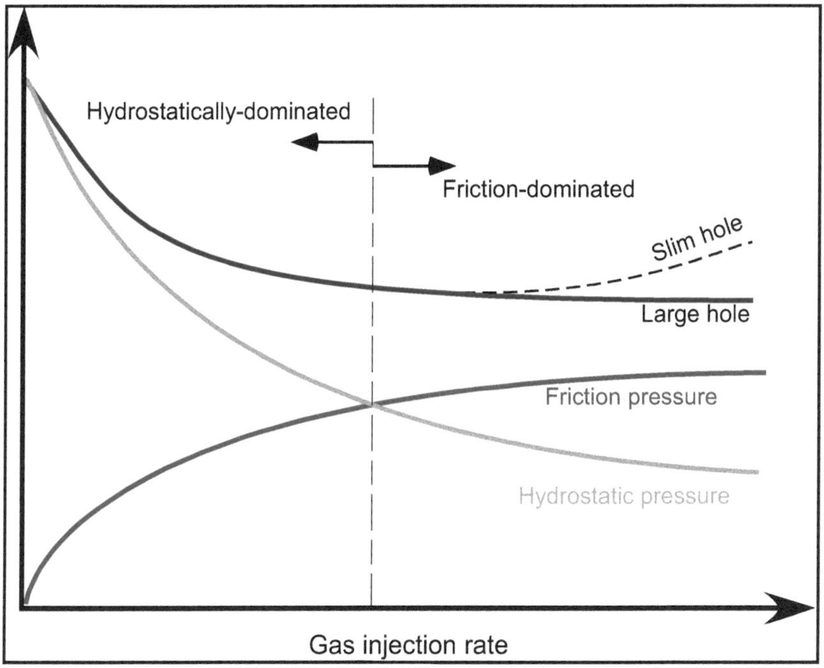

Figure 3–2 *Hydrostatic and friction dominated regimes (Nas, 2006)*

3.1.5 Gaseated Fluids in Horizontal Wells

In long horizontal wells, the increase in friction loss, or equivalent circulating density (ECD), makes the wellbore pressure at the toe of the well higher than at the heel. There is no point in injecting additional gas at the point where the well is starting to be horizontal because gas will not further reduce the hydrostatic pressure in the horizontal section. In a long and flat well, a simple decision has to be made whether the pressure is going to be controlled at the heel or at the toe of the well. There is a practical limit on how long a lateral can be drilled and remain underbalanced. In actual practice, it is difficult to calculate the ECD difference in gaseated horizontal systems because the hole is a long separator, and even with consistent agitation from drillpipe rotation the liquid and gas systems separate.

3.2 Advantages and Concerns of Gaseated Systems

Flexibility and simplicity make gaseated systems an attractive choice for reducing bottom-hole pressure. Reduced wellbore pressure can eliminate the problem of lost circulation and differential sticking, and can significantly increase drilling rate. It has also become increasingly evident that reducing overpressure against the reservoir can reduce reservoir damage.

The flexibility to use almost any liquid or liquid system makes it practical to find a liquid base that is compatible with sensitive formations, and one that limits the problems of contamination to the mud system. If a constant circulation technique is not being used, it is difficult to maintain a close balance on the bottom-hole pressure during a drilling connection-trip cycle, and so it is important to also have a minimal-damaging fluid for the overpressure surges.

The incremental cost of a gaseated system varies with the task. In its simplest form, the extra cost is the sum of the gas cost plus the cost of a rotating control device, a simple separator, and a string float. The cost of the gas is generally the greatest cost, but if very close control of pressure is required, the choke and separator system cost can be significant.

One kind of drilling method cannot be used to drill all wells. Every method has its own constraints and limitations. Gaseated systems eliminate some concerns of conventional drilling. However, the nature of gaseated system brings some other issues to the design process.

3.2.1 Advantages of Operating Gaseated Systems

3.2.1.1 Reservoir Protection

Underbalanced drilling (UBD) is considered a drilling method to protect the reservoir by reducing formation damage during the operation. A well-designed UBD operation reduces or eliminates problems associated with solid and fluid invasion into the formation such as pore plugging, phase trapping, clay reaction, fluid incompatibility, and the formation of emulsions. UBD does not eliminate all sources of formation damage. Therefore, the main benefit from the UBD operation is the reduction of formation damage attributable to solids and fluid invasion.

3.2.1.2 Reduces or Eliminates Lost Circulation

The first applications of gaseated systems in the industry were to prevent non-productive time (NPT) by eliminating lost circulation.

3.2.1.3 Eliminates Differential Sticking

Differential pressure sticking is the result of thick mud cake and excessive annular pressure. In a well-designed UBD, the annular pressure is less than formation pressure and no mud cake is present. Therefore, the conditions for differential sticking are eliminated.

3.2.1.4 Increases Rate of Penetration and Bit Life

In a formation with a very low rate of penetration, UBD can generally be applied to improve penetration rate. In drilling with three-cone bits, higher bottom-hole pressure holds cuttings down against the bottom of the wellbore (chip hold down pressure). The phenomenon increases the time to clean the bottom-hole and drill new rock. UBD eliminates chip hold down and increases the rate of penetration as illustrated in Figure 3–3. The actual complete explanation is more complex than this, but the total effect is that the cuttings that remain under the bit repress drilling rate. The drilling rate increase is not as pronounced with PDC or drag type bits because of their different cutting effect.

Roller Cone bit life is expected to be higher in UBD and MPD than conventional drilling. In UBD the bit is exposed to less stress and low-solids nonabrasive mud. UBD increases the ROP, then lower weight on bit (WOB) is required to achieve desired rate of penetration and this leads to higher bit life.

Higher rate of penetration and longer bit life reduces the number of drill bits and trip time to change the bit, and therefore improves the economics of the operation.

3.2.1.5 Reservoir Evaluation

During underbalanced drilling, pay zones can be detected immediately after penetrating the formation by measuring and observing fluid at the blooie line or after the separator. Formation fluid can be monitored at the surface to identify and study pay zones. Single or multi rate drawdown tests are achievable during drilling operation for well test purposes to estimate reservoir productivity.

3.3 Challenges with Operating Gaseated Systems

3.3.1 Cost

Gaseated drilling is normally more expensive on a daily basis than conventional drilling, especially in remote locations. In addition to conventional operating costs, a rotating control device, compressors, separators, flare lines, storage tanks for oil if it is encountered, more personnel, and more space is required, imposing a higher operational cost. Offshore locations and the presence of sour gas also increases the cost. The extra cost may be justified by advantages of a gaseated system. Costs can be reduced by integration of the extra equipment and services that are required to drill underbalanced. This is further discussed in Chapter 11, Equipment and Equipment Integration.

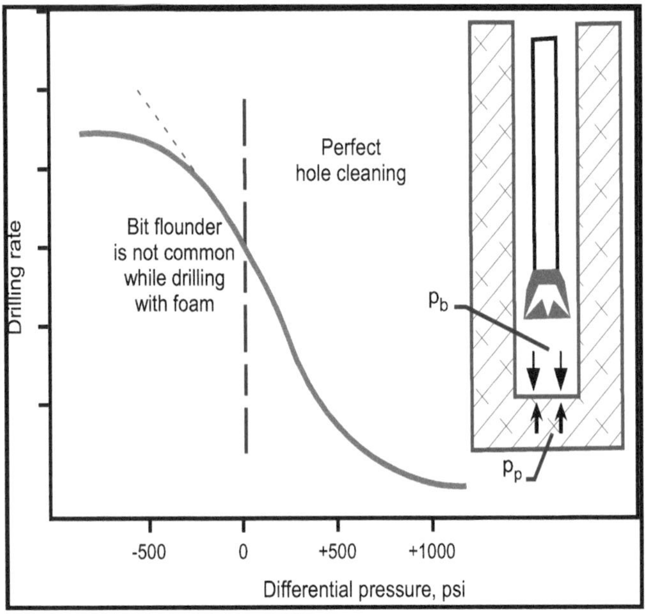

Figure 3–3 *Pressure drop versus ROP*

3.3.2 Pressure Surges

The gaseated system is unstable. The gas and liquid separate by gravity and require mixing to keep them combined. The instability of a gaseated system induces pressure surges. The gas migrates up hole to form a large gas bubble zone while the fluid falls down the hole to form a solid fluid column below the gas bubble.

As the gas reaches the surface and escapes through the choke, the upper wellbore is then full of a fluid column that causes increased bottom-hole pressure and keeps the gas below the fluid column compressed. As the fluid column circulates out through the choke and reduces wellbore pressure, the gas expands and further unloads the hole, reduces bottom-hole pressure, and the cycle repeats. Cycle time can be between one minute to twenty minutes.

Pressure surges causes formation damage and wellbore instability problems. Pressure surging can be controlled during drilling by a combination of pipe rotation, velocity, and surface back pressure held on the return annulus. During drilling, an impressed surface pressure of 5 to 15 atmospheres (70–220 psi) keeps the gas compressed enough that with the upward flow of fluid, the system stays mixed.

The injected gas separates during connections when gravity causes the gas to move upward and liquid to displace downward. The rate of separation depends on the size of gas bubbles and viscosity of the fluid. Large bubbles move upward faster than small bubbles. This means that keeping the well pressurized with the 5 to 15 atmospheres (70–220 psi) of surface back pressure to minimize gas bubble size is important. Increased viscosity of the liquid phase slows down the gas-liquid separation but makes it more difficult to separate gas from liquid at the surface as well as increasing the circulating density (ECD).

To minimize pressure surges on connections, extra gas can be injected just before the connection to dry the upper section of drillpipe. After the connection, the extra gas provides a boost that reduces much of the connection based pressure surge (see Figure 3–4).

3.3.3 Other Challenges

- **Fractures:** In presence of large wide fractures, the well fluid displace into the fractures and cause a continual low level lost return situation, which will turn into a low level well kick on a connection. The source of this problem is gravity displacement of the drilling fluid and flow back when the pump is turned off (see Figure 3–5).
- **Imbibition:** Capillary forces within the reservoir can cause fluid imbibitions, where liquids are "sucked" into the reservoir even though the wellbore is underbalanced. To minimize the fluid imbibition, annulus pressure should be less than formation pressure by the value of capillary pressure, and the liquid drilling phase should be the non-wetting phase of the reservoir. Imbibition can be measured from cores in the laboratory (Guo and Ghalambor, 2006).
- **Periodic kill:** It may be required unless the pipe is stripped/snubbed in and out or a down-hole valve is used. Going overbalanced to kill the well can damage the formation or be ineffective due to lost circulation.

Other challenges of UBD operations:

- Corrosion is a problem associated with the use of air because of oxygen introduction in hot down-hole environment. It is discussed in Chapter 13.
- Surface fires and explosion can occur if hydrocarbons are presented with oxygen.

- Vibration of drillstring occurs because aerated drilling fluid does not support the pipe fully as in the case of conventional drilling fluids.
- Friction factor is sometimes higher in aerated fluids compared to conventional fluids, thus resulting in increased torque and drag.
- Proper hole cleaning might be a problem in aerated drilling fluids resulting in stuck pipe and increase pressure drop.
- Conventional mud pulse (MWD) signals are attenuated in aerated drilling.

3.4 Flowing Hydrostatic Pressure Prediction

The gaseated system is very flexible and the volumes that can be used are equally flexible. The upper limit of liquid injection rate is set by either the bottom-hole pressure requirement or by ECD in the upper hole. The frictional response (ECD) to increased liquid velocity due to expanding gas in the upper part of the hole is not quite linear and depends primarily on the wetted perimeter and annular area.

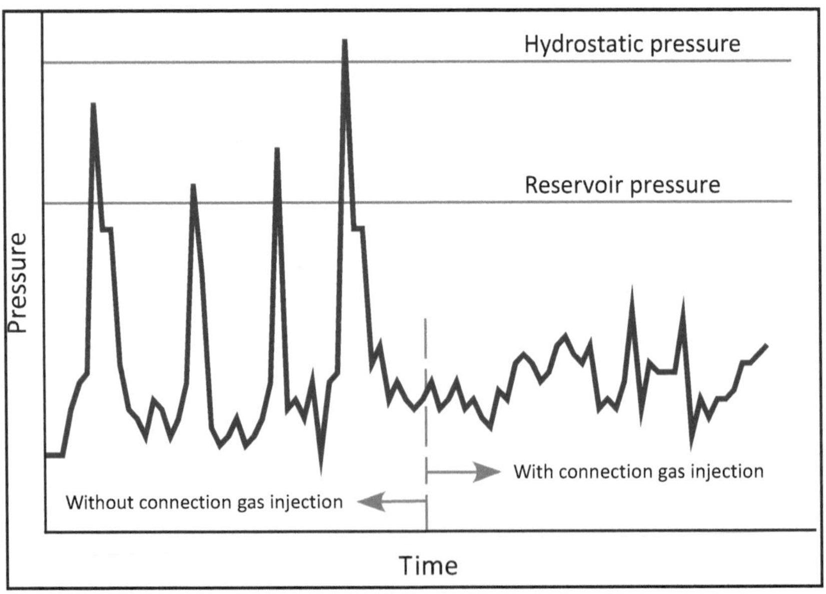

Figure 3–4 *Minimization of slugging by extra gas circulation before connection (Bennion et al., 1998)*

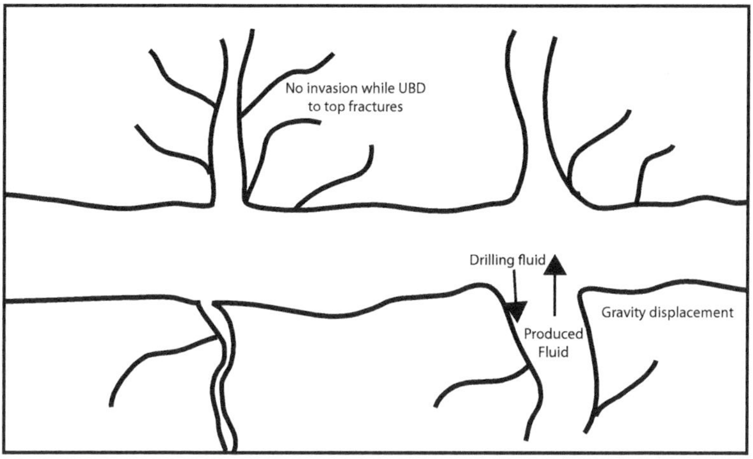

Figure 3–5 *Gravity induced invation in UBD operations (Bennion et al., 1998)*

Poettmann and Bergman (1955) developed one of the earliest models for the hydrostatic pressure of gaseated fluids at static condition. Figure 3–6 illustrates a chart developed by Poettmann and Bergman for hydrostatic pressure of gaseated system at 100°F.

The limitation of this method is that it does not consider dynamic conditions such as hydraulic friction and the effects of fluid segregation in the annulus. The gas volume required to reach the bottom-hole pressure reduction is low by a factor of 2 or 3.

To estimate the bottom-hole pressure during gaseated drilling, use one of the available models, Table 3–1 or correlations available for multiphase flow such as available mechanistic models (Shaoham, 2006), (Hasan and Kabir, 2002), (Guo and Ghalambor model, 2002), Hagedorn and Brown correlation, and the Beggs and Brill correlation, etc. If the gaseated system is assumed as a homogenous fluid system, consider that the gas-liquid mixture behaves as a power-law fluid because the gas-liquid mixture is in high turbulence. A simple friction pressure loss in the annulus for power-law is

$$\Delta P = \Delta L\ f\ \rho\ V_a^2 / \left\{21.1\left(D_h - D_p\right)\right\} \tag{3.1}$$

where

f = The Fanning Friction Factor obtained after calculating Reynolds number

ΔP = pressure drop due to friction, psi

ΔL = Length being evaluated, ft

ρ = fluid density, ppg

u_a = plastic viscosity of fluid

V_a = Mean annular velocity, ft/sec

D_h = Hole diameter, in.

D_p = Drillstring diameter, in.

3.5 Operations-Basic Gaseated Fluids

The following suggestions and ideas may be at variance with the written procedures of any particular operator or service company. The important part of these procedures is to try to understand the problem that is presented and be prepared to act if equipment goes down or the procedures are leading into a larger problem.

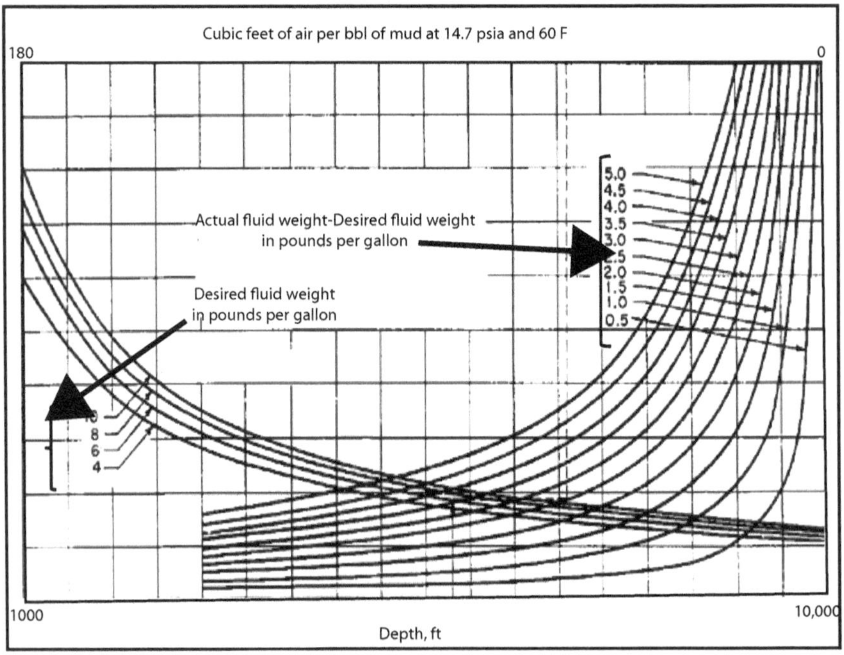

Figure 3–6 *Air requirements for reducing average mud density at average fluid temperature of 100°F (Poettmann and Bergman, 1955)*

3.5.1 Gas and Liquid Volumes

In gaseated systems, start with a liquid volume that would give an annular velocity of 120 ft/min (36.6 m/min). Perhaps a better statement would be to start with a liquid volume that will clean the hole. The liquid part of the system is generally considered to have the greatest influence on cleaning out from under the bit and on general hole cleaning, as well as on drilling motor operation. In a general sense this is true and is why planning begins with the common industry annular velocity.

Add to the liquid volume, the volume of gas required to reduce the bottom-hole pressure and to force the system into the more stable friction dominated regime. This generally requires the use of a computer model of gaseated systems. In the friction dominated regime, the system is more sensitive to volume changes than gas volume changes and tends to act more like a liquid system.

The final volumetric requirements are a compromise between:

- Bottom-hole pressure requirements
- The requirement for a friction dominated regime
- A sum of fluid and gas volumes at the motor pressure that is adequate to run the drilling motor in an effective part of the torque curve
- Hole cleaning, especially in a horizontal bore
- The volume and pressure required against the available compressors or source of natural gas. Both peak and operating pressures need to be within compressor or gas line operating range

3.5.2 Gas Injection Rate

Gas injection rate can be almost anything and depends upon:

- Reduction in bottom-hole pressure
- Onset or effect of the friction dominated regime
- Annular backpressure which keeps the gas compressed and reduces the quality. In small hole sizes, the friction dominated regime may produce all the backpressure necessary to stabilize the surging
- Viscosity of the fluid phase
- Cost

3.5.3 Compressors

The best practice is to use the complete output from a compressor and not have to bypass gas. Bypassing gas works opposite from need since when well pressure rises, generally it means more gas is required, but at a higher pressure the bypass orifice will release more gas. If gas is bypassed, be sure the bypass is upstream of the gas meter so volume to the well can be properly metered.

3.5.4 Liquid Hold Up Volume of Liquid in the Hole

The annular liquid hold up versus gas injection rate and drill-string liquid hold up versus gas injection rate graphs are used to calculate total fluid in the well as shown in Figure 3–7.

3.5.5 The Liquid or "Mud" System

The liquid system is the basis for the gaseated system. Almost any drilling fluid can be used that would not damage the reservoir, but would inhibit shale and resist temperature or contamination. In practice, most fluid systems are water, diesel oil, or a light synthetic oil. The liquid system that is most desirable has low fragile gel strength to facilitate the release of the gas at the flow line or separator. In horizontal or high angle wells, the most desirable fluid would also have a good carrying capacity for drill cuttings.

High viscosity in the liquid phase makes it difficult to remove the gas at the surface. In most large surface holes, the mud is sent to a large mud pit where the gas will have time to work out of the fluid.

The most common fluid is water, with a controlled pH and with anti-corrosion chemicals. The limitation to water (or oil) is its limited carrying capacity. It is difficult to clean a washed-out hole with water. It is also a problem to clean the 50° to 70° casing curve in high angle wells with pure water or oil. In a horizontal well, the dune effect is significant.

The lifting capacity of the water may be improved by using thixiotropic drilling fluids like an XC polymer or Mixed Metal Oxide systems. These systems will improve carrying capacity and reduce the dune effect without a significant increase in the APL. Other polymers or drilling fluid systems, like low lime mud, can also be used, providing they do not exceed the ECD beyond the program limits. However, almost any increase in viscosity of the liquid system increases the difficulty of releasing the gas from the fluid at the pits.

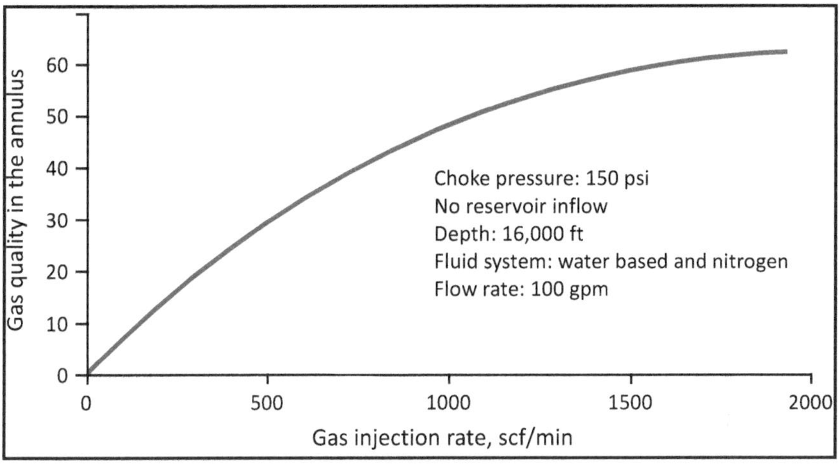

Figure 3–7 *Percentage of gas by volume in the annulus versus gas injection rate (Nas, 2006)*

The obvious exception to low gel strength is when drilling large shallow surface holes where the viscosity and gel strength of the fluid can be deliberately high to contain the gas and keep it from breaking out and bypassing the liquid. Typically, the mud is a bentonite-water mixture and has enough surface containment provided that most of the air can work its way up and out of the system.

Synthetic oil or diesel oil is commonly used in small diameter completions or re-entries within a reservoir. When oil is used, the gas is normally nitrogen, but natural gas has also been used. The oil/liquid phase gaseated fluids have low viscosities and limited carrying capacity, but they will adequately clean gauge or near gauge holes.

Viscosifying additives and the presence of polymers increases the risk of producing emulsion and foam. Reducing the amount of viscofying additives and polymers, and adding defoaming agents minimizes the risk of emulsion and foaming.

3.6 General Limits of Gas and Fluid Volumes

3.6.1 Gas Limits

There are limits to how much gas can be added to the system to reduce the bottom-hole pressure.

In a gaseated system, the liquid needs to be in the continuous phase. The maximum gas quality at the surface is normally limited to about 80% to avoid pressure surges and loosing cutting transport capacity near the surface.

With increasing gas additions, the increased velocity of the displaced fluid causes friction, or APL, that becomes greater than the pressure reduction from adding more gas to the system. This is a common practical limit to gas volume (the friction dominated regime).

3.6.2 Limits to Liquid Volume

The liquid being pumped down the drillpipe cleans the bit and runs the down-hole motor. The liquid is the primary hole cleaner. The lower limit on liquid injection is:

- Hole cleaning
- The drilling motor

The upper limit on liquid injection rate is:

- The friction dominated regime
- The drilling motor

Figure 3–8 illustrates a general planning procedure. The operating region should be within the reservoir pressure target, limitations of the down-hole motor, cutting transport, and wellbore stability. The flow should be in the friction dominated region to ensure smoother operation.

3.6.3 Back Pressure

Back pressure is one of the keys to maintaining a smooth pressure regime. The annular back pressure increases bottom-hole pressure in a non-linear manner.

The tendency of the gas and liquid to separate becomes severe at about 80% quality (percent of gas in the system at a point). This is the reason for using back pressure. Above 80% quality, even when circulating:

- The liquid and gas separate
- Alternate slugs of air and gas cause pressure surges in the hole
- Gas becomes the continuous phase and cuttings cannot be lifted out of the hole. The cuttings fall back increasing the surging effect and occasionally will stick the pipe

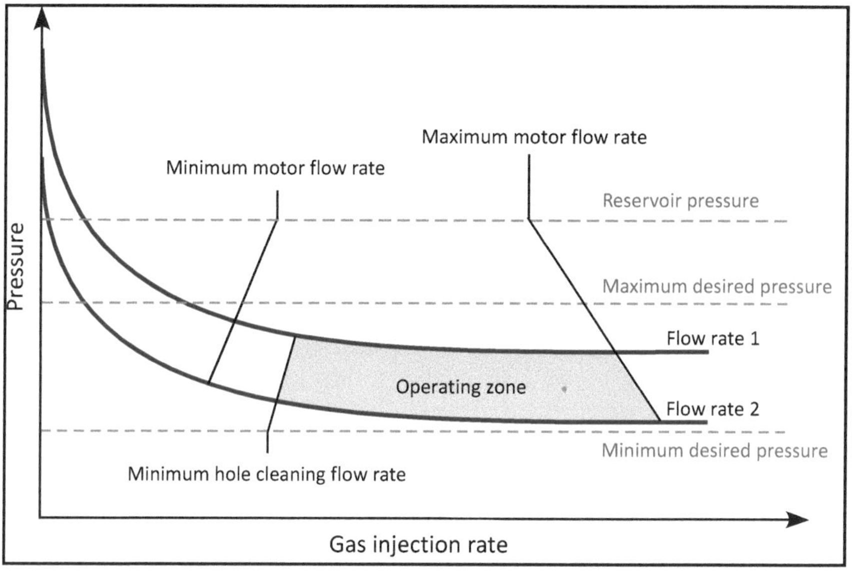

Figure 3–8 *Operating window for bottom-hole pressure (Nas, 2006)*

Calculate from one of the computer programs the back pressure that will be needed to control surging, about an 80% quality, in the friction dominated regime, and re-correct the gas volume to the proper down-hole pressure (default is 7 atm. of back pressure, ~100 psi, or 700 kPa). Some iteration may be required for a final program. On a practical basis, there needs to be several solutions. Implementation at the rig may require some modification of the results from the model. In small hole sizes, the friction dominated regime may produce all the backpressure necessary to stabilize the surging.

3.6.4 Motor Constraints

The motor is placed on the bottom until proper standpipe pressure is built up (approximately 400 psi). As drilling is initiated, pump pressure decreases and increases with changes in motor torque. The equivalent liquid rate through the motor needs to be calculated to ensure sufficient torque. Consult the down-hole motor provider for the limits of gas and liquid injection rates versus torque.

Motors can stall out and this can be signaled by a rapid increase in pump pressure. It is sometimes hard to detect a motor stall when using compressible fluids. When the pressure increases, the driller assumes that the motor has stalled so he will pick it up off the bottom, but this will cause the release of gas from the drillstring which can cause the motor to over-speed by exceeding its maximum flow rate (Nas, 2006). The use of small jets in the bit will help mitigate the over-speed problem, but at the expense of higher compressor pressure.

3.7 Solids Control Equipment

Solids control equipment is not specifically part of these discussions. However, gaseated systems require recycling of the drilling fluid. The purposes of maintaining the underbalance are to protect the reservoir, avoid differential sticking, and increase the drilling rate. These requirements may be at risk if the solids in the drilling fluid are allowed to build up and increase the APL. Additional books in this series and many other references discuss the problems of inadequate solids control (Robinson, 2003).

3.8 Methods of Gas Injection

The problems of surging and pressure variations on trips and connections has lead to a number of different techniques for gas injection into the system. These include:

- Drillpipe injection
  - Drillpipe injection using a jet sub
  - Drillpipe injection using constant circulating sub
- Parasite string
- Dual casing strings, or concentric casing strings
- Remote well injection

3.8.1 Drillpipe Injection of Gas

Injection of gas down the drillpipe in conjunction with the drilling fluid is the original and simplest method for gaseating a system. The gas is injected into the fluid being pumped through the standpipe. Final mixing of the fluid and gas takes place at the bit where the gas is at the smallest volume and available energy (pressure drop) is at the greatest.

The advantages of drillpipe injection:

- It is simple.
- The total volume of fluid and gas is available at the motor, bit, and to clean the hole.

There are, however, some challenges to drillpipe injection:

- If the gas contains oxygen, and it maximizes the chance of corrosion, especially in the hotter higher pressured bottom-hole environment.
- It is more difficult to control pressure surging and re-aerate the annulus after long connections or trips.
- In high angle horizontal wells, the wellbore is a long separator that allows the liquid and gas to separate. This makes hole cleaning and controlling pressure surges more difficult.
- Connections take longer because the compressed gas needs to bleed out of the drillpipe before the drillpipe can be unscrewed.
- Mud pulse MWD cannot be used.

3.8.1.1 Drillpipe Injection Using a Constant Circulating Sub

Circulating subs are available to make up on the drillpipe. These may be sourced through several service companies and manufacturers (see Figure 3–9). This reduces the potential for pressure surges and tends to keep the gaseating operation in steady state flow. The use of the sub does not require any change in hole design, well head or major modifications to the rig (Ridley et al., 2011).

3.8.2 Drillpipe Injection of Gas Using a Drillpipe Jet Sub

One of the big problems with drillpipe injection is the buildup of pressure due to a long column of (non-gaseated) mud in the annulus after trips and long connections. As the column of mud is pumped up the annulus, it keeps the gaseated fluid below compressed, which increases bottom-hole pressure. When the gas finally gets near the surface, it expands and unloads the hole, which abruptly decreases bottom-hole pressure. In areas of lost circulation, the pressure increase can be so severe that total lost circulation occurs and the hole will not circulate.

Field operations faced with this problem developed the drillpipe jet sub. It is normally a 3 ft (1 m) sub of drill collar stock drilled for a bit jet with a pinned-in string float above the jet. The bit jet is usually about 11/32 in. (8.73 mm) and connects the inside of the drillpipe with the annulus. It is typically sized to let out about 20% of the drillpipe fluid to the annulus. The jet sub appears to preferentially pass a greater percent of gas. The sub is usually placed at 2,500 ft to 3,000 ft (760 m to 900 m), just inside the surface casing. In some

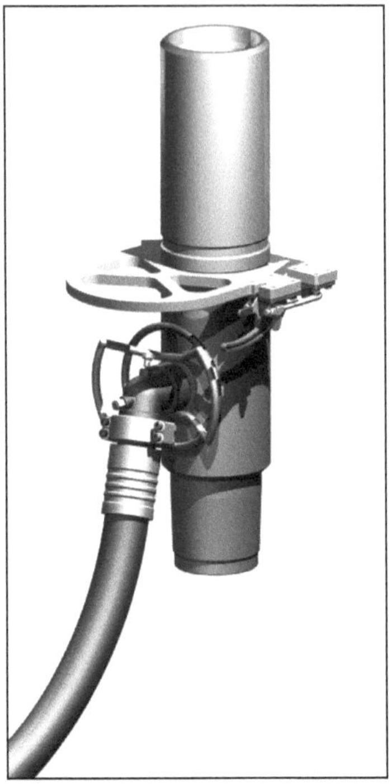

Figure 3–9 *Constant circulating sub "Non Stop Driller" with quick connect*

cases, two jet subs have been placed in the string at about 2,000 ft (600 m), and again at 3,500 or 4,000 ft (1,060 m to 1,200 m).

The advantages of the jet sub:

- On connections, the gas from the sub helps in gaseating the column of the fluid above the jet sub and limits the wellbore pressure surge.
- On trips, it starts the aeration of the system early and helps prevent lost circulation by limiting the initial pressure surge.
- It does not require pre-planning and can be put in the string at any time with no drillstring modification.

The problems with the jet sub:

- It passes about 20% of the circulating fluid that is then not available to the bit or motor.

- It works in reverse, when the annulus pressure is the lowest the jet sends the most mud to the annulus and when the annulus pressure is the highest, it represses fluid from the jet.
- It has the same problem as drillpipe injection with horizontal holes.
- Mud Pulse MWD cannot be used.

A string float should always be run as part of the sub, just above the jet. If there is a pressurized wellbore, when the jet sub is pulled through the rotating head there will be a brief release of the pressure trapped between the jet float and the next lower string float or the bit float. The drillpipe stand containing the jet sub can be stripped through the annular preventer to keep the pressure release below the floor and into the flow line.

3.8.3 Gas Injection with a Parasite Tubing String

A parasite string was originally an external string of 2 3/8 in. to 3 1/2 in. (73 mm to 90 mm) tubing strapped to and run with the surface casing, but some later strings have been 1.5 in. (36.75 mm coiled tubing). The string and a fitting was welded into the surface casing about a joint above the bottom. The bottom of the tubing string included a burst valve and a non-return valve (NRV). Gas is injected into the tubing and exits above the base of the surface pipe to gaseate the mud from about 3,000 ft (1,000 m) to the surface.

It was mathematically observed that until the gas bubbles from the bottom of the well reached about 3,000 ft (1,000 m), very little expansion took place. If the gas were to be injected near the base of the surface pipe, typically at 2,500 ft to 3,000 ft, (750 m to 1,000 m), the gas would displace almost the maximum amount of drilling fluid and so reduce the bottom-hole pressure about as much as practical.

The advantages of the parasite tubing string:

- Gas can be injected during a connection and eliminate surging and lost returns on a connection.
- Gas can be injected during most of a trip and avoid trip surge.
- It allows the use of a mud pulse MWD.
- It limits any oxygen corrosion to the cooler upper part of the hole.

The challenges with a parasite tubing string:

- A larger hole at the surface is required to accommodate the tubing.
- Modifications to the wellhead are required to allow for passing the tubing.
- The risk of crushing or damaging the tubing exists while running the tubing.
- The tubing has a tendency to put the casing off-center and does not allow for moving the casing while cementing. Being off-center and not being able to move the casing while cementing affects the quality of cement bond.
- A burst plate and NRV needs to be used in the tubing string to prevent cement from filling the tubing.
- With a tubing parasite string, gas injection is independent of rig pump operation and aerates the upper section of the hole on an independent basis during connections and trips. However, the control of bottom-hole pressure is presumed to be a steady state operation with the mud pumps running. With the pumps off, the full drilling volume of air will unload too much of the hole. When the mud pump is off, the gas volume needs to be reduced to about one half to one third to maintain a steady bottom-hole pressure.
- Since the gas is left on during connections and while tripping in the hole, a greater overall volume of gas is used. This is only significant if metered natural gas or cryogenic nitrogen is used, since compressors work on a day rate basis.

3.8.4 Concentric Casing String (Sometimes Called a Dual Casing String)

The concentric casing technique requires a size larger casing through which is run an inner flush joint casing string, leaving a secondary or "false" annulus for gas injection.

Provision must be made for left hand torque from the drillstring. The inner string may be run and hung off as a liner allowing the upper section to be retrieved after drilling. It may also be run with a left hand thread without any down-hole hardware to eliminate any constriction in the annulus. Above the end of the flush joint string is a perforated or slotted short joint that allows the gas from the secondary annulus to enter the drillpipe annulus.

The concentric casing string was developed to resolve the problems with drillpipe injection and the parasite string. The concentric casing solves surging problems on connections since it is always injecting gas and trip surges are of less magnitude. Running a concentric casing is easier for the crew, and can be run around the curve and run deeper than a parasite tubing string. Other advantages of parasite casing injection method:

- The main or outer string of casing can be reciprocated or rotated while cementing.
- The inner casing string can be run as a tie-back liner and retrieved to remove the slotted sub, or removed for reuse.
- It lends itself to the use of the Deployment Valve or Down-hole Safety Valve.
- Mud Pulse MWD can be used.
- With a dual casing string, gas injection is independent of rig pump operation and aerates the upper hole on an independent basis during connections and trips. However, the control of bottom-hole pressure is presumed to be a steady state operation with the mud pumps running. To manage the bottom-hole pressure during connections when the mud pump is off, the gas volume needs to be reduced to maintain a steady bottom-hole pressure.
- One of the better uses is constant circulation by diverting the mud pump output along with the gas to the false annulus.

Challenges with concentric tubing:

- It requires a larger hole.
- Wellhead needs modification, or a "B section" added to hang the string.
- If it is run as a liner, the hang-off assembly constricts the annulus.
- In wells where pressure might occur, the surface pipe has to have well pressure integrity because well pressure might back up through the slots.
- The false, or outer annulus, has a large storage capacity and the stored energy from the compressed gas in the annulus can cause surging of its own accord. If surging is allowed to start to unload the hole, the stored energy may unload the entire upper hole, which can cause a significant negative pressure

surge followed by a positive pressure surge until the false annulus builds up enough pressure to flow gas again. The dual string annulus has a storage volume, and when the annulus pressure is released it must be bled slowly to keep from unloading the hole. The smaller the volume of gas and the lower the pressure in the "false" annulus, the easier it is to control the surging.

- The inner string may be centralized, but it must handle the reactive torque from the drill sting. The inner string may be set on packer slips at the bottom or it may have a left hand thread, in which case it can be hung from the "B section." The exact set up depends upon the commercial availability of the hang off and packer equipment used in conjunction with the slotted sub or ported collar and Deployment or Down-hole Safety Valve.
- Slotted or perforated area is required to pass the gas (see Section 2, Using Concentric Casing with Gaseated Systems, Principles and Examples, for a further discussion).

3.8.5 The Dual Well System

A system patented by CDX Corp (Zupanick and Rial, 2006) used for drilling coal bed methane involves some of the same technologies. A vertical hole is drilled into the reservoir about 200 ft (60 m) beyond the proposed end of the landing point of the curve. The directional hole then intersects the vertical hole in the reservoir and continues on to drill one or more laterals as shown in Figure 3–10. If several curved holes intersect the vertical hole, the vertical hole can be an efficient point for production, while the curves are abandoned.

The system is gaseated from the vertical hole while drilling the directional hole. The process is similar to drilling with a dual casing string, but since the point of gas entry is in the horizontal section of the well, several hundred feet from the surface annulus, more care must be taken not to unload the hole.

3.9 Well Kicks (Gas, Oil, or Water Flows)

The onset of significant gas, oil, or water flows may tend to change the bottom-hole pressure by either unloading the hole (gas) or increasing the wellbore pressure (water and oil).

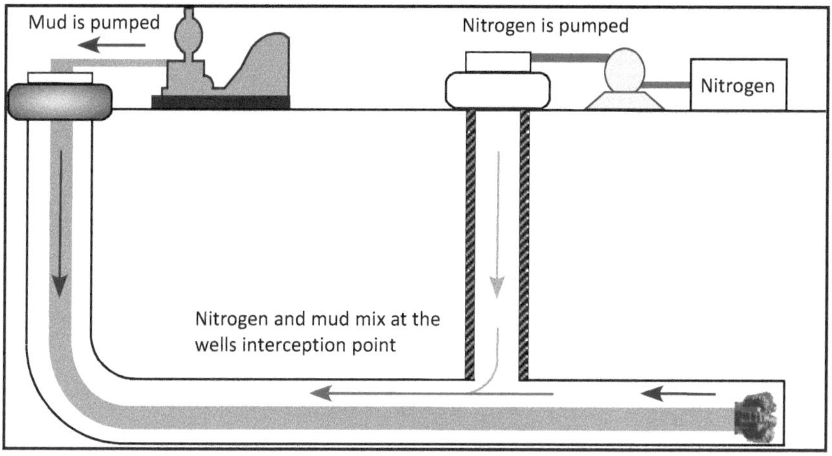

Figure 3–10 *The system is gaseated from the vertical hole while drilling the directional hole (Zupanick and Rial, 2006)*

3.9.1 Gas Flow

Often, increasing gas flow has little effect on bottom-hole pressure due to the effect of the friction dominated regime. In remote locations, or where there is no infrastructure, the surface gas has been flared. In other circumstances, gas is sent to the sale line to offset drilling costs and restrict pollution. There are situations where gas flow intentionally has not been suppressed because it will cause lost circulation (such as Austin Chalk and other fractured formations).

With a gas flow, it is possible to increase the surface pressure to suppress the flow, and then increase the fluid volume ratio to keep the flow suppressed and return to a normal surface backpressure.

There is the option with gas flows to use the gas from the well and shut down or reduce the surface injection volume of gas. Economically this makes sense. The injected gas may be a small enough volume (due to pressure) at the drilling motor that the loss of the gas down the drillpipe doesn't make much difference in motor operation. However, it is important to check motor throughput with any change in drillpipe injection values.

3.9.2 Water or Oil Flow

Water or oil flows will increase the bottom-hole pressure by increasing the ratio of liquid to gas; the influx may be self limiting. The choke pressure may be increased to repress the subsurface flow, while bottom-hole pressure is increasing from the increased liquid fraction of the gaseated system.

Oil and water kicks dilute the drilling fluid system and increase surface separation and disposal problems. It may be desirable to limit (suppress) some or much of the flow of a well fluid. There are, however, exceptions to this where oil is sent to the sale line or water is needed to combat lost returns. This is a "what if" problem that needs to be part of the drilling program.

3.10 Operational Concerns and Challenges

3.10.1 Pressure Surges

Pressure surges are caused by separation of the gas and liquid where the liquid flows down and the gas rises up until there is a large interval of gas at the top of the well followed by a solid column of fluid as shown in Figure 3–11. As the gas is unloading, the pressure drops because the hole is partly emptied. Then as the liquid is pumped up, usually with a load of cuttings, it keeps the gas below it compressed and the bottom-hole pressure builds until the liquid column is short enough to let the gas below it unload the top of the hole and complete the cycle.

The function of annular back pressure is to keep the gas compressed to where it is less than about 80% of the total volume of gas and fluid at the top of the hole. The planned back pressure required to control surging will almost always require some "tweaking" at the well site. The default value of 100 psi (700 kPa) is a good starting pressure. A choke pressure gauge is adequate for this purpose if it reads in 10/20 psi (or 100 kPa) increments that will show enough change to efficiently monitor the annulus pressure.

Most of the surging problems with gaseated systems occur at the start of the well. To avoid operating problems, at the start of a gaseated operation before drilling into the formation, the well should be circulated with the liquid and gas rates in the following plan:

- Plan for at least two hours of circulating to balance the system.
- Use an increase in back pressure to stop surging.
- Balance back pressure and bottom-hole pressure with more gas or less liquid.
- Both the drillpipe pressure and the annulus pressure should be constant.

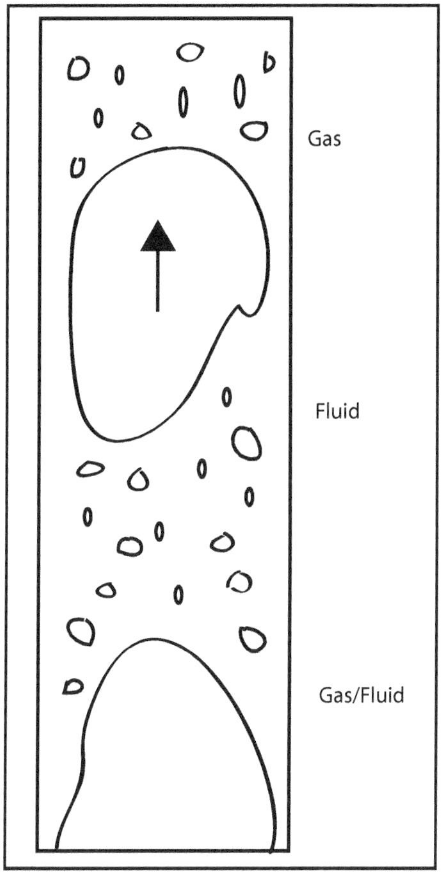

Figure 3–11 *Separation of gas and fluid*

- Use the minimum back pressure with pipe rotation to determine the final pressure.
- Then practice connections until the crew and system work smoothly.

See Section 2, Using Concentric Casing with Gaseated Systems, Principles and Examples, for a further discussion of the surging problem with dual casing strings.

3.10.2 Unloading the Casing

To unload a hole full of mud or water, pump the gaseated system until the pressure rises too high for the compressors. Then bypass the compressors and pump just liquid until the pump pressure goes down enough to restart the air. Repeat this procedure until the hole is completely gaseated, then establish a steady state flow before drilling ahead.

This technique pressures up the hole, so if the cement has already been drilled, consider using about three stages for going in the hole instead of going to the bottom and unloading all at once. Staging in is only important if the open formation might lose circulation or cave under repeated pressure surges.

3.10.3 Connections

If hole stability or reservoir damage due to pressure surging is not a consideration, connections can be made in a normal manner. The use of the drillpipe jet sub can limit the worst of the surging, but constant circulation with either a consistent circulating sub or a dual drillstring is even a better choice.

Major reservoir influxes can occur, due to loss of annular friction pressure, when pump and compressors are turned off if adequate choke pressure is not maintained (see Figure 3–12). When drilling is then resumed there will be an unstable system until the influx is circulated out of the hole.

3.10.4 Stripping in Underbalanced Operations

As a matter of safety and good practice, careful consideration should be given to whether or not to strip out of the hole. The general objection to stripping out of the hole is that it increases trip time and wears out the sealing element in the rotating control device. Trip time and the replacement of the RCD element are significant costs subject to criticism if the well is not producing hydrocarbons. In many cases, the wellbore pressure is being controlled to avoid lost circulation or to increase drilling rate and it appears that there will be little danger of gas in the annulus. So in practice, when drilling outside the reservoir, the use of the rotating head is a judgment call that depends upon the local circumstances and the operator's policies.

The important difference to remember is that there is no heavy column of mud as a first barrier against gas reaching the surface. Therefore it is prudent to keep the rotating head active where gas is in evidence.

The critical point in most stripping operations is when the bottom-hole assembly arrives at the surface and the rotating head needs to be set aside. With a partly empty hole, it is a good practice to

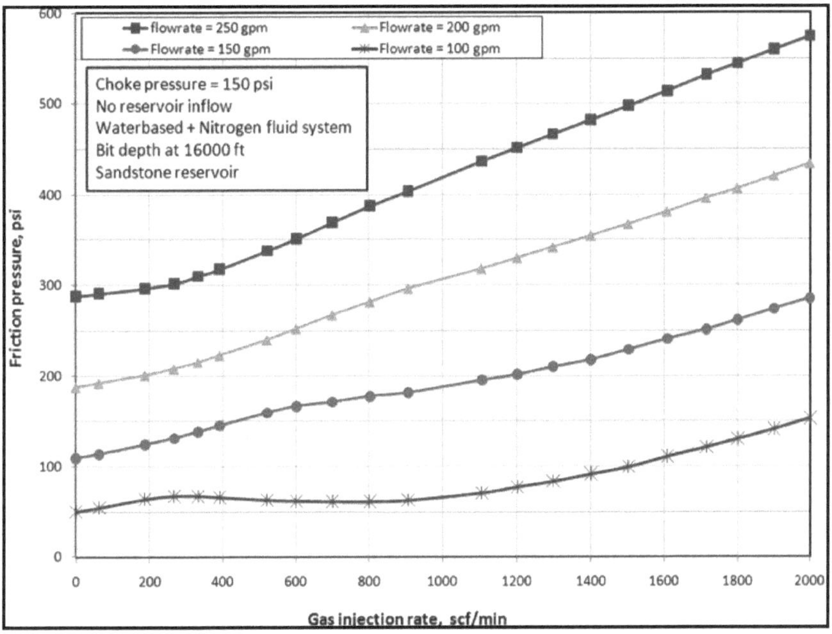

Figure 3–12 *Gas injection versus friction pressure (Nas, 2003)*

be prepared to use the annular preventer and ram spacing to pass the last elements of the BHA.

The use of stabilizers at the collars and bit and with a drilling motor needs to be planned so that it is possible to pass the stabilizer blades between the rams and annular or between the ram sets if the wellbore becomes pressurized. The ability to strip the BHA out and back in the hole is the back-up barrier to failure of a floating mud cap or a down-hole casing (deployment) valve.

3.10.5 Pipe Light

Pipe light is discussed in Chapter 1.

3.10.6 Snubbing

Snubbing is positive control of the drillpipe, but it comes with a high cost in time for rig mounted rams, and a higher daily cost in rig time and equipment with the full snubbing unit. However, it is the ultimate solution for high pressure and/or H_2S problems.

Snubbing is discussed in Chapter 6.

3.11 Questions

1. What is the single greatest technical challenge with gaseated systems?

2. What basic surface and downhole equipment would be needed for a vertical gaseated (land) well drilled into a light oils sandstone reservoir?

3. If the well is flowing, what steps need to be taken to make a connection.

4. What would be the first estimate for water and air volumes if you want to reduce bottom hole pressure 295 psi (3,000 kPa) from the pressure provided by a static column of fresh water? You want to be at or in the onset of the friction dominated regime. Depth is 12,000 ft, hole size is 6.25 in. (159 mm) using 4 ½ in. (114 mm) drill pipe and ignoring drill collars for this example.

5. List the various methods of injection of gas in a gaseated system.

THEORETICAL AND MATHEMATICAL QUESTIONS

6. Give a general equation for bottom hole pressure during gaseated drilling in a steady state.

7. List the various flow patterns that might be present in an aerated system that includes a horizontal leg.

8. A vertical well is planned for aerated mud drilling. The TVD is 12,000 ft and hole size is 6 in. Casing is cemented at 6,000 ft. Drill string is made of 11,500 ft, 3.5 in. drill pipe with 2.602 in. ID and the rest is 5 in. × 2.25 in. drill collar. Drill bit has 3 × 16 nozzles. The well is planned for liquid injection rate of 300 gpm, gas injection rate of 1,000 scfm, and 100 psi back pressure. The rate of penetration is limited to 60 ft/hr. Ignore fluid influx into the wellbore, and temperature gradient of 0.01°F/ft, and 0.3 in. and 0.0018 in. for wellbore and steel roughness. Use Guo's model to:

   (a) Calculate gas fraction and fluid velocity at the surface

   (b) Calculate pressure, gas fraction, and fluid velocity at casing seat, top of drill collar and bottom hole

(c) Calculate pressure drop at the nozzles

(d) Calculate standpipe pressure

9. Using the data from the previous problem, calculate bottom hole pressure and stand pipe pressure for:

   (a) Water influx of 5 bbl/hr

   (b) Gas influx of 100 scfm

10. Calculate minimum required annular velocity for hole cleaning criterion if cuttings larger than 0.5 in. are not expected.

11. Compute the required air injection rate to give a BHP of 2,497 psi while drilling 8 ½ in. hole at 6,000 ft and circulating 8.6 ppg at 350 gpm. State all assumptions necessary.

3.12 Answers

1. The single greatest challenge in gaseated systems is control of pressure surges or control of surging.

2. Basic mechanical surface and down-hole equipment for a gaseated system drilling vertically into a depleted sand stone reservoir at 9,000 ft (3,000 m).

   Nitrogen generators or a gas source, compressors and boosters

   Rotating control head

   Bit and string floats

   Separator

   Chokes and Manifold System

   Flare and flare line

3. To make a connection with the above well when using drillpipe injection of the gas.

   1. Pick up the pipe and circulate to clear the cuttings off bottom.

   2. Turn off the mud pump and at the same time.

3. Shut in the well at the choke, but maintain about 200 psi choke pressure.

4. Load (dry) the upper drillpipe with air.(No more than 5 minutes) The nitrogen from the drillpipe will also give a "boost" to nitrogen in the annulus when circulation is started and keep a solid slug of fluid from forming.

5. Bypass the compressors or gas to the flow line jet.

6. Bypass the nitrogen trapped in the standpipe and drillpipe out the standpipe manifold to the open part of the flowline, or simply blow it off in a safe manner.

7. Make the connection.

8. Pick up the pipe and turn the nitrogen back into the drillpipe.

9. Start the mud pump.

10. Open the rotating head-flowline valve when the pump pressure builds up to within 100 psi (700 kPa) of the drilling pressure.

4. The first estimate for liquid and air volumes to reduce the bottom-hole pressure by 295 psi (3,000 kPa) in a 6.25 in. hole (158.75 mm) while in the onset of the friction dominated regime would be 200 gpm of water and 1,500 cfm of air.

5. The various methods of injecting gas into a gaseated system are:

   Drillpipe injection

   Drillpipe injection with a jet sub

   Parasite tubing lines

   Dual casing string

   Dual well injection (only as a matter of general interest)

6. What is a general equation for bottom home pressure with an aerated system in a steady state while drilling.

$$P_{bh} = P_{hyd} + P_f + P_{acc} + P_{surf}$$

where

P_{bh} = Bottom-hole pressure

P_{hyd} = Hydrostatic pressure

P_f = Frictional pressure loss

P_{acc} = Acceleration pressure

P_{surf} = Surface back pressure

7. List the various flow patterns that might be present in an aerated system that includes a horizontal leg.

   1. Turbulent flow
   2. Bubbly flow
   3. Stratified flow
   4. Slug flow

8. (a) (b)

Reference point	Pressure, psi	Gas fraction, %	Velocity, ft/s
Surface	100	78	15.65
Casing seat	1,843	67	8.4
Top of Drill collar	4,392	10.5	6.5
Bottomhole	5,111	5	12.6

(c) Pressure drop at the nozzles calculates to 224 psi

(d) Standpipe pressure is 3,240 psi

9. (a) Bottom-hole pressure = 5,618 and standpipe pressure = 4,179

   (b) Bottom-hole pressure = 5,019 and standpipe pressure = 3,189

10. u_{sl} = 1.42 ft/sec and u_{tr} = 0.63 ft/sec therefore minimum annular velocity for hole cleaning (u_{min}) is 2.05 ft/sec.

11. 11.13 scf/bbl

3.13 References

Bennion, D.B., Thomas, F.B., Bietz, R.F., Bennion, D.W. "Underbalanced Drilling: Praises and Perils," *SPE Drilling & Completion*, Vol. 13, No. 4, 1998, pp. 214–222.

Brantly, J.E. *History of Oil Well Drilling*, Gulf Publishing Company, Houston, TX, USA, 1971.

Doan, Q.T., Oguztoreli, M., Masuda, Y., Yonezawa, T., Kobayashi, A., Naganawa, S., Kamp, A. "Modeling of Transient Cuttings Transport in Underbalanced Drilling (UBD)," *SPE Journal*, Vol. 8, No. 2, June 2003, pp. 160.

Dupriest, F. E., Koederitz, W. L. "Maximizing Drill Rates with Real-Time Surveillance of Mechanical Specific Energy," SPE 92194 presented at IADC/SPE Drilling Conference, Amsterdam, February 23–25, 2005.

Griffith, P. "Multiphase Flow in Pipes," *Journal of Petroleum Technology*, March 1984, pp. 361–367.

Gucuyener, I.H. "Design of Aerated Mud for Low Pressure Drilling," SPE 80491 presented at SPE Asia Pacific Oil and Gas Conference and Exhibition, Jakarta, April 15–17, 2003.

Guo, B. and Ghalambor, A. *Gas Volume Requirements for Underbalanced Drilling Deviated Holes*, PennWell Publishing Company, Tulsa, OK, USA, 2002.

Guo, B., Sun, K. and Ghalambor, A. "A Closed Form Hydraulics Equation for Predicting Bottom-hole Pressure in UBD with Foam," SPE 81640 presented at the IADC/SPE Underbalanced Technology Conference and Exhibition, Houston, TX, USA, March 25–26, 2003.

Guo, B., Ghalambor. A. "A Guideline to Optimizing Pressure Differential in Underbalanced Drilling for Reducing Formation Damage," SPE 98083 presented at the International Symposium and Exhibition on Formation Damage Control, Lafayette, LA, USA, February 15–17, 2006.

Guo, B., Song, S., Chacko, J., and Ghalambor, A. *Offshore Pipelines*, Elsevier, Oxford, UK, 2005.

Hannegan, D. and Divine, R. "Underbalanced Drilling–Perceptions and Realities of Today's Technology in Offshore Applications" SPE 74448 presented at the IADC/SPE Drilling Conference, Dallas, TX, USA, February 26–28, 2002.

Lage, A.C.V.M. and Time, R.W. "Mechanistic Model for Upward Two-Phase Flow in Annuli," SPE 63127 presented at the SPE Annual Technical Conference and Exhibition, Dallas, TX, USA, October 1–4, 2000.

Li, J., Walker, S. "Sensitivity Analysis of Hole Cleaning Parameters in Directional Wells," *SPE Journal*, Vol. 6, No. 4, December 2001, pp. 356.

Lourenço, A.M.F., Martins, A.L., Andrade, P.H. Jr., Nakagawa, E. Y. "Investigating Solids-Carrying Capacity for an Optimized Hydraulics Program in Aerated Polymer-Based-Fluid Drilling," SPE 99113 presented at the IADC/SPE Drilling Conference, Miami, FL, USA, February 21–23, 2006.

Meldey, J., George H., Maurer, W. and Garkasi, A. "Use of Hollow Glass Spheres for Underbalanced Drilling Fluids," SPE 30500 presented at the Annual Technical Conference and Exhibition, Dallas, TX, USA, October 22–25, 1995.

Nas, S., "Introduction to Underbalanced Drilling," Weatherford Private Publication Ref: APR-WUBS-WFT-001, 2006.

Ogena, M. S., Gonzales, R. C., Palao, F., Toralde, J. S., Bayking, E. "Aerated fluids drilling used in Philippines field to minimize well interference while infill drilling," *Drilling Contractor Magazine*, March/April 2007, pp. 76.

Perez-Tellez, C., Smith, J.R., Edwards, J.K. "A New Comprehensive, Mechanistic Model for Underbalanced Drilling Improves Wellbore Pressure Prediction," *SPE Drilling & Completion*, Vol 18, No 3, September 2003, pp. 199.

Poettmann, F.H. and Bergman, W.E. "Density of Drilling Muds Reduced by Air Injection," *World Oil*, August 1995, pp. 97–100.

Ramalho, J. and Davidson, I.A. "Well-Control Aspects of Underbalanced Drilling Operations," SPE 106367 presented at the IADC/SPE Asia Pacific Drilling Technology Conference and Exhibition, Bangkok, Thailand, November 13–15, 2006.

Rehm, Bill. "Don't Overlook Aerated Mud," *Oil and Gas Journal,* December 1965.

Rehm, B., Schubert, L. "How to control surface and bottom-hole pressure during UBD—unbalanced drilling—Statistical Data Included," *World Oil*, Vol. 222, No. 3, March 2001, pp. 48–55.

Rehm, B. *Practical Underbalanced Drilling and Workover*, Petroleum Extension Service, University of Texas, Austin, TX, USA, 2002.

Ridley, K. et al. "Continuous Circulation Reduces NPT," *World Oil*, March 2011.

Rommetveit, R., Sævareid, O., Lage, A.C.V.M., Guarneri, A., Georges, C., Nakagawa, E., Bijleveld, A. "Dynamic Underbalanced Drilling Effects are Predicted by Design Model," SPE 56920, Offshore Europe Oil and Gas Exhibition and Conference, Aberdeen, September 7–10, 1999

Saponja, J. "Challenges with Jointed-Pipe Underbalanced Operations," *SPE Drilling & Completion*, Vol 13, No 2, June 1998, pp. 121.

Saponja, J. "Canadian Experience Underscores Importance of Fluids," *Underbalanced Drilling Technology*, September 2003.

Sunthankar, A. A., Miska, S., Kuru, E., Kamp, A. "New Developments in Aerated Mud Hydraulics for Horizontal Well Drilling," SPE 62897 presented at the SPE Annual Technical Conference and Exhibition, Dallas, TX, USA, October 1–4, 2000.

Teichrob, R.R. "Low-Pressure Reservoir Drilled With Air/N2 in a Closed System," *Oil and Gas Journal*, March 1994, pp. 80–89.

Teichrob, R., Baillargeon, D. "Part 1—Principal Components of Conventional Underbalanced Drilling (UBD) Packages, Including the Flow Back, Injection and Data Acquisition Systems," *World Oil*, March 2000.

Tian, S., Medley, G.H., Stone, C.R. "Optimizing Circulation While Drilling Underbalanced," *World Oil*, Vol. 221, No. 6, June 2000.

Zhou, L., Ahmed, R.M., Miska, S.Z., Takach, N.E., Yu, M., Saasen, A. "Hydraulics of Drilling with Aerated Muds under Simulated Borehole Conditions," SPE 92484 presented at the IADC/SPE Drilling Conference, Amsterdam, The Netherlands, February 23–25, 2005.

Zhou, L. "Hole Cleaning During UBD in Horizontal and Inclined Wellbore," SPE 98926 presented at the IADC/SPE Drilling Conference, Miami, FL, USA, February 21–23, 2006.

Zupanick, J.A. and Rial, M., Nov. 14, 2006, *Method and System for Recirculating Fluid in a Well System*, US patent #7,134,494.

Section 2
Using Concentric Casing with Gaseated Systems, Principles and Examples

Paco Vieira, Weatherford Services, U.S., LP

3.14 UBD—Concentric Casing Gas Injection

Gasified fluids are commonly used in underbalanced and low head drilling due to their advantages in respect to other drilling fluid systems.

These advantages include:

- A wide range of equivalent circulating density (ECD)
- Ease of ECD control by simple adjustment of gas or fluid
- Almost any fluid and gas can be used with minimum adjustment to properties

Several gas injection techniques are used to mix the gas and liquid phases, such as parasite string, concentric string, drillpipe gas injection, and drillpipe injection with a jet sub. Each one of them has advantages and disadvantages. The selection of the appropriate gas injection system is a key piece for the design of a controlled pressure drilling operation that requires the use of multiphase drilling fluid systems. This discussion is limited to the use of a concentric casing system for gas injection, the most flexible of the techniques. Figure 3–13.

The use of concentric casing gas injection has the benefit of allowing the use of conventional equipment, and is an easy way to handle the drillpipe during connections.

However, inappropriate control of the concentric casing gas injection system can generate fluctuations on the gas injection pressure that will lead to variations of the volumetric gas flow rate that passes to the drilling annulus. Irregular gas flow to the drilling annulus creates variations in the ECD that significantly affect the optimal execution of an underbalanced or low head drilling operation.

An improved gas flow technique has been successfully applied to drill depleted reservoirs for UBD applications in the Middle East and North Africa region. The following sections describe the main concepts behind this gas injection technique and its results in UBD applications among the North African and Middle East countries.

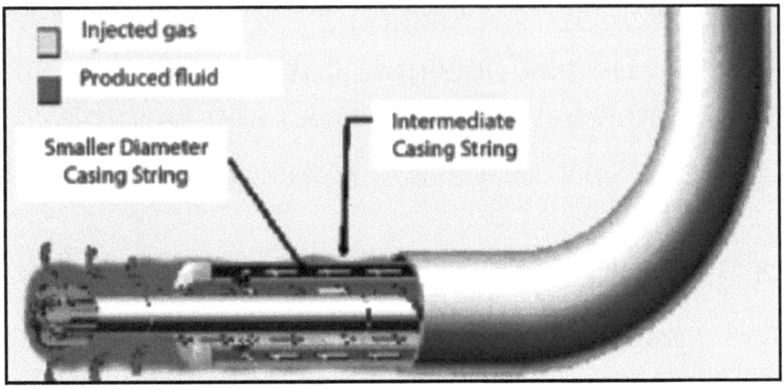

Figure 3–13 *Micro-annulus injection*

3.15 First Applications

In the initial stages of using concentric casing gas injection, services and operating companies start to experience a considerable amount of pressure fluctuation due to the "accumulator effect." When injecting a compressible fluid into the microannulus created by the two concentric casing strings, the gas will start being accumulated and increasing the micro-annulus pressure until the pressure in the microannulus at the injection ports depth is greater than the pressure in the main annulus at the injection ports. At that moment gas will start entering the main annulus at a volumetric rate that is not necessary the same as that being injected into the microannulus at the surface.

The injection rate at the down-hole ports increases as the annular pressure is reduced by aerating the annulus. This continues until the micro-annulus pressure is lowered enough that it no longer overcomes the drilling fluid pressure in the annulus. The cycle then repeats itself. This is known in the industry as the "accumulator effect".

3.16 Options to Mitigate the Pressure Fluctuations

Once the problem was identified, different options were identified as a solution to minimize or eliminate it:

- Achieve critical flow through the ports
- Control the microannulus pressure constant through applying annular back pressure
- Use of gas lift valves technology
- Drillpipe injection with a jet sub

3.16.1 Critical Flow Concept

One way to control the accumulator effect that appears obvious is to design the ports size so critical flow is achieved through the injection ports; this implies that the volumetric gas passing through the injection ports is not dependent of the downstream conditions after the ports. For compressible flow, it is possible for the velocity in the ports to reach the speed of sound or the sonic velocity in the fluids. Consequently, if the fluid reaches sonic velocities within the ports, the flow behavior become independent of the conditions downstream of the ports. This condition is called "critical flow." If the maximum fluid velocity of the gas at the ports is less than the sonic velocity, then the flow is called "sub critical" flow. Thus the prediction of the sonic velocities and the boundaries between critical and subcritical conditions is necessary to describe the flow behavior of the gas through the ports. Figure 3–15 shows the dependence of the gas flow rate through the ports on the ratio between the downstream to upstream ports.

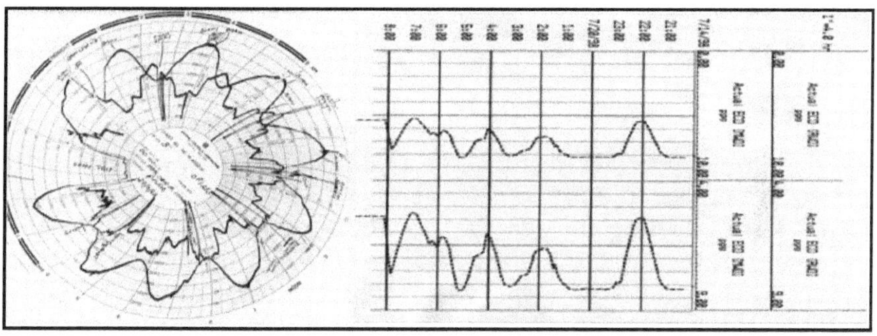

Figure 3–14 *Pressure surges due to gas/liquid separation*

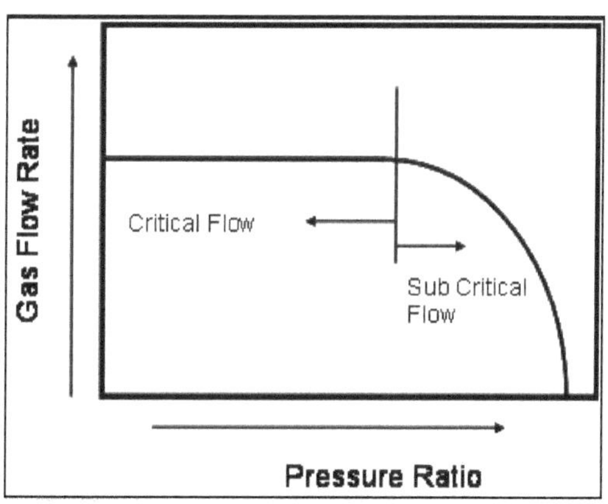

Figure 3–15 *Flow through restrictions*

For gas injection using concentric string, it is necessary to guarantee a continuous gas volumetric flow rate through a ports flow area, even if pressure variations are expected due to the different possible multiphase flow patterns on the upstream conditions (drilling annulus).

Flow Calculations are shown in Example 3-1.

Example 3–1 Calculation for Critical Pressure Ports

To calculate the theoretical appropriated ports flow area, first we need to estimate the pressures in the upstream and downstream of the injection ports in order to determine the critical pressure ratio, Figure 3–16.

The critical-pressure ratio is defined as the ratio between the downstream and upstream pressure in the ports. For a gas the critical pressure ratio is also.

$$y_c = \frac{P_2}{P_1} = \left(\frac{2}{k+1}\right)^{\frac{k}{k-1}} \tag{3.2}$$

The pressure downstream of the ports (P_2) can be estimated with the equivalent circulating density at the injection point assumed by using a steady state multiphase hydraulic simulator. The pressure upstream of the ports (P_1) can also be calculated using a single-phase gas hydraulic simulator and it will be the indicator of the injection pressure requirements at surface. The ratio of specific heats for a gas (k) is given by the following equation

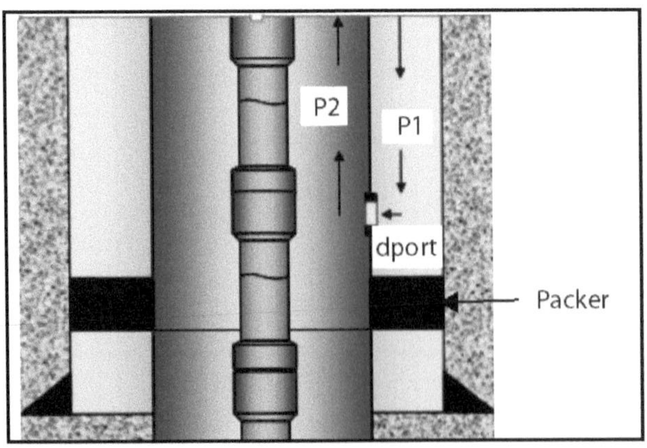

Figure 3–16 *Pressure at the injection ports*

$$k = \frac{C_p}{C_v} \tag{3.3}$$

For Air and other diatomic gases, k is approximately 1.4 and the critical ratio becomes 0.53. Values of hydrocarbons gases are typically between 1.25–1.3. The Bernoulli equation is combined with an isentropic equation of state resulting in the following equation

$$q_{sc} = \frac{C_n \times P_1 \times d_{ch}^{\ 2}}{\sqrt{\gamma_g \times T_1 \times Z_1}} \sqrt{\left(\frac{k}{k+1}\right)\left(y^{\frac{2}{k}} - y^{\frac{k+1}{k}}\right)} \tag{3.4}$$

where

$$C_n = \frac{C_s \times C_d \times T_{sc}}{P_{sc}} \tag{3.5}$$

Table 3–1 indicates the Constant and Units for Eqs. (3.2–3.5).

Knowing the critical pressure ratio, we can estimate the range of pressure capacity that booster at the surface had to have in order to inject the gas in the annulus seeking to maintain critical flow thought the restriction. There is no better way to explain the concept than developing a practical example. Data for the example is provided in Figure 3–17.

The minimum pressure at the upstream of the port should be equal or greater than the critical pressure to create supersonic flow through the port. From Eq. (3.2):

$$y_c = \left(\frac{2}{k+1}\right)^{\frac{k}{k-1}} = \left(\frac{2}{1.4+1}\right)^{\frac{1.4}{1.4-1}} = 0.53$$

$$P_1 = \frac{P_2}{y_c} = \frac{P_2}{0.53} = 3774 \text{ psia}$$

Z factor at 3,774 psia and 200°F is 0.81.

Table 3–1 Constants and Units for Eqs. (3.2–3.5)

Symbol	Constant & Units
Q_{sc}	Mscf/D
d_{ch}	In
P	Psia
T	°R
C_s	27.611
C_d	0.865
P_{sc}	14.696 psia
T_{sc}	519.68°R
C_n	844

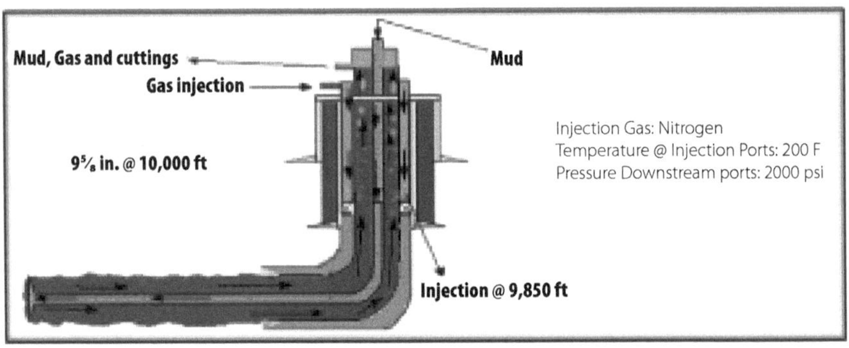

Figure 3–17 *Example of microannulus*

$$d_{ch} = \sqrt{\frac{q_{sc}\sqrt{\gamma_g \times T_1 \times Z_1}}{Cn \times P_1\sqrt{\left(\frac{k}{k+1}\right)\left(y^{2/k} - y^{\frac{k+1}{k}}\right)}}} = \sqrt{\frac{1728\sqrt{0.97\times 660\times 0.81}}{844\times 3774\sqrt{\left(\frac{1.4}{1.4+1}\right)\left(0.53^{(2/1.4)} - 0.53^{\left(\frac{1.4+1}{1.4}\right)}\right)}}} \quad (3.6)$$

- As can be observed from the calculation result, (d_{ch} = 0.25 in. or 6.35 mm). If 4 ports are installed, then the diameter of each port should be 0.125 in. The equivalent diameter for the total flow area (TFA) in the ports is small, leading to the risk of plugging by debris.
- The pressures required (about 3,800 psi) are beyond normal compression limits for the available drilling compressors.
- The ports TFA are designed for a specific flow rate, giving no room for changes.
- There are also references that, as a rule of thumb, the flow ports TFA have to be less than 1% of the main annulus flow area in order to achieve stability in the system. In the previous case, the outer casing is 9 5/8 in., 53.5 lbm/ft (ID = 8.535) and the inner casing is 7 in., 35 lbm/ft (ID = 6.004 in.). Applying this concept to the previous case, the recommended equivalent diameter is 0.5 in. or 12.7 mm (see Figure 3–18).

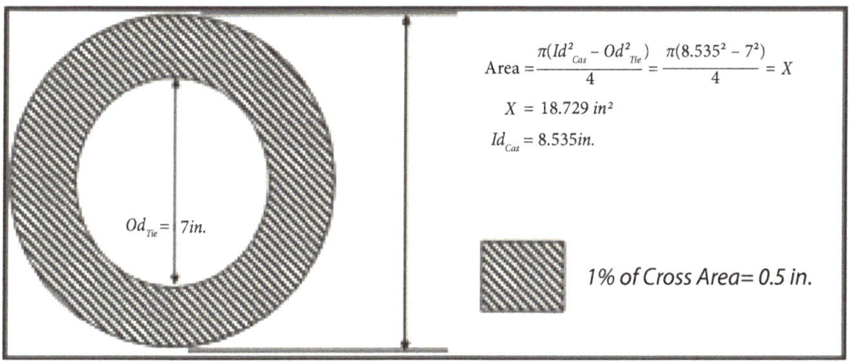

Figure 3–18 *Area calculation*

Without entering in the discussion of the proper TFA port calculation the experience in the Middle East has shown that rather than designing a port TFA to achieve critical flow, it is more effective and practical to control the pressure downstream of the ports through applying annular back pressure and keeping the injection annulus pressure as constant as possible. This requires a continuous monitoring by the UBD engineer and a precise annular back pressure control.

3.16.2 Micro-Annulus Pressure Control

A better and more practical approach is to control the pressure in the main annulus, which in turn limits pressure variations in the microannulus. This allows more flexibility in the down-hole pressure port size.

Annular pressure control is performed by applying annular back pressure through the underbalanced drilling (UBD) choke. When UBD operations are performed with the concentric gas injection, and the injection ports TFA are not designed for critical flow, the pressure control requires monitoring by the UBD engineer and choke operator. Utilizing this type of choke control, the choke operator will need to continuously monitor the microannulus pressure and adjust the choke position to control the pressure accordingly at the value indicated by the UBD engineer.

The appearance of automatic choke systems for managed pressure drilling operations (MPD), changed the ways of controlling the pressure that can be utilized for UBD operations with concentric gas injection (see Figure 3–19). The automatic MPD choke systems can be set to adjust the annular back pressure based on set controlled pressure parameters like standpipe pressure or choke pressure, allowing a more precise control of the microannulus pressure and minimizing the pressure fluctuations that the system creates for the called "accumulator effect."

3.16.3 Injection Port Sizes

Controlling the gas volumetric rates at the injection ports by applying back pressure will not depend on the ports TFA. However, a

Figure 3–19 *Automatic choke systems*

TFA has to be designed. Several criteria are utilized for the design of a pre-perforated productions liner and also applies for the design of the injection ports. Lab tests have shown that pre-perforated casing tubular with perforation phases 0-60-90-120-180 with a maximum of 16 perforations/ft of 0.37–0.75 in. in diameter will have minimum impact in the overall resistance of the tubular. A rule of thumb that satisfies this criteria is to have a minimum of three ¾ in. hole phase 90 degrees from each other.

A default starting surface pressure in the annulus is 150 psi or 10 atm.

3.17 Middle East and North Africa Experience

3.17.1 Oman

Since 2002, more than 50 wells have been drilled in several fields in Oman utilizing concentric multiphase fluids with concentric casing gas injection. The drilling techniques, reservoir production and operational parameters will determine the optimum degree of underbalanced drilling which will be strictly controlled during all phases of the operation. Accurate control of the nitrogen injection through concentric casing has been achieved by annular backpressure control (see Figure 3–20).

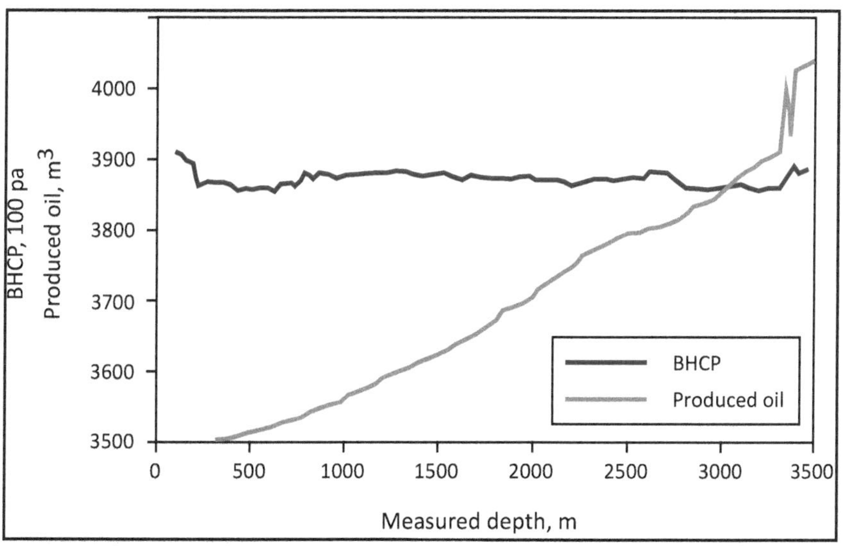

Figure 3–20 *UBD Oman – pressure control*

3.17.2 Algeria

The Hassi Messaoud field, discovered in 1956, is the largest oilfield in Algeria covering about 2,000 km^2 and currently including over 1,000 wells.

Production is from thick, Cambrian age sandstone at a vertical depth of approximately 3,400 m. This sandstone is an extremely hard and, abrasive candidate for UBD technology with the objectives of improving productivity, minimizing formation damage, improving the rates of penetration and eliminating drilling problems. Drilling the wells in Hassi Messaoud in underbalanced conditions assures the maximum well deliverability with the highest efficiency (see Figure 3–21).

Based on the reservoir pressure and depth, a gasified fluid system is required in order to achieve a controlled bottom-hole pressure drawdown leading to stable underbalanced drilling conditions. More than 30 wells have been drilled successfully underbalanced utilizing concentric casing gas injection. The total days of drilling the horizontal production section went from 27 days to an average of 12 days per well (see Figure 3–22). Pressure has been controlled, mitigating the "accumulator effect" by controlling annular surface back pressure while injection up to 2,400 SCFM of nitrogen (see Figure 3–23).

3.17.3 Libya

Underbalanced drilling technology was utilized to mitigate formation damage caused by solids and liquid invasion of the reservoir on Sabah; Fidaa and Sarir Fields utilizing concentric multiphase fluids

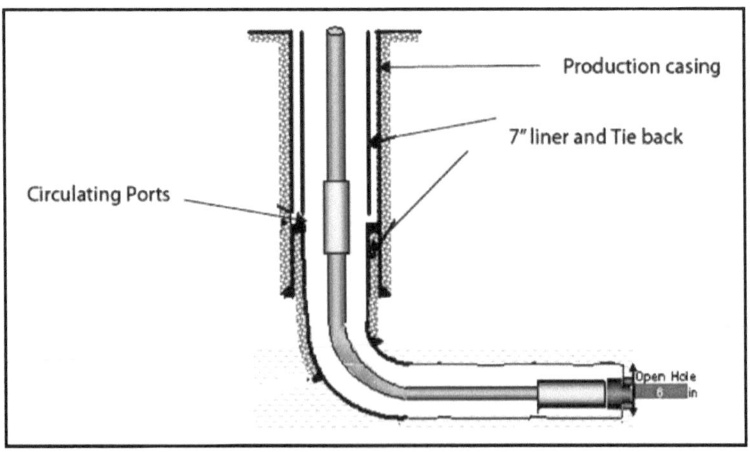

Figure 3–21 *Hassi Messaoud underbalanced assembly and casing points*

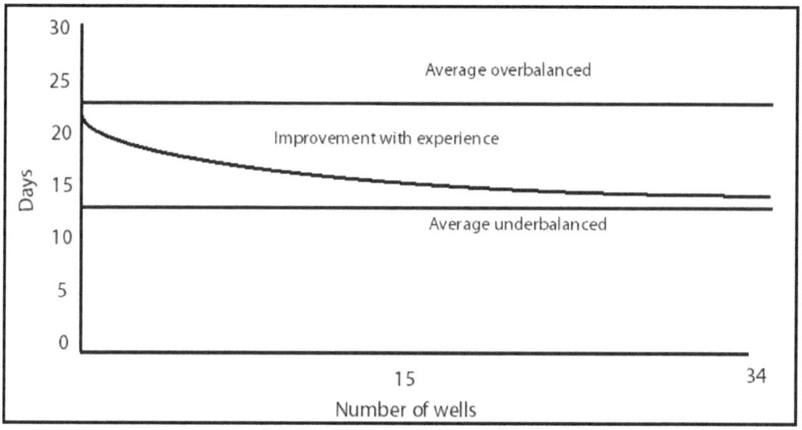

Figure 3–22 *Reduction in drilling days with experience*

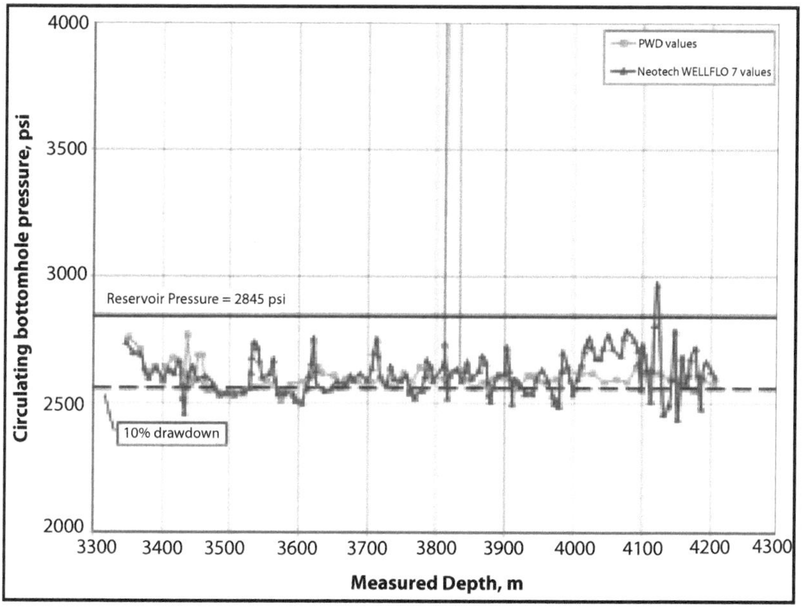

Figure 3–23 *Hassi Messaoud well, effect of controlling back pressure*

with concentric casing gas injection since 2005 (see Figure 3–24). Reservoirs where UBD technology was applied have similar depletion conditions and rock properties that made them good candidates for the application of the technology.

The most recent applications were performed in the Sarir C-Main field. This field was put on production in 1966. Three wells were drilled in an undeveloped fault block, at the northeastern part of Sarir C-Main,

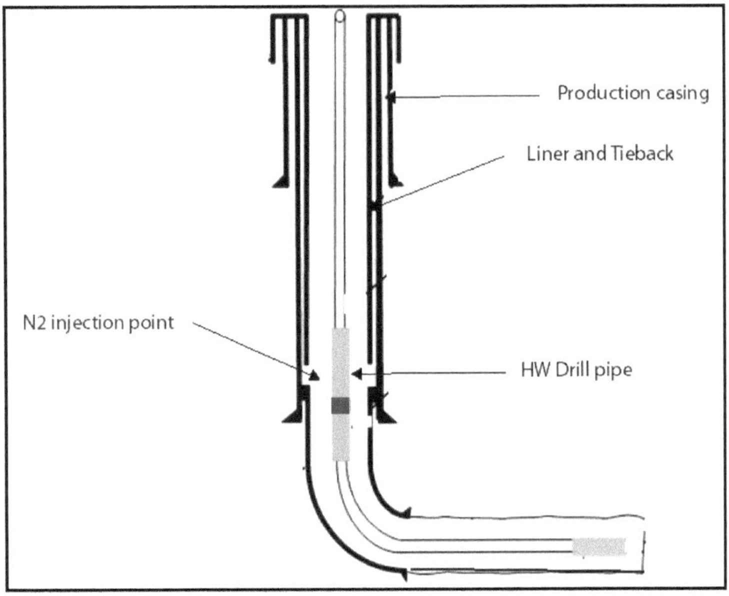

Figure 3–24 *Libya Sarir main underbalanced casing points*

to encounter an oil-filled member, which consists of sandstone, fine to medium grained, from the Lower Cretaceous age. The drilling techniques, reservoir production, and operational parameters determined the optimum degree of underbalanced, which was strictly controlled during all phases of the operation injection up to 2,400 SCFM of nitrogen, utilizing concentric casing gas injection technique.

3.18 References

Doron P., Granica D. and Barnea D. "Slurry Flow in Horizontal pipes—Experimental and Modeling" *Int J. Multiphase Flow,* Vol. 13, No. 4, 1987, pp. 535–547.

Kamp A.M., Rivero M. "Layer Modeling for Cutting Transport in Highly Inclined Wellbores" SPE 53942 presented at the 1999 SPE Latin and Caribbean Petroleum Engineering Conference, Caracas, Venezuela 21–23 April 1999.

King, G.E. *An Introduction to the Basics of Well Completions, Stimulations and Workover,* George E. King, Tulsa, Oklahoma, 1995.

King, G.E., "The Effect of High-Density Perforating on the Mechanical Crush Resistance of Casing," SPE 18843 presented at the SPE Production Operations Symposium, Oklahoma City, OK, USA, 1989.

King, G.E., "Casing Crush Resistance Loss to High-Density Perforating—Casing Test," SPE 20634 presented at the SPE Annual Technical Conference and Exhibition, New Orleans, LA, USA, 1990.

Section 3
Field Cases for Gaseated UBD Systems

Abdullah Al-Yami, Texas A&M University

3.19 Underbalanced Drilling Long-Term Performance (Kimery and McCaffrey, 2004)

This section compares different field cases of wells drilled underbalanced to offset wells drilled conventionally. The comparison was done by using decline analysis and economic techniques.

3.19.1 Case 1

Elkton Formation-Harmattan East Field: the formation is dolomitized carbonate with a permeability of 0.1 to 5 mD, porosity of 6 to 12%, initial water saturation of 11 to 30%, gross pay of 8.7 to 32.8 meter, and an initial pressure of 12.3 to 21.6 MPa.

The wells were initially drilled vertically then hydraulically fractured. Drilling horizontal wells was done later to maximize the production. However, drilling horizontal wells overbalanced did not show any improvement compared to vertical wells even after stimulation. Therefore, it was decided to use underbalanced drilling to drill horizontal wells, and the first horizontal well was drilled. The initial production rate was higher by 24% than conventionally drilled wells.

3.19.2 Case 2

Glauconitic Formation: Garden Plains Field: the formation is sandstone with a permeability of 0.1 to 1.2 mD, porosity of 12 to 27%, initial water saturation of 32 to 64%, gross pay of 1.7 to 23.5 meter, and an initial pressure of 5.4 to 8.7 MPa.

The horizontal well increased initial production rate by an average of 118% compared to wells drilled conventionally.

3.19.3 Case 3

Pekisko Formation-Three Hills Creek Field: the formation is clean limestone. It is coarsely crindoidal and fragmental to fine-grained with a permeability of 0.25 to 5 mD, porosity of 4.5 to 11%, initial water saturation of 20 to 30%, gross pay of 1.7 to 10.3 meter, and an initial pressure of 3 to 12 MPa. Horizontal underbalanced wells increased the initial production rate by an average of 238% more than conventional wells. The conventional wells were drilled vertically and stimulated (hydraulic fracturing or acid fracturing).

3.19.4 Case 4

Gething X pool-Kaybob Field: the area is a highly heterogeneous, fluvial-incised vally. "The lithology consists of conglomeratic, coarse to fine-grained facies" with a permeability of 0.07 to 4.2 mD, porosity of 10.5 to 19.7%, initial water saturation of 23 to 47%, gross pay of 1.9 to 11.5 meter, and an initial pressure of 11.9 to 15.0 MPa. The horizontal underbalanced wells showed an increase in initial production rate by an average of 254%.

3.19.5 Negative Field Cases

If the reservoir is not suitable for underbalanced drilling and/or formation damage resulted due to operational techniques, then we will not have good performance of wells drilled underbalanced. An example is found in Section 3.19.6.

3.19.6 Case 5

Cardium Formation-Ansell Field: the formation is sandstone with a permeability of 0.05 to 1.7 mD, porosity of 9.5 to 13%, initial water saturation of 17 to 37%, gross pay of 5 to 19 meter, and initial pressure of 15.4 to 21.9 MPa. Initial production rate from conventionally drilled wells was higher by 27%: the Cardium formation is not suitable for horizontal drilling because of its low permeability. Fracturing gave better results than horizontal drilling. Also, underbalanced drilling operations were not optimized to prevent formation damage.

3.20 Production Enhancement-Brunei (Sarssam et al., 2003)

The Rasau field has a sandstone reservoir (1–100 md) that formed from a shallow marine environment. This resulted in stacked laterally sequence formation. The producing formation is at 4,900–6,500 ft. The oil is light (40° API) with different gas cap sizes. Produced oil and gas were used to save cost on operation and minimize formation damage. Wells were drilled at a high angle through the target reservoir, intersecting with many thin layers of reservoir to maximum exposure. However, this lead to great exposure of shallow formations.

Underbalanced drilling was initiated in an attempt to minimize formation damage. A skin factor up to 100 was recorded in wells drilled conventionally. The highest skin factor was recorded for wells that were drilled with unconditioned drilling fluid and wells that were completed as open hole with wire wrapped screens.

A special tank and a solid-separation equipment were required. All rubber in the surface well control, separation system and down-hole motors needed to be resistant against the field oil used. Electromagnetic measurement while drilling was used to provide continuous transmission to surface.

3.20.1 Field Case 1

The pipe got stuck after initiation of two-phase underbalanced drilling. The pipe was released but got stuck several times later. Then the bottom-hole assemble was pulled up and run again to establish underbalanced condition. However, the pipe got stuck one more time before reaching underbalanced conditions. This time, oil and brine circulation and jarring was done to release the stuck pipe. Debris was recovered on the surface due to possible whipstock and window damage. It was not known where the underbalanced conditions or the use of brine caused the hole to collapse, which caused pipe stuck problems. A second whipstock was installed above the first whipstock and the formation was drilled overbalanced. However, when an attempt was done to drill underbalanced (with still 200 psi overbalanced), the hole began to cave and the operation was abandoned. The hole was displaced to water-based mud and drilled conventionally. A good reason to justify this is the high degree of depletion and the high amount of shale in this formation.

3.20.2 Field Case 2

The target reservoir was not completely depleted from the original hydrostatic pressure. So its risk level is reduced in terms of wellbore stability problems. A 7 in. casing was installed and cemented. Then the float shoe was drilled, and drilling was switched to underbalanced mode to drill the 6 1/8 in. hole section. Field oil was used in a single-phase to establish a bottom-hole circulating pressure of 1,950 psi. While tripping out of the hole to change EM-MWD tool, the dual flapper float valve failed causing hydrocarbon flow up the drillpipe. Instead of killing the well and risking the underbalanced conditions, the drillpipe was isolated with a through-tubing inflatable bridge plug to trip out in underbalanced mode. Underbalanced drilling was resumed but large solids were observed at the surface, which is an indication of wellbore stability problems.

After drilling to TD and pulling the bit to the shoe, a three rate production test was performed. The well productivity index was 16.8 BOPD/psi. the GOR was 968 scf/B and there was no water production. A skin factor of zero resulted, and the range of flowing bottom-hole pressure was 1,870–1,930 psi.

While the well is in underbalanced condition, an inflatable plug (PIP) was installed in the casing by using wireline. Then drillpipes were used to install wire wrapped screens and liner hanger/packer assembly into the inflatable plug. The PIP was deflated and pushed down, the linger hanger/packer was placed, and the assembly was stung out of the liner packer.

Around 100 bbls of brine were lost to the formation while displacing the hole to brine above the knock-out isolation valves due to a failure in the valves. An inflatable plug was set inside the linger hanger, the drillpipe was pulled out of the hole, and the completion tubing was installed. The inflatable plug was then retrieved using wireline.

3.20.3 Field Case 3

This field case was also a horizontal well producer where a high percentage of shale was expected. A 7 in. casing was installed in the shale just above the sand formation. Drilling conventionally was stopped just above the shoe, then drilling switched to underbalanced using field oil and injected gas to achieve 1,770 psi. The reservoir pressure was 1,870 psi. The well was shut in for a few hours to repair the gas compressor. However, an increase in the static bottom-hole pressure was observed with 2,000 psi. Then underbalanced drilling was resumed using flow drilling without the need to inject any more gas. The target bottom-hole pressure was changed to 1,850 psi.

After drilling to TD and pulling the bit to the shoe, a three rate production test was performed. The well productivity index was 9.9 BOPD/psi. The GOR was 478 scf/B, and there was no water production. A skin factor of zero resulted and the range of flowing bottom-hole pressure was 1,800–1,870 psi.

Great amount of solids were observed during washing and back reaming. The same above procedure to install the screen was repeated, however, the PIP was not pushed all the way to TD. It was pushed 100 ft above the TD. After confirming that the KOIV was holding pressure, the hole was displaced to brine. However, this resulted in reduction of productivity index from 9.9 to 3 BOPD/psi.

Table 3–2 UBD Trial Well Description

Trail Well	Length Drilled (ft)	Net-to-Gross (TG) (%)	Reservoir Pressure (% of original)	Description	Relative Stability Risk
Well 1 (RS34ST1/2)	1702	50%	30–40%	Reservoir pressure highly depleted. Shale content moderate to high.	1 (high)
Well 2 (RS40)	1257	80%	90%	Reservoir pressure close to virgin hydrostatic (no depletion). Shale content low.	3 (low)
Well 3 (RS39)	407	40%	90%	Reservoir pressure close to virgin hydrostatic (no depletion). Shale content moderate to high.	2 (med)

Table 3–3 UBD Well Productivity

Trial Well	Productivity Index Before Completion (bbl/d/psi)	Productivity Index After Completion & Hook-up (bbl/d/psi)	Reduction in PI (%)	Comments
Well 1 (RS34)	N/A	N/A	N/A	UBD unsuccessful due to probable wellbore collapse. Well drilled conventionally.
Well 2 (RS40)	16.8	6.5	61%	Reduction in PI likely caused by loss of 100 bbl brine to the reservoir due to KOIV failure, autopacking, and plugging of the wire-wrapped screens.
Well 3 (FS39)	9.9	3.0	70%	Reduction in PI likely caused by prolific reservoir sand (unable to get screens to TD) plus autopacking and solids plugging of the wire-wrapped screens.

A couple of recommendations were given:

- Evaluate more options of completion, such as using expandable sand screen to support the wellbore and prevent screen plugging, and also investigate underbalanced gravel-packing.
- Side tracking or running the casing shoe in the sand instead of the shale formation to minimize borehole stability problems.

3.21 Fractured Carbonates (Murphy et al., 2006)

Reservoir characterization was performed by underbalanced drilling the formation and gathering and observing surface and bottom-hole flowing parameters in real time by using pressure while drilling sub.

Tight Shuaiba carbonate can be damaged easily when exposed to overbalanced conditions. To prevent this, a down-hole isolation valve was installed for drilling and even for completion to avoid any potential to kill the well when tripping out of the hole. The down-hole isolation valve was installed as deep as possible in the concentric casing. The concentric casing was used instead of a drillpipe for gas injection. The use of the down-hole isolation valve also prevented fluid from flowing to the surface when there was no pipe installed.

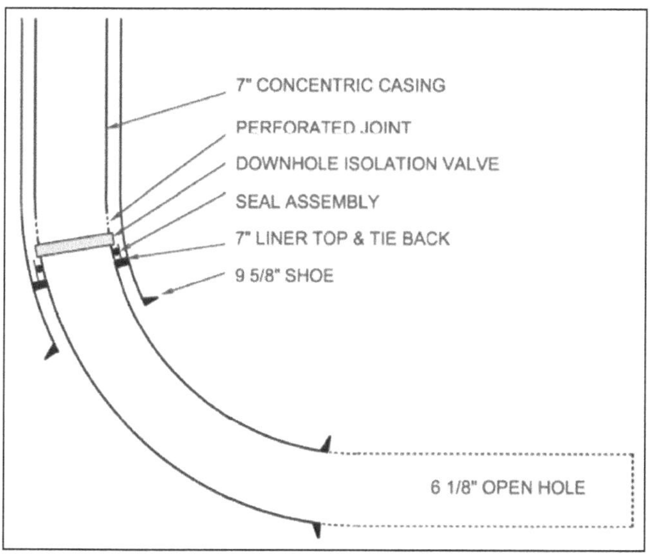

Figure 3–25 *Schematic showing UBD well design for each well*

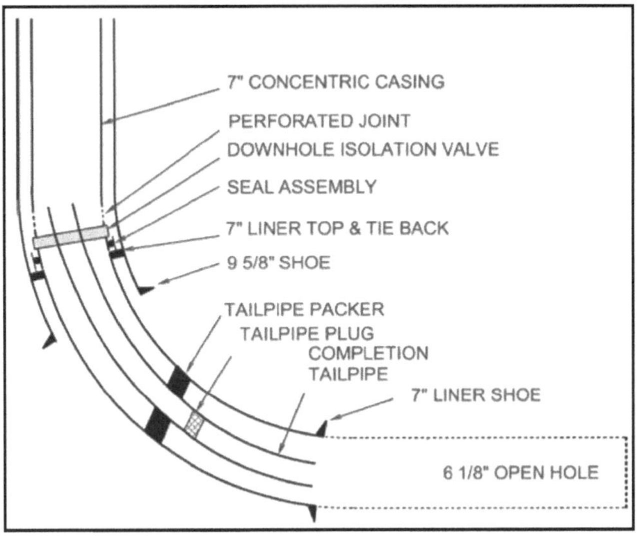

Figure 3–26 *Schematic showing typical UBD well when completed*

Completion (packer and tail pipe) was run on drillpipe while the down-hole isolation valve was closed. They were set in the liner below the concentric casing polished bore receptacle. The packer, with its preset plug, isolated the reservoir which allowed the removal of the down-hole isolation valve and concentric casing. Next, production tubing was installed in the well by stabbing into the tail pipe.

3.21.1 Field Cases

The down-hole isolation valve failed due to the damage of the sealing flapper. The well had to be killed before tripping the drillpipe. Killing the well in this way damaged the reservoir. A clear reduction in production rate was observed.

An enlarged borehole diameter prevented the use of SET. Better practices were developed to solve this problem, such as using a bit size of 6 1/8 in. instead of 6 in., using a motor with reduced bit to bend length to reduce the bend required, reducing the amount of flow test which reduced the time in open-hole, reducing circulation time before well testing, and minimizing drawdown. However, no actual reasons were given behind over-gauge hole in underbalanced drilling for this formation.

Drilled solids gathered behind the concentric casing, and while retrieving the concentric casing, the solids fell and accumulated on the plug installed in the tail packer. This incident occurred in three wells out of four. A coil tubing unit was used to remove the solids from the plug in order to remove it successfully. To prevent this from happening again, the "on/off connector of the packer assembly was extended to a point just above the perforated joint in the concentric casing. The result was the plug was recovered on the first attempt."

3.22 Comparison of Conventional and Underbalanced Drilling (Culen et al., 2003)

Saih Rawl Field has under-pressure oil wells. These wells are usually completed by an electric submersible pump. Underbalanced drilling was reintroduced to drill a well with three laterals. Three of the laterals were drilled overbalanced and stimulated, and the remaining two laterals were drilled underbalanced.

The Shuaiba reservoir in Saih Rawl has a tight formation with 1–10 mD. The formation is carbonate with 26% porosity. Average reservoir temperature is 74°C. The initial water cut is 40%. The oil has a viscosity of 1.7 cp at 30°C. There is no H_2S reported.

Saih Rawl wells have been drilled and stimulated conventionally with weighted mud. Each lateral was displaced to water. However, dynamic loss was reported when displaced to water after the stimulation operation.

Production logging tool (PLT) logs run on the Saih Rawl drilled wells showed that flow only came from 10% of the lateral. The reason for this was believed to be formation damage. So a decision was made to use underbalanced drilling to minimize this problem. A schematic of the well is shown in Figure 3–27.

Field natural gas was used through a concentric liner. The natural gas compressor was attached to the stand pipe to provide underbalanced mode. Siah crude oil was used for the drilling operation.

Due to HSE and training prerequisites, the first three laterals were drilled overbalanced while the remaining two legs were drilled underbalanced.

3.22.1 Lateral 1 and 2 Drilled Overbalanced

The lateral 1 was drilled in 4.95 days to 1,493 meters. The lateral was acidized and then displaced to water. Total dynamic losses of 485 m^3 were recorded. The whipstock was then installed inside the casing and a window was milled for lateral 2.

The lateral 2 was drilled in 5.79 days to 1,632 meters. The lateral was acidized and displaced to hole with total dynamic losses of 366 m^3.

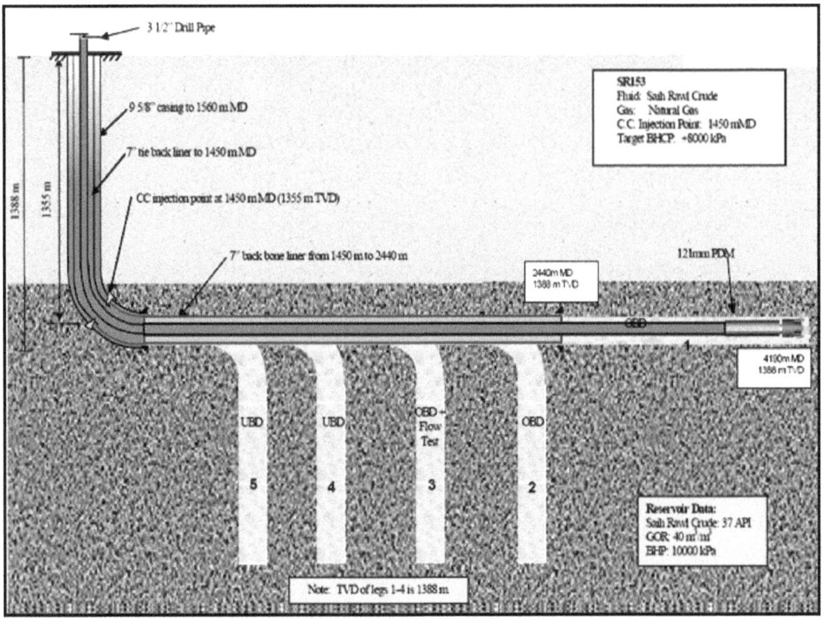

Figure 3–27 *Schematic of Siah Rawl underbalance wells*

3.22.2 Lateral 3 Drilled Overbalanced

In this lateral, instead of flushing with water after stimulation, the lateral was displaced to oil to avoid the dynamic losses reported earlier in the previous two legs. The operation was a success and production was quicker in this lateral because there was no dynamic loss. The skin factor measured from this later showed a negative value, which supported this good practice of stimulation.

3.22.3 Lateral 4

Underbalanced drilling was activated by pumping oil at 500–600 gpm and natural gas at 60–90 m^3/min through a concentric casing. The pumped oil and injected gas aid in overcoming the hydrostatic pressure of the oil in the annulus. The oil in the annulus then flowed into the injection sub at the concentric casing at 1,450 ft. Once the flow was pseudo-steady and the wellhead pressure was about 3,000 kPa, the oil injection through the concentric casing stopped and gas injection was resumed. Oil was produced while drilling with traces of water, and a total of 1,340 meters were drilled. Figure 3–28 shows underbalanced drilling improved the productivity index by more than 80% for the last three lateral sections.

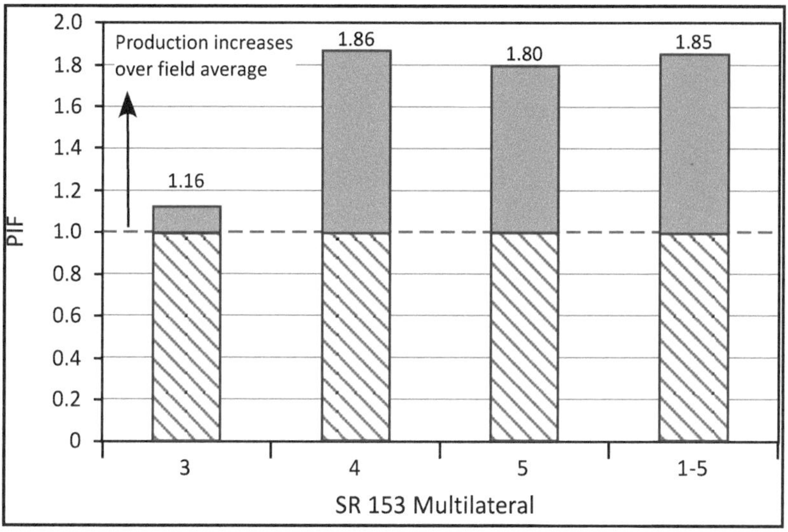

Figure 3–28 *Productivity improvement of SR153 over field average*

3.23 UBD North Sea (Ramalho et al., 2003)

Nimer field has two high permeable reservoirs (±1 Darcy) named Amin and Al Khlata. The formation is sandstone and it contains medium gravity crude with 21° API and 300–500 cP. Horizontal wells with sand screen and artificial lift methods did not show good performance from these wells, so underbalanced drilling was initiated in this field for this reason.

Crude oil with gas injection was selected to drill underbalanced. The gas selected was deoxygenated air for economic reasons. Sayala crude was selected because of its low density (0.78 sg) and low viscosity (30 cP), flammability of 0–5°C, and auto ignition of 293°C. A concentric annulus (9 5/8 in. × 7 in.) was selected for gas injection. The concentric annulus was selected because of the high amount of gas required for injection and because of the incompatibility of mud pulse telemetry measurement while drilling. Electromagnetic telemetry was considered but was evaluated as uneconomical for this operation.

A down-hole deployment valve was used as a part of the 7 in. tie-back concentric casing. Cuttings monitoring in the crude oil was done to minimize potential formation damage.

It was expected that once the formation oil was produced and used for drilling the cutting problems would be minimized. However, sliding problems and increased torque problems were observed. The

ROP was reduced, short wipe trips were made, and circulation time was increased prior to connection to help in cuttings removal. More cuttings were also produced, which was then related to the hole stability problems the first two wells were drilled at an earlier TD than planned.

One more problem was bottom-hole pressure fluctuations. The Nimer field was complex and it had a static reservoir pressure that varied throughout the field. Static bottom-hole pressure measurements were not easy because of the beam lift system. During the drilling, static pressure was measured after drilling 20 meters of reservoir, and then the drawdown was applied with respect to that measurement. It was observed that the measured static pressure varied in the toe, midsection, and the heel of the hole.

The following corrective actions to reduce NPT:

- An 11 1/16 in. RCH was used instead of the 13 5/8 in. RCH to make it more convenient to switch from conventional mode to underbalance by just removing the adapter nipple and installing the pack off and bearing.
- 20 4¾ in. drill collars were added above the 3½ in. drillpipe to improve drilling performance.
- To improve cutting transport a 65/35 of Sayyala and produced Nimer crude was used to initiate drilling, and this improved hole cleaning.
- When tripping pipe, the well was circulated clean, but the gas injection was resumed and the well flowed while stripping back to the shoe. At the shoe, the well was allowed to produce by itself with resumed circulation (1 ½ hole volume was produced) to remove all drilling fluid used.

3.24 Kuwait Fractured Dolomite (Vieira et al., 2007)

The Shuaiba formation was a fractured dolomitic limestone reservoir, and also a lost circulation formation. The lost drilling fluid in the Shuaiba formation resulted in severe formation damage. Underbalanced drilling was initiated in this formation to minimize lost circulation problems and to evaluate the potential of hydrocarbons in the Shuaiba formation. The Kuwait Oil Company believed that potential resources in the Shuaiba formation were not utilized because of possible failure associated with drill stem test due to the highly fractured formation. The Shuaiba, with 7.1–8.9 ppg formation pressure, was not a high pressure formation.

Since the Shuaiba formation is a depleted reservoir, a two-phase drilling fluid was proposed. Water and cryogenic nitrogen were used in the fluid phase because of the short period of this project. If the maximum gas volume fraction is 20%, then conventional mud pulse tools (MWD/LWD/PWD) are used, otherwise the electromagnetic tools will be utilized. Kuwait Oil Company considered using the conventional mud pulse tools so the volume of gas was below 20%. This was done by increasing the pressure in the drillstring by selecting a suitable nozzle size (1×12; 2×9).

Drilling underbalanced was started with 400 gpm of 7.0 ppg water and 1,200 scfm of nitrogen. The PWD recorded 3,150 psi, then formation water of 9.2 ppg was produced and PWD recorded an increased pressure of 3,650 psi. The amount of nitrogen was reduced from 1,200 scfm to 1,000 scfm to minimize nitrogen consumption, and PWD recorded an increase of 3,900 psi. Traces of hydrocarbons were recorded and water production was approximately 4,900 bbl/day during underbalanced drilling.

In the underbalanced drilling, nitrogen and water were injected in different ratios. The minimum annular liquid velocities were between 162–230 ft/min.

3.25 UBD Mexico Fractured and Depleted Formations (Tellez et al., 2009)

Concentric casing nitrogen injection was used to drill a high angle well. The use of such a method allowed the use of conventional mud pulse MWD/LWD tool and at the same time kept the bottom-hole circulating pressure as desired. The use of electromagnetic tools is limited at a high temperature.

A temporary tie-back was set on top of an intermediate liner, and gas was injected through ports in the tie-back microannulus to tie back drillpipe annulus as the well was drilled. In this case the well was drilled conventionally with only fluids pumped inside the drillpipe to minimize the effect of nitrogen on the conventional mud pulse.

In concentric casing nitrogen injection, the slug phenomena might occur due to gas compressibility in concentric annulus. First nitrogen was injected in the annulus, then mud and nitrogen were pumped inside the concentric annulus to decrease injection pressure and returned fluid.

3.26 Thailand—Down-Hole Deployment Valve
(Timms et al., 2005)

When tripping out of the hole during underbalanced operations, the well might be killed first which may cause formation damage. Another option is to use a subbing unit or trip out while the well is still flowing; both scenarios impose safety risks. This section discusses the development of a down-hole valve to close and isolate the well at the required depth. The use of the valve will eliminate the use of a snubbing unit since the reservoir pressure can be balanced below the valve location. The down-hole valve is run with the casing and is controlled using control lines from the surface. This section discusses the use of the down-hole valve in Thailand.

Snubbing takes a longer time than conventional tripping and requires up to 8 people to handle the operation in 24 hours.

A 7 in. casing was set at the top of the reservoir and a 6 1/8 in. hole section reservoir was drilled underbalanced and completed with 4 1/2 in. slotted liner. The down-hole valve is a full bore casing shut-off valve run with the casing, and can allow down-hole tools entry when in open position. It is a flapper valve providing a 5,000 psi metal to metal seal and is available in different sizes and works in temperatures up to 300°F.

The down-hole valve was opened by movement of an internal sleeve when operated hydraulically. The fluid was pumped to the valve by a ¼ in. tubing hydraulic lines clamped to the casing by cross-coupling clamps (one line for opening and the other one for closing). The down-hole valve could be designed to be retrieved when used in concentric casing or cemented in place for permanent installation. The casing was cemented using an inner string cementing technique to prevent risk of the valve's damaging.

A 5000 ps rated down-hole valve was set in the well which saved 7.5 days on tripping the bit and a string of perforated liner.

3.27 Underbalanced Experience in Libya (Safar et al., 2006)

Sabah G55 was the first well drilled as an underbalanced horizontal well in Libya. The reservoir (Beda C) was depleted with an estimated pressure of 1,050 psi. A total of 62 wells have been drilled; lost circulation was experienced in part of this reservoir due to its good porosity (up to 35%).

One of the 62 wells was drilled horizontally (G53). Severe lost circulation occurred when drilling the top of Beda C with viscous diesel. The drilling fluid was changed to water-based mud, however, the circulation was lost completely and only 580 ft of horizontal section was drilled.

Underbalanced drilling was initiated in G55 to drill from 5,961 ft MD to 7,628 ft MD. The objectives were to minimize formation damage, minimize lost circulation, characterize the reservoir, and improve rate of penetration. The well was drilled conventionally and a casing was run and cemented at 5,011 ft MD. Then a 6 in. horizontal well was drilled in the reservoir with two phases (nitrogen and Sabah crude oil) to 7,331 ft MD. The BHA consisted of a 6 in. polycrystalline diamond compact (PDC) bit, a 4 ¾ in. positive displacement motor with 1.5° bent housing. A tapered drillstring was used with a 5 in. drillpipe above 7 in. liner and 3.5 in. drillpipe and heavyweight drillpipe in the liner and open-hole. EMD and gamma ray tools were used for geo-steering and pressure measurements.

The horizontal well drilled underbalanced resulted in higher production, lower water cut and gas-oil ratio than the horizontal well drilled unconventionally.

The well was killed several times to change the RCD elements, which discontinued the underbalanced drilling process. Drillpipe ring grooves were the main reason to cut the RCD elements. The alignment of the derrick and BOP stack was not done properly.

3.28 Massive Lost Circulation in Libya (Safar et al., 2007)

The major problem with the Facha reservoir was enormous lost circulation and differential sticking, especially in horizontal well sections. Underbalanced drilling was initiated mainly to eliminate lost circulation and minimize formation damage. This is the second well that was drilled underbalanced after the G55 well as mentioned above. The field had a salt creeping problem which required the use of a higher collapse resistance casing. However, this type of casing was not available, and to avoid the salt creeping problem two stringers (9 5/8 in. casing and 7 in. liner) were used to isolate the salt beds.

Several scenarios were investigated to find out the best method to inject gas. The first method was done by utilizing a drillpipe gas injection system without tieback liner. The nitrogen will be injected into the standpipe manifold and mixed with crude oil and then pumped down the pipe. The advantages of such methods are that less gas equipment is used, lower bottom-hole pressure can be achieved, and no rig time will be lost due to running or retrieving of the tieback. However, the disadvantages are increased in bottom-hole circulating pressure when breaking the circulation connection. The conventional LWD will be affected because of the presence of gas (no more than

15% volume fraction of gas should be presented). And finally, hole cleaning will be a problem when fluid velocity goes below the minimum recommended value above the top of liner (TOL).

The second option was to use drillpipe gas injection system utilizing a tie back to enhance the cutting removal. However, rig time will be lost to install and retrieve the 7 in. tieback and install an extra wellhead.

Concentric casing gas injection was used utilizing a 7 in. tie back to surface with a perforated interval of 38 ft above the top of the liner. Several benefits were realized with concentric casing injection, such as possibility of using conventional LWD since one-phase drilling fluid is pumped inside the drillpipe.

A BOP stack and RCD were installed before starting to drill underbalanced. The hole was displaced to crude oil and concentric casing annulus was displaced with nitrogen. The circulating pressure was stabilized with 200 gpm of oil and 1,300 scfm of nitrogen. The first lateral section was 2,053 ft. The RCD elements were replaced 11 times due to the G drillpipe ring groove. The second lateral was 1,517 ft. The underbalanced drilling program resulted in 700-950 BOPD while drilling. A total of 10,918 bbl of oil was produced and sent daily to the Fidda field station.

After completing the underbalanced drilling, the crew was not able to revive the 7 in. tie back due to cuttings accumulation and settling in the 9 5/8 in. × 7 in. annulus below the perforated interval. To avoid this, the perforated interval should be located 5 ft above the TOL instead of 38 ft. The ROP was controlled to be below 50 ft/hr to avoid loading the annulus with cutting and to maintain bottom-hole pressure below 1,900 psi.

3.29 References

Culen, M.S., Harthi, S. and Hashimi, H. "A Direct Comparison Between Conventional and Underbalanced Drilling Techniques in the Saih Rawl Field, Oman," SPE 81629 presented at the IADC/SPE Underbalanced Technology Conference and Exhibition, Houston, TX, USA, March 25–26, 2003.

Kimery, D., McCaffrey, M. "Underbalanced Drilling in Canada: Tracking the Long-Term Performance of Underbalanced Drilling Projects in Canada," SPE 91593 presented at the IADC/SPE Underbalanced Technology Conference and Exhibition, Houston, TX, USA, October 11–12, 2004.

Murphy, D., Al-Busaidi, R., Wind, J., Davidson, I., Mykytiw, C., Kennedy and Arsenault, L. "Applications of Underbalanced-Drilling Reservoir Characterization for Water Shutoff in a Fractured Carbonate Reservoir—A

Project Overview," *SPE Drilling and Completions (SPEDC)*, September 2006, pp.153–157.

Ramalho, J., Medeiros, R. and Davidson, I. "The Nimer Story: Reservoir Exploitation Using UBD Techniques," SPE 81623 presented at the IADC/SPE Underbalanced Technology Conference and Exhibition Houston, TX, USA, March 25–26, 2003.

Safar, H., Azhary, S., Hijazi, A., Qutob, H., Chopty, J. and Pham, C. "Dual Lateral Horizontal Well Drilled Underbalanced in a Depleted Reservoir in Libya—Case Study," SPE 107307 presented at the IADC/SPE Middle East Drilling Technology Conference and Exhibition, Cairo, Egypt, October 22–24, 2007.

Safar, H.A., Majdoub A.A., Azhary, S.A., Qutob, H., Chopty, J. and Obeidat, H. "The First Horizontal Underbalanced Well Drilled in Libya: A Case Study," SPE 101073 presented at the IADC/SPE Indian Drilling Technology Conference and Exhibition, Mumbai, India, October 16–18, 2006.

Sarssam, M., Peterson, R., Ward, M., Elliott, D., McMillan, S. "Underbalanced Drilling For Production Enhancement in the Rasau Oil Field, Brunei," SPE 85319 presented at the IADC/SPE Middle East Drilling Technology Conference and Exhibition, Abu Dhabi, UAE, October 20–22, 2003.

Tellez, C.P., Urbieta, A., Lupo, C., Castellanos, J.M., Ramirez, O., Puerto, G., Bedoya, J., Gabaldon, O., Beltran, J., Castiblanco, G. "MPD Concentric Nitrogen Injection Used to Drill a Successful Horizontal Well in Fractured and Depleted Mature Reservoir in Mexico South Region," SPE 122982 presented at the Latin America and Caribbean Petroleum Engineering Conference, Cartagena, Colombia, May 31–June 3, 2009.

Timms, A., Muir, K. and Wuest, C. "Down-hole Deployment Valve-Case History," SPE 93784 presented at the Asia Pacific Oil & Gas Conference and Exhibition, Jakarta, Indonesia, April 5–7, 2005.

Vieira, P., Larroque, F., Saleh, A.M., Ismael, H., Qutob, H.H. and Chopty, J.R. "Kuwait Employs Underbalanced Drilling Technology To Improve Drilling Performance While Simultaneously Evaluating the Reservoir," SPE 106672 presented at the Offshore Europe Conference, Aberdeen, Scotland, UK, September 4–7, 2007.

Section 4
Two-Phase Flow Modeling

Arash Haghshenas, Boots & Coots

3.30 History of Two-Phase Flow Modeling

Drilling with gaseated systems requires understanding the nature of the multiphase flow in the wellbore. Multiphase flow is studied in various engineering disciplines and industrial applications such as chemical engineering plants, nuclear engineering, and piping. Generally in these applications multiphase flow prevails in a tubing or pipe. However, in gaseated systems the multiphase flow prevails in the annulus as well as drillstring. The multiphase flow in pipe is extended to annular flow in the wellbore by applying the concept of hydraulic diameter.

Early approaches of determining pressure profile in the wellbore were empirical correlations. Initially correlations developed for vertical wells by Poettman and Carpenter, Griffith and Wallis, Duns and Ros, Hagedorn and Brown, and Orkiszewski. In the 1960s directional drilling and completion emerged, the available correlation for vertical flow was unreliable and the industry was looking for a correlation for inclined wells. Beggs and Brill developed the first multiphase flow correlation for inclined wells in 1973.

Alongside the development of empirical correlations, many researchers focused on the nature and mechanistic modeling of the multiphase flow. The work of Taitel et al. (1980) and Barnea et al. (1985) improved the concept of mechanistic modeling and defining flow patterns in the medium. Following the improvement in flow pattern recognition, Ozon et al. and Hasan and Kabir improved the accuracy of earlier mechanistic models. A review of the early mechanistic models and correlations was published by Brill (1987).

A brief review of the literature indicates that empirical correlations and mechanistic models are used extensively to predict multiphase flow behavior. The exact approach of multiphase modeling includes solving the Navier-Stokes equation for all phases simultaneously. However, the exact solution of Navier-Stokes is limited to simple systems (Taitel,1996), and no significant success is reported in the literature for oil and gas application.

The science of multiphase flow in the petroleum industry evolved for predicting production rate of oil and gas wells. Initial software used for gaseated and mist drilling were production software which simulated the drilling operation as water, oil, and gas production from a reservoir. Experimental set-ups in the research institutes and fields were used to create multiphase flow correlations. The main short coming of correlations is that their application is limited to the experimental set-up condition and should not be extended to other conditions. Among empirical correlations, only the Hagedorn and Brown correlation showed reasonable results for a wide range of application. The original Hagedorn and Brown correlation was modified through the time; the modified form is one of the most popular correlations. Hasan and Kabir (2002), Gomez et al. (1999), Shoham (2006) and other authors justified the accuracy and applicability of the Hagedorn and Brown correlation by an extensive number of experiments.

3.30.1 Two-Phase Procedure

In this section, different methods of pressure calculation for multiphase fluid are briefly discussed. The Hagedorn and Brown correlation and the Hasan and Kabir (2007A) model are explained at the end of this section. The Hagedorn and Brown correlation is reported to have good accuracy compared to other models, including mechanistic models. The general Hagedorn and Brown model is modified for annular flow conditions by applying the Ansari model (Ansari, 1994). However, in this discussion, the Hasan and Kabir (2007A) model is used because of its simplicity, and the calculated results are similar to the other mechanistic models.

3.30.2 Mechanistic Models

Researchers developed mechanistic models to describe the behavior of multiphase fluids. Mechanistic models are simplified models with defined dimensionless parameters. The purpose of dimensionless parameters is to generalize the solution to different conditions and applications (Taitel, 1994). However, some parameters of the mechanistic models are estimated through experimental data and correlations. The advantages of mechanistic models is that they can be applied to different conditions rather than the experimental set-up with less loss of accuracy compared to the empirical correlations.

Takács (2005) described the development of the mechanistic models that includes three different stages:

1. Homogeneous mixture modeling
2. Recognition slippage without considering the different flow patterns
3. Recognition of slippage and flow patterns

Early mechanistic models assumed a homogeneous mixture and used the concept of single-phase fluids to calculate the pressure profile in the wellbore. They correlated the pressure profile with Reynolds number and friction factor parameter. The next modification to the mechanistic models was considering the effect of gravity, which causes different phases with different densities to move at different velocities. The velocity difference between the different phases creates extra shear stress at contact layers between the phases and results in extra pressure loss. Also, the slippage affects the liquid holdup in the wellbore. Later, different flow pattern and liquid holdup were recognized during multiphase flow. Different friction factors were associated to the different flow patterns, which improved the accuracy of the mechanistic models.

Determining the flow pattern is the first and most important step in pressure calculation using mechanistic models. Next, liquid holdup and friction factor are determined according to the prevailing flow pattern. At the end, frictional pressure drop is calculated and added to the hydrostatic pressure to calculate the pressure at a given point. To apply the mechanistic models the well should be divided into several sections to track the flow pattern and pressure profile in the wellbore.

To determine the fluid pattern; flow rates, pressure, and temperature should be known at a given point. For drilling application either the bottom-hole or the surface condition is known and flow pattern prediction can be started from that known point. For underbalanced drilling purposes, it is easier to start from the surface with known operational condition and then calculate bottom-hole pressure or any other desired point in the wellbore. The procedure of calculating pressure in the presence of multiphase flow is given by Shoham (2006) and is shown in the Figure 3–29.

3.31 Gaseated Flow

3.31.1 Flow Patterns

When two-phases are present in the system, flow patterns in the system depend on the gas/liquid velocities, inclination angle, and fluid properties. The mechanism of energy and momentum transfer for each flow pattern differs from other flow patterns. Thus it is important to know the flow pattern to study flow behavior in the system. Two-phase flow patterns are widely studied and can be predicted. However, three and four-phase flows that prevail during UBD and MPD are not well understood. Different flow patterns are observed for horizontal and vertical pipes and discussed separately.

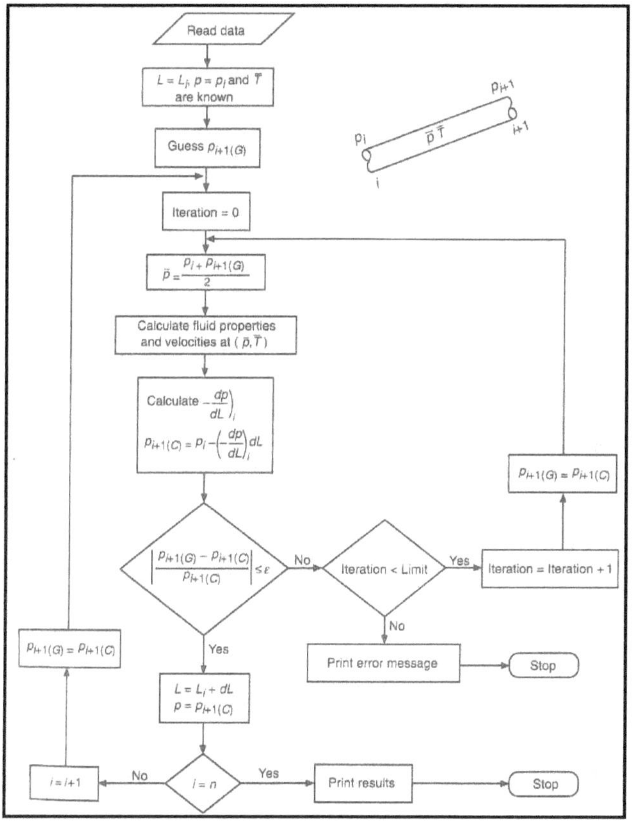

Figure 3–29 *Computational flow chart for average properties of fluids in one segment of pipe or annulus (Shoham, 2006)*

3.31.2 Vertical Flow Patterns

Flow regimes observed in a vertical pipe are shown in Figure 3–30. Flow patterns observed in a vertical pipe are:

> **Bubble flow**—Bubble flow occurs at low gas injection rates. Gas bubbles discrete in the continuous liquid phase and flow in a spiral form in the middle of the pipe.
>
> **Slug flow**—Increasing the gas injection rate, large bubbles coalesce with smaller bubbles, generating larger bubbles. When large bullet-shaped bubbles are present in the pipe, the flow is slug flow and the bubble is a Taylor bubble. Taylor bubbles do not have enough pressure to support the liquid

phase. Therefore, liquid slips down the Taylor bubble and collect under the bubble until the next Taylor bubble arrives.

Churn flow—Churn flow pattern is chaotic slug flow. Churn flow occurs at higher gas injection than slug flow. Taylor bubble is deformed and the slug has an oscillatory motion.

Annular flow—At high gas injection rates, gas flows as a continuous phase in the middle of pipe. Some liquid droplets entrain the gas phase and increase the phase reaction between the gas and liquid.

Mist flow—At very high gas injection rates, gas overcomes liquid-pipe tension force and removes the liquid film from the pipe. In this case, gas phase is the continuous phase and liquid phase is present as dispersed droplets.

Flow patterns observed in the annulus are the same as flow patterns in pipes. However, the geometry of annulus, eccentric geometry, and presence of cutting beds shifts the expected flow pattern boundaries in the system.

For safety reasons and operational concepts, surface choke pressure is adjusted to control bottom-hole pressure and gas velocity in the annulus. Choke handling affects mainly gas velocity and changes flow pattern in the wellbore.

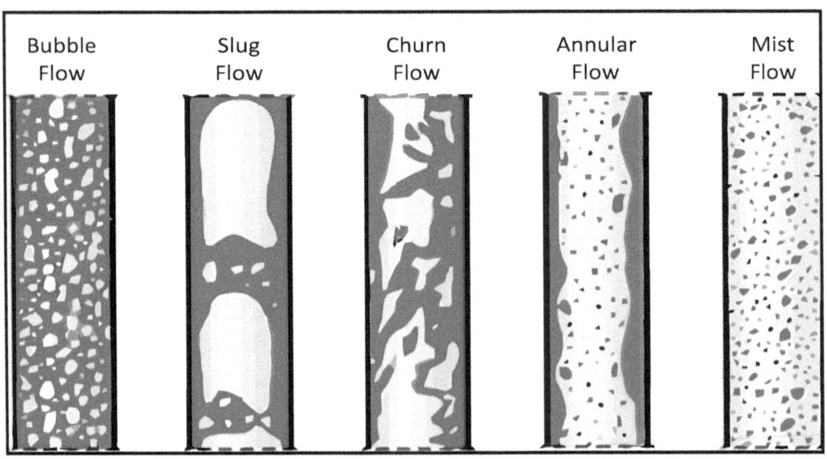

Figure 3–30 *Flow patterns in vertical flow (Griffith, 1984)*

3.31.3 Horizontal Flow Patterns

Flow regimes observed in a horizontal pipe depend on gas and liquid velocities. Several classifications are defined in the literature for horizontal flow as shown in Figure 3–31.

Flow patterns discussed here are the most widely accepted and observed in UBD operation.

Dispersed bubble flow—At very low gas and high liquid injection rates, gas moves in the pipe as small bubbles. This flow pattern is dispersed bubble flow. Gas bubbles do not have the same size. Due to lighter density, gas bubbles tend to move on the upper section of pipe.

Stratified smooth flow—At low liquid and gas velocities in a horizontal pipe, gas and liquid separate and gas moves on the top and liquid on the bottom; the liquid-gas interface surface is smooth. This flow pattern is stratified smooth flow.

Stratified wavy flow—ncreasing the gas injection from stratified smooth flow adds turbulence to liquid-gas interface surface, causing a wavy interface. This flow pattern is stratified wavy flow.

Slug flow—If gas further increases from stratified wavy flow, the wavy motion of interface increases till it reaches the upper side of the pipe and blocks the gas continuity in the system. This flow pattern is slug flow. In slug flow, the flow regime is not uniform in the pipe. Gas phase moves in separate pockets which are separated by columns (slugs) of the liquid phase.

Annular flow—If gas injection rate further increases, gas moves as a core in the pipe surrounded by liquid. Due to gravity, liquid phase is thicker at the bottom.

Based on experiments, flow pattern maps are developed to predict flow patterns under a specific condition. Some of these maps are very simple, using superficial velocities to predict the flow pattern, like Mandhane, as shown in Figure 3–32; some of them use series of dimensionless parameters for mapping, like Taitel and Dukkler's (Guo et al., 2005).

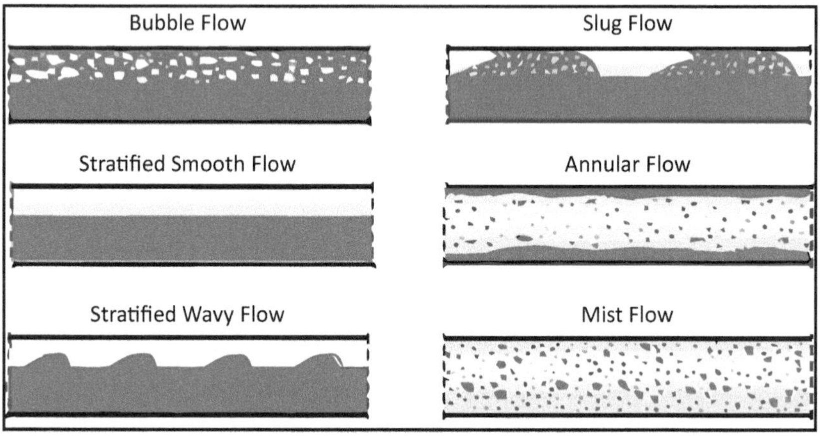

Figure 3–31 *Flow patterns in horizontal flow*

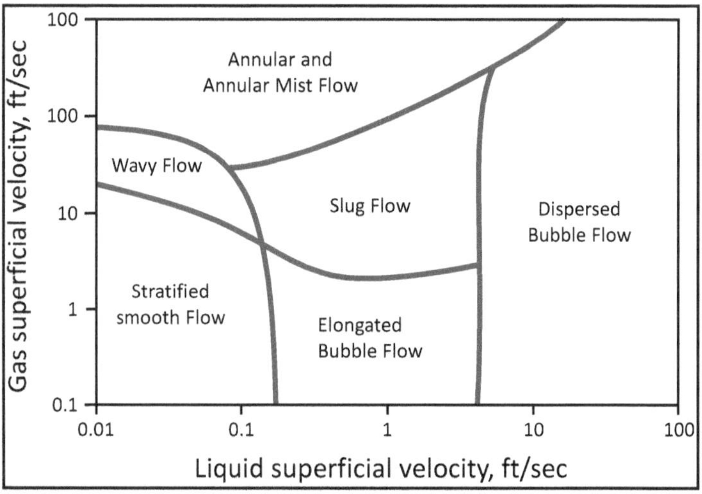

Figure 3–32 *Mandhane flow pattern*

3.32 Hydraulic Modeling

Modeling is a decision making tool used to predict drilling conditions under a specific condition. Operational challenges can be studied before the operation to develop an optimum program to save time and cost. Several models are presented in the literature regarding pipe flow. For oil industry application, wetted diameter and friction factor are modified to simplify the annular flow to pipe flow and improve the accuracy.

Models of UBD (and MPD) hydraulics have a unique complexity of wellbore geometry, borehole roughness, temperature profile, drillpipe rotation, and solids in the fluid. All of these conditions are simplified to be able to develop an applicable model.

3.32.1 Empirical and Mechanistic Models Developed for Oilfield Applications

Empirical models are valid over a limited range of gas and liquid injection rates and under specific conditions. Recently, researchers focused to develop mechanistic models valid for a wide range of gas and liquid injection rates.

Rommetveit et al. (1999) mentioned that bottom-hole pressure during UBD drilling can be expressed as:

$$P_{bh} = P_{hyd} + P_f + P_{acc} + P_{surf} \tag{3.7}$$

where

P_{bh} = Bottom-hole flowing pressure

P_{hyd} = Hydrostatic pressure

P_f = Frictional pressure loss

P_{acc} = Acceleration pressure

P_{surf} = Surface back pressure

Hydrostatic pressure is function of gas and liquid density and gas volume fraction. Density of gas is a function of temperature and pressure. Gas volume fraction is a function of gas and liquid flow rates at the annular condition. Temperature and pressure do not significantly affect water density and are usually neglected, except for oil-based muds.

Frictional pressure loss in the system depends on several factors which are not completely understood. Friction pressure drop is strongly a function of flow pattern regime, gas and liquid velocities, and wetted diameter.

Acceleration pressure measures the amount of pressure drop in the system due to acceleration of fluid in the system. This type of pressure drop is not present in single-phase systems.

Gas and liquid mixing causes an extra pressure drop inside the gasified mud. This pressure drop depends on flow pattern regime. As this pressure drop is not well understood, it is considered an extra frictional pressure drop.

Surface back pressure is extra pressure on the system to control gas liquid ratio (GLR) at the surface to maintain stability of the gasified mud; it is also a tool to control the bottom-hole pressure.

Modeling methods described in this section are considered immiscible gas/liquid systems. Temperature and pressure are simplified for a water base but are also important in oil base systems. In systems like natural gas and oil-based fluid, as gas dissolves into the liquid phase flash calculations need to be added to determine gas and liquid volumes under bottom-hole conditions.

All pressure calculations are dynamically related to each other and depend on time and system condition. When a change is introduced into the system, it affects all of the pressure calculations. The result depends on the way that all components influence each other. For example, increasing gas injection rate may increase or decrease the bottom-hole pressure.

Different flow pattern regimes have different behaviors and friction factors. The first consideration is determining flow pattern. In order to select the flow pattern, it is necessary to have fluid properties, temperature, and pressure. Fluid properties depend on temperature and pressure. Temperature is assumed equal to the formation's temperature. Pressure is the unknown. The procedure is to assume a pressure and determine if the assumed value is correct. Therefore, numerical analysis is needed.

The first step is to determine flow pattern. Several equations are available for this purpose. Flow pattern depends on inclination angle. Zhou et al. (2005) proposed boundaries to determine flow pattern considering the inclination angle. In practice, annular flow is not expected during gasified operation. It may occur very close to the surface at high gas injection rates and relatively low liquid injection rates. Slug and bubble flow are dominant flow regimes in the annulus. Stratified flow prevails in horizontal and highly inclined sections. Flow pattern models indicate different flow patterns at a given condition. Hasan and Kabir (2007B) recommend the following sequence of annular flow patterns in the vertical section of the hole:

1. Churn flow
2. Slug flow
3. Bubble flow

If the wellbore pressure calculation starts from the surface and initially assumes annular flow, and the flow pattern condition does not match, try to find the flow pattern by going down the above hierarchy. For example, it is not possible for the flow pattern changes to go from churn to annular while going from the surface to the bottomhole, so the next step would be to impose the churn flow pattern on the calculations. The sequence of flow pattern remains until the wellbore geometry changes, as at the casing seat or at the top of the collars.

For more details on mechanistic modeling readers are directed to Hasan and Kabir (2002), and Shoham (2006).

3.32.2 Hagedorn and Brown Correlation

The Hagedorn and Brown correlation initially developed for vertical wells with 1 in. to 2 in. (25–50 mm) tubing sizes. This correlation does not consider liquid holdup or flow pattern in the tubing and annulus, which simplifies the application of this correlation. The Hagedorn and Brown correlation is developed for vertical flow and care must be taken for deviated sections. However, some modifications were added to the original correlation to improve the accuracy of the model.

Table 3–4 Hagedorn and Brown Correlation

$N_{lv} = 1.938\, v_{sl} \sqrt[4]{\frac{\rho_l}{\sigma_l}}$	liquid velocity number	(3.8)
$N_{gv} = 1.938\, v_{sg} \sqrt[4]{\frac{\rho_l}{\sigma_l}}$	gas velocity number	(3.9)
$N_d = 10.1\, d \sqrt{\frac{\rho_l}{\sigma_l}}$	pipe diameter number	(3.10)
$N_l = 0.157\, \mu_l \sqrt[4]{\frac{1}{\rho_l\, \sigma_l}}$	liquid viscosity number	(3.11)
$C_{Nl} = Exp\{-4.895 - 1.0775 \ln(N_l) - 0.80822[\ln(N_l)]^2 - 0.1597[\ln(N_l)]^3 - 0.01019[\ln(N_l)]^4\}$		(3.12)
$N_f = \left(\frac{V_{ld}}{V_{gd}^{\;0.575}}\right)\left(\frac{p}{14.65}\right)^{0.1} C_{Nl}\left(\frac{10^6}{N_d}\right)$		(3.13)

Table 3–4 Hagedorn and Brown Correlation (cont'd)

$$\left\{ \begin{array}{ll} \dfrac{V_{gd} N_l^{\;0.38}}{N_d^{\;2.14}} \le 0.01 & \psi = 1 \\ \dfrac{V_{gd} N_l^{\;0.38}}{N_d^{\;2.14}} > 0.01 & \psi = \dfrac{1.80}{\left(1 + \exp(3.47 - 99.4x)\right)^{0.276}} \end{array} \right\} \tag{3.14}$$

$$f_l = \psi \exp\{(-3.6372) + 0.8813(\ln(N_f)) - 0.1335(\ln(N_f))^2 + 0.018534(\ln(N_f))^3 - 0.001066(\ln(N_f))^4\} \tag{3.15}$$

$$f_g = 1 - f_l \tag{3.16}$$

$$\mu_m = \mu_l^{\;f_l} + \mu_g^{\;(1-f_l)} \tag{3.17}$$

$$C_l = \frac{ql}{ql + qg} \tag{3.18}$$

$$\rho_n = C_l \rho_l + (1 - C_l)\rho_g \tag{3.19}$$

$$v_m = v_{sl} + v_{sg} \tag{3.20}$$

$$\mathrm{Re} = 1488 \frac{\rho_n v_m d_{hyd}}{\mu_m} \tag{3.21}$$

$$f = 0.184(\mathrm{Re})^{-0.2} \tag{3.22}$$

$$\rho_s = f_l \rho_l + (1 - f_l)\rho_g \tag{3.23}$$

$$\left(\frac{dp}{dz}\right)_f = f \frac{\rho_{n^2} v_m^{\;2}}{288 g_c \rho_s d} \tag{3.24}$$

$$\left(\frac{dp}{dz}\right)_h = \frac{\rho_s}{144} \tag{3.25}$$

$$\left(\frac{dp}{dz}\right)_t = \left(\frac{dp}{dz}\right)_f + \left(\frac{dp}{dz}\right)_h \tag{3.26}$$

where

C_l = non-slip holdup
d = pipe diameter, in.
f = friction factor
f_l = liquid fraction
f_g = gas fraction
N_d = pipe diameter number
N_{gv} = gas velocity number
N_l = liquid viscosity number
N_{lv} = liquid velocity number
$\left(\frac{dp}{dz}\right)_f$ = frictional pressure, psi/ft
$\left(\frac{dp}{dz}\right)_h$ = hydrostatic pressure, psi/ft
$\left(\frac{dp}{dz}\right)_t$ = total pressure change, psi/ft
μ_l = viscosity of the liquid phase, cP
v_{sg} = gas superficial velocity, ft/sec
v_{sl} = liquid superficial velocity, ft/sec
υ_m = mixture velocity, ft/sec
ρ_n = non-slip density of the mixture, lb/cu.ft
ρ_m = mixture density, lb/cu.ft
ρ_{ns} = no-slip mixture density, lb/cu.ft
σ_l = interfacial tension, dyne/cm

Three diagrams were developed to determine the required variables. First, the dimensionless factors are calculated for the given pressure and temperature. Then the factors from the graphs, parameters, y, CN_l, and ε_l are determined. Eq. (3.12) through Eq. (3.15) are fitted estimations for the graphs in the Figure 3–33 (Hasan and Kabir, 2002).

Modifications are applied to the Hagedorn and Brown correlation to improve the accuracy of the model. The first correction is to check for f_l to be greater or equal to C_l. If f_l is less than C_l then the value of C_l

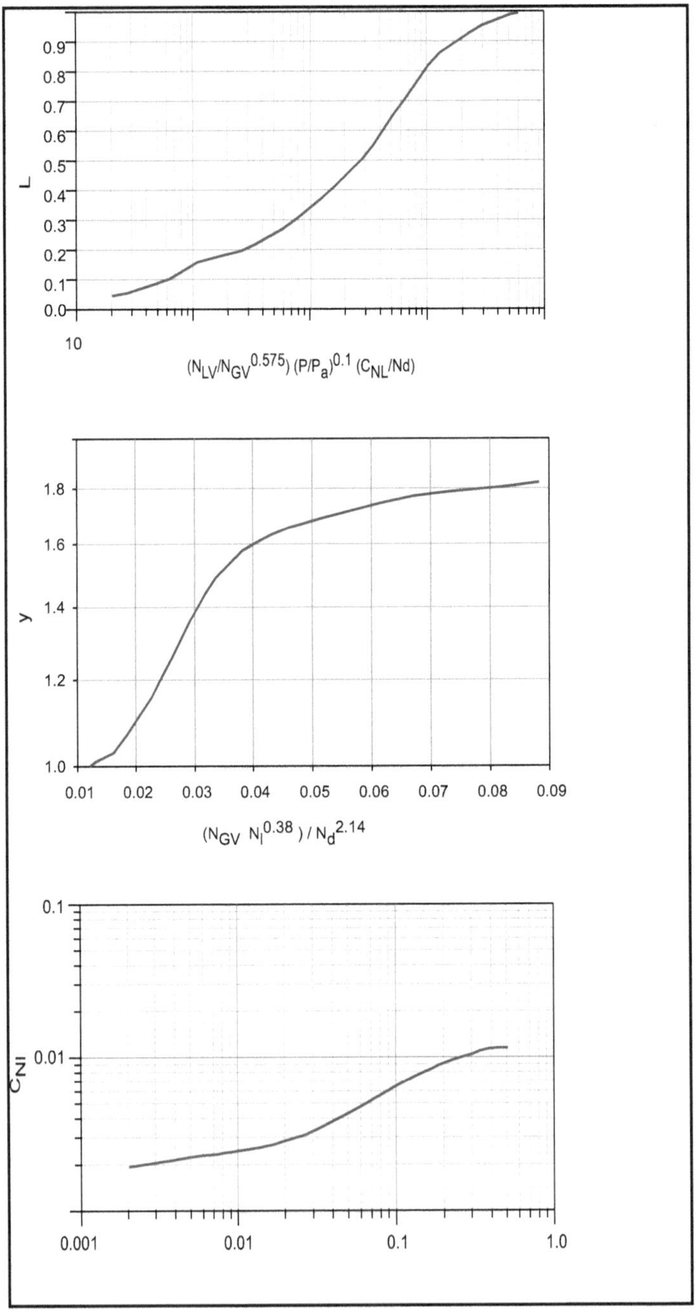

Figure 3–33 *Hagedorn-Brown diagrams (Gábor Takás)*

is used for f_l. Other modifications include using different correlations for annular and bubble flow conditions. For example, the Griffith correlation can be used for bubbly flow and the Duns and Ros correlation for annular/mist flow.

Example 3–2

Calculate the pressure gradient for the given condition in Table 3–4. The wellbore is 8.5 in. and the drillpipe OD is 3.5 in.

Table 3–5 Input Data for a Given Wellbore Geometry and Fluid Injection Parameters

Q_g		4	ft^3/sec
Q_l		2	ft^3/sec
ρ_l		65	lbm/ft^3
ρ_g		1.5	lbm/ft^3
μ_g		5	cp
$\mu\lambda$		0.02	cp
σ	70 dyne/cm	0.154	lbm/sec^2
Pressure		300	psi

Solution:

Area	0.327	ft^2
Hydraulic diameter	0.42	ft
Area	0.3272492	ft^2
V_{sg}	12.22	ft/sec
V_{sgl}	6.11	ft/sec
V_{ld}	11.63	
V_{gd}	23.26	
N_d	48.58	
N_l	0.1144	
ψ	1	
C_{NI}	0.0024	
N_f	127.26	

f_l	0.375
f_g	0.625
μ^m	0.1588 cp
C_l	0.33
ρ_n	22.667 ibm/ft^3
ρ_s	25.33 ibm/ft^3
v_m	18.33 ft/sec
Re	1.6×10^6
f	0.01054
$(fp/dz)_f$	0.019 psi/ft
$(fp/dz)_h$	0.176 psi/ft
$(fp/dz)_t$	0.195 psi/ft

3.32.3 Guo's Closed-Form Hydraulics for Gasified Systems

Several equations are available to calculate pressure during multiphase flow, considering different flow patterns and transient flow between flow regimes. These equations are complex. Assuming steady state condition, Guo et al. (2004) developed a closed-form hydraulic model for gaseated fluid by assuming the gasified mixture as a non-slip homogenous mixture.

$$dp = \gamma_m \left(1 + \frac{f\,u_m^{\,2}}{2\,g\,D_h}\right) dh \tag{3.27}$$

where γ_m is density of gasified mixture in lb/ft$_3$. The final derivation results in a series of equations that are solved using trial-error. Table 3–6 summarizes all the equations.

Table 3–6 Summary of the Equations Required for Guo's Model

$$a = \frac{0.0014\,d_b^{\,2}\,S_s\,ROP + 0.2\,W_m Q_m + 1.44\,S_f Q_f + 0.019\,S_g Q_{go}}{T\,Q_{go}} \cos(\theta) \tag{3.28}$$

Table 3–6 Summary of the Equations Required for Guo's Model (cont'd)

$$b = \frac{0.033\,Q_m + 0.023\,Q_f}{TQ_{go}} \tag{3.29}$$

$$c = \frac{9.77\,T\,Q_{go}}{A} \tag{3.30}$$

$$d = \frac{0.33\,Q_m + 0.22\,Q_f}{A} \tag{3.31}$$

$$GLR = \frac{4.07(T + T_s)Q_{go}}{(P_s + P)\left(\frac{Q_m}{7.48} + \frac{5.615\,Q_f}{60}\right)} \tag{3.32}$$

$$\bar{e} = \frac{e_p\,D_i + e_b\,D_o}{D_i + D_o} \tag{3.33}$$

$$f = F_{LHU}\left[\frac{1}{1.74 - 2\log\left(\frac{2\,\bar{e}}{D_h}\right)}\right] \tag{3.34}$$

$$e = \frac{f}{2\,g\,D_h\,\cos(\theta)} \tag{3.35}$$

$$m = \frac{c\,d\,e}{1 + d^2 c} \tag{3.36}$$

$$n = \frac{c^2\,e}{(1 + d^2\,e)^2} \tag{3.37}$$

$$144\,b\,(P - P_S) + \frac{1 - 2\,b\,m}{2}\ln\left|\frac{(144\,P + m)^2 + n}{(144\,P_S + m)^2 + n}\right| - \frac{m + b\,n - b\,m^2}{\sqrt{n}}\left[\tan^{-1}\left(\frac{144\,P + m}{\sqrt{n}}\right) - \tan^{-1}\left(\frac{144\,P_S + m}{\sqrt{n}}\right)\right] = a\left(1 + d^2 e\right)L \tag{3.38}$$

where

A = Flow passage area, in^2

d_b = Bit size, in

D_h = Hydraulic diameter, ft

D_i = Inner diameter, ft

D_o = Outer diameter, ft

e_b = Tubular roughness

e_p = Tubular roughness

$\bar{e}$ = Average wall roughness, ft

F_{LHU} = Friction factor correction due to liquid hold-up

g = Gravity acceleration, 32.2 ft/sec^2

L = Length of section, ft

P = Pressure at the bottom of section, lb/ft^2

P_s = Pressure at the top of section, lb/ft^2

Q_f = Formation-fluid influx rate, bbl/hr

Q_{go} = Gas injection rate, scf/min

Q_m = Mud injection rate, gpm

ROP = Rate of penetration, ft/hr

S_f = Specific gravity of formation-fluid influx, water = 1

S_g = Injecting gas specific gravity, air = 1

S_s = Specific density of solids, water = 1

T = Absolute temperature in the wellbore, °R

T_s = Absolute temperature at the surface, °R

W_m = Mud weight, ppg

θ = Inclination angle from vertical

The Moody friction factor is valid for single-phase flow and results in a very low pressure drop for multiphase flow system. Friction factor is corrected for the liquid hold-up in a system by a correction factor called F_{LHU}, which is a correlation based on average GLR of the system as given in Eq. (3.34).

The friction factor is a function of the average roughness in the wellbore. Guo et al. (2004) recommend values of 0.00015 ft for steel tubular and 0.004 ft for wellbore. If tubular and wellbore roughness is available, use them instead of the given values. The average roughness in the wellbore is estimated using Eq. (3.33).

3.32.4 Pressure Drop through Bit Nozzles

Mechanical energy balance is a common way to calculate bit nozzles pressure drop. For two-phase flow, the pressure drop equation is given by Eq 3.28 as given by Gucuyener (2003). Iteration is needed to calculate the nozzle pressure drop using Eq 3.22.

$$1.08\times10^{-4}\left(\frac{u_N}{C_N}\right)^2+\frac{\varphi_g}{7.48\,\rho_g}P_{bh}\,\mathrm{Ln}\left(\frac{P_{bh}}{P_{bh}+P_{bit}}\right)-\frac{\varphi_l}{7.48\,\rho_l}P_{bit}=0 \quad (3.39)$$

$$u_N=2.4\left(Q_g+0.1337Q_l\right)/A_N \quad (3.40)$$

$$A_N=\sum_{i=1}^{N}d_{Ni}^2/1303.8 \quad (3.41)$$

$$\varphi_g=\frac{Q_{go}}{Q_{go}+13.09\;Q_l} \quad (3.42)$$

$$\varphi_l=\frac{13.09\;Q_l}{Q_{go}+13.09\;Q_l} \quad (3.43)$$

where

u_N = Velocity in ft/sec

C_N = Nozzle discharge coefficient, ($C_N = 0.95$)

φ_g, φ_l = Gas and liquid mass fractions

ρ_g, ρ_l = Gas and liquid density, ppg

P_{bh} = Bottom-hole pressure, psi

P_{bit} = Pressure drop through nozzles, psi

A_N = Total nozzle area, in^2

Q_{go} = Gas injection rate, scf/m

Q_g = Gas flow rate at bottom-hole condition, ft^3/min

Q_l = Liquid injection rate, gpm

3.32.5 Pressure Drop in the Drillpipe

In the drillpipe, pressure and velocity are higher than in the wellbore. Therefore, the main dominant flow is bubble flow, which makes pressure profile calculations easier. The flow of gas and liquid can be assumed as a homogenous mixture, and the slippage between the phases can be ignored, reducing the multiphase flow calculation to a single-phase fluid flow.

The Griffith correlation is one of the most widely used for bubble flow calculation in the industry to calculate the frictional pressure loss for bubble flow. The Griffith model assumes a slippage velocity (drift-flux) between the gas and the liquid phase. The velocity of the fluid in the pipe is related to the superficial velocities of gas-liquid phases. Then the density and the viscosity of the liquid phase are used to determine the Reynolds number and frictional pressure drop. The relationship between the liquid velocity and the superficial velocities is given as

$$E_L = 1 - \frac{1}{2}\left[1 + \frac{V_m}{V_s} - \sqrt{\left(1 + \frac{V_m}{V_s}\right)^2 - 4\left(\frac{V_{sg}}{V_s}\right)}\right] \tag{3.44}$$

The value of V_s is 0.8 ft/sec or 0.24 m/sec in SI unit. When the E_L is calculated, the velocity of the in-situ velocity of the liquid phase in the tubing is calculated as

$$V_l = \frac{V_{sl}}{E_L} \tag{3.45}$$

The Reynolds number is then calculated using the properties of the liquid phase.

$$Re = \frac{\rho_l V_l D}{\mu_L} \tag{3.46}$$

The friction factor can be calculated using the Reynolds number. For Fanning friction factor, the frictional pressure drop is given in Eq. (3.29). Note that the Moody friction factor is 4 times greater than Fanning friction factor.

$$\Delta P_f = \frac{2 f \rho_l V_l^2}{g_c D} L \tag{3.47}$$

The hydrostatic pressure of the gas and liquid mixture is calculated using the mixture density of the gas and liquid.

$$\Delta P_h = \rho_m g L \tag{3.48}$$

3.32.6 Cutting Transport Models

Cutting transport modeling provides important results to study the effect of different operational parameters. Several attempts have been published in the literature to study the cutting transport phenomenon, Zhou (2008), Doan et al. (2003), Li and Walker (2001), Lourence et al. (2001), and Guo and Ghalambor (2002). These attempts either consider minimum required velocity or a mechanistic model for cutting removal. In minimum required velocity models, gasified fluid velocity is determined to move the cuttings out of the hole. These models are appropriate for vertical wells. For inclined and horizontal wells the minimum required velocity is multiplied by an empirical factor to ensure hole cleaning.

Mechanistic models determine the sweeping effect of gasified fluid on the cuttings. Mechanistic models allow studying the formation of cutting beds in horizontal and inclined sections. Properties of a gasified system are a function of pressure and temperature. Therefore, cutting transport study should be combined with hydraulics studies to determine required values.

Gas and liquid injection rate should be enough to carry cuttings out of the hole. Minimum cutting carrying velocity is usually used to assure hole cleaning. Minimum velocity is the sum of slip velocity and cutting generation velocity as shown in Eq 3.26. If gaseated fluid velocity is higher than the minimum velocity, cuttings are carried out of hole properly. High and low fluid velocities cause problems. Low fluid velocity does not clean the hole properly and high fluid velocity erodes the wellbore and increases the diameter.

$$u_{min} = \alpha \left(u_{sl} + u_{tr}\right) \tag{3.49}$$

Coefficient α, is an empirical field coefficient to correct the equation for inclined and horizontal sections. For vertical wells $\alpha = 1$, and for inclined and horizontal wells usually the value of 2 is considered for α in calculations.

Slip velocity calculation depends on the flow regime. Turbulent flow for gasified water-based fluids and transitional to turbulent flow for gasified oil-based fluids are expected in the wellbore. Also, as drag force of fluid on cuttings is minimal in a turbulent condition, slip velocity is higher. Therefore, turbulent flow is considered for slip velocity calculation. For turbulent flow regimes (Re > 2,000), slip velocity is estimated using

$$u_{sl} = 5.35\sqrt{\frac{d_s\left(\rho_s - \rho_m\right)}{\rho_m}} \tag{3.50}$$

where

u_{sl} = Slip velocity, ft/sec

d_s = Cuttings diameter, ft

ρ_s = Cuttings density, lb/ft^3

ρ_m = Density of gaseated fluid, lb/ft^3

Slip velocity depends on cuttings size, density of fluid, and density difference between fluid and cuttings. Slip velocity increases as either/both density or size of cuttings increases. Increasing the density of gaseated fluid reduces slip velocity. Cuttings generated velocity depends on the volume rate of generated cuttings and optimum cuttings concentration in the system. Volume of generated cuttings depends on wellbore diameter and rate of penetration. Cutting concentration is adjusted in value to prevent cuttings accumulation in the wellbore. Guo and Ghalambor (2002) recommend 4% cuttings concentration in the system.

$$u_{tr} = \frac{\pi d_b^2}{4 C_p A}\left(\frac{ROP}{3600}\right) \tag{3.51}$$

where C_p is the concentration of cuttings.

For a specific gas/liquid injection rates, required minimum velocity can be adjusted by controlling the rate of penetration or density of gaseated fluid. For specific gas and liquid injection rates, density is function of pressure. Increasing the gas or liquid injection rate does not mean higher density or lower slip velocity. Effects of surface back pressure, gas, and liquid injection rate should be studied before introducing any changes to system.

3.33 Questions and Answers

Questions and answers covering this section are found in Section 3.11 and Section 3.12, page 138 and 139.

3.34 References

See Chapter 3.13, page 142.

CHAPTER 4

Foam Drilling

Bill Rehm, Drilling Consultant
Amir Paknejad, Add Energy, LLC

4.1 Introduction to Foam Drilling and Workover

This chapter discusses the use of foam as an underbalanced drilling and workover fluid. The chapter contains definitions of foam and foam properties, how to build a foam, common field procedures, a discussion of foam agents, and the theory of foam flow. Commonly accepted limits to foam drilling are also discussed, as are examples of using foam under extreme conditions.

4.1.1 Foam Description

Foam is a low density system that has the advantage of having a high lifting and hole cleaning capacity that can be combined with a very low fluid flow. A foam system also acts to aid in reducing lost circulation because the bubbles in the system expand as they enter the lower-pressured lost zone.

Foam has been variously described as

- an agglomeration of bubbles surrounded by a thin liquid film,
- a dispersion of gas in a liquid, and
- an emulsion of a gas in a liquid.

Whichever is the best description, a foam bubble is surrounded by a "skin" of polar molecules that tend to keep the bubble of gas dispersed evenly throughout the liquid.

The term "emulsion" will be used in this discussion. While it is a simple and easy to understand term, on a physical chemistry basis "emulsion" is not technically correct.

The bubble of gas can be air, nitrogen, or natural gas. The gas type is generally chosen for economics, compatibility, and/or safety. The liquid phase is almost always water, but oil has also been used as explained in Section 2, page 234.

Foam more closely approaches plug flow than any other standard flow pattern. However, since the composition or percent of gas varies with pressure, the flow pattern may start to approach a streamline or Bingham-plastic fluid. Many of the computer models match actual well conditions by using the power law calculations (Section 4, page 255).

The unique characteristics of the foam system make it very attractive for drilling re-entries, horizontal holes, large surface holes, and work-over operations. The characteristics of the foam system are:

- High lifting capacity independent of velocity
- Stability in very low density systems (minimal pressure surging)
- Resistance to lost returns
- Corrosion protection

A foam system can reduce the bottom-hole pressure in a shallow well to as little as 20% of a column of water or to as little as 50% of a column of water in a 10,000 ft. hole. Pressure reduction comes from displacing water in the hole with "bubbles" of gas. As the hole gets deeper than about 8,000 ft. (2,500 m), the gas is so compressed that the pressure gradient starts to approach that of the water (or oil) continuous phase.

The use of a drilling motor with foam requires higher foam volume to help run the motor, so the bottom-hole pressure reduction will be less than with non-motor applications.

The foam system is dependent upon a good quality foaming agent. As defined here, a foaming agent is not just a surface tension lowering detergent such as that used in mist drilling, but is a material that causes a much higher film strength around the gas bubbles. The difference between foam and mist can be shown in Figure 4–1. The foaming agent and foam stiffeners can be controlled to give foam that will remain in a foamed state during a trip, and break upon being released into the pit.

Another important part of the foam system is the water. While foams can be made with almost any water, the best and most economical foam is made with "drinking" quality water.

The incremental costs of a foam system are the cost of the foaming agent, corrosion chemicals, compressors, rotating control

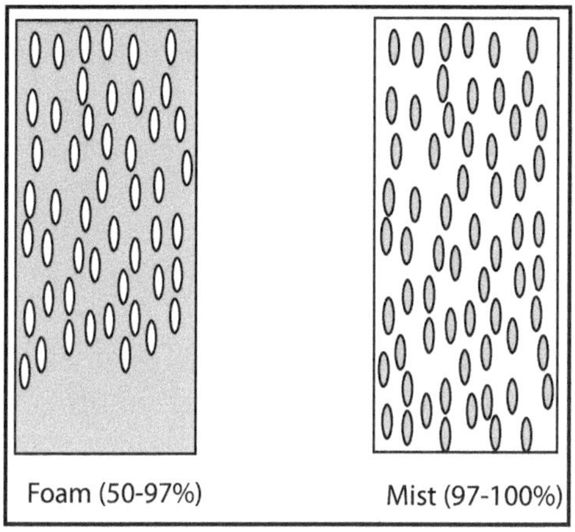

Figure 4–1 *Difference between foam and mist system (Medley, 1998)*

device (RCD), and separator system. All manner of surface systems have been used with foam that varies from foam confined by a RCD and expelled from the flow line to an earth pit, to complex BOP systems, chokes, and manifolds with pressurized separators and complete solids control systems.

4.1.2 General Operational Ideas with Foam Systems

The foam system is a reasonably flexible and pressure-stable system within the limits that it remains as an "emulsion" of gas in a liquid. Some basic considerations are:

- The systems start to be temperature sensitive at about 200°F (100°C). Foam systems will work at much higher temperatures but the tendency to break down during a long trip increases with higher temperatures. Some special foam has been reported to operate at 275°F or 135°C. Foam has great insulating properties and when combined with the heat absorption caused by the expansion of the gas, it keeps the circulating foam system relatively cooler than the wellbore.
- The quality of the system (% gas) must never be so high that the liquid is not in the continuous phase (air by volume should be less than 85–90% on a practical operating basis). When air reaches the continuous phase, the cuttings are no longer lifted out of the hole and the system starts to surge.

- The lifting capacity of the field system starts to degrade when the quality drops below about 40% (laboratory results show 52%).
- The best foam systems are made with good quality water. Any change from "drinking quality" water requires chemical treatment and/or more foaming and extending agents. If you can't drink the water, it's going to be expensive to make good foam.
- Foam is more stable under pressure than when in the pits, so it is possible after a little field experimenting with the chemical system to make foam that is stable on connections and trips, but breaks in the flow line and pits.
- Single pass foam should break by itself in the pit. A one to three day drilling or workover job is ideal for a "single pass" foam. Some of the water can be recovered from the waste pit and be reused and minimum extra equipment is needed.
- Recyclable foam should break with an alcohol spray (or with a pH change). Good solids control equipment and precise chemical addition is critical. One of the big problems with a recycle system is that as the solids in the water build up, they tend to act as a foam breaker or more properly, a foam inhibitor. Hydrophobic solids that break foam (fine carbon particles) and hydrophilic solids (clays) that make foam more stable are a complex subject. So, the solids should be kept to a minimum in recyclable foams.
- The critical part of a good consistent foam system is the precise measurement of the foam agent and other chemicals that go into the water stream.
- Separators have their place in foam drilling as a breaker area for recyclable foam, and as a safety system if a gas reservoir is being drilled, or if natural gas is being used as the gas phase. Many "single pass" foam drilling operations do not use a separator.
- In areas where corrosion is known to occur, corrosion can be controlled by treating out the oxygen in the make up water, the use of a corrosion inhibitor and by closely monitoring the iron content at the flow line. In properly designed foam, the air is encapsulated by the foaming agent and oxygen is limited in its ability to pass through the walls of the air bubble.

- Foam also encapsulates small quantities of down-hole gases, and in the case of CO_2 and H_2S, tends to limit their corrosive action. However, the action is not always complete and the gasses are still released at the surface.
- Foam tends, in theory, to have a slightly higher annular pressure drop than an aerated water system, but the difference is hard to observe.

4.2 History of Foam Systems

The first clear use of deliberately made foam systems appears to be drilling the large emplacement holes for nuclear testing in Nevada in the USA in the 1960s. The holes were up to 15 ft (4.6 m) in diameter and 5,000 ft (1,500 m) deep. In these emplacement holes, the water table was at about 2,000 ft (600 m) and the rhyolite rock had been shattered by previous explosions. Reverse circulating with a very 'Stiff Foam' was the only way the holes could be drilled. Since the circulation time was so long—sometimes a day or more—the foam had to resist gravity draining of the liquid phase. This was accomplished with the addition of bentonite and CMC (sodium carboxymethylcellulose) to the mixture of water and foaming agent. The foam remained stabile in the reserve pit in a desert atmosphere for a year or more before the pits could be closed. This mixture gave the early foam systems a reputation of being hard to de-foam.

In the following years, Stan Hutcheson with Chevron Oil Company in California used a better foaming agent without CMC or bentonite to successfully clean sand out of production wells. Without the bentonite and CMC the foam would break in the pits and still provide the lifting capacity required.

While foam is used in many drilling operations, workover and cleanout operations still provide the greatest number of foam jobs.

4.2.1 Definition of Foam Terms

Foam is an "emulsion" of gas in water (or occasionally oil). The foaming agent forms a chemical coat around each gas bubble and keeps it from merging with other gas bubbles.

It is common field practice in some areas to call foam "mist drilling." That is a major error in nomenclature. Mist is water injected into air where air is in the continuous state. Mist uses a detergent to lower the surface tension of the water to help break the water into fine droplets.

Foam is an "emulsion" of gas in water where the water is the continuous state. A foaming agent decreases the surface tension of the water but also contains bipolar molecules that form around each small bubble of air to make a "skin" that restricts the coalescing of foam gas bubbles and makes the system more stable.

Wet foam is used to describe a foam with a high percentage of water; it breaks easily and feels "wet." In actual field practice, the wet foam needs more foaming agent or a foaming enhancer or there simply is too much water and not enough gas.

Dry Foam is used to describe a foam with a low percentage of water and a higher treatment of foaming agent. It is more persistent and will support a higher percentage of air to water.

Stiff Foam is an old expression that was developed to define a foam made with a polymer (typically CMC) and bentonite, as well as a foaming agent. It was very stiff and persistent and could remain in the foam state for days or weeks. The term is now often used to define a regular drilling foam.

Stabile Foam is an old expression used to differentiate a foam that used only chemical foaming agents without the use of bentonite and a polymer.

Quality is the percent by volume of gas in the gross mixture at a *particular point or pressure*. The term can be expressed as a percent, decimal, or number, i.e., 70%, 0.70, or 70. The quality is a function of the volume of air injected into a certain volume of liquid, and the pressure at the point of interest.

The upper limit of quality with *water as the continuous phase* is a quality of 90 to 98. Above that gas percentage, the gas becomes the continuous phase. In the laboratory, it is possible to develop 98 quality foam, but in the wellbore with field water and field foam agents, at about 85 or 90 quality the gas may start to be the continuous state and the water becomes discontinuous—essentially a mist. At the point where gas starts to become the continuous phase the system starts to surge into alternate slugs of water and gas. The foam loses its lifting capacity and drops the cuttings.

The lower limit of efficient foam based on lifting capacity is reported from the literature to be a quality of 52. Anecdotal field reports from vertical holes indicate that foam has a very good lifting capacity as it goes towards a quality of 30. Figure 4–2 shows relative lifting force versus foam quality. A further discussion of field foam stability is found in Section 4.37, page 266.

Ratio—Also known as the surface ratio of volumes, is the ratio of injected gas at standard temperature and pressure (STP) to liquid in a common measurement system (standard cubic feet or liters), where the liquid volume is always expressed as 1. A ratio of 200/1 would mean 200 sft^3 of gas injected to every ft^3 (7.48 gal) of water (in metric this would still be 200L gas to 1 Liter of water). The ratio is a tool for dealing with an air or gas compressor output at surface conditions.

Texture—The term "texture" has been proposed to cover the viscosity and flow characteristics of foam (Guo, 2003). At present, most of these measurements have not been established, nor are there any simple reproducible field tests. Foam texture depends upon the pressure and temperature to a much larger degree than do standard single-phase drilling fluids. It is also a function of the foaming agents and the induced shear.

Half-Life—A standard mixing test is performed and a column of foam is developed. When half the liquid used in the test has separated from the foam, the time is noted as "half-life."

Solids—In drilling terminology, "solids" refers to drill solids in the drilling fluid. (Water analysis uses the term "solids" to refer to the ion content of the water.)

Yield Stress (of bulk foams)—This term is not directly used in this chapter but contains, or defines, why foams plug fractures so efficiently as well as containing some of the concept of why foam lifts cuttings so well (Gajbhiye, R.N., Kam, S.I., 2010).

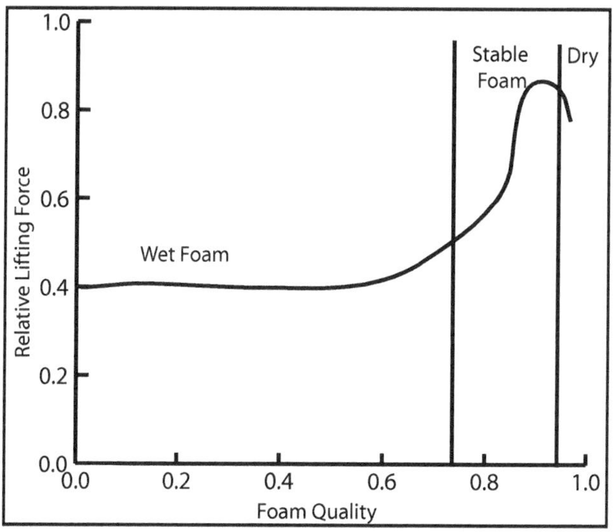

Figure 4–2 *Foam lifting capacity (Bayer et al., 1972)*

4.2.2 Requirements for Good Foam

Three basic conditions are required for good foam:

1. Start with clean "drinking" quality water. The more stray ions (solids) that are in the water, the more treating and chemical agent the water requires. If there is poor quality water, treat it and clean it before adding the foaming agents.

2. The foaming agents must have the proper concentration on a consistent basis. (Foaming agents vary widely in concentration and content. They are considered proprietary and field operations generally have no idea of any of the specifications.)

3. There must be enough energy put into the system to break the gas up into discrete small bubbles so that they can be encapsulated by the chemical agent. The final and best mixer is a good pressure drop (at least 200 psi or 1,400 kPa) at the bit.

4.3 Advantages of Foam Systems

There are advantages to foam systems:

- Foam systems display little in the way of pressure surging, with minimal overpressure damage to reservoirs and formations.
- Bottom-hole pressure can be reduced to below that of gaseated fluids.
- The system has a greater lifting capacity than any other drilling fluid (see Figure 4–3).
- It reduces or stops lost circulation.
- It permits very high drilling rates because of foam ability to clean under the bit and clean the annulus.
- As a cleanout fluid, it uses very little water with low annular velocity.
- Properly made foam can reduce or limit drillpipe corrosion.

From these advantages come the following:

- Better reservoir protection (from surges and pressures)
- No differential sticking

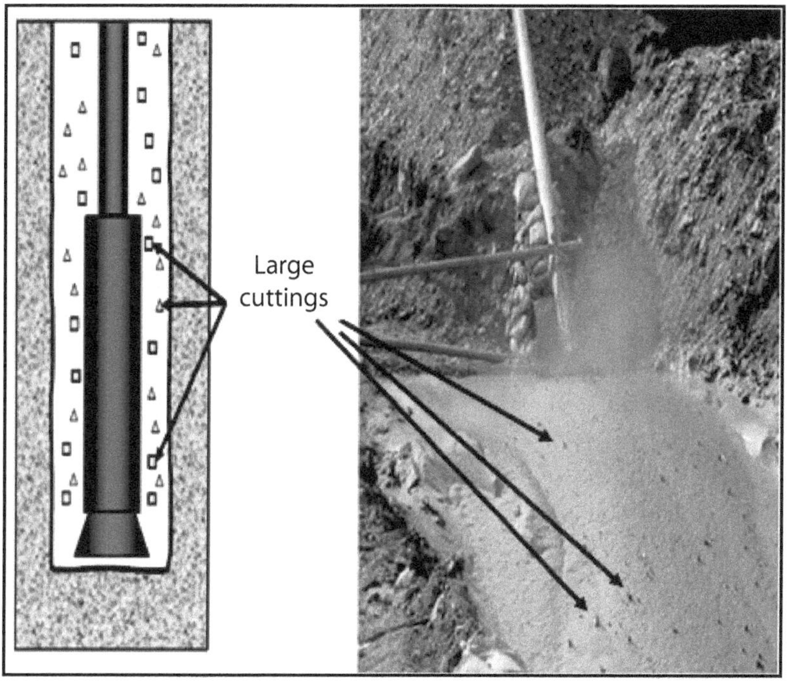

Figure 4–3 *Large cuttings cleaning in foam drilling (Medley, 1998 and Nas, 2006)*

- Higher drilling rates with better hole cleaning and lower wellbore pressure
- Excellent cleanout of cuttings or frac sand with low annular velocity
- Low ECD (equivalent circulating density)

4.3.1 Stabile System

Foam systems do not have pressure surges like gaseated systems, nor do they collapse like gas or gas mist systems when the compressors are stopped. Foam is a continuous stabile system that acts more like a mud system than an air system.

The nature of foam, small bubbles stabilized in a continuous liquid phase, makes a hydraulically stabile system. When the gas is sheared into the water through a surface mixer or through the bit (and motor) the gas is further dispersed as fine bubbles with each bubble surrounded by a chemical skin. As long as the bubbles in the annulus are kept under a pressure of three to six atmospheres, they remain small enough to resist floating upward, expanding, and coalescing. While coalescing of the bubbles eventually takes place, pressure on the system keeps the foam relatively stable.

The combined system of bubbles of gas in water tends to act as a single-phase system—a sort of thickened drilling mud. (This is not actually true, the bubble system is not truly locked into position in the water, but the operational effect is much the same.) The system does not surge because the gas does not separate from the water and form slug flow.

Formation and reservoir protection is enhanced by the stability of the system so foam systems can be run with tighter down-hole pressure tolerances than gaseated systems.

4.3.2 Wellbore Pressure Reduction

The pressure at any point in the wellbore is the sum of the hydrostatic pressure and the friction pressure, plus a small increment of internal friction and acceleration value. The stability of foam makes it possible to reduce pressure in the hole and maintain it on a consistent basis. Some important notes:

- Minimum hydrostatic pressure is achieved with a greater gas content than gaseated systems. This can be done without inducing separation of the gas and water.
- Because foam systems have such a great lifting capacity, they can be used with low annular velocities which limit the friction pressure. Motor drilling requires higher flow volumes in the foam, and this will limit the extreme end of pressure reduction.
- Drilling rate is enhanced by reduced pressure against the formation. Foam is an excellent "power drilling" fluid.

4.3.3 Lifting Capacity

Foam systems have a much higher lifting capacity than any other drilling fluid. This appears to be the result of the foam structure where the bubbles are held in a flexible structure. Unlike other mud systems which depend upon annular velocity to overcome slip velocity, foam has a very low slip velocity. This is especially noticeable in milling operations where the steel shavings are brought to the surface at close to calculated lag time. Another good example is foam cleanouts of sand or frac sand, where foam at the flowline appears to carry close to

40% by volume of sand. Foam lifting capacity is particularly good for cleaning in a horizontal hole where it limits dune formation.

4.3.4 Lost Circulation

Foam resists lost circulation. In a zone of lost returns the pressure is lower than in the wellbore. When foam enters a lost zone, the bubbles expand as the pressure is reduced, and eventually plug the zone. Foam will stop quite massive losses, but there is a limit to the permeability or fracture size. In very large open fractures or vugular zones, the bubbles have no place to lodge and the system will only limit lost circulation by reducing the wellbore pressure (see Figure 4–4).

4.3.5 High Drilling Rate

There are two parts to drill rate, the instantaneous penetration rate, and the distance actually drilled or "made good" during a 24 hour period.

Instantaneous penetration rate is increased significantly by underbalanced conditions. In general, the greater the under balance the higher the drill rate. The consistency and lack of surging make foam "power drilling" a very efficient procedure. The cleaning ability of the foam under the bit represses the onset of bit flounder and so allows higher drill rates. Instantaneous drill rate is particularly high in horizontal holes. High drilling rate is much more pronounced with cone or conventional bits than it is with PDC or drag type bits.

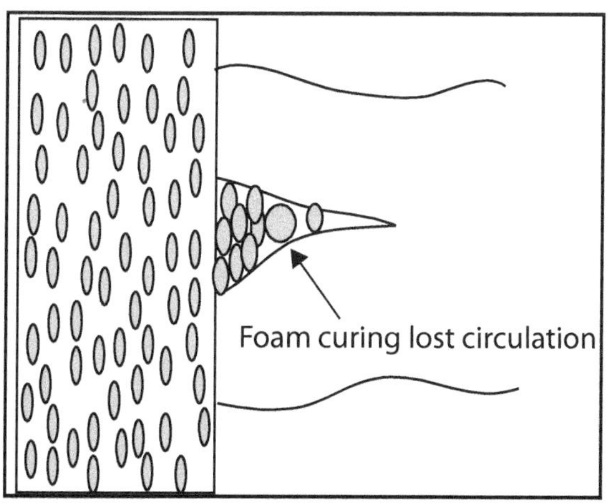

Figure 4–4 *Curing lost circulation zone with foam*

The second part of the drill rate (rate made good) depends more upon the cleaning ability of the drilling fluid. Circulating, washing and reaming are non-productive time (NPT). The high lifting and cleaning ability of foam reduces the time required to clean the hole. NPT reduction from circulating and washing the hole is especially apparent in foam drilled horizontal or high angle bore holes.

4.4 Challenges and Technical Limits with Foam Systems

4.4.1 Cost

Foam systems are primarily water systems with either air or nitrogen injected into the system. Mechanical equipment includes compressors, drillstring floats, a rotating control device, a separator, and a flare. The gas source (compressor, etc) is a major cost in this group. The separator system costs can be minor with a simple atmospheric separator or major with closed system separators. These mechanical costs are similar for all of the underbalanced systems. One of the major costs with a foam system is the foaming agents and other allied chemicals for stabilization and corrosion control. Costs vary with location and hole size, but chemical usage is higher with down-hole temperatures above 200°F (93°C), or impure water sources. It is not unusual to have chemical costs of 2,000 USD/day in an 8,000 ft. 6 in. (2,500 m, 152 mm) hole. Foam chemicals are semi-commodities so with large projects, the chemical costs can be significantly reduced with bulk purchase.

4.4.2 Hot Holes

With most foaming agents, operating costs start to get higher as the temperature increases above 200°F or 100°C. While bottom-hole temperatures can be considerably higher and the foam can tolerate higher temperatures while circulating, very hot wells are not normally satisfactory candidates for foam systems. This is particularly evident on trips where the foam at the bottom of the hole collapses and the bubbles tend to migrate in a manner similar to gaseated systems.

However, there has been at least one "high" temperature water-base foam developed that will operate at 275°F (135°C) and tolerated connections, but it has to be replaced after a trip (Weiss et al., 1997). The Oleofoam (oil continuous phase foam) discussed in Section 4.36, page 264, is proposed to be relatively stable to 400°F or 200°C.

4.4.3 Foam Breakdown from Other Conditions

Once foam is properly "emulsified" (sheared through the bit at a high pressure drop), it is quite stabile and will resist immediate breakdown in hot salt or acid water flows and oil flows. The foam will generally hold together while circulating under pressure. However, it tends to separate with time during trips. In the pit, contaminated foam will break quickly because light oil, hot mineral, or acid water are good defoamers. (heavy oils have a minimal effect on the destabilizing foam). More foaming agent concentration is required for recycling in case of oil or water influx.

Some formation waters, especially acid water below a pH of 5, and high solids (i.e., high ion content) water make it difficult to maintain strong and stable recyclable foam, primarily because of the continual build up of acid or ions when the foam water is re-circulated.

The simple solutions to these problems:

- Decrease the influx with additional surface pressure, or
- Increase bottom-hole pressure by increasing the water volume in the foam and making it denser,
- Add stabilizers to the foam, and
- Do not try to re-cycle the foam; instead, make it a single pass system.

4.5 One Pass Systems or Disposable Foam

The majority of foam systems are used with workover or cleanout operations. The foam is generally discarded at the flow-line because on a short operation with intermittent circulation, it is cheaper and more convenient to temporarily store the foam in a pit and dispose of the resulting water than it is to recycle the water and chemicals (see Figure 4–5).

Properly built foam for this type of operation can be very persistent, or can be developed to start breaking down at the flow line and return to 90% water within 15 minutes. An alcohol spray can be used if the foam does not break quickly. Water usage is 20 to 30 gal/min (75 to 115L/m) in a six inch hole (152 mm). Even with motor drilling, water quantities are small, about 100 gal/min (379lpm). Most of the foam agents are biodegradable and present in very small concentrations. Disposal of the foam water depends upon the local environmental rules.

Figure 4–5 *Disposable foam*

4.6 Recycle Foam

There are three types of defoaming methods for recycling. They can be roughly classified as mechanical, natural and chemical methods.

The mechanical method requires additional equipment and appears to have a slow defoaming rate. There have been a number of attempts to defoam mechanically with parallel plates and/or vacuum action in separators, but the systems have only been partly successful.

Natural defoaming type requires a longer time for defoaming and utilizes an open circulation system. The foam formula is optimized to match foam circulation time from the wellbore by controlling half-life. By doing this, proper hole cleaning can be provided and the foam will still break in an extended pit system. Recycling can be achieved by cleaning the water of drill cuttings and adding more foaming agents. The problem is with the buildup of fines over several circulations which degrade the foaming capability. This method was used in Iran for stiff foam, saving 1/3 of drilling fluids cost (Wan et al., 2010).

Chemical defoaming uses alcohol sprayed in at the blooie line (note below that Weatherford often utilizes another chemical method). The most common recyclable foams are systems where the foaming agents are carefully metered and the system is defoamed at the flow line by an alcohol spray. The alcohol evaporates in the separator and over the shale shaker to a constant base solution of water with a small percent of alcohol. The foam is then reconstituted with new foaming agents and recycled.

When a low carbon alcohol is sprayed on the foam it will integrate into the liquid film of the foam and reduce the surface tension of a part of the original stabilizers in the liquid film of the bubble. When re-blending the foam base liquid, the alcohol defoamer will increase the viscosity and will increase the foam liquid film strength; alcohol will lose its defoaming character and the foam fluid can be refoamed (Wan et al., 2010).

Other commercial foaming agents have been used, and it is possible that a better agent than alcohol can be obtained.

Weatherford has a unique recyclable foam (Transfoam) that is defoamed by changing the system from basic (high pH) to acid (low pH), Figure 4–6. Acid is added at the flow line, which causes the foam to immediately break. The makeup water is then run over the shaker and through the solids control equipment. It is then chemically treated with foaming agents and corrosion inhibitors as required, and returned to a basic system with caustic soda or a similar basic material, at which point it will again make a foam. The chemical makeup of the foam system uses three materials which will foam in a high pH environment, but interfere with each other in an acid environment causing the foam to immediately go "flat."

Metering of the foaming agents in all methods of recycle is important so that the system will return to a "standardized" system that fulfills the foaming and defoaming requirements, as well as limiting the cost of the system.

Recyclable foams are not necessarily less expensive than a throwaway system because of the additional cost of solids control equipment and separators. Chemical costs may be less in a recycle system, but in general the recycle costs are close to the "throw away" cost. The difference is in the disposal problem.

4.7 Basic Design of Foam Systems

4.7.1 The Method of Bottom-Hole Pressure Reduction

Foam systems reduce bottom-hole pressure by displacing some of the liquid in the hole with gas bubbles, so in effect, the hole is partly empty. The greatest actual liquid displacement results from the gas expansion as it rises in the upper part of the annulus from 3,000 ft

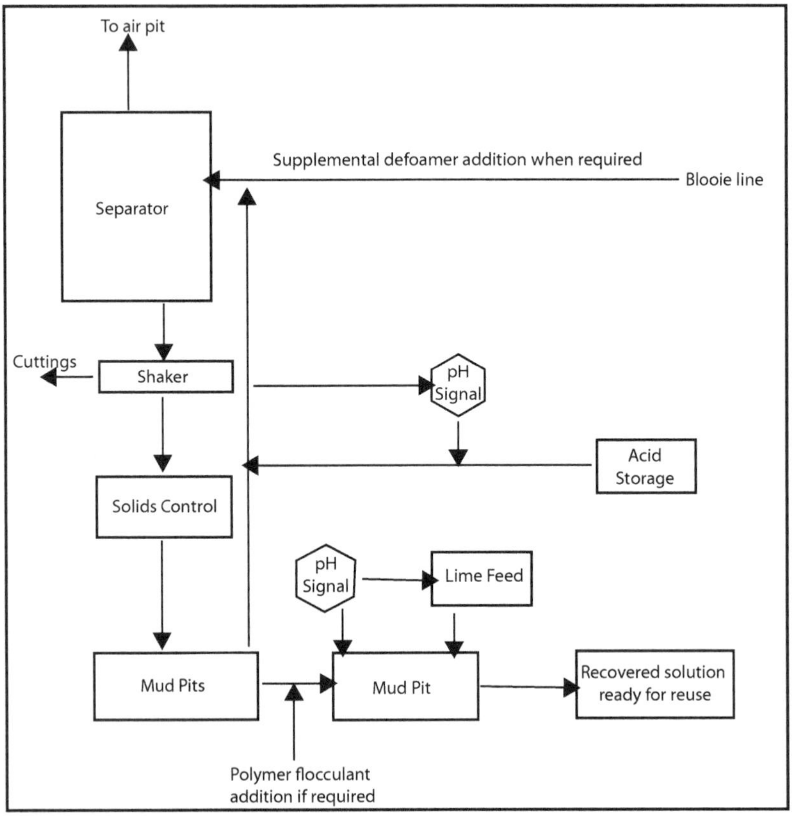

Figure 4–6 *Transfoam recyclable foam system*

(1,000 m) to the surface. The bottom-hole pressure reduction is not intuitive because of the non-linear effect of gas expansion and bore-hole friction.

Foam systems generally use between 5:1 to 500:1 ratio (gas at standard conditions: liquid), where liquid volume is defined as 1. The ratio has no operating relationship to quality, which is a measure of the actual percent gas in the system under pressure. A typical foam system in a rotary drilled vertical 6 in. (152 mm) hole uses 1,000 scfm (28 m^3) of gas with 25 gpm or 3.3 ft^3 /min (95 l/m) of liquid, or an injected gas to liquid ratio of 300/l. The nature of gas compression is that at the bottom of the annulus the gas is compressed to a 60 quality, or the original 300/1 injected ratio of gas to liquid is changed to where the gas is only 60% of the total volume (1.5:1). As the foam goes up the annulus and pressure is reduced, the gas expands to where it is finally close to 90% of the volume (9:1). It then reaches the ratio of 300/1 as it is released to atmospheric pressure, Figure 4–7.

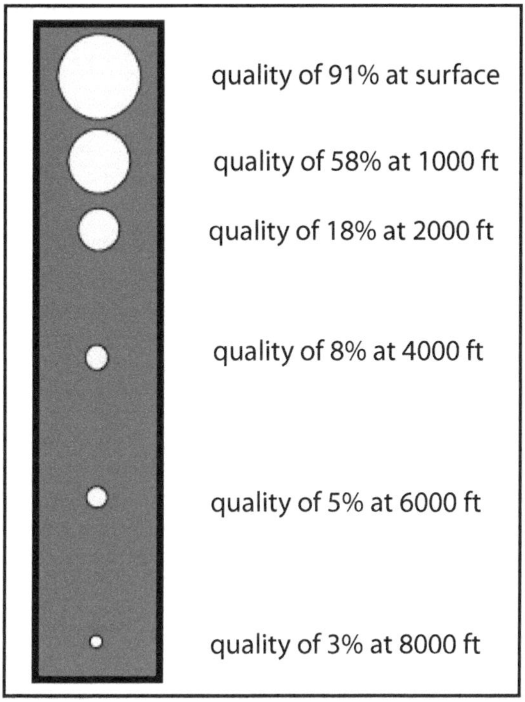

Figure 4–7 *Quality is the percent of gas by volume in an interval (Medley et al., 1998).*

The quickest and easiest way to control bottom-hole pressure on a temporary basis is to change the choke pressure. For small increases or decreases in pressure this is satisfactory. However:

- Higher surface pressure makes connections and trips more difficult and disproportionately increases the pressure at the base of the casing.
- Decreasing the surface pressure to below about 100 psi (680 kPa) may cause the foam to break into gas/water slugs in the upper annulus, so changing the liquid and gas volumes is a more satisfactory solution.

However, there is not a direct linear relationship between surface pressure increase and bottom-hole pressure increase. The bottom-hole pressure increase is greater than the surface pressure increase because of compression of gas in the annulus. The drillpipe pressure is not as reliable as in well control operations because there is gas in the drillpipe (as discussed in greater detail in Section 1.12, Well Control in

Underbalanced Drilling). Drill pipe pressure is a direct reading of bottom-hole pressure, but it is also a function of the gas volume that is (compressed) in the drillpipe.

On a practical basis, a surface pressure increase is used to limit a liquid flow or gas flow. Surface pressure would be decreased to limit lost circulation or find the onset of gas or liquid flows. This can be either a mathematical correction or a "wait and see."

The nature of foam makes it hard to quickly increase or decrease the bottom-hole pressure. When using a choke to temporarily increase the BHP, the foam system must be compressed. This depends upon the pump output and the hole volume. To decrease bottom-hole pressure the foam must be de-compressed by releasing surface pressure. It is not uncommon to take up to an hour to permanently compress or decompress a foam filled hole so as to stabilize a change in bottom-hole pressure. The change in bottom-hole pressure cannot be quickly measured by a change in the drillpipe pressure, as in a well control scenario, because of compression of gas in the drillpipe and the long lag time for compression/decompression. A model of lag time can simplify some of these uncertainties.

The more permanent solution to change bottom-hole pressure is to change the injected liquid volume or liquid ratio. This will take a full circulation, normally more than an hour. It is always best to first model the change since the solution may not be intuitive.

Increasing the gas ratio or decreasing the liquid volume decreases the bottom-hole pressure since the gas replaces some of the liquid in the hole. Increasing the gas volume may lead to the friction dominated regime where increasing the gas volume does not decrease bottom-hole pressure (although this is not normally a problem with foam systems). Decreasing the liquid volume may affect the drilling motor and directional capabilities, so motor throughput volumes need to be considered.

4.7.2 The Effect Fluid and Gas Volumes on Hole Cleaning and Motor Operation

Foam is characterized by larger, cleaner, and sharper bit cuttings than would be seen when drilling with water or a low viscosity mud. Foam, with its structured form, appears to displace cuttings from under the bit and carry them to the surface with increased efficiency.

Reduced bottom-hole pressure leads to increased drill rate, and this is very evident in foam systems. Because foam has a great cleaning and holding capacity, drilling rates have to be pushed very high to see any significant fill up on connections. Drill rate limits from bit flounder (recognized by the inability to drill faster with increased bit weight or rotary speed) and bit whirl are the common limits to drilling rate with foam systems.

Horizontal and high angle holes see a reduction in hole cleaning when sliding using a bent housing motor because it is difficult to clean the hole without pipe rotation. However, the nature of foam limits settling and significantly limits the formation of dunes on the bottom of the hole. This makes foam holes easier and quicker to clean than conventional drilled holes when rotation is started again.

The significant hole cleaning problems found with a single-phase or gaseated system that occurs in the 50° to 70° section of the hole are seldom a problem in foam systems because of the different flow pattern.

The best hole cleaning is done with foam with a quality of 50% to 90%. Within that range of quality there is little apparent effect on field operations. Above 90% quality there is the chance that the foam will change to a gas continuous phase and lose its lifting capacity. Below a quality of 50% foam appears to show a decreased lifting capacity. The lower limit of experience, with acceptable low foam qualities in vertical holes, seems to be in the range of 35 to 40 quality.

Foam systems generally operate on a surface ratio of 50/1 to 500/1 which yields a down-hole quality of 52–90%. The lower ratio is typical of straight slim holes and workovers where the high end of the ratio generally follows a motor in a horizontal hole. This seems like an inverse statement but motor operation needs the higher ratios because of extra liquid in the system required to run the motor needs to be balanced with more gas.

Bottom-hole pressure is one of the controlling factors in the relative amount of gas and water to be used. Gas compression is the reason for the large range of ratios. As more water is added to the system, a disproportionate amount of gas is needed to maintain a low bottom-hole pressure. As more water is added, more foaming agent is needed to keep the system from becoming too wet.

In the **hydrostatic** regime, more gas reduces bottom-hole pressure. Almost all foam systems will operate in the hydrostatic regime. The chapter on gaseated fluid (Chapter 3) describes how the **friction dominated** regime controls the upper limit of gas injection in gaseated systems. Increasing velocity with addition of more expanding gas finally causes the friction loss in the upper part of the annulus to overcome any reduction in bottom-hole pressure. The friction dominated regime is basically controlled by controlling the liquid volume in the system. Foam is not as challenged by the friction dominated regime (as are gaseated systems) because foam systems do not need as high an annular velocity to clean the hole and use less liquid and more gas in the system. Friction domination can occur when using a wet foam with motor drilling in a slim hole.

Water volume with a foam/rotary system is much lower than required for foam/motor operations. With drilling motors, the liquid volume is increased to the minimum required for high torque motor operations. The gas volume is increased to keep the desirable ratio or quality for bottom-hole pressure. The sum of the compressed gas and the liquid makes up the motor operating volume required for maximum torque.

Example 4–1 Motor Calculation

Motor Requirement for a certain 4 ¾ in. (120 mm) motor, from the motor table:

175 gpm (660 lpm) liquid equivalent flow for desired torque and speed

Assume 100 psi (689 kPa) pressure drop at the motor

Hole

6 ½ in. (165 mm) hole

Depth 10,000 ft (3,048 m)

Bottom-hole Pressure, 2,680 psi, (18,477 kPa) 5.1 ppg (611kG/m^2) equivalent

Surface Back Pressure, 100 psi (689 kPa)

Desired foam quality at surface, 90%

Foam Requirement

135 gpm of liquid (512 l/m)

810 scfm of air or nitrogen (23 m^3/minute)

This will give a total equivalent volume of 175 gpm at the bit

Points to be made about this example:

1. There will be variations in the answer due to different models, flow assumptions, and BHA assumptions.

2. Pressure drop across the motor will range from 100 psi to 800 psi during drilling.

3. The answer is not intuitive!

There appears to be some slippage with foam (and gaseated) systems that limits maximum motor torque on the test bed. In field practice it is hard to determine if this is occurring and if it has any effect on the drilling.

4.7.3 The Limits to Water and Gas Injection Volumes

4.7.3.1 The Limits to Water Injection Volume

Foam is normally controlled to stay within a quality of 52% to 90%. Within those limits the system has superb lifting capacity. A lower limit of water injection in 4 ¾ in. to 6 ½ in. holes is in the range of 20 to 30 gpm. This range is generally proportional with annular area. (see Figure 4–8). This liquid volume with the gas required to stay in the quality range will lift steel milling cuttings or up to 40% by volume frac sand.

The upper limit of water injection volumes is controlled by the bottom-hole pressure requirement and motor.

4.7.3.2 The Limit of Gas Injected Volume

Maximum gas volume injection is limited by the ability to keep foam flow in the upper part of the hole. Too much gas by volume (above about 90%) will change the system to a gas continuous phase and force mist, annular, or slug flow. There is a limit to the annular surface pressure that can be used to compress the upper gas based on the required bottom-hole pressure.

There have been cases where gas flows have been encountered where the gas/water ratio at the surface has exceeded the suggested limit of 500/1 with no apparent ill effects, but a higher impressed surface pressure was used. In the case of a large gas flow that is being sent to the sale line, much or all of the injected gas can be stopped.

A lower general limit to gas injection volume is the desire to keep the ratio or percent of gas in the bottom-hole no lower than 50%. However, quality levels of 40% in vertical holes have been recorded with no apparent problem with lifting capacity.

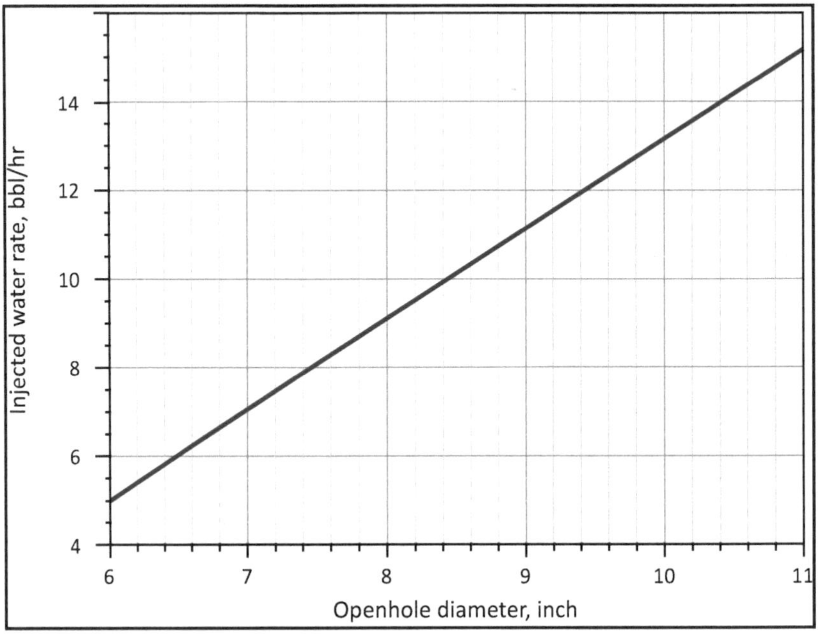

Figure 4–8 *Required water volume versus hole size*

4.7.4 Stabilizing the Foam System

In operations, a surface pressure of 50–150 psi (350–1,000 kpa) normally is enough to reduce the near-surface quality of the foam to prevent revision to an air continuous system and the accompanying surging. While the rotating head (RCD) causes a small back pressure, this is normally not enough and a foam well generally needs to be choked. In workovers, the combination of the rotating head or snubbing packer and small diameter flow-line can often produce adequate back pressure.

Trips require a chemical modification of the foam system if the foam column is to be kept full. More of the primary foaming agent does not increase the life of foam once a good foam system has been developed. Foam enhancers will increase the life of foam so that the column will stand with minimum shrinkage during a trip. There is a later discussion in this chapter about the use of chemical agents.

Foam breaks when the skin around the gas bubble becomes too thin. This occurs when the gas bubble gets too large or when the skin thins by gravity drainage. Foam stability depends on the size of the gas bubble and the tenacity of the film around the bubble. Figure 4–9

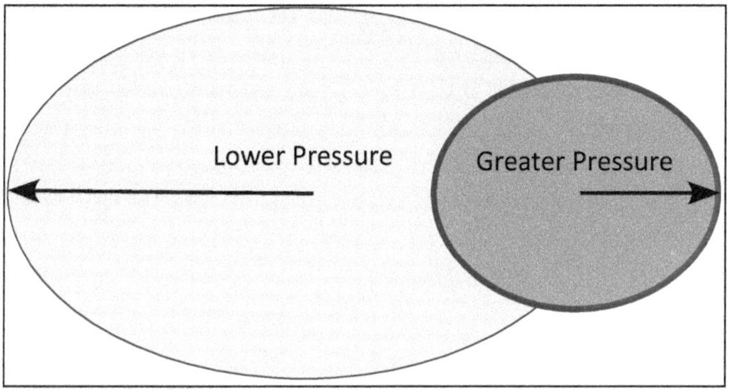

Figure 4–9 *Pressure redistribution of foam bubbles (Roberson, 1992)*

shows pressure redistribution of foam bubbles and how the skin becomes thinner with increased bubble size.

Defoaming occurs naturally at the surface or in the pits with time due to gravity thinning of the skin. Defoaming can be accelerated by the use of an alcohol spray. Much of the alcohol evaporates over the shaker and leaves only a small residue in the water. There are other materials that can accelerate defoaming such as aluminum stearate, fine silica dust, or light oils. When considering the end use of the water and other environmental and safety considerations, as well as cost, an alcohol spray is generally the best defoamer.

4.7.5 Lost Circulation and Foam

Foam is one of the best lost circulation materials. To cause lost circulation, the pressure in the wellbore has to be higher than the pressure in the fracture or lost zone. As foam flows into the fracture, the pressure decreases and the bubbles in the foam increase in size until they block the channel. The time element where the loss is blocked depends upon the stability of the foam. However, foam under pressure is much more stable than foam in the pits.

In a related system, Afrox, lost circulation material is made by inducing very small bubbles with an emulsifying agent in the mud pits and pumping the controlled bubbles into the lost zone. The bubble volume is so small that there is only a minimal effect on the pumps. The bubbles expand in the lower pressured lost zone and plug it.

4.7.5.1 Foam in Karst Topography or with Extreme Lost Circulation

Stiffened foam with CMC or bentonite has worked in Southeast Asia in the Karst topography by reducing the loss of fluid and allowing partial returns. This approaches some of the common practices of drilling with floating mud caps or drilling with partial returns. The stiffened foam has a long half-life, often approaching days, and provides flow resistance into fractures or small caverns. The constant circulating sub "Non-Stop Driller" has helped the process by avoiding shutting the foaming down during connections. With extreme lost circulation, the force required to start circulating may be enough to cause total losses (Zwager, 2011).

4.7.6 Controlling the Half-Life and Lifting Capacity of Foam

Foam drilling suffers from lack of field measurement of foam properties. Foam is sensitive to pressure so any measurement of properties needs to take place under pressure, or at least be interpolated to a pressure condition. All of the field drilling mud viscosity measurement systems are irrelevant to foam properties.

The standard foam measurement is "half-life" in Figure 4–10. The half-life measures the persistence (texture) of the foam under atmospheric pressure. Half-life is a reasonable screening process for the efficiency of a foaming agent with different water samples; it also gives some idea of the difficulty of breaking foam in the pits. It says nothing about the lifting capacity or ECD of the foam in the hole, or how persistent the foam is under pressure during a connection or trip. However, some work with workover rigs seems to indicate that a foam half-life of greater than 12 minutes indicates that the foam will be persistent during a trip (Evans, 2002).

The "half-life" test uses a blender (not a mixer) to make foam from 100 ml of water and whatever foam agents are to be added (generally on the order of ½–1%). The blender is run for 30 seconds, which makes foam approaching a 90 quality foam (90% air). The foam is immediately poured into a 1,000 ml graduate. The half-life is the time it takes for 50 ml of water to accumulate in the bottom of the graduate. The height of the foam column is also sometimes a help when analyzing the foam.

Foam persistence (texture) as measured by the half-life test, can only marginally be increased with more primary foaming agent. With a foam that is persistent to the flowline while drilling, the addition of more foaming agent appears to have little practical effect in the field except for making the foam fluffier in the pit. However, an increase in foamer seems to thicken the skin on the foam bubble and decrease oxygen corrosion or corrosion from gasses entering the system.

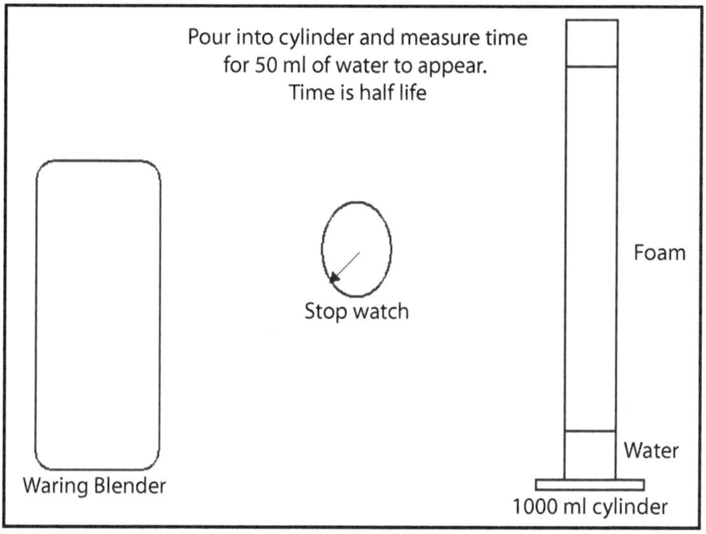

Figure 4–10 *Standard half-life test*

The addition of a polymer or foam stiffener will:

- Increase foam persistence when under pressure during a connection or trip
- May increase the lifting capacity
- May increase the annular pressure loss

Of the preceding effects, the most overlooked is the first. In considering procedures during a connection or trip, the most important item is that the flow-line must be closed and the foam kept under about 10 atmospheres of pressure to keep the quality of the foam low enough, about 85%, in order to keep the liquid at the continuous state.

Foam breaks when the skin around the gas bubble becomes too thin. This occurs when the gas bubble gets too large (which is why the foam is kept under pressure), or when the polymers in the foam skin thin by gravity drainage. Degradation of the film happens when the temperature is too high. Defoamers upset or destroy the film.

Polymers or foam stiffeners should be used with a light hand to avoid excessive annular pressure loss (APL), and to avoid too much persistence at the surface.

Polymer addition to the foam water can be pilot tested by adding the polymer to the water in 0.125 lb/bbl (0.35 kg/m^3) elements and measuring the apparent viscosity and yield point as done with a normal drilling mud, or with the XC polymer, measuring the very low shear rate viscosity. The treated water and foaming agents should then be tested using the half-life test. The test is imprecise, but it will give some indication of where to start with polymer treatment to obtain the stability required for drilling or trips. The final proof is a test in the hole.

Make-up water should always be tested with a half-life test. To test the effect of down-hole contamination, first build the foam and add the contaminate using the lowest blender speed, then test for half-life.

4.8 Water, Gas, and Chemical Agents

4.8.1 The Liquid Foam Base

4.8.1.1 Fresh Water as a Foam Base

Almost all foam systems use water as the continuous phase of the system. The best foam systems use 'drinking quality" water. Any increase in salts or materials in solution in the make-up water will increase the cost of the foam and decrease the foam stability. Contamination from a water hauling truck can require the addition of more foam agents. In areas where there is significant corrosion in an oil field, it is almost imperative to start with a clean water base with minimal ionic (solids) or bacterial content to control corrosion.

4.8.1.2 Brackish Water as a Foam Base

In many desert areas, brackish water is available as a foam base. There are specific foam agents that can make a foam with a brackish water base. All possible hardness (Na^+, Ca^{++}) of the water needs to be treated out with caustic or soda ash. However, be careful about using excessive soda ash ($NaCO_3$) as it will build up excessive bicarbonates (HCO_3^-) that seem to lead to increased corrosion The brackish water foaming agents will deal with chloride content, but excessive anions increase the treating cost.

4.8.1.3 Oil as a Foam Base

See Section 4.27, page 245.

4.9 Foaming Agents and Foam Extenders

Foaming agents are considered proprietary materials by the vendors, and it is difficult to ascertain the materials and concentrations. In general, most commercial foaming agents are mixtures of various chemicals; for example, foaming agents used in cold weather areas must have antifreeze added. There are no standards for foaming agents. The only test is the "half-life" test which, while it tells little about the action of the foaming agent under pressure, is a good screening method for the various foamer concentrations and the effect of the available water. Table 4–1 and Table 4–2 show common foaming agents and common foam extenders.

Idealized foam structure for different qualities is shown in Figure 4–11. Optimizing the foam fluid additives can result in a more stable foam as shown in Figure 4–12 (Hutchins and Miller, 2005).

It is left to the drilling engineer or foam drilling supervisor to figure the persistence (the texture) and lifting capacity of the foam. While this sounds like a very marginal operation, in actual practice these items can be observed and modified quickly in field operations.

Table 4–1 Common Foaming Agents[a]

COMMON PRIMARY FOAMERS	
Ethoxyl Alcohol Ether Sulfates (Anionic)	High Cost
Widespread Applicability	
Excellent solubility	
Good thermal stability	
Comment: Foam efficiency drops off rapidly when make-up water salinity goes above 40,000Cl^-. Oil tolerance can be enhanced by adding diphenyl oxide disulfonates. Addition of more primary foamer past about a "10" minute half-life will not increase foam pesistence (texture).	
Alpha Olephin Sulfonates (Anionic)	Poor brine solubility
Relatively low cost	
Excellent thermal stability	
Outstanding hydrocarbon tolerance	
Comment: Good for fresh water applications. Make sure you have drinking water quality. More AOS beyond a good foam does not improve foam strength.	
COMMON FOAM ADDITIVES	
Amino-Propyl Betaines (Amphoteric)	Expensive
Excellent stability	Best use is as a foam booster
Good thermal stability	
Good brine solubility	
Fair yield	

Table 4–1 Common Foaming Agents[a] (cont'd)

Alkanol Amides (Cationic)	Incompatible with most Anionic foamers
Good thermal stability	A foam booster
Comment: Be careful with this material. Be sure to pilot test.	
Sodium Sulfosuccinates (Anionic)	Poor brine solubility
Common usage	A foam stabilizer
Good freshwater solubility	
Good thermal solubility	
Alkyl-Phenol Ethoxylates (Non-ionic)	Poor foam quality
Good surfactant	A foam stabilizer
Soluble in all water	
Good thermal stability	
Compatible with other additives	
Comment: Not much of a foamer, usually used as a water wetting agent.	

[a] The table of common foaming agents is just that—common foaming agents. There are other proprietary agents that are used by the various service companies that are not included in the table.

Table 4–2 Common Foam Extenders or Stiffeners[a]

X-C Polymer (Slightly Anionic)	May need a biocide in some areas
Organic polymer, a sugar that degrades to an alcohol	Expensive
Thixiotropic, stiffness reduces with velocity	Hard to mix
Extends and stabilizes the foam with minimal increase in ECD	Should test with Low Shear Rate values
Comment: One of the best of the foam extenders. Can make a foam by itself.	
HEC (Anionic)	Not thixiotropic
"Organic Polymer" degrades with time	Increases ECD
Good foam stiffener	
Inexpensive	
Comment: Common and easy to use.	

Table 4–2 Common Foam Extenders or Stiffeners[a] (cont'd)

CMC-Carbymethylcellulose (Anionic)	Inexpensive
Stiffens the foam	Tends to make the foam too persistent
Comment: Be careful not to overdo the material.	
Bentonite (Anionic)	Too persistent
Makes the foam very stiff and persistent	
Cheap	
Comment: Works well with a throw away foam for a large surface hole. Used in conjunction with CMC. Foam in pits is very persistent and will last for weeks.	

[a] Pilot test before using, too much can make the foam unmanageable at the surface.

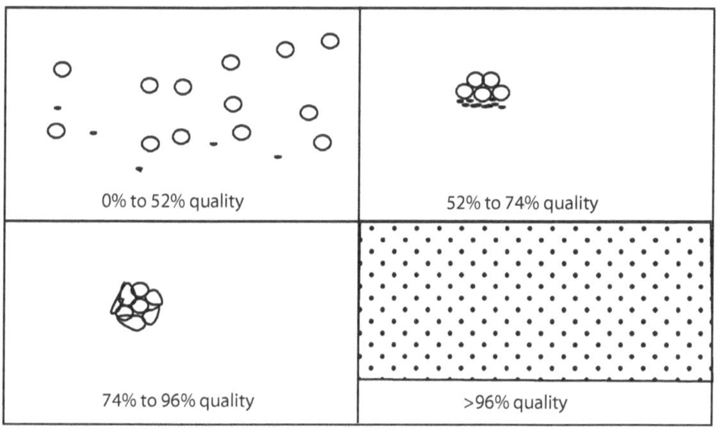

Figure 4–11 *Different foam structure for different qualities (Hutchins and Miller, 2005)*

4.10 Trips and Connections

The key is the persistence of the foam column. The foam should be stiff enough to remain whole during a connection when a 2–3 atm pressure is kept on the annulus.

After a connection and as soon as the standpipe is up to drilling pressure and the flowline is opened, there should be foam returns with minimal heading (see Figure 4–13) and no more than two minutes until there is a steady state flow (see Figure 4–14).

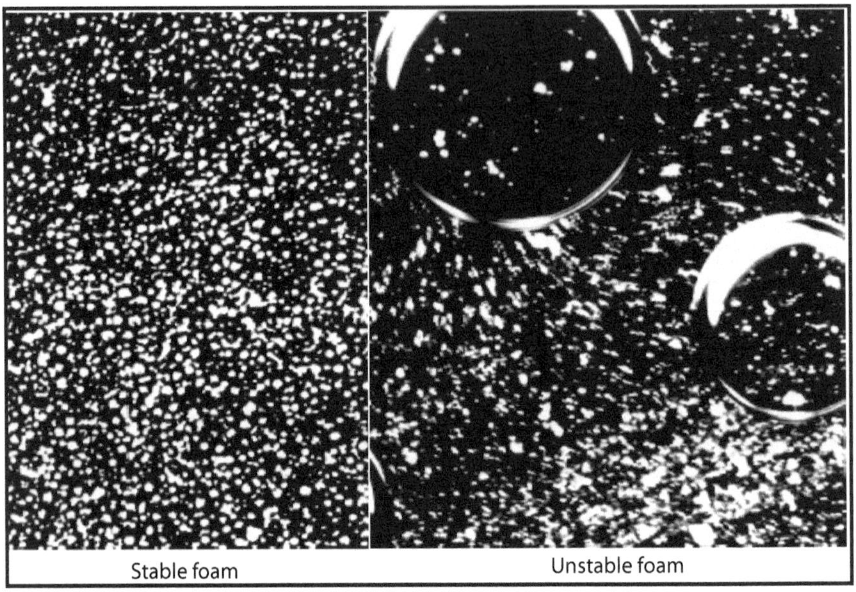

Figure 4–12 *Effect of foam fluid additives on the stability of the foams (Richard and Matthew, 2005)*

Figure 4–13 *Foam heading*

Figure 4–14 *Stable foam*

Maintaining a column of foam during a trip may not be practical if it requires stiffening the foam with an additive. This will have to be added and pumped for a full circulation. The critical point is how much pressure surge or lost returns on a trip is acceptable. To minimize the pressure surge on a trip requires a full column of foam at all times. A full column of foam requires stripping all the way out of the hole and back in while maintaining up to 10 atm of annular pressure. If the foam column drops, it may be possible to foam up on the way in the hole. However, once the foam breaks, it forms a slug of water that represses air expansion and creates a pressure increase in the hole until it is circulated out. It is more difficult to make persistent foams with a bottom-hole temperature above 212°F (100°C) because of partial breakdown in foaming agents over a long period of time during a trip.

The more persistent foam (the stronger the texture) has a higher annular pressure loss. The increase in APL depends on the foam additives or polymer used as a stiffener. At present, the only good way of measuring this is with a bottom-hole pressure gage on an MWD.

Table 4–3 Making a Connection

Making a Connection with a Foam System

1. Circulate and pass at least one tool joint
2. Pipe can be set on the slips or at item #6
3. Turn off the liquid feed pump (mud pump), and injection pumps (if used)
4. Shut in or choke the flow line but maintain an annular pressure of 2 or 3 atmospheres
5. Blow the drillpipe dry to below the string float—depending upon the location of the string float, this will be when the standpipe pressure rises 100 psi (700 kPa)
6. Bypass the compressor
7. Blow down the standpipe and drillpipe gas pressure through the standpipe manifold
8. Make the connection
9. Put the liquid, and gas back on line
10. Do not open the flow line until the pressure builds to near normal circulating pressure
11. Drilling can be started after the pressure starts to build. Caution! This is a field derived or drilling policy point. In slow drilling, there is little danger of cuttings so build up and drilling can start as soon as the pressure starts to rise. In very fast drilling such as in coal bed methane holes, wait until there is full circulation before drilling
12. The survey point depends upon the type of survey instruments in use

Table 4–4 Trips

The purpose of the tripping procedure is to try to keep the bottom-hole pressure constant. The objections to this procedure are:

1. It requires stripping the pipe. Stripping creates wear on the rotating head
2. If the pipe is slugged with water, it creates some extra BHP
3. The annular pressure is released when the BHA or HWDP reaches the BOP. Release the surface pressure when the BHA arrives at the BOP and then when going back in the hole, foam up with pressure at the bottom of the casing or at about 3,000 ft. (1,000 m)

Making a Trip with a Foam System

Policies or practices may change this procedure, which is for the purpose of a guide in this manual:

1. Circulate the hole clean and pass tool joints
2. Pipe can be set on the slips or at item #6
3. Bypass the compressors
4. Displace 50% of the drillpipe with water (to pull dry pipe)

Table 4–4 Trips (cont'd)

Making a Trip with a Foam System (cont'd)
5. Turn off the liquid feed pump (mud pump) and injection pumps, if used; and
6. Shut in the flow line but maintain the normal surface pressure of 2 or 3 atmospheres
7. Release any drillpipe gas pressure through the standpipe manifold (as a safety measure)
8. Strip pipe Do not fill the hole Watch for any increase in annular pressure
9. The annular pressure will normally be released and the hole opened when the heavy weight pipe or the BHA arrives at the surface. Wait a few minutes at this point to be sure there is no flow from the hole
10. Going back in the hole, stop and foam up the hole until there are full foam returns at 3,000 ft; (1,000 m, the kick off point for a horizontal hole, or/and at the end of the casing
11. Strip back to bottom
12. Circulate a full column of foam

4.11 Questions

1. List three advantages to a foam system over other fluid and air systems.

2. Gaseated mud and air mist systems are unstable because the gas and air separated very easily. Why is foam, which contains both fluid and gas, a stabile system?

3. A foam system may start to become unstable above a quality of 90. Why?

4. List the steps to making a connection and to start drilling again.

5. What would be minimum nitrogen and water volumes required to rotary drill (no motor) with a foam in an 8 ½ in. hole 8,000 ft. deep (2,400 m) while maintaining a bottom-hole pressure equal to a 0.28 psi/ft gradient?

6. If the influx rate is zero and injection GLR is 7 scfm per gpm, will you have stable foam without back pressure? If not, how much is the required back pressure?

7. Given the following table, plot the ECD for depth ranging from 1,000 to 10,000 ft.

Liquid Injection Rate (Ql):	300	gpm
Total Depth (H):	10000	ft
Annulus OD:	7.875	in
Drill String OD:	4.50	In
Inclination Angle:	45.00	Deg
Surface Temperature:	520	R
Mud Weight (W_m):	8.4	ppg
Gas Specific Gravity (S_g):	1	air=1
Formation Fluid Specific Gravity (S_f):	1	water=1
Geothermal Gradient (G):	0.01	°F/ft
Formation Fluid Influx Rate (Q_f):	20	bbl/hr
Inj GLR:	6.5	scf/gal
Backpressure (Ps):	14.7	psia
Gas Injection Rate (Qo):	1950	scfm
Liquid Weight:	62.90	lb/ft3
Cross Sectional Area (A):	32.80265	in^2
Hydrolic Diameter (dH):	3.375	in
Foam Quality at Surface :	0.978926	

8. In an 8.75 in. hole, 400 gpm is a normal mud rate. What should be your initial rate for foam drilling?

4.12 Answers

1. Listed in the text are five major advantages to foam systems plus some other comments:

   Foam systems display little in the way of pressure surging, with minimal overpressure damage to reservoirs and formations.

   Bottom-hole pressure can be reduced to below that of gaseated fluids.

   The system has a greater lifting capacity than any other drilling fluid.

   It reduces or stops lost circulation.

   It permits very high drilling rates because of foam ability to clean under the bit and clean the annulus.

From these advantages come better reservoir protection (from surges and pressures), no differential sticking, higher drilling rates, and much better hole cleaning.

2. Foam, which contains both fluid and gas is stable because the gas is emulsified in the fluid. Each gas bubble is surrounded by a chemical skin.

3. A foam system starts to become unstable above a quality of 85 because the gas has expanded so much that the skin around the bubble breaks and the gas becomes the continuous phase while the water becomes the discontinuous phase.

4. The steps to making a connection and then commence drilling again are:

    1. Circulate and pass a tool joint
    2. Turn off the mud pump and foam pump
    3. Shut in or choke the flow line maintaining an annular pressure of 2 or 3 atm
    4. Blow the pipe dry to below the upper string float
    5. Bypass the compressors
    6. Blow down the standpipe drillpipe
    7. Make the connection
    8. Put the foam back into the drillpipe
    9. Do not open the flow line until the pressure builds towards drilling pressure
    10. Start to drill

5. The minimum nitrogen and water volumes required to rotary drill (no motor) with a foam in an 8 ½ in. hole 8,000 ft deep (2,400 m) to obtain 2,200 psi (15,000 kPa) on bottom would be about 50 gpm (190 L) of liquid and 2,500 scf/m (70.8 m^3) gas.

6. No, because GLR max is 4.32 scfm per gpm. The required back pressure is 24 psia.

7.

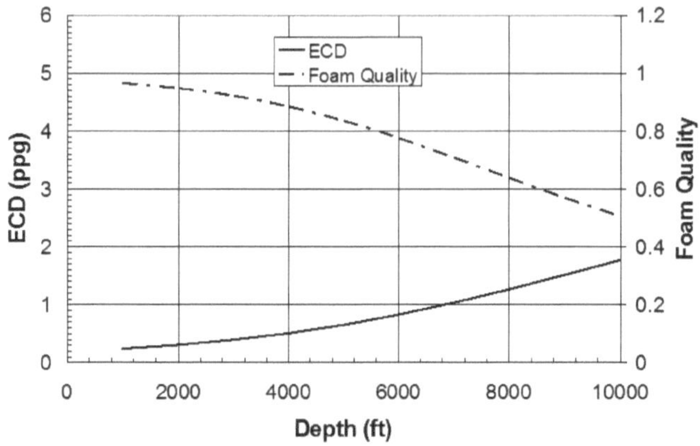

8. Begin with 1/10 of conventional mud rate required to give a 120 ft/min (40 m/min) annular velocity. Start with 40 gpm as an initial try.

4.13 References

Able, L.W., Bowden, J.R., Campbell, P. J. *Firefighting and Blowout Control,* Wild Well Control, Inc., USA, 1994.

Amoco Production Company, *Drilling Fluids Manual,* Amoco Corporation, 1994.

API, *Underbalance Drilling Operations,* API Recommended Practice 92U, First Edition, API, Washington, DC, USA, 2008.

Beyer, A.H., Millhone, R.S., and Foote, R.W. "Flow Behavior of Foam as a Well Circulating Fluid," SPE 3986 presented at the SPE Annual Fall Meeting, San Antonio, TX, USA, October 8–11, 1972.

Brantly, J.E. *History of Oil Well Drilling,* Gulf Publishing Company, Houston, TX, USA, 1971.

Chafin, M., Medley, G., Rehm, W. *Underbalanced Drilling and Completion Manual,* Maurer Engineering for the DEA 101 Project, 1998.

Clearwater, Inc. "Underbalanced Drilling Fluids, (Air, Mist, Foam and Mud)," Product Information, Pittsburgh, PA, 1996.

Evans, T., Protreat Technologies, Denver, CO, USA, Personal conversations from 2002–2005.

Gajbhiye, R.N., Kan, S.J. "Characterization of Foam Flow-in Horizontal Pipes by Using Two-Flow Regime Concept," *Chemical Engineering Science,* 66, No. 8, 2011, pp. 1536–1549.

Guo, B., Miska, S. and Hareland, G. "A Simple Approach to Determination of Bottom-hole Pressure in Directional Foam Drilling," proceeding of the ASME-ETCE Conference, Houston, TX, USA, January 25–February 1, 1995.

Guo, B., Sun, K., Ghalambor, A. "A Closed Form Hydraulics Equation for Predicting Bottom-hole Pressure in UBD with Foam," SPE 81640 presented at the IADC/SPE Underbalanced Technology Conference and Exhibition, Houston, TX, USA, March 25-26, 2003.

Hall, D.L. and Roberts, R.D. "Offshore Drilling with Preformed Stable Foam," SPE 12794 presented at the SPE California Regional Meeting, Long Beach, CA., USA, April 11–13, 1984.

Hutchins, R. and Miller, M.J. "A Circulating-Foam Loop for Evaluating Foam at Conditions of Use," *SPE Production & Facilities*, 20, No. 4, 2005, pp. 286–294.

Krug, J. and Mitchell, B.J. "Charts Help Find Volume, Pressure Needed For Foam Drilling," *Oil and Gas Journal*, February 1972, pp. 61–64.

Li, Y. and Kuru, E. "Optimization of Hole cleaning in Vertical Wells Using Foam," *Energy Sources, Part A: Recovery, Utilization and Environmental Effects*, 31, No. 1, 2009, pp. 1–16.

Lyons, W.C., Gao, B., and Seidel, F.A. *Air and Gas Drilling Manual,* Second Edition, McGraw Hill, New York, NY, USA, 2001.

McLennan, J., Carden, R., Curry, D., Stone, C.R., and Wyman, R. *Underbalanced Drilling Manual,* GRI Ref No. 97/0236, Gas Research Institute, Chicago, IL, USA, 1997.

Medley, G.H., Stone, R.C., Colbert, W.J., and McGowen III, H.E. *Underbalanced Operations Manual*, Signa Engineering Corp., Houston, TX, USA, 1998.

Nas, Steve. "Introduction to Underbalanced Drilling," Weatherford Private Publication Ref: APR-WUBS-WFT-001, 2006.

Rehm, W., Schubert, J., Haghshenas, A., Paknejad, A. and Hughes, J. *Managed Pressure Drilling,* Gulf Publishing Company, Houston, TX, USA, 2008.

Rehm, W.A. *Practical Underbalanced Drilling and Workover*, Petroleum Extension Service, University of Texas, Austin, TX, USA, 2002.

Robertson, L. *Reduced Pressure Drilling Systems*, Bachman Drilling and Production Specialties Inc., Oklahoma City, OK, USA, 1992.

Wan, L., Meng, Y., Li, Y., Wang, J., Shu, X., and Zeng, Q. "The Study of the Circulation of Drilling Foam," SPE 131068 presented at the SPE International Oil & Gas Conference and Exhibition, Beijing, China, June 8–10, 2010.

Zwager, D., Personal Communication, 2011.

Section 2
Field Foam Properties

Reuben Graham, Weatherford International

4.14 Introduction

Foam drilling has a lot of "ifs" and "buts." and there are differences of opinion about some of the methods of dealing with foam systems. Foam stability, half life, and quality are areas where different experience has lead to different opinions. The properties of a foam will be quite different if it is being used for milling on a workover, as a drilling fluid in a surface hole or in a deep hot well, or as a solution to severe lost circulation as in an area of karst topography.

In order to keep a free flowing discussion in about foam systems, Reuban Graham was invited to discuss some of the gray areas where opinions, data, models, and experience are often at variance.

4.15 Quality (GVF) and Foam Stability

One of the most confusing terms in foam drilling is "quality." Quality has nothing to do with how good the foam is but rather is simply a term applied to gas volume fraction (GVF), which is the actual percent of gas in the mixture at a certain point in the annulus. For many years, there has been a concept in the foam engineering/modeling circles that the dividing line between foam and mist was at a GVF of 96%. To those that have seen foam pile up on location several feet high, that concept seems ridiculous. However, to those that have observed a blooie line discharging a rain-like stream it seems to be accurate. We will examine some factors not commonly taken into consideration.

The manner in which the foam forms and the subsequent shear it undergoes determines the viscosity and strength of the foam. The mixing of water, foaming agent, and air together does not necessarily result in foam unless the mixing action provides sufficient shear to generate a good bubble structure. The guiding rule that is the smaller the bubbles, the stronger the foam film will hold together. Some experts (Gao, 2003) have referred to this as "texture."

However, the amount of shear experienced in the wellbore (bit nozzles etc.) will have a big effect on the texture or strength of the foam, yet not change the quality (GVF). Once the small bubbles form through adequate shearing and mixing with an adequate foaming agent, the foam is in a state of continuity or stability, so it will not immediately "break" at a quality of 96% or even 99%.

In time, the foam "breaks" as the liquid will drain out of the system. However, this should not happen until the foam is out of the well or ceases to be in a turbulent mixture.

There are two general approaches to foam stability (persistence, foam strength, or texture).

1. In the main body of Chapter 4, the general approach is to make a foam that breaks easily at the pit. The down side of a less robust foam is that it is necessary to keep the foam pressurized to allow the bubbles to remain small which keeps the foam stabile.

2. The general approach in this section is that the foam is more robust, which eliminates the requirement for keeping the foam under pressure. The down side is that more effort is required to break and recycle the foam.

Measurement of foam strength in advance of well operations is an imprecise art. The best practical measurement available is the "half-life" test (see Figure 4–15). The foam, when tested, should have a half-life between 7 and 15 minutes depending upon the experience of the operator and the connection requirements. The longer half-life tends to make the ECD higher but pressure surges smaller. This is one of the gray areas with foam because there is no reliable field measurement.

4.16 Carrying Capacity and Settling

There is very little settling of cuttings from a well constituted foam system. This is particularly evident when foam is used as a milling fluid or as a cleanout fluid after a fracture operation. When milling, the heavy steel cuttings come to surface very close to lag time.

In cleanout operations, up to 30% to 50% by volume frac sand (proppant) is carried to the surface with a very low settling rate. In both of these extreme conditions, there is little to no fill-up. The viscosity curve, shown Figure 4–2 with lifting capacity of foam is in the main body of Chapter 4.

In horizontal and high angle holes, there is duneing if the pipe is not rotated, but the dunning effect is smaller than with single-phase fluid or two-phase fluids as in gaseated drilling fluids.

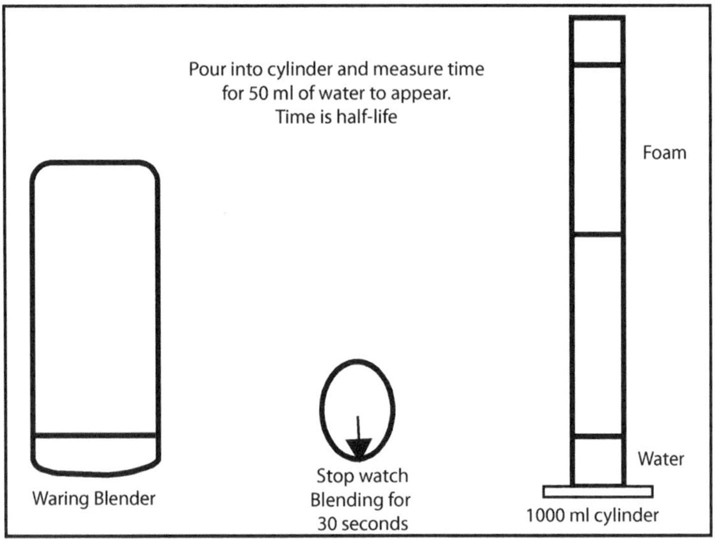

Figure 4–15 *The half-life test*

4.17 Water and Oil Dilution

In a properly constituted and sheared foam, large volumes of water or oil can be carried to the surface without upsetting (contaminating) the foam. An easy way to tell good carrying capacity is at the blooie line; if the flow is steady without pressure spikes and heavy surging, then the foam is adequate to remove the water. If the reverse is true then the foam needs to be made more robust or the pressure needs to be increased to limit the water flow. It is common for a well-designed foam to "puff" or have minor surging at the blooie line.

4.18 Modeling and Field Results

Critical to modeling is foam stability. Models must first make the assumption that the foam is properly constituted and sheared since there is no method of measuring the "texture." Stating that foam "breaks" at 96 or 99% GVF gives a nice point that can be used to simplify modeling and explain how foam works. The fact that it just is not true can be a little troublesome because it introduces some areas of uncertainty. Since the extent or strength of the restraining force of the film around the gas bubble is unknown, it is very difficult to consider it in model calculations. Hence there is a "gray" area, requiring a handling of the mathematics in a different manner. The gray area is where many factors influence the stability of the systems and whether it acts as a single-phase or a two-phase fluid. Anything that affects the strength of the bubble film (including chemicals, salts, oil, cuttings and fine solids, shear, influxes, temperature, and pressure) can put the foam in either state.

Some models have been successful in modeling foam partly because they do not consider foam to break until it gets to 99.5% (which is used for friction and carrying capacity). The model always considers the system to be single-phase. The model may also consider a reduction in friction when the GVF reaches 97.5%. Other models make different assumptions or allow the modeler to make an assumption on stability. Modeling normally does not consider the possibility of multi-phase flow in foam.

The basic interest in the field is in the overall performance of the foam in the hole, not an isolated bubble. There may be some foam breakage along the sides of the casing, but the main point is that the vast majority of the system is still acting as a common single-phase fluid that does not break until it has some quiescent settling time.

Considering that the single biggest problem with foam is breaking the foam, and the popular position that foam breaks is at 96% GVF, it is obvious that these two positions clash. Otherwise, the addition of gas to the foam would break it. The strength of the liquid film must be reduced, damaged or destroyed to release the air. That is the purpose of surface tension reducing chemicals to break foam (alcohols, etc).

This is not to say that the quality or GVF has no meaning. Rather, the pressure in the bubble must be sufficient to stretch the bubble liquid film to the point where it will break and the gas escapes. Even when conducting a foam test in a lab, it is noticeable that the bubbles at the top of the column slowly lose their liquid film and grow in bubble size until they eventually break. They normally do not break until long after they have reached a GVF of 96%.

4.19 Two-Phase Foam

One common event for foam-drilled wells is the separation of gas and foam. When the GVF increases too rapidly while ascending to the surface, you may see gas breakout of the foam carrying the cuttings out the blooie line. This situation is difficult to model, particularly where polymers are involved.

A robust foam does not break apart as it comes to the surface (see Figure 4–17), but creates a build up of shaving cream-like foam as it comes out of the blooie line. However, where there is a small amount of liquid, it appears the foam cannot hold all of the air and separates

Figure 4–16 *Example of mist flow (Weatherford U.S., LP)*

Figure 4–17 *Robust foam at the blooie line (Weatherford U.S., LP)*

into air and foam (the foam carries the cuttings and the air carries the foam). (Existing video tapes show that the foam carries the cuttings but the gas is flowing faster than the foam) The foam is often loaded with cuttings and hence, there is the familiar puffing at the blooie line. The air is moving faster than the foam from the break out point to the surface. This action then is two-phase foam of air and foam. This action is more common where there is little water in the foam (say less than 20/25 gallons/min) as more liquid could hold greater quantities of air/gas. There is some evidence of the gas/water separation, however, little is reported (possibly because of preconceived ideas). Modeling does not address this two-phase foam action, as current models predict either it breaks into air and liquid or holds together as a single phase.

4.20 Testing

Currently no standard test for measuring foam properties at the wellbore condition is available. Typical oilfield practice requires pouring a 100 ml liquid sample into a blender and blending it on high speed for 30 to 60 seconds. After sufficient blending, pour the foamed fluid into a 1,000 ml cylinder and measure the height and half-life, as shown in Figure 4–15.

During any foam operation, it is critical to make this test and record the results with the observed results of the foam operation. The rig crew normally does this, thus necessitating training sessions for both the foam tests and the corrosion tests as requested by the corrosion engineer.

However, there is a problem with this foam test procedure. The resultant mixtures do not contain the same amount of gas (GVF) or represent the foam behavior under pressure. Flow loop testers have long known that foam under pressure has a much longer half-life than at atmospheric pressure. One of the real needs for foam drilling is the development of testing procedures that will reflect how the foam performs in the drilling situations.

4.21 Circulating Time Tests

The practice of measuring circulating time for wells has long been in use with mud programs. The trick with foam is to inject the trace material (normally concrete dye) into the foam stream as a concentrated slug. In addition, run the test without a break in the injection process. The primary purpose of these tests is to validate the model used for the design of the operation by matching the compressible flow system.

Figure 4–18 depicts the recommended system to get the dye in concentrated doses. The following two diagrams represent typical results.

Figure 4–19 and Figure 4–20 show a change in color. To calculate circulating time pick the first sample to show a change in color. This process is to calibrate the model in use for the design of the operation.

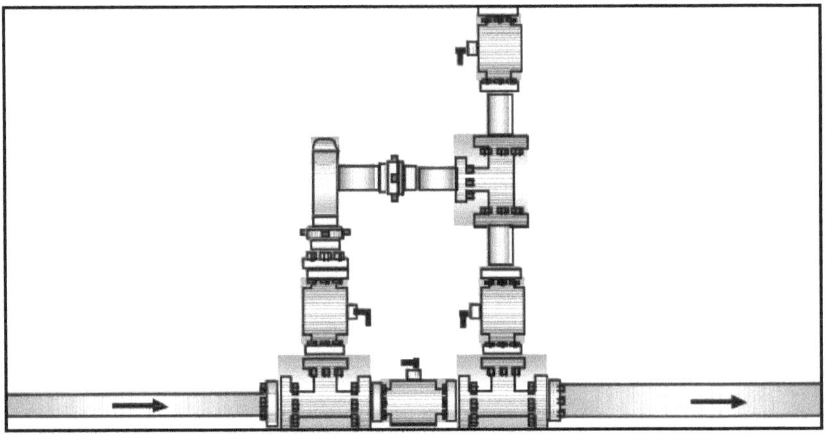

Figure 4–18 *Dye injection manifold*

Figure 4–19 *Dye color change (Weatherford U.S., LP)*

Figure 4–20 *Dye color change*

4.22 Operational Considerations

The first operational consideration in using foam as a drilling media is whether to use a one-pass system or to recycle the foam. Many factors enter into this decision. Probably the major one is environmental concerns. Recycling allows minimum left over fluid, especially for long jobs. However, recycling is more complex than a one-pass system.

For a one-pass system, the foam is pumped in the well and out to a pit or other storage system. Using proper care, the liquid can be recycled via a natural de-foaming process and generally accomplished using the rig's steel pits or with external pits. As the foam comes out of the rotating control device, inject the defoamer and extra water (as required). This action breaks the foam (breaking is a 95% action, it never breaks into just water and gas). To counter this fact it is a good idea to keep a low level in the first pit (no more than ¾ full. The broken foam will have a fluff-like top. Most of the defoamers contain alcohol so the top will also contain more alcohol. That is why the first pit is not stirred (let the alcohol build up). This practice makes it easier to re-foam the liquid. It is often necessary to dump produced water out of foam systems, normally the second pit. Normally it is necessary to pump the liquid from the first pit to the second pit (DO NOT FLOW OVER THE TOP). It may seem as if these precautions are unnecessary, but they are important in keeping the foam job running smooth. In addition, a good practice involves using a desilter to clean the foamed liquid because the foam carries little pieces of dirt and rock that potentially interferes with further cutting carrying capacity. When recycling, the make up surfactant should be less than 0.4%, but let the foam tests dictate how much to use. Figure 4–21 is a suggested scenario for pumping foam.

4.23 Chemicals

Even using cryogenic nitrogen or natural gas, utilize the corrosion protection chemicals in the system. Do not let two or more companies treat for corrosion as they may create a chemical imbalance that can quickly destroy the tubulars. The rule is when conducting a drilling operation with foam, run a good corrosion program and frequent tests.

In discussing the part chemicals play in foam drilling, it is fair to say that it is very important in the ultimate success or failure of the job. It is important to use chemicals that are compatible. Most of the foaming agents are anionic, but there are some cationic materials. In addition, some materials are just not compatible.

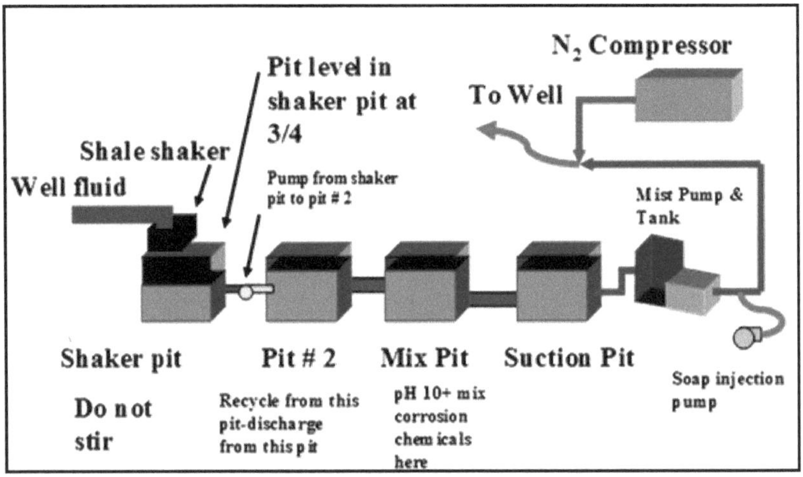

Figure 4–21 *Foam surface system*

It is important to match the chemicals to the formations encountered. Some formations are sensitive to water, some are sensitive to pH, some are sensitive to velocity and all are sensitive to pressure surges. Proper connection practices will minimize down-hole pressure surges during drilling. The best way to avoid using the wrong chemicals is to test samples of the formations for sensitivity prior to drilling operations.

It is not a good practice to use polymers as a matter of course in foam. They increase half-life by increasing foam persistence, but they also increase the friction loss. Connection and tripping procedures that are improved with more persistent foam are of prime importance in drilling a trouble free well with foam, but increased friction loss is detrimental. Too much polymer also makes it difficult to defoam the system. To ensure compatibility, test the foam chemistry and the formation fluids in the lab prior to the job.

4.24 Pumping Surfactant and Chemicals

While many operators inject the surfactant and chemicals on the downstream (high-pressure side) of the pump, the most economical and accurate method involves injecting the surfactant and chemicals on the suction side of the pump. The proportioning pumps sometimes lose accuracy under varying pressure conditions.

4.25 Jet Subs

Jet subs are tools run in the drillstring to allow a portion of the pumped stream to exit the drillstring with the percentage of exiting fluid controlled by the size of the jet and the differential pressure, Figure 4–22 and Figure 4–23.

When making connections, bottom and upper column, the liquid must be aerated so the hydrostatic will not force the gas into the water zone. If required, run the jet subs out of the casing and into the open hole because under normal flow conditions they typically exhaust 10% in normal flow and should not damage the wellbore. It is very likely the above case requires two jet subs, thus enabling drilling through the water zone to the next casing point.

4.26 Hammers and Motors

Present day practice routinely combines foam drilling with hammers and motors. The hammer usage is prevalent in big holes, and the motor usage is prevalent in deviated wells. Most of the hammer operators have computer programs for guidance of air volumes and liquid volumes, but they generally only apply to shallow wells. Hammers or motors are power units. The hammer gets its power from a piston striking the bit; the harder it strikes the more power. The annulus

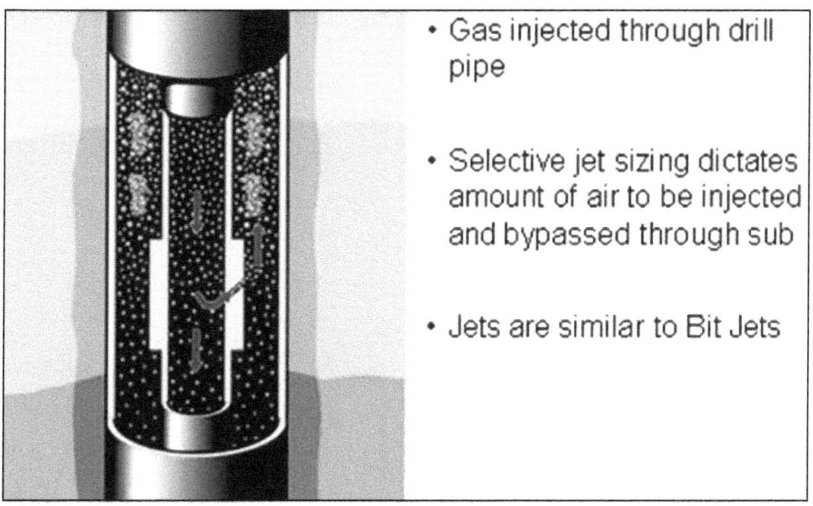

Figure 4–22 *Jet sub (Weatherford U.S., LP)*

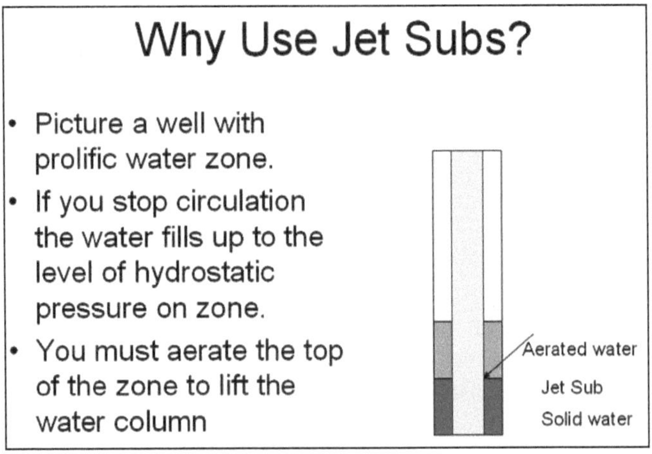

Figure 4–23 *Why use jet subs*

pressure adversely affects the hammer efficiency, and they work best in a low pressure environment.

The motor gets its power from a fluid turning the rotor. The more force the fluid transmits to the rotor the more powerful the rotor. The motor is sensitive to the viscosity and density of the fluid passing through, thus requiring greater volumes for air than water. However, the foam is viscous and acts more like water than air.

Section 3
Oil-Based Foam Drilling Fluid

Olusegun M. Falana, Weatherford International

4.27 Introduction

There are five basic drilling fluids employed in underbalanced drilling (UBD): liquid, air, mist, aerated, and foam[1]. When the choice of a drilling fluid is liquid or aerated, there are two alternatives for base fluid: one is water or aqueous, and the other is oil or non-aqueous base fluid. The focus of this section is on the first and novel oil-based foam drilling fluid, *OleoFoam HT,* developed by Weatherford International.

In order to fully reap the benefits of UBD, mist and foam are the primary fluids used when faced with operations in environments with less than 4.0 pounds per gallon equivalent pore pressure. The benefits of properly executed UBD operation with mist or foam include the ability to control fluid loss and differential sticking, increasing rate of penetration, and the minimization of formation damage. Reducing inflow minimizes surface handling cost, flaring cost and environmental impact[2], which are also important requirements. Foam is the fluid system of choice for efficient hole cleaning, employment of foam with all its advantages in UBD operations is dwarfed by operational costs in water sensitive formations, mainly because until now, existing foam drilling fluid systems are all water-based.

Water sensitive formations with bottom-hole temperatures greater than 250 °F with or without an influx of contaminants (H_2S, brine, etc.) constitute a hostile environment for water-based foam. The oil-based foam system, OleoFoam HT, was developed to extend the full spectrum of foam drilling benefits to water sensitive reservoirs (Kippie et al., US 2008/057556; Kakadjian et al., US 2010/0000795 A1). OleoFoam HT is a recyclable foam system compatible with H_2S scavengers, and is stable up to 450°F (230°C). The system has been successfully applied in hostile environments.

The retrograde condensate of the Smackover Carbonate and the Norphlet Sandstone in the Gulf Coast of the United States exemplifies such an extreme environment. The OleoFoam HT system was used for an operator in the area with impressive results. Not only was the job completed ahead of schedule, but there were instantaneous rates of

1. The fluids are covered in Chapters 3 through 5.
2. Details on Foam Drilling Systems are presented in Chapter 4.

penetration as high as 165 ft per hour, or 50 m per hour, compared to the 4 ft per hour, or 1.2 m per hour, previously experienced with a water-based foam system. A bit was saved in the operation, and most importantly, there were impressive improvements in production (Sepulveda et al., 2008).

The fact that the system is adaptable permitted the use of an environmentally benign and readily biodegradable base fluid (HF 1,000, an oxygenate mixture produced by Sasol) in the drilling of a well in Houston, Texas. Specifically, the fluid was designed to offer double the benefits of providing the ability to drill through sensitive formation and meet environmental requirement. The customer saves at least one million dollars on casing alone.

The San Juan field in New Mexico is one more example where OleoFoam has offered unparallel opportunity to complete a viable well. Up until the operation that involved the use of OleoFoam, there was no record of successful completion of a well in the field because of a prevalence of a highly unstable formation (Sepulvida et al., 2008).

4.28 Development of Oil-Based Foam Drilling Fluid

Oil-based foam drilling fluid (OBFDF) was developed to meet UBD demand for extreme foam drilling for hostile environment conditions. In most practical situations, a suitable OBFDF must meet some necessary criteria for effective hole cleaning, optimum rate of penetration, or successful operation. The following criteria are established from Weatherford practices; ultimately fluid system must be formulated in conformance with outcome of flow modeling of the specific drilling operation:

- It must exhibit foam height (FH) and half-life (HL) higher than 130 mL and 3 min, respectively. These properties measure foamability and stability of the resultant foam in a system. Foam height is the volume of foam generated by mixing 100 mL base liquid and foaming solution for 1 minute with a Hamilton Beach blender (ASTM D3519). Foam half-life measures how long it takes for 50 mL of liquid to break from the foam. Both properties are measured at atmospheric conditions. Although FH and HL requirements will be job dependent, preferably both must be high for a successful operation and optimum UBD benefits.
- The fluid must be compatible with common additives such as scale and corrosion inhibitors.

- Fluid must be compatible with down-hole tools; i.e., non-deleterious to seals on Kelly Hose, Mud Pumps, RCDs, and the like.
- The system must be robust: adaptable to any desired oil-base fluid (diesel, native crude, synthetic oil, etc.), be thermally stable [≤ 250°F (120°C) but preferably at ≥ 400°F (200°C)] and withstand cuttings, brine, condensate, or other common contaminants.
- The fluid should be recyclable to ease fluids handling and minimize chemical usage.
- The fluid should be compatible with refinery processes to allow selling of makeup fluids and produced hydrocarbons.

4.29 OleoFoam HT System

The *OleoFoam HT* System was developed to preserve the aforementioned advantages of drilling with foam, gain the ability to operate in higher temperatures, and to maintain permeability in water sensitive formations. The system utilizes a foamer (*OleoFoam C*), a viscosifier (*OleoVis*) and other additives that are job specific. For example, the presence of H_2S will necessitate inclusion of a scavenging additive.

4.29.1 Oil Foamer

Initial hurdle was to develop a foamer that could meet the first four or all the desirable criteria. Additionally, a suitable foamer must be cost effective and devoid of toxicity properties of existing fluorocarbons, which are costly and known to foam hydrocarbons. Characteristically, oil base foams are unstable as they readily dissipate with intangible half-life; in part, because surface tension is much lower in oil (~21 dynes cm^{-1}) than in water (~72 dynes cm^{-1}). Stable foam, when base fluid or continuous phase is crude oil, diesel, mineral oil, alpha olefins and/or ester based oils (etc.), posed a complex chemical problem. Generations of oil foams that would be stable at high temperatures added to the complexity of the problem.

Fortunately, the silicone foamer in *OleoFoam C*, successfully met the preset criteria. The surface active agent is a silicone oligomer (polydimethylsiloxane, DC 1250) available from Dow Corning. Thus, OleoFoam C offers stable foam properties with desirable FH and HL values even at ultralow foamer concentrations (Figure 4–24 and Table 4–5). In the foam tests presented in Table 4–5, 1.0% of the foamer was

Figure 4–24 *OleoFoam HT in red diesel as base fluid*

Table 4–5 Properties of OleoFoam HT System (Recyclability, Robustness, & Thermal Stability)

System (Base Fluid)	3.5% Sea Water	3.0% KCL	Condensate	Crude Oil	Heat Stability °F (24hr)	Foam Height (mL)	Half Life (min)	Ave Recycle (X10)	Fan 35A (Pv,Yp)
OleoFoam HT (Neat Diesel)						170	4:00	200/4:15	11, 2
	30%					170	4:00		
		20%				160	4:00		
			20%			180	4:00		
				20%		160	4:00		
OleoFoam HT (Neat Diesel)					450	160	4:00		
Crude Oil (36 API, Neat)						180	5:00		
OleoFoam HT (Neat HF 1000)						170	4:00	200/4:15	5, 2
	30%					160	2:30		
		20%				160	2:30		
			20%			160	2:30		
				20%		190	3:00		
OleoFoam HT (Neat HF 1000)					450	160	3:40		

[1]OleoFoam C and OleoVis are of equal concentration of 1.0 % of solution of each in the respective fluids.
[2]HF 1000 is a completely biodegradable base fluid.

employed, which amounts to 500 *ppm* of the active agent despite the presence of contaminants in the continuous phase.

4.29.2 Defoamer

A silicone-based defoamer, *OleoBreak*, was also identified to be compatible with the foamer. The defoamer is also a silicone oligomer or polydimethylsiloxane (DC 200 Fluid, 50 centistokes) from Dow Corning. Consequently, the system can be taken through a series of foam-defoam-refoam processes and still retain desirable foam properties. A test on the recyclability of the system is summed up in Table 4–5. In one case (row 3 column 9), red diesel was used as the continuous phase, while in the second case a biodegradable oxygenate mixture, HF 1000 (row 14 column 10) from Sasol was employed. A complete test process involves foam-defoam; then the same is repeated 9 more times for 10 cycles. Apparently, considering initial foam height of 170 mL, an average height of 200 mL and a 4 min 15 sec half-life, foam properties improve over the 10 cycles suggesting the defoamer to be a profoamer because no more foamer was added. Incidentally, most common defoamers like alcohols, mixture of alcohols, ethers, hydrocarbon, or other combinations thereof are incompatible with either the silicone-based foamer or viscosifier in the system.

4.29.3 Viscosifier

There are several oil soluble polymers (including organophilic clay) in the market, but most are unfit for foam drilling fluids formulation. For instance, organophilic clay will increase solid composition of the fluid system with a resultant increase in equivalent circulation density in foam drilling operations. Therefore, the clay is undesirable in foam drilling fluids. Initially, Weatherford utilized a hydrocarbon gelling technology (US 2008/0257556 A1) that involves generation of a phosphate complex as a viscosifier resulting from reaction of the phosphate with a suitable amount of initiator. Now, desirable properties are found with a hydrogenated *Styrene-Isoprene copolymer* available from Infinium. Employment of a star-like oil soluble block copolymer, *OleoVis*, which forms micelles at high temperatures (Shultz, 1991) and increases viscosity of the continuous phase (hence decreasing drainage), allows the flexibility to modify foam properties for drilling conditions like hole cleaning and bottom-hole pressure control.

The result of the influence of the block copolymer on a 55% N_2 foam system using a tubular recirculating rheometer (see Figure 4–25) is given in Figure 4–26.

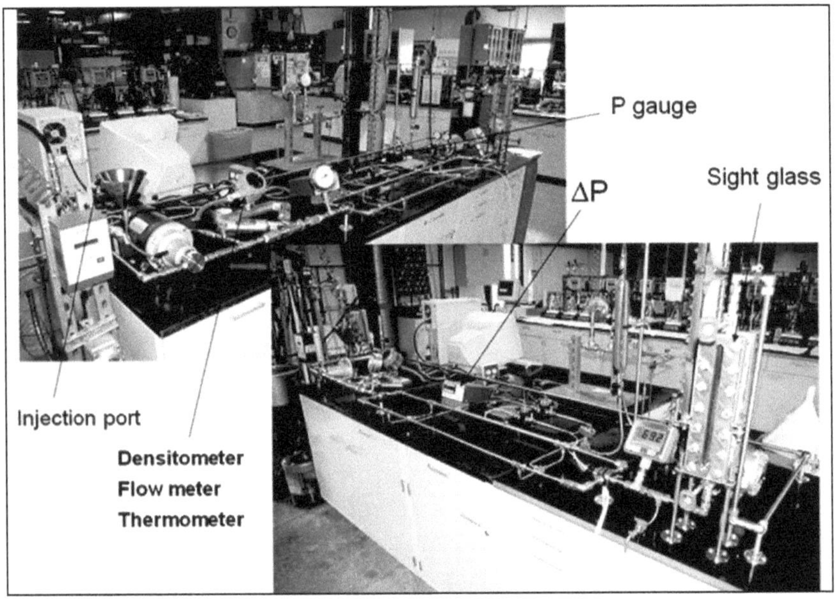

Figure 4–25 *Sections of recirculating rheometer for study of foam properties at high pressure and room temperature*

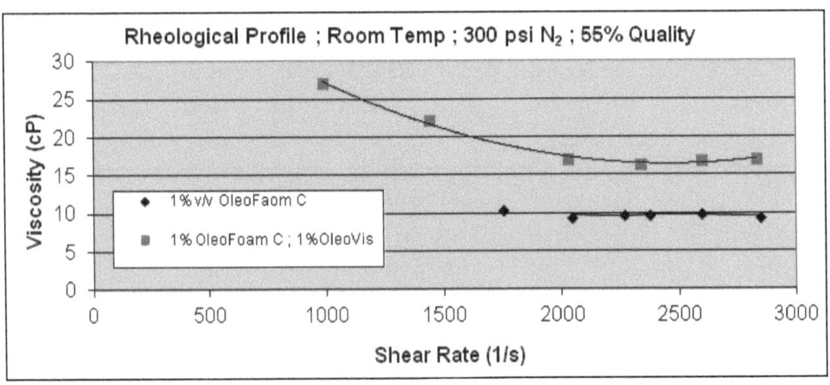

Figure 4–26 *Rheological profiles of OleoFoam HT Systems with and without OleoVis*

4.30 Features

4.30.1 Recyclability

OleoFoam HT system is suitable for a recycle, non-recycle or stable foam, mist, or two-phase operations. The system is reusable from well to well and can be refined for fluid recovery.

4.30.2 Robust

The robustness of the fluid is demonstrated by its ability to retain functionality in the presence of 100% crude (e.g., 36 API), brine Table 4–5, and cuttings. It is thermally stable up to 450°F (230°C, rows 14 and 21, Table 4–5).

4.30.3 Versatile

OleoFoam HT is versatile with diesel, local crude, internal olefins or with similar readily available makeup fluid substitutes.

4.30.4 Corrosion Resistant

Typically, there is little problem with the corrosion of down-hole tools when oil-based fluids are used. In some hostile environments with noxious gas or brine, additional inhibiting additives are needed. OleoFoam HT offers similar protection from down-hole corrosion and is compatible with most oil soluble inhibiting additives.

4.30.5 Environmentally Friendly

A completely environmentally friendly system can be built by using a biodegradable and nontoxic makeup fluid like HF-1,000 among others produced by Sasol.

4.31 Challenges

One challenge that has been encountered while drilling with OleoFoam HT is the inability to drill in the presence of more than 30% condensate. Thus, unloading wells with condensate remains problematic.

Lately, further research has shown that the system is selective in its ability to produce stable foam with condensate as base fluid. In view of unique characteristics of sources, some condensates have compositional differences sufficient to elicit different foam

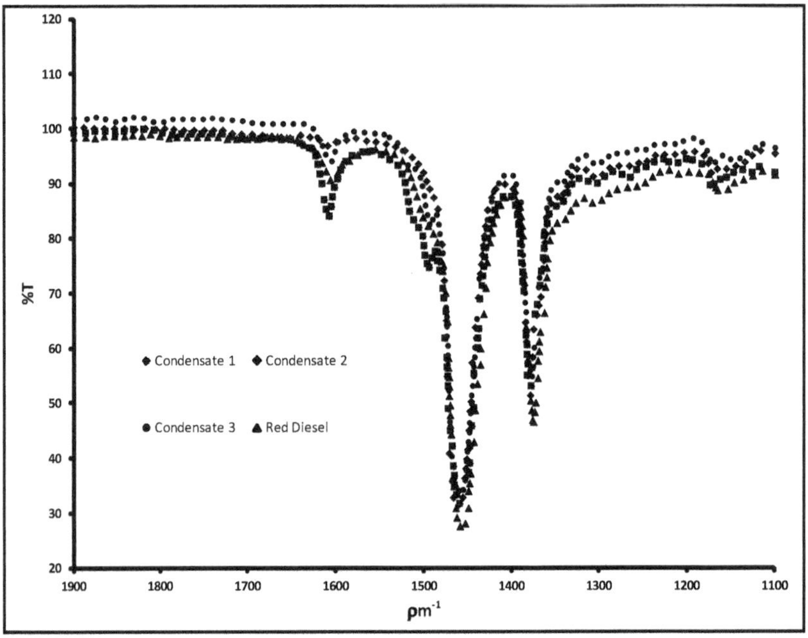

Figure 4–27 *FT-IR spectra of condensates from three different sources and US number 2 Red Diesel*

properties with a given foamer. In fact, condensates do have different compositions that are source specific as evidenced in the segment of Infra-Red Fourier Transform spectra of red diesel and three condensates from three different sources shown in Figure 4–27. Noticeably, *Condensate-2* (C-2) has little or no absorptions at 1,607 and the shoulder at 1,495 wave numbers (cm-1) that are characteristics of a $\nu_{C=C}$ ring stretches (suggesting presence of aromatics), which are relatively profound in *Condensate-1* (C-1). Not surprisingly then, OleoFoam C produces stable foam in C-1 as well as "red diesel"; while another foamer was found to be compatible with C-2. Thus, C-1 contains significantly more aromatic "character" than C-2, and is respectively foamed by aromatic compatible foamer (OleoFoam C, OFC) and non-aromatic compatible fluorocarbon foamer. Given non-aromatic containing base fluid or condensate, the surface active agent of choice used in a proprietary formulation is an environmentally benign Fluoroaliphatic Polymeric Esters with average molecular weight of at least 1,000,000 grams per mole (WO 2008/089391 A1, WO 2008/089386 A2). The surfactant is available through the 3-M Innovative Properties Company in Saint Paul, Minnesota. Accordingly, Weatherford has a second foamer which has been found to offer desirable foam properties where the proprietary silicone foamer system fails. Yet, the need to run FT-IR as a diagnostic tool for condensate classification and selection of proper foamer remains imperative.

One additional challenge is that the effect of ingredients in surfactants, viscosifier or base fluid on down-hole assembly must be known. Oftentimes, the nature of seals on down-hole assemblies are not disclosed; down-hole conditions in one well might render a base fluid usable while the other renders some undesirable. Indeed, not every base fluid with desirable foam properties is suitable for drilling operations. In essence, local light crude that shows great foam properties might have deleterious effect on the seals of down-hole tools. Hence, it is paramount to investigate the effect of local base fluids on seals before their adoption.

4.32 Conclusion

Weatherford has developed series of foamers and agents that extend the use of drilling foams into high temperature and water sensitive zones. When used with a nitrogen gas system, oil-based foam fluid has the low density, lifting capacity, and stability of a water-based foam with the basic advantages of oil-base drilling fluid. Field results have proven robustness of the foam fluid system, including its resistance to contamination, as well as its ability to increase drilling rates significantly in water sensitive formation.

Not only can diesel or crude now be foamed with desirable foam properties, but dissimilar condensates can be selectively foamed and/or lifted depending on target drilling operation. Undoubtedly, the introduction of the high temperature oil-based foam system will extend the use of this unique fluid into areas where it was previously impossible to use foam.

4.32.1 Acknowledgement

OMF wishes to thank WFT for support and permission to present information on the OleoFoam HT System. Thanks go to Mike DiBiasio, Jhon Sepulveda, Shirley Glen; Frank Zamora, Sarkis Kakadjian, Ed Marshall and others who have made contributions that culminated in the offering of the OleoFoam HT System.

4.33 References

ASTM International. "Standard Test Method for Foaming Properties of Surface-Active Agents," ASTM D-1173-07.

Kippie, D. and Zamora, F., 2008, *Non-Aqueous Foam Composition for Gas Lift Injection and Methods for making and Using Same*, US Patent 0257556.

Kakadjian, S.R., Falana, O.M., DiBiasio, M.A., Marshall, E. and Zamora, F., 2010, *Enhanced Oil-Based Foam Drilling Fluid Compositions and Method for Making and Using Same*, US Patent 0000795 A1.

Schulz, D.N. and Glass E. *Polymers as Rheology Modifiers,* ACS Symposium Series, Vol. 462, American Chemical Society, Washington, DC, USA, 1991, pp. 260–265.

Sepulvida, J. et al. "Oil-Based Foam and Proper Underbalanced Drilling Practices Improve Drilling Efficiency in a Deep Gulf Coast Well," SPE 115536 presented at the SPE Annual Technical Conference and Exhibition, Denver, CO, USA, September 21–24, 2008.

Section 4
Foam Rheology

Amir Paknejad, Add Energy, LLC

4.34 Introduction

The concept of foam rheology was first discussed by Sibree[1] in 1934. He found that the apparent viscosity of foam was higher than that of each its constituents. He also found that foam shows Newtonian behavior below its critical shear-stress value and shows plug flow behavior above its shear-stress.

The study of fire-fighting foam behavior by Grove et al.[2] produced one of the earliest, most enlightening papers on foam rheology. By measuring the effects of pressure, shear stress, and foam quality on the apparent viscosity, they indicated that at high shear rates the apparent viscosity is independent of shear rate. They also found that at constant pressures viscosity highly depends on and varies directly with foam quality.

Fried's[3] results with a modified rotational viscometer were also consistent with the previous works. He found that foam viscosity decreased with decreasing foam quality and increased with increasing tube diameters.

The Raza and Marsden[4] study of foams with qualities ranging 0.7–0.96 stated that foam shows pseudoplastic behavior below critical flow rates and plug flow behavior above that. They found that the critical flow rate is dependent on both foam quality and tube diameter. They did not correct their experiments for slippage at the tube wall and for the compressibility of the foam.

David and Marsden[5] corrected the experimental results for both semi-compressibility and slippage at the wall. They concluded that corrected viscosity for both slippage and compressibility is independent of foam quality. They also found that slip coefficient increased with shear stress, but the corrected apparent viscosity still increased with the tube diameter.

4.35 Mathematical Concepts

4.35.1 Einstein

Einstein's[6] theoretical development was the first mathematical treatment of rheological problems in foam. He considered the foam as a suspension of solids in liquid. Based on the energy balance criteria, and considering no slip at the surface of the particles, his two-phase viscosity for the foam quality ranging from 0 to 0.45 is given by

$$\mu_F = \mu_l \left(1 + 2.5\ \Gamma\right) \tag{4.1}$$

where

μ_F = Foam viscosity

μ_l = Base liquid viscosity

Γ = Foam quality

Hatschek[7] developed a similar foam rheology model based on Stoke's law for foam quality ranging from 0 to 0.74 as

$$\mu_F = \mu_l \left(1 + 4.5\ \Gamma\right) \tag{4.2}$$

He also described the viscosity of foam for qualities between 0.74 and 0.99. It was based on conservation of energy during interference, deformation, and packed bubbles within the foam flow boundary. His proposed model for the foam viscosity is

$$\mu_F = \mu_l \left(\frac{1}{1 - \Gamma^{\frac{1}{3}}}\right) \tag{4.3}$$

This equation is only applicable in high shear rates where the foam viscosity is almost independent of shear rate.

4.35.2 Mitchell

Mitchell[8] developed his model based on Rabinowitsch's theory. Considering the foam as a Bingham plastic fluid, the plastic viscosity and the apparent yield stress were correlated with the foam quality and the shear stress-shear rate equation was presented as

$$\left(\tau - \tau_y\right) = \mu_p \cdot \gamma \tag{4.4}$$

He also proposed two empirically derived equations for foam viscosity. The equations of foam viscosity for foam qualities from 0 to 0.54 and 0.54 to 0.97 are expressed respectively as

$$\mu_F = \mu_l (1 + 3.6\Gamma) \tag{4.5}$$

$$\mu_F = \mu_l \left(\frac{1}{1 - \Gamma^{0.49}} \right) \tag{4.6}$$

The graphical representations of viscosity-quality relationship for these pioneer methods are shown in Figure 4–28 and Figure 4–29. Further experiments showed that combining the plastic viscosity and the yield point was significantly more reliable than using the plastic viscosity alone.

In contradiction with previous models that pertained only to capillary tubes, Beyer et al.[9] were the first to present a model based on both laboratory and pilot-scale tests with horizontal pipes. They followed Mooney's procedure to correlate slip velocity with liquid volume fraction and wall shear stress. The authors stated that the accuracy of their model in large diameter wells may be increased by accounting for liquid buildup. Using yield shear stress obtained in pilot-scale experimental data, the viscosity is given as

For 0.02< *LVF* <0.1;

$$\mu_o = \frac{1}{(7200LVF + 267)} \tag{4.7}$$

For 0.1< *LVF* <0.25;

$$\mu_o = \frac{1}{(2533LVF + 733)} \tag{4.8}$$

where

LVF = Liquid Volume Fraction

μ_o = Bingham Viscosity

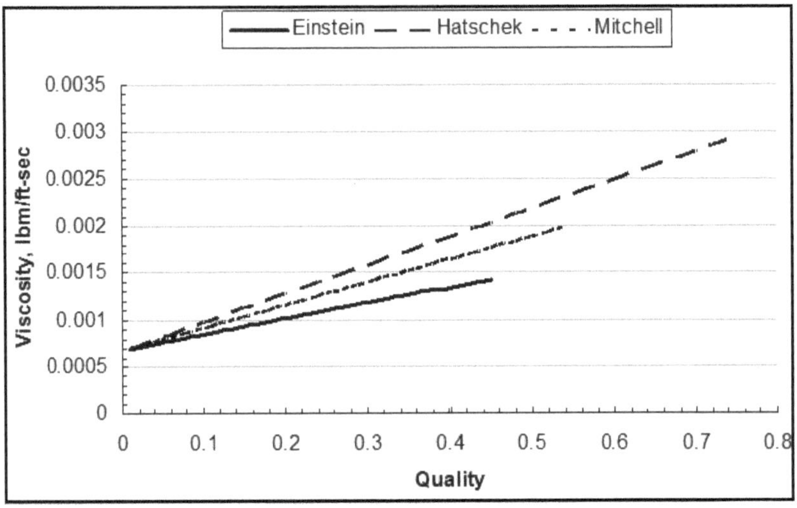

Figure 4–28 *Viscosity quality relationship (Einstein, 1906)*

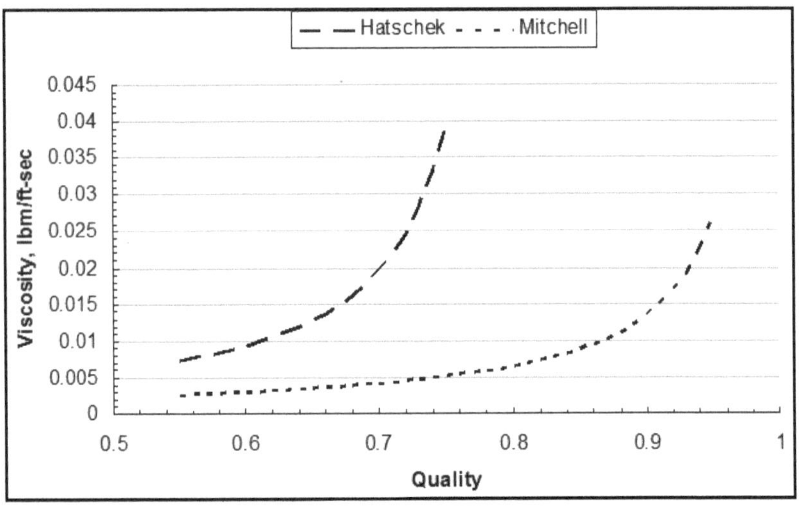

Figure 4–29 *Viscosity quality relationship (Mitchell, 1979)*

4.35.3 Blauer et al.

Blauer et al.[10] proposed to use effective viscosity, density, average velocity, and pipe diameter to calculate the Reynolds number and Fanning friction factor for foam fluids. They found that the relationship between the Reynolds number and Fanning friction factor for foam was the same as that of a single-phase fluid. They assumed that foam behaves like a Bingham plastic fluid with no slippage in a laminar flow region. The foam plastic viscosity and yield stress were experimentally determined as a function of foam quality, and are presented in Figure 4–30 and Figure 4–31.

Their proposed equation for effective viscosity of Bingham plastic foam is given as

$$\mu_e = \mu_P + \frac{g_c \tau_y D}{6v} \tag{4.9}$$

where

μ_e = Effective viscosity

μ_e = Plastic viscosity

g_c = 32.2

τ_y = Yield stress

D = Tube inside diameter

v = Velocity

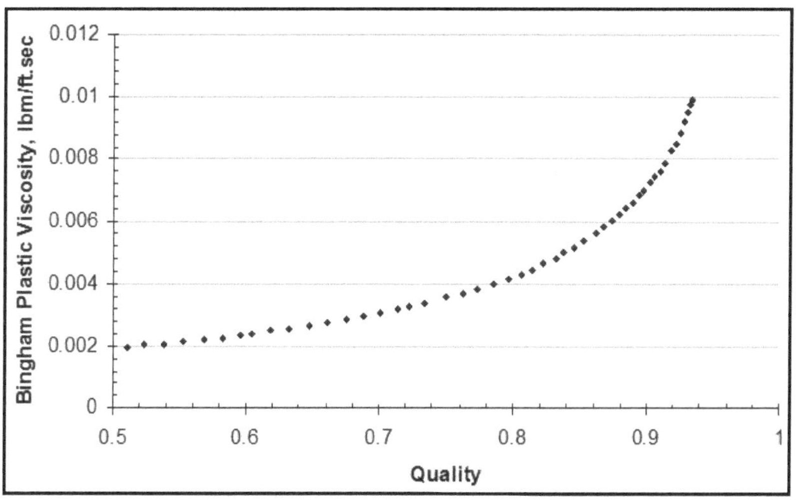

Figure 4–30 *Bingham plastic viscosity versus quality*

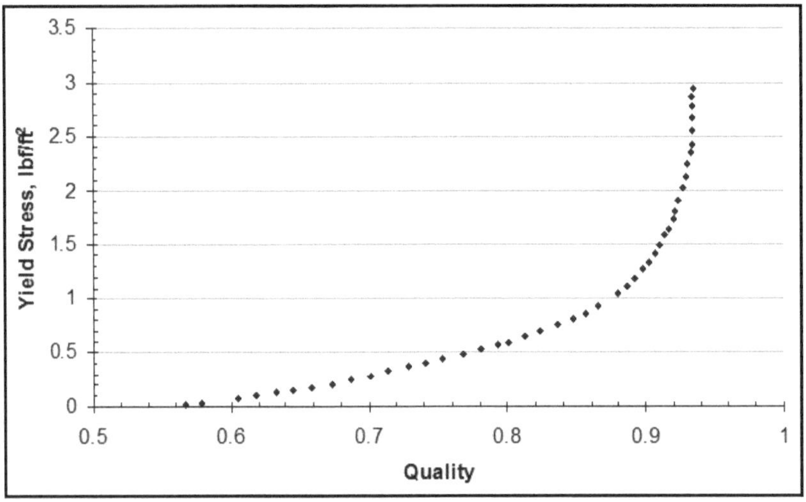

Figure 4–31 *Yield stress versus quality*

4.35.4 Reidenbach et al.

Reidenbach et al.[11] performed experimental work with water foams using nitrogen as an internal phase. They proposed a Herschel-Bulkley model to describe a laminar foam flow through pipes. They found that substitution of CO_2 for N_2 as the internal phase gives similar laminar rheology. A modified scale-up relationship was used to describe compressibility in turbulent flow. They defined a correlation for apparent viscosity that is used as a Newtonian viscosity in standard pressure drop calculations. Apparent viscosity is defined as

$$\mu_a = \tau_{yp}\left(\frac{8v}{d}\right)^{-1} + K\left(\frac{8v}{d}\right)^{n-1} \tag{4.10}$$

where

μ_a = Apparent viscosity

τ_{yp} = True yield point stress

d = Pipe inside diameter

v = Bulk velocity

K = Consistency index

Based on the experimental measurements performed by Martins et al.[12], the parameters of power law model were correlated with foam quality as

$$K = 0.0813\left[\frac{1-\Gamma}{\Gamma}\right]^{-1.591}$$

$$n = 0.8242\left[\frac{1-\Gamma}{\Gamma}\right]^{0.5164}$$

As flow behavior and consistency index were obtained, according to the definition of the apparent viscosity for the Power law fluids, the apparent viscosity can be calculated as

$$\mu_a = K\gamma^{n-1}$$

where

$$\gamma = \left(\frac{2n+1}{3n}\right)\cdot\left(\frac{12v}{D_H}\right)$$

γ = Shear rate, $\sec^{-1}$

The flow consistency index and the flow behavior index correlations can also be presented as it is shown in Figure 4–32.

4.35.5 Sanghani

Sanghani [13] set up an experiment with a concentric annular viscometer to closely simulate actual wellbore conditions. He found that the power law model was statistically superior to the Bingham model in correlating their data. He concluded that foam is pseudoplastic at low shear rates and Bingham plastic at high shear rates. He also provided experimental data to correlate pseudoplastic parameters K and n, as a function of foam quality as below:

$$K = -0.15626 + 56.147\Gamma - 312.77\Gamma^2 + 576.65\Gamma^3 + 63.960\Gamma^4 - 960.46\Gamma^5 - 154.68\Gamma^6 + 1670.2\Gamma^7 - 937.88\Gamma^8 \quad (4.11)$$

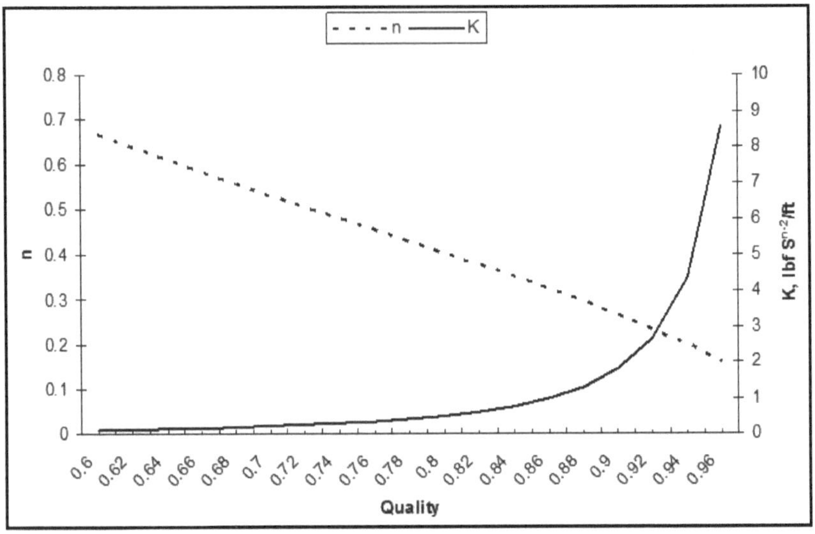

Figure 4–32 *Flow consistency index and flow behavior*

$$n = 0.095932 + 2.3654\Gamma - 10.467\Gamma^2 + 12.955\Gamma^3 + 14.467\Gamma^4 - 39.673\Gamma^5 + 20.625\Gamma^6 \tag{4.12}$$

and

$$\mu_e = K\left(\frac{2n+1}{3n}\right)^n \left(\frac{12v_f}{D_H}\right)^{n-1} \tag{4.13}$$

where

D_H = Hydraulic diameter

The flow consistency index and the flow behavior index correlations can also be presented as it is shown in Figure 4–33.

4.35.6 Ozbayoglu et al.

Recently Ozbayoglu et al.[14] conducted a rheological study for foam. They concluded that wall slip effect is not negligible and should be considered to establish the true flow behavior of foam in pipes. By using the Rabinowitch-Mooney equation, the rate of shear at the tube wall is given as follows:

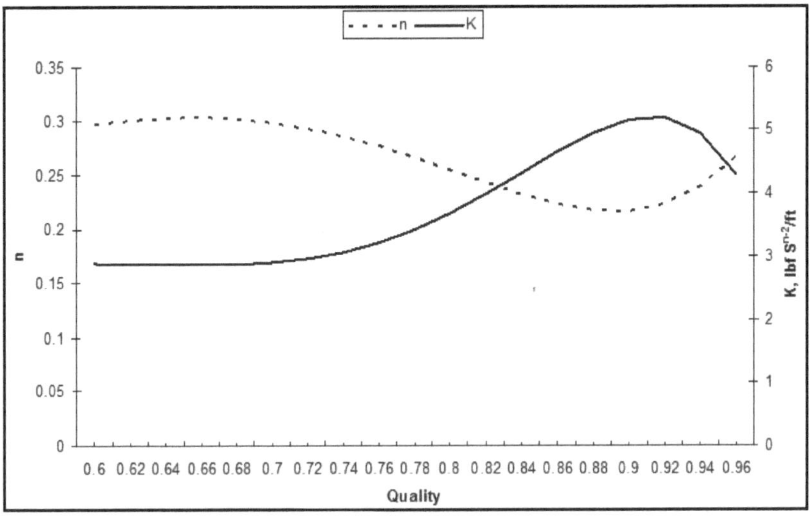

Figure 4–33 *Index correlations*

$$\gamma_w = \left(\frac{8v}{D}\right)\frac{3n+1}{4n} \tag{4.14}$$

where

$$n = \frac{d\ln(\tau_w)}{d\ln\left(\frac{8v}{D}\right)} \tag{4.15}$$

and

$$\tau_w = \frac{\Delta P}{L}\cdot\frac{D}{4} \tag{4.16}$$

τ_w = Wall shear stress

In their method, the Newtonian shear rate was corrected for the slip effect at the wall and $\ln(\tau_w)$ was plotted versus

$$\ln\left(\frac{8v}{d}\right)$$

Then, n was determined for different shear rates and, using curve-fitting and statistical analysis, the rheological properties were determined based on the plot of (τ_w) versus

$$\frac{3n+1}{4n}\left(\frac{8v}{d}\right)$$

Their experimental data indicated that foam rheology was best characterized by the power law model for foam qualities ranging 0.7 to 0.8, and the Bingham plastic model gives better fit for higher foam qualities.

4.36 Foam Quality

Foam quality is the volumetric ratio of gas-phase to gas/liquid-phase. The foam quality index Γ can be defined as

$$\Gamma = \frac{V_g}{V_g + V_l + V_f} \tag{4.17}$$

where

V_g = Gas volume

V_l = Liquid volume

V_f =Formation fluid influx volume

The foam quality is a function of the pressure in the annulus. Most foams are stable when foam quality is between 0.6–0.97. Low-quality foam contains more liquid than a high quality foam. The foam bubbles of low-quality foam, also known as "wet foam," are found to be spherical and evenly dispersed. The foam bubbles of high-quality foam, also known as "dry foam," have a polyhedral structure with thin separating films.

In stable foam drilling operations, the lower limit of foam quality is usually found at the bottom of the annulus and the higher limit at the top of the annulus.

The foam quality at the bottom of the annulus is controlled by a valve placed on the flow line at the top of the annulus. Since the back pressure upstream of the valve can be set to a desired value, the control of the foam quality at the top enables us to determine the foam quality at any position in the annulus (particularly at the bottom of the annulus).

Qualities greater than 0.97 cause the continuous cellular foam structure that entraps the gaseous phase to become unstable, and the foam to turn into mist. When quality is less than 0.6, gas forms isolated bubbles that are independent of the liquid-phase and two phases can move with different velocities which breaks down the foam structure. To make the Eq. (4.17) more applicable, the ideal gas law is used as

$$\frac{P_s V_s}{T_s} = \frac{PV}{T} \tag{4.18}$$

where

P = Pressure at any point, psi

P_s = Pressure at surface, psi

T = Temperature at any point, °R

T_s = Temperature at surface, °R

V = Gas volume at any point, ft^3

V_s = Gas volume at surface, ft^3

Which can be rearranged as

$$V = \frac{P_s T}{P T_s} V_s \tag{4.19}$$

Substituting Eq. (4.19) into Eq. (4.17) yields

$$\Gamma = \frac{\frac{P_s T}{P T_s} Q_{gs}}{\frac{P_s T}{P T_s} Q_{gs} + Q_l + Q_f} \tag{4.20}$$

where

Q_{gs} = Gas injection rate

Q_l = Liquid injection rate

Q_f = Formation fluid influx rate

Gas-liquid ratio (*GLR*) is defined as

$$GLR = \frac{Q_{gs}}{Q_l} \tag{4.21}$$

In order to maintain the surface foam quality at the desired value, we need to set an appropriate gas-liquid injection ratio at the surface. Based on the maximum allowable foam quality at surface without backpressure, the *GLR* is calculated as

$$GLR = \frac{\Gamma_{\max}}{1-\Gamma_{\max}}\left[\frac{1}{7.48}+\frac{5.615Q_f}{60Q_l}\right] \tag{4.22}$$

By substituting Eq. (4.21) and Eq. (4.22) into Eq. (4.20), the correlation for foam quality at any point can be expressed as

$$\Gamma = \frac{\frac{P_sT}{PT_s}GLR}{\frac{P_sT}{PT_s}GLR+\frac{1}{7.48}+\frac{5.615Q_f}{60Q_l}} \tag{4.23}$$

4.37 Foam Specific Weight

The specific weight is defined as the weight per unit volume of a material. The specific weight of a mixture would be the weighted average of the components. The specific weight of foam can be expressed as

$$\gamma_f = \Gamma\gamma_g + (1-\Gamma)\gamma_l \tag{4.24}$$

where

γ_g = Specific weight of gas phase

γ_l = Specific weight of liquid phase

The ideal gas law is used to account for pressure and temperature dependency of the gas specific weight

$$\gamma_g = 2.7 \times \frac{S_g P}{T} \tag{4.25}$$

where

S_g = Specific gravity of gas

Introducing Eq. (4.25) into Eq. (4.24) results in

$$\gamma_f = \gamma_l - \left[\gamma_l - 2.7 \times \frac{S_g P}{T} \right] \times \Gamma \tag{4.26}$$

4.38 Foam Velocity

The liquid and gas injection rates are controlled by the operator. In event of formation influx, the formation's fluid influx rate can be measured at the surface. Knowing these values, the foam velocity can be expressed as

$$v_f = \frac{Q_g + Q_l + Q_f}{A} \tag{4.27}$$

where A = Cross sectional area of flow path.

Introducing Eq. (4.21) and Eq. (4.22) into Eq. (4.27), the velocity of foam at any point is given by

$$v_f = \frac{\dfrac{P_s T}{P T_s} GLR + \dfrac{1}{7.48} + \dfrac{5.615 Q_f}{60 Q_l}}{A} \left(\frac{144}{60} Q_l \right) \tag{4.28}$$

In foam drilling, unlike the conventional mud drilling, the velocity of the fluid varies with depth. The foam velocity profile is highly affected by the variations of pressure and temperature along the wellbore. Generally, at constant rate and uniform wellbore geometry, the foam velocity decreases with depth.

4.39 Foam Friction Factor

Determining friction factor is crucial for foam flow calculations. Assuming that stable foam flow falls into a laminar flow regime, the theoretical approach for the Moody friction factor is expressed as a function of Reynolds number:

$$f = \frac{64}{\mathrm{Re}} \tag{4.29}$$

with Reynolds number calculated as

$$N_{\mathrm{Re}} = \frac{\bar{v} D_H \bar{\rho}}{\mu_e} \tag{4.30}$$

where

f = Moody friction factor

N_{Re} = Reynolds number

$\bar{v}$ = Average foam velocity

$\bar{\rho}$ = Average foam density

μ_e = Effective viscosity

It has been reported in several cases that the friction factor given by Eq. (4.29) has been too high. Guo et al.[15] developed an empirical correlation derived from two-phase flow regimes that gives good results for foam flow in conditions commonly encountered in foam drilling. This empirical approach, which uses the weight flow rate, is expressed as

$$w = 0.0765 S_g Q_g + 8.33 S_l Q_l \tag{4.31}$$

where

w = Mass flow rate

$$D\rho v = 0.02173 \frac{w}{D_H} \tag{4.32}$$

$$f = 4 \times 10^{1.444 - 2.5 \log(D\rho v)} \tag{4.33}$$

For the Reynolds numbers ranging 2,000–4,000, transitional flow conditions, the Colbrook-White equation can be used to calculate the friction factor. This equation can not be solved directly and must be solved by trial and error.

$$\frac{1}{\sqrt{f}} = -2Log\left(\frac{\varepsilon}{3.7 \cdot D} + \frac{2.51}{N_{\mathrm{Re}} \cdot \sqrt{f}}\right) \tag{4.34}$$

The Churchill equation is another empirical equation developed for the calculation of the friction factor in transitional flow conditions. The Churchill equation is represented as

$$f = 8 \cdot \left(\left(\frac{8}{N_{\mathrm{Re}}}\right)^{12} + \frac{1}{(A+B)^{1.5}}\right)^{\frac{1}{12}} \tag{4.35}$$

where

$$A = \left(2.457 \cdot Ln\left(\frac{1}{\left(\frac{7}{N_{\mathrm{Re}}}\right)^{0.9} + \frac{0.27 \cdot \varepsilon}{D}}\right)\right)^{16} \tag{4.36}$$

$$B = \left(\frac{37530}{N_{\mathrm{Re}}}\right)^{16} \tag{4.37}$$

When the Reynolds number is greater than 4,000, turbulent flow regime, the smooth pipe friction factor can be calculated using a Blasius type equation as

$$f = 0.184(N_{\mathrm{Re}})^{-0.2} \tag{4.38}$$

At large Reynolds numbers, the friction factor is independent of fluid viscosity and only depends on the roughness of the pipe. The Nikuradse equation for fully rough flow can be used to calculate the friction factor as

$$\frac{1}{\sqrt{f}} = 1.74 - 2Log\left(\frac{2 \cdot \varepsilon}{D}\right) \tag{4.39}$$

Jain came up with a non-iterative method that covers all ranges of smooth, transition, and fully rough flow. The Jain equation is represented as

$$\frac{1}{\sqrt{f}} = 1.14 - 2Log\left(\frac{\varepsilon}{D} \cdot \frac{21.25}{N_{\text{Re}}^{\ 0.9}}\right) \tag{4.40}$$

4.40 Cuttings Removal Phenomena

In foam drilling operations, volumetric injection rates of gas and liquid must meet the requirements of cuttings transport capacity and foam stability.

In contrast with air drilling, the minimum kinetic energy criterion for cuttings transport should not be used in foam drilling. The mixture of foam and cuttings can not be treated as a homogeneous system. Significant differences between foam and cuttings velocities may exist in the annulus. Therefore, the minimum drag force criterion is the best applicable option. Considering the laminar foam flow in normal foam drilling operations, the following equation is suggested by Moore[16] to estimate cuttings settling velocity in foam:

$$v_{sl} = 1.56 \frac{D_s \left(\rho_s - \rho_f\right)^{0.667}}{\rho_f^{\ 0.333} \mu_e^{\ 0.333}} \tag{4.41}$$

where

v_{sl} = Cuttings settling velocity

D_s = Cuttings equivalent diameter

ρ_f = Foam density

μ_e = Effective foam viscosity

The equation above is proposed for non-Newtonian power law fluids.

However, the required cuttings transport velocity depends on how fast the cuttings are generated by the drill bit and the amount of cuttings concentration allowed in the borehole during drilling.

When solid particles are settling down in a steady and still fluid of lower density, they first accelerate under the force of gravity and then decelerate due to the increasing drag force between the cuttings and the fluid. After a certain time, the velocity of the cutting reaches a constant value, known as terminal velocity. The cuttings terminal velocity is given by Guo and Ghalambor[17]:

$$v_{tr} = \frac{\pi d_h^{\,2}}{4C_pA}\left[\frac{R_p}{3600}\right] \tag{4.42}$$

where

v_{tr} = Terminal velocity

d_h = Hole diameter

C_p = Cuttings concentration

A = Annulus area

R_p = Rate of penetration

Finally, the foam velocity required to transport the cuttings to the surface is

$$v_{foam} = v_{sl} + v_{tr} \tag{4.43}$$

In drilling with conventional incompressible drilling fluids, it is possible to calculate frictional and hydrostatic pressures separately and then to determine the overall pressure drop. In drilling with foam, such a method is not possible because the frictional and hydrostatic pressure components influence each other.

4.41 Background and History

Krug and Mitchell[18] were the first to develop charts for finding volume and injection-pressure requirements for foam drilling operations. They developed a numerical model based on modified Buckingham-Reiner equations. They assumed incremental pipe lengths with fixed pressure drop along the wellbore and pressure-dependent variables were averaged over each pressure increment.

Beyer et al.[9] developed a finite-difference approach for predicting injection pressures during foam circulation. Their model was found to be valid for foam qualities greater than 0.75. For foam qualities below 0.75 they proposed that frictional gradients be interpolated between that of 0.75-quality foam and the value for flowing water of the same velocity.

Blauer et al.[10] proposed a method in which an effective Newtonian viscosity and foam density are calculated, and hydraulic analysis proceeds as if the fluid were an incompressible Newtonian fluid. They assumed gas-phase density to be negligible. To revise this assumption, Blauer and Kohlhaas[19] proposed another model that iterates on length. To monitor the changes in quality, velocity, viscosity, and pressure, the wellbore was divided into length segments. The real gas law was used to account for gas compressibility.

Okpobiri and Ikoku[20] developed an iterative semi-empirical method for the prediction of frictional losses caused by the solid-liquid phase of foam. Foam flow in this method was mathematically modeled as Ostwald-de Waele power law fluid and solutions were obtained by iterating on pipe/annulus segments length.

Guo et al.[21] presented a trial and error method to estimate the frictional and hydrostatic pressures. They coupled the pressure components through the pressure-dependent fluid density. Foam was treated as a power law fluid.

Lord[22] was the first who used a different and more sophisticated approach. He developed an equation of state based on the real gas law and mass balance conservation. He assumed an average friction factor for the entire system and numerically solved the mechanical energy balance equation for compressible fluid flow. To produce good results with this method, one needs to estimate the best appropriate value of friction factor.

Sporker et al.[23] proposed a new two-phase flow equation to modify Lord's solution. Instead of the real gas equation of state they suggested the use of the virial equation. They solved the differential mechanical energy balance equation and presented an explicit expression for pressure loss of foam flow.

Liu and Medley[24] modified the previous work with accounting for formation fluids influx.

Gardiner et al.[25] proposed an alternative approach by combining the "volume equalized power law" model and the Hagen-Poiseuille equation. They also derived an explicit expression for foam flow pressure losses.

Guo et al.[15] recently presented a closed form hydraulics equation for predicting bottom-hole pressure for foam drilling operations. Their analytical model couples the frictional and hydrostatic pressure components in vertical and inclined wellbores. This newly developed model foam is considered to have the power law behavior.

Numerical methods seem to be the best approach when dealing with compressible fluids. Different numerical solutions used to model the foam hydraulics are summarized below:

- Explicit
  - Differential mechanical energy balance equation for compressible fluid flow
  - Volume equalized power law model and Hagen-Poiseuille equation
- Implicit
  - Closed form hydraulics equation
- Iterative
  - Ostwald-de Waele power law fluid equation
  - Modified Buckingham-Reiner equations
- Finite-Difference Approach

4.42 Steady State Foam Flow

Before going into further details, it is necessary to introduce the basic governing equation of steady state foam flow. Consider foam flow in an inclined conduit section under steady-state conditions. Based on the first law of thermodynamics, the governing equation for flow can be presented as

$$dP = \gamma_f \left[Cos(\theta) + \frac{f v_f^{\,2}}{2gD_H} \right] dL \tag{4.44}$$

where

P = Pressure

γ_f = Specific weight of foam

f = Moody friction factor

v_f = Foam velocity

D_H = Conduit hydraulic diameter

L = Conduit length

θ = Inclination angle

g = 32.2

As it was mentioned, there are several methods proposed to solve the equation above. In this chapter, two different methods will be discussed in detail. Among the different numerical methods, the finite-difference method (FDM) has been acknowledged as one of the most powerful methods for compressible flow analysis. Solving the Eq. (4.44) by the FDM method is rather complex and should be incorporated into a computer code.

The following are some basic assumptions which were made to simplify the solution procedure for the pressure loss calculations along the wellbore:

- It is assumed that compressible gasses can be approximated by the ideal gas law.
- Calculations are based on a constant wellbore geometry.
- The term bottomhole is applied for the total depth just above the drill collars.
- Foam rheology is characterized by the power law model.
- Wellbore temperature is assumed to be equal to the formation temperature.
- Gas expansion right below the drill bit (Joule-Thomson effect) is ignored.
- The minimum drag force criterion is used for cuttings transport analysis.
- The trouble free cuttings concentration is assumed to be 4%.
- Formation fluid influx, if any, is considered as an incompressible fluid which is known as water.

The procedure used in a FDM solution can be explained by the following steps:

Step 1: To maintain the surface foam quality at the desired maximum value, gas-liquid injection ratio (GLR) at the surface should be set to an appropriate value. Based on the maximum allowable foam quality and no formation water influx, GLR at surface pressure (no backpressure at the choke) is calculated.

Step 2: To measure the foam properties at the top of the interval using a preset foam quality at the surface, foam specific weight and hydrostatic pressure are calculated. Knowing the pressure at the bottom, the foam quality and the foam specific weight at bottom are calculated. By taking the average for the foam quality and the foam specific weight over the length, the new hydrostatic bottom-hole pressure is calculated. This procedure is repeated several times to obtain an accurate hydrostatic bottom-hole pressure.

Step 3: To determine the viscosity, foam consistency index K and flow behavior index n are calculated based on the average foam quality calculated in Step 2.

Step 4: To calculate the required velocity to transport the cuttings to the surface, terminal velocity and cuttings settling velocity should be determined. Considering the fact that cuttings settling velocity and viscosity are a function of each other, the values of each is determined by iteration. Once the required minimum foam velocity is determined, the liquid injection rate is calculated. Next, Reynolds number and Moody friction factor are calculated based on the actual foam velocity and viscosity. Knowing the flow properties, the total hydrostatic and frictional pressure at the bottom is calculated.

Step 5: Using the pressure at the bottom obtained in Step 4, the quality and specific weight of foam at the bottom are calculated. Then the average pressure between the surface and the bottom is calculated. The average pressure of the interval is then used to calculate the average quality, specific weight, viscosity, velocity, and Moody friction factor. Consequently, a new bottom-hole pressure is obtained. At this point, Step 4 and Step 5 are repeated several times to update the values used in each step.

Step 6: Foam flow has entered the next depth increment of the wellbore. The foam properties at the top of the new depth interval are replaced by the calculated flow properties at the bottom of the previous interval. Next, Step 4 and Step 5 are repeated to calculate the properties at the bottom.

Step 7: To ensure successful removal of the cuttings, Step 4 is repeated for the latest interval and new liquid injection rate is calculated. Using the new injection rate, Step 3 through 6 are repeated for the entire intervals.

Step 8: The foam quality at the bottom obtained in Step 7 is checked to make sure that it is not lower than the desired minimum foam quality. If the foam quality is lower than the required minimum foam quality, it means back-pressure is needed at the surface choke. The minimum required back-pressure to maintain the foam quality greater than the minimum limit is calculated. The back pressure is then replaced by the latest value and Step 2 through 8 are repeated.

Step 9: The Step 2 through 8 are repeated until the total depth is reached. Note that the calculated bottom-hole pressure is based on the minimum required back-pressure at the choke. This means that no lower values of back-pressure can be used. Further information on calculations and equations used in each step can be found in Sanghani, V.[13] The summary of the procedure is shown in Figure 4–34.

A simpler method of solving Eq. (4.44) is to directly calculate the average value of each parameter and then substitute the results into the main equation. Example 4-2 is presented to describe this solution approach in more detail.

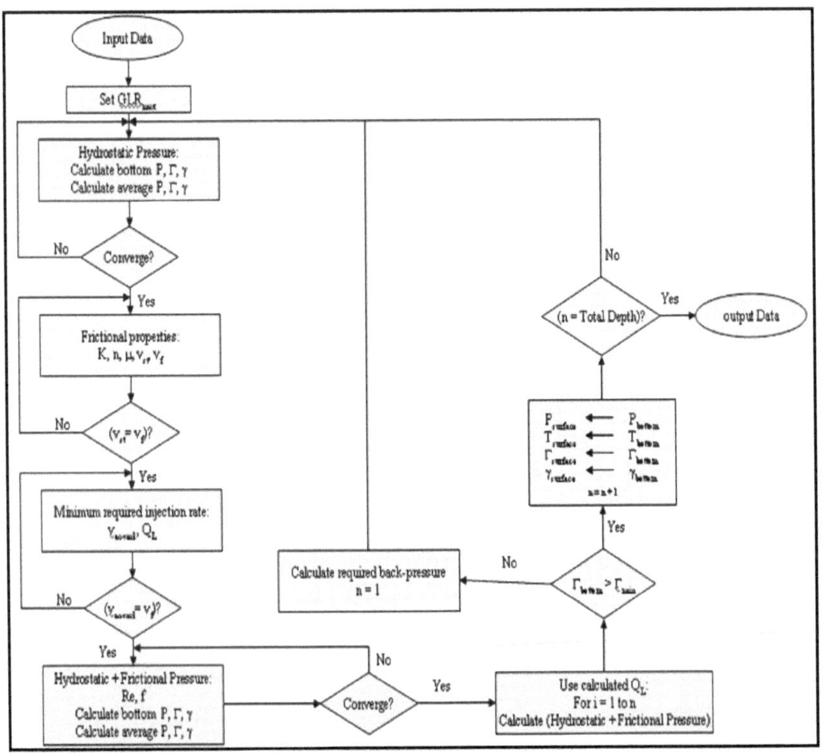

Figure 4–34 *Model calculation procedure*

Example 4–2

Table 4–6 Example 4-2 Input Data

Well and Drill String Data	
Total depth, ft	10000
Hole diameter, in	7.875
Wellbore inclination	0
OD of drill pipe, in	4.5
Length of drill pipe, ft	9500
Length of drill collars, ft	500
OD of drill collars, in	7.5
Operative Data	
Surface temperature, °F	80
Surface pressure, psia	14.7
Thermal gradient, °F/ft	0.01
Maximum quality @ surface	0.95
Minimum quality @ bottom	0.7
Average cuttings concentration	4%
Average cuttings size, in	0.5
Cuttings specific gravity	2.7
Formation water influx, bbl/hr	0
Rate of Penetration (ROP), ft/hr	60
Surface choke back-pressure, psi	100

Using the minimum required injection rates, calculate static bottom-hole pressure and dynamic bottom-hole pressure while drilling a well described in Table 4–6.

Solution

Beginning with basic calculation:

$$Annulus\ Area = \frac{\pi}{4}\left(7.875^2 - 4.5^2\right) \longrightarrow A_{Ann} = 32.8\ in^2$$

$$Hydraulic\ Diameter = \frac{7.875 - 4.5}{12} \longrightarrow D_H = 0.28\ ft$$

$$Surface\ Temperature = 460 + 80 \longrightarrow T_s = 540\ ^oR$$

$$Bottom-hole\ Temperature = (460 + 80) + (9500 \times 0.01) \longrightarrow T_b = 635\ ^oR$$

Part A

Based on the maximum allowable foam quality and no formation water influx, using Eq. (4.22), *GLR* at surface pressure (no backpressure at the choke) is calculated as

$$GLR = \frac{0.95}{1-0.95}\left[\frac{1}{7.48}+0\right] \longrightarrow GLR = 2.54 \ {}^{scf}\!/_{gal}$$

Assuming a liquid specific weigh of 62.4 lb/ft^3, the specific weight at the surface, using Eq. (4.26), would be calculated as

$$\gamma_{surface} = 62.4 - \left[62.4 - 2.7\times\frac{1\times100}{540}\right]\times0.95 \longrightarrow \gamma_f = 3.6 \ {}^{lb}\!/_{ft^3}$$

To calculate the bottom-hole pressure, first just calculate the hydrostatic pressure with no frictional pressure loss. Assuming a constant specific weight along the wellbore, the resulting bottom-hole pressure is calculated to be

$$P_{bottom} = 3.6\times9500 + 14.7\times144 \longrightarrow P_{bottom} = 48555 \ pcf \longrightarrow P_{bottom} = 337 \ psi$$

Based on the calculated bottom-hole pressure, using Eq. (4.23), the foam quality at the bottom is calculated to be

$$\Gamma_{bottom} = \frac{\dfrac{100\times635}{337\times540}\times2.54}{\dfrac{100\times635}{337\times540}\times2.54 + \dfrac{1}{7.48} + 0} \longrightarrow \Gamma_{bottom} = 0.87$$

And the resulting specific weight at the bottom would be

$$\gamma_{bottom} = 62.4 - \left[62.4 - 2.7\times\frac{1\times337}{635}\right]\times0.87 \longrightarrow \gamma_f = 9.43 \ {}^{lb}\!/_{ft^3}$$

Taking an average over the entire length yields

$$\gamma_{average} = \frac{9.43+3.6}{2} \longrightarrow \gamma_{average} = 6.51 \ {}^{lb}\!/_{ft^3}$$

Now, using the average specific weight, the new bottom-hole pressure is calculated to be

$$P_{bottom} = 6.51 \times 9500 + 14.7 \times 144 \longrightarrow P_{bottom} = 76265\ psf \longrightarrow P_{bottom} = 529\ psi$$

Based on the new bottom-hole pressure, the quality and specific weight are recalculated as

$$\Gamma_{bottom} = \frac{\frac{100 \times 635}{529 \times 540} \times 2.54}{\frac{100 \times 635}{529 \times 540} \times 2.54 + \frac{1}{7.48} + 0} \longrightarrow \Gamma_{bottom} = 0.81$$

$$\gamma_{bottom} = 62.4 - \left[62.4 - 2.7 \times \frac{1 \times 529}{635}\right] \times 0.81 \longrightarrow \gamma_f = 13.8\ {}^{lb}\!/_{ft^3}$$

Again, taking an average over the entire length yields

$$\gamma_{average} = \frac{13.8 + 3.6}{2} \longrightarrow \gamma_{average} = 8.69\ {}^{lb}\!/_{ft^3}$$

After several iterations the final bottom-hole properties are calculated to be

$$P_{bottom} = 898\ psi$$
$$\Gamma_{bottom} = 0.71$$
$$\gamma_{bottom} = 20.6\ {}^{lb}\!/_{ft^3}$$

Part B

Now that the average values of pressure, quality, and specific weight are calculated, using these values, the frictional flow properties are to be determined. First, a liquid injection rate needs to be assumed. The liquid injection rate should be assumed high enough to ensure the successful cuttings removal. However, it could not be higher than certain limits opposed by the pumps, compressors, wellbore conditions, economical issues, and more. The exact value, known as the minimum required injection rate, which satisfies the cuttings removal criteria, would be obtained by several iterations. To proceed with the solution, a liquid injection rate of 50 gal/min is proposed as the first guessed value for this parameter.

The average pressure, quality, and specific weight are:

$$P_{bottom} = 499 \;\; psi$$
$$\Gamma_{\;bottom} = 0.83$$
$$\gamma_{bottom} = 12.1 \;\; {}^{lb}\!/\!{}_{ft^3}$$

Using Eq. (4.13), the fluid's average velocity is calculated as

$$v_f = \frac{\frac{100 \times 635}{499 \times 540} \times 2.54 + \frac{1}{7.48} + 0}{32.8} \left(\frac{144}{60} \times 50 \right) \longrightarrow v_f = 2.68 \;\; {}^{ft}\!/\!{}_{sec}$$

Assuming that foam behaves as a power law fluid, using Eq. (4.12) and Eq. (4.11) proposed by Sanghani[13], the power law constants n and K are calculated as

$$n = 0.095932 + 2.3654(0.83) - 10.467(0.83)^2 + 12.955(0.83)^3$$
$$+14.467(0.83)^4 - 39.673(0.83)^5 + 20.625(0.83)^6 = 0.237$$

$$K = -0.15626 + 56.147(0.83) - 312.77(0.83)^2 + 576.65(0.83)^3 + 63.960(0.83)^4$$
$$-960.46(0.83)^5 - 154.68(0.83)^6 + 1670.2(0.83)^7 - 937.88(0.83)^8 = 4.17$$

Substituting in Eq. (4.13):

$$\mu_e = 4.17 \cdot \left(\frac{2(0.237) + 1}{3(0.237)} \right)^{0.237} \left(\frac{12 \times 1}{0.28} \right)^{(0.237-1)} \longrightarrow \mu_e = 0.133$$

Knowing the effective viscosity, the Reynolds number is calculated:

$$N_{Re} = \frac{2.67 \times 12.1 \times 0.28}{0.133} \longrightarrow N_{Re} = 68.5$$

The friction factor for a laminar foam flow, given by Eq. (4.29), is calculated as

$$f = \frac{64}{68.5} = 0.934$$

Knowing the required flow parameters, the total bottom-hole pressure (the sum of the hydrostatic and frictional pressure loss) can be calculated by Eq. (4.44)

$$P = 20.6 \times \left[1 + \frac{0.934 \times 2.67^2}{2 \times 32.2 \times 0.28} \right] \times 9500 + 100 \times 144 \longrightarrow P = 1195 \ psi$$

This, yields to a quality of

$$\Gamma_{bottom} = \frac{\frac{100 \times 635}{1195 \times 540} \times 2.54}{\frac{100 \times 635}{1195 \times 540} \times 2.54 + \frac{1}{7.48} + 0} \longrightarrow \Gamma_{bottom} = 0.65$$

and specific weight of

$$\gamma_{bottom} = 62.4 - \left[62.4 - 2.7 \times \frac{1 \times 1094}{635} \right] \times 0.67 \longrightarrow \gamma_f = 25.1 \ {lb}/{ft^3}$$

Calculating the average values

$$\gamma_{average} = \frac{25.1 + 3.6}{2} \longrightarrow \gamma_{average} = 14.3 \ {lb}/{ft^3}$$

$$\Gamma_{average} = \frac{0.95 + 0.65}{2} \longrightarrow \Gamma_{average} = 0.80$$

and P_{ave} = 647 psi.
These average values are used to recalculate the flow parameters as

$$v_f = \frac{\frac{100 \times 635}{647 \times 540} \times 2.54 + \frac{1}{7.48} + 0}{32.8} \left(\frac{144}{60} \times 50 \right) \longrightarrow v_f = 2.18 \ {ft}/{\sec}$$

$$n = 0.095932 + 2.3654(0.80) - 10.467(0.80)^2 + 12.955(0.80)^3$$
$$+14.467(0.80)^4 - 39.673(0.80)^5 + 20.625(0.80)^6 = 0.254$$

$$K = -0.15626 + 56.147(0.80) - 312.77(0.80)^2 + 576.65(0.80)^3 + 63.960(0.80)^4$$
$$-960.46(0.80)^5 - 154.68(0.80)^6 + 1670.2(0.80)^7 - 937.88(0.80)^8 = 3.69$$

$$\mu_e = 3.69 \cdot \left(\frac{2(0.254)+1}{3(0.254)}\right)^{0.254} \left(\frac{12\times 1}{0.28}\right)^{(0.254-1)} \longrightarrow \mu_e = 0.149$$

$$N_{\mathrm{Re}} = \frac{2.18\times 14.3\times 0.28}{0.149} \longrightarrow N_{\mathrm{Re}} = 58.7$$

$$f = \frac{64}{58.7} = 1.09$$

Once the frictional flow parameters are determined, the bottom-hole pressure would be calculated:

$$P = 14.3\times\left[1 + \frac{1.09\times 2.18^2}{2\times 32.2\times 0.28}\right]\times 9500 + 100\times 144 \longrightarrow P = 1314\ psi$$

After several iterations the bottom-hole parameters are determined as

$$P_{bottom} = 1383\ psi$$
$$\Gamma_{bottom} = 0.62$$
$$\gamma_{bottom} = 27.5\ {}^{lb}\!/_{ft^3}$$

To check for sufficient lift capacity, based on the calculated bottom-hole parameters, the frictional flow parameters at the bottom are calculated as

$$v_f = \frac{\frac{100\times 635}{1383\times 540}\times 2.54 + \frac{1}{7.48} + 0}{32.8}\left(\frac{144}{60}\times 50\right) \longrightarrow v_f = 1.28\ {}^{ft}\!/_{sec}$$

$$n = 0.301\ and\ K = 2.87 \longrightarrow \mu_e = 0.208$$

Note that for the calculation of minimum required rate and its corresponding velocity, instead of dealing with average properties, all the calculations would be performed at bottom-hole conditions. The reason for that lies in the fact that velocity decreases with depth. Hence, the lowest cuttings carrying capacity of the fluid would occur at the bottom of the well.

The cuttings settling and terminal velocity are given by Eq. (4.41) and Eq. (4.42) respectively

$$v_{sl} = 1.56 \times \frac{0.5 \times (2.7 \times 62.4 - 27.5)^{0.667}}{27.5^{0.333} \times 0.208^{0.333}} \longrightarrow v_{sl} = 0.986 \ {}^{ft}\!/_{sec}$$

$$v_{tr} = \frac{\pi (7.875)^2}{4 \times 0.04 \times 32.8} \left[\frac{60}{3600} \right] \longrightarrow v_{tr} = 0.619 \ {}^{ft}\!/_{sec}$$

Finally, the foam velocity required to transport the cuttings to the surface is

$$v_{required} = 0.986 + 0.619 = 1.61 \ {}^{ft}\!/_{sec}$$

Comparing the minimum required velocity and actual velocity at the bottom of the well shows that: $v_{required} \rangle \ v_{actual}$ hence, the first assumed value for the liquid injection rate does not meet the minimum lift requirements. The minimum required rate can be found by trial and error. In this specific case, the minimum required injection rate would be 66 gal/min, which corresponds to a bottom-hole pressure of 1,413 psi.

It should be noted that the solution presented in Example 4–2 gives an approximate estimate of the bottom-hole conditions. The fact is that taking an average value for a property, over the entire length of the wellbore, would be an inherently inaccurate approach. To clarify this, using the FDM solution approach, the same example is worked out and the results are presented in a graphical form. In the developed solution, the entire length of the wellbore is divided into segments, and flow properties in each segment are calculated numerically. Under the same surface back-pressure of 100 psi, the minimum required injection rate is calculated to be 68 gal/min, which shows about 3% increase, compared to the 66 gal/min estimated by the first method. This injection rate would correspond to a bottom-hole pressure of 1,493 psi. This means that the first method underestimates the bottom-hole pressure by 5%.

The pressure profile along the wellbore is presented by Figure 4–35. As it is shown, in spite of non-linearity of the pressure profile, it is fairly acceptable to assume a linear trend.

The major error lies in averaging the velocity value. Figure 4–36 represents the velocity profile along the wellbore calculated by both methods. It is clearly obvious that the velocity trend, computed by the FDM, is more parabolic rather than linear. Hence, when applying the first method, assuming a linear velocity trend could be found as the main source of error.

Furthermore, the accuracy of the frictional pressure drop calculations is strongly affected by the validity of the selected rheological model. Some uncertainties about the foam rheology still exist; more investigation on rheological properties of the different types of foams would increase the accuracy of the frictional pressure drop calculations.

For instance, if using another power law model such as the Martins et al.[12] rheology model, the calculated bottom-hole pressure would increase by more than 20% (1,825 psi). Likewise, using a different rheological model such as Blauer's Bingham plastic rheological model would yield a bottom-hole pressure of 1,020 psi. That, compared to the power law model, underestimates the bottom-hole pressure by 30%.

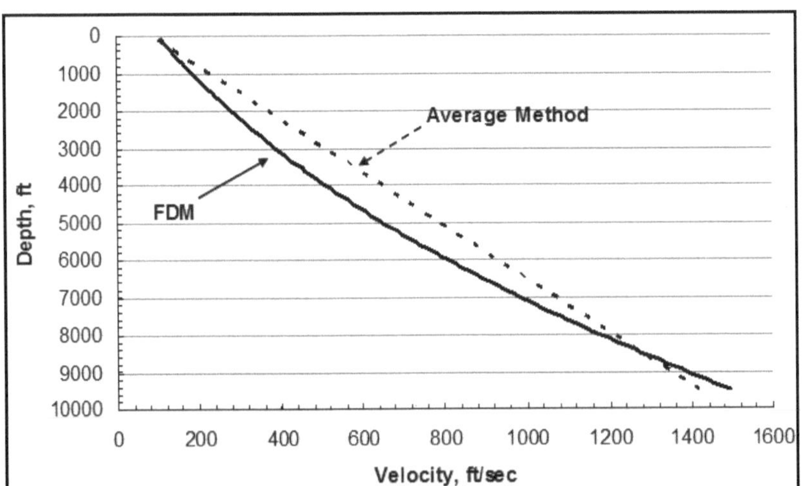

Figure 4–35 *Foam pressure profile*

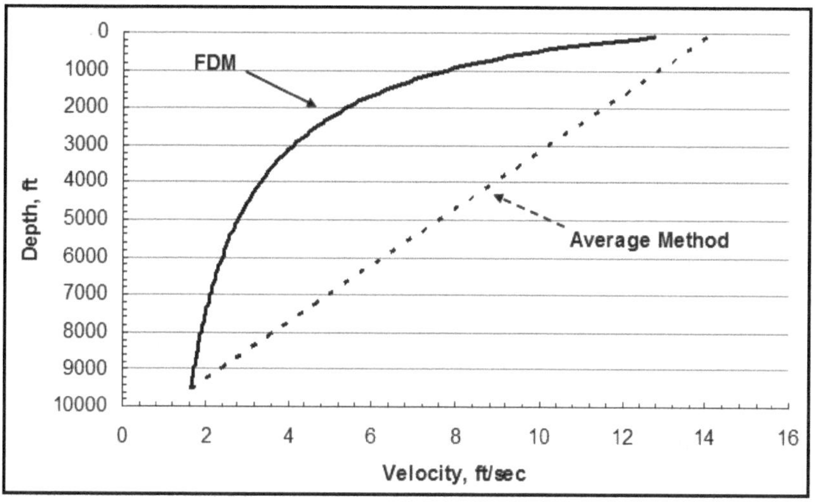

Figure 4–36 *Foam velocity profile*

4.43 Importance of Surface Back-Pressure

The back-pressure could be defined as the pressure within a system caused by fluid friction or an induced resistance to flow through the system. The necessary back-pressure is often created and controlled by a valve that is set to operate under the desired range of conditions. In foam drilling operations, a surface choke is used to create the required pressure to maintain the foam quality at the bottom greater than the minimum limit.

To investigate the importance of the surface back-pressure, different values of back-pressure were introduced into Example 4–2. The results presented in Figure 4–37 shows how, at a constant injection rate, both the bottom-hole pressure and the foam quality at the bottom of the wellbore are affected by the surface back-pressure. As the back-pressure increases, because of the resulting increase in foam quality, the bottom-hole pressure begins to drop. However, when a certain limit is reached, the frictional pressure would prevail over the hydrostatic pressure drop caused by the increase in foam quality. From this point the bottom-hole pressure is more dominated by the frictional pressure rather than the hydrostatic pressure, and increasing the back-pressure increases the bottom-hole pressure.

Figure 4–38 shows how the minimum required injection rates would be affected by the surface back-pressure. As it is shown, increasing the back-pressure decreases the minimum injection rate required for the successful removal of the cuttings. Compared to the

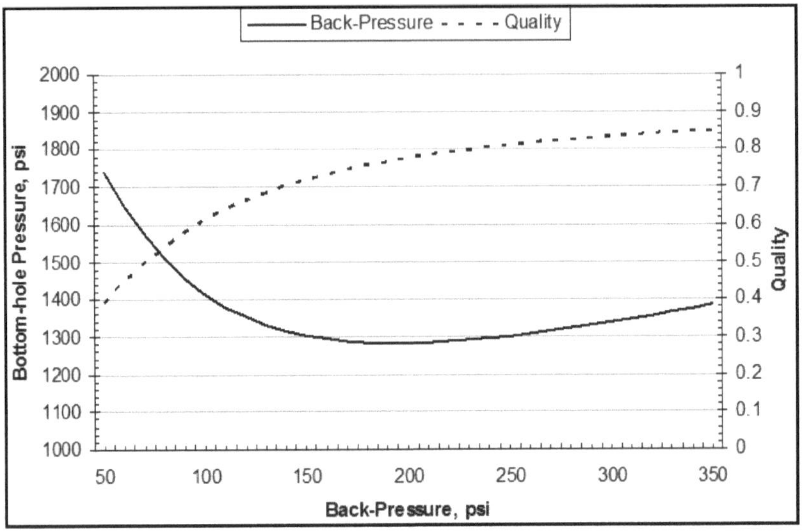

Figure 4–37 *Foam—impressed surface pressure versus quality*

liquid injection rate, the gas injection rate would decrease dramatically. The pressure profile along the wellbore is highly affected by both the surface back-pressure and the injection rate. Finding the best combination of back-pressure and injection rate is not an easy task, and many factors should be taken into account.

4.44 Cuttings Transport in Horizontal Wells

The knowledge of cuttings transport phenomena in horizontal wells is very limited. Using experimental approaches, Iyoho[26] and Larsen[27] developed mechanistic models to investigate the cuttings transport phenomenon in inclined wells. Based on forces acting on a particle, and considering three different modes of cuttings movement, a two layer model was developed by Clark and Bickham[28]. Another two layer model, which was defined by a layer of stationary bed and a layer of pure fluid flow above that, was developed by Gavignet and Sobey[29]. Later, Nguyen and Rahman[30] considered an additional layer of fluid-solids dispersion and developed their three layer model. All these models required simultaneous solutions for the same number of equations and unknowns. Their studies were later expanded to include cuttings transport with multi-phase fluids for inclined and horizontal

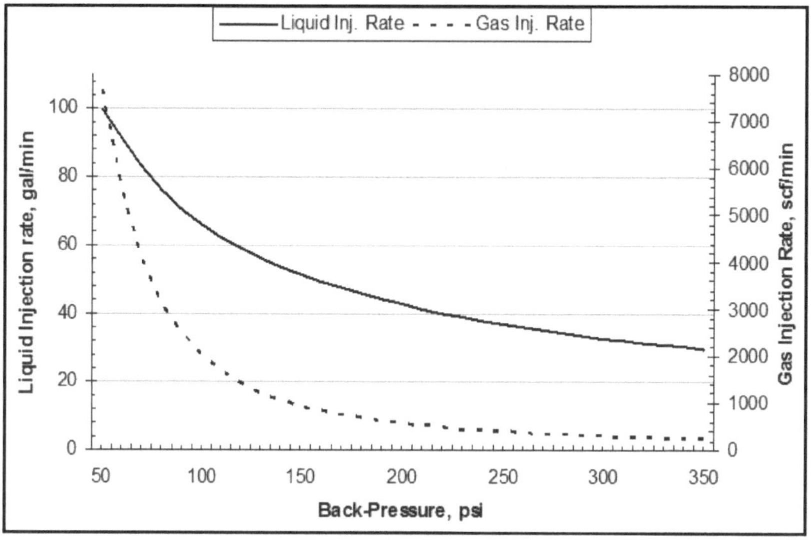

Figure 4–38 *Injection rate versus impressed surface pressure to hold a bottom-hole pressure*

wells. Saintpere et al.[31] were the first to analyze the foam cuttings transport capability in inclined and horizontal wells. Based on the experiments, Martins et al.[12] proposed comprehensive curve fitting functions to predict the cuttings bed erosion capacity in horizontal wells as a function of foam quality and Reynolds number. Using the principles of mass and linear momentum conservation, Ozbayoglu et al. developed a three layer model discussed in Example 4–3.

Example 4–3 Three Layer Model

The horizontal section of the wellbore can be divided into three distinct layers. These three layers, from top to the bottom, are described as:

First layer: Flowing fluid region

Second layer: Flowing mixture of fluid and solids

Third layer: Stationary bed of solids

Note that the wellbore would be divided into longitudinal grids with a specific length. Within each grid, the pressure, specific weight, velocity, fluid properties and so on are assumed to be constant. Within the first layer, the variations of flow properties are ignored.

Also, the slip between the fluid and the cuttings in the second layer is taken into account. The cuttings in the third layer are assumed to be uniformly compacted, thus porosity would be constant. The graphical representation of these layers is shown in Figure 4–39.

Mass Balance

Assuming a constant penetration rate and injection rate, the fluid phase mass balance for a single grid can be written as

$$v_1 A_1 \rho_{fluid} + v_2 A_2 \rho_{fluid}\left(1 - C_{c2}\right) = \bar{v} A_{wall}\left(1 - C_c\right)\rho_{fluid}$$

And mass balance for the solid phase would be written as

$$v_2 A_2 C_{c2} \rho_c = \bar{v} A_{wall} C_c \rho_c$$

where the slurry velocity at the second layer

$$v_2 = v_{2fluid} - \frac{v_{slip} \rho_c C_{c2}}{\rho_{suspension}}$$

And in-situ slurry density and transport velocity of the solids are determined by

$$\rho_{suspension} = \rho_{fluid}\left(1 - C_{c2}\right) + \rho_c C_{c2}$$

$$v_T = v_{2f} - v_{slip}$$

Also, the average velocity and the feed cuttings concentration are defined as

$$\bar{v} = \frac{Q_f + A_{bit} ROP}{A_{wall}}$$

and

$$C_c = \lambda \frac{ROP}{\bar{v}\left[1 - \left(\frac{d_i}{d_o}\right)^2\right]}$$

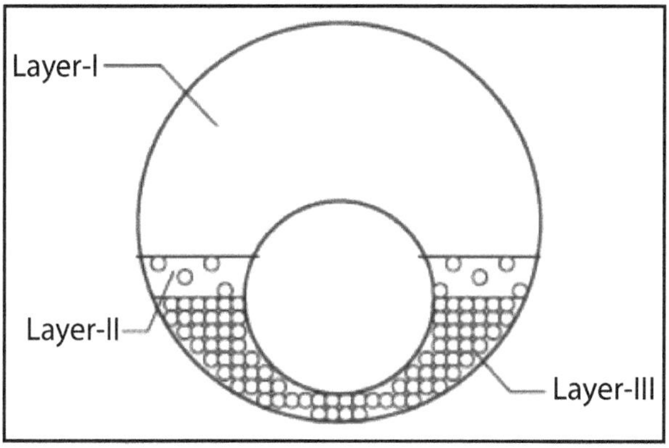

Figure 4–39 *Three layers of flow*

where

$$\lambda = \frac{m_{total} - m^*_{bed}}{m_{total}} = \frac{A_{bit} - A^*_{bed}(1-\varphi_{bed})}{A_{bit}}$$

Momentum Balance

Based on Figure 4–40, the free body diagram of the wellbore, the force balance for the first layer can be written as

$$\Delta PA_1 - \tau_{1\text{-}2}S_{1\text{-}2}\Delta L - \tau_{1\text{-}wall}S_{1\text{-}wall}\Delta L - \rho_f gA_1\Delta L \sin\alpha = 0$$

The second layer's force balance

$$\Delta PA_2 + \tau_{1\text{-}2}S_{1\text{-}2}\Delta L - \tau_{2\text{-}3}S_{2\text{-}3}\Delta L - \tau_{2\text{-}wall}S_{2\text{-}wall}\Delta L - \rho_s gA_2\Delta L \sin\alpha = 0$$

And in the third layer

$$\Delta PA_3 + \tau_{2\text{-}3}S_{2\text{-}3}\Delta L - f_{3\text{-}wall}\rho_{bed}A_3\Delta L \cos\alpha = 0$$

where the shear stress would be defined as

$$\tau_{i\text{-}j} = f_{f_{i\text{-}j}} \frac{\rho_i (v_i - v_j)^2}{2}$$

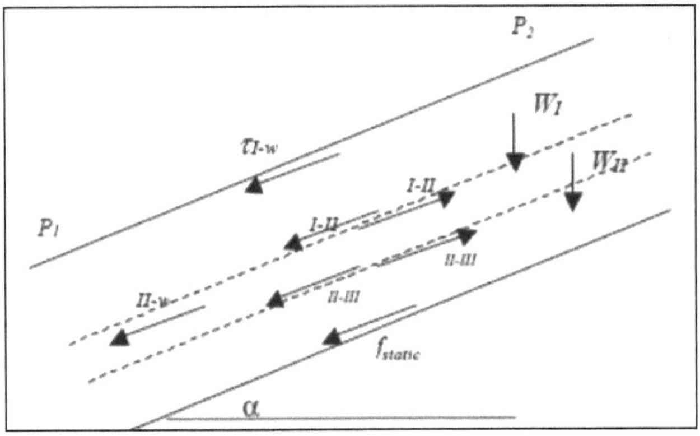

Figure 4–40 *Free body diagram*

The surfaces where the shear takes place are indicated by the indices i and j.

Friction Factors

The static friction factor between the cuttings bed and the wellbore

$$f_{3-wall} = 0.617\left(\frac{A_{bed}}{A_{wall}}\right)^{0.252}$$

The friction factor between the layer boundaries

$$\frac{1}{\sqrt{2f_f}} = -0.86\ln\left(\frac{\frac{d_c}{d_{hyd}}}{3.7} + \frac{2.51}{N_{Re}\sqrt{2f_f}}\right)$$

Friction factor between the first and the second layer

$$\frac{1}{\sqrt{2f_{f'}}} = -0.86\ln\left(\frac{\frac{d_{chr}}{d_{hyd}}}{3.7} + \frac{2.51}{N_{Re}\sqrt{2f_{f'}}}\right)$$

and

$$f_f = f_{f'}\left(\frac{d_{chr} - d_c}{2d_c} + 1\right)$$

where the characteristic diameter is defined as

$$d_{chr} = d_c \sqrt[3]{\frac{C_{c2}}{C_{cbed}}}$$

Finally, the viscosity of the cutting-fluid mixture at the second layer

$$K'_{suspension} = K'\left(1 + 2.5C_{c2} + 10.05C_{c2}^2 + 0.00273e^{16.6C_{c2}}\right)$$

Slip Velocity

$$v_{slip} = \sqrt{\left(\frac{4}{3} \cdot \frac{d_c g}{\left(C_{lift} \sin\alpha + C_{drag} \sin\alpha\right)}\right)\left(\frac{\rho_c - \rho_f}{\rho_f}\left(\xi(\alpha,\beta,\theta)\right)\right)}$$

where (ξ) would depend on the compaction behavior of the cuttings that generate the bed.

In-Situ Cuttings Concentration

$$C_{c2} = \frac{2N_C^2\left(C_{C-bottom} - C_{C-top}\right) + C_{c-top}h_2^2 v_Y^2 - \dfrac{\left(C_{c-bottom} - C_{c-top}\right)h_2 v_y\left(2N_c + h_2 v_y\right)}{e^{\frac{h_2 v_y}{N_c}} - 1}}{h_2^2 v_y^2}$$

Solution Procedure

The solution approach would deal with a total of twelve equations including:

- Two equations from the conservation of mass
- Three equations from the conservation of momentum
- Five equations due to the interface between each layer and corresponding wellbore section
- An equation from the slip between the cuttings and the foam
- An equation which describes the moving cuttings concentration in the second layer

The unknowns for this set of equations are:

- Pressure drop in the axial flow direction
- Thickness for each layer
- Fluid flow velocities in the first and the second layers
- Cuttings slip velocity in the second layer
- Average cuttings concentration in the second layer
- Interfacial stress among each layer, as well as corresponding wellbore section

Having a total of twelve unknowns, a simultaneous solution of these twelve equations for these twelve unknowns is required. The next example shows how the above equations would be used to solve a problem.

4.45 Pressure Drop across Bit Nozzles

A nozzle is a device that causes the interchange of kinetic and internal energy of a fluid as a result of change in cross section of a flow conduit. Nozzles are often installed into the orifices at the drill bit for better removal of cuttings. Pressure loss across the nozzles is usually very significant. There is no universal equation for predicting pressure loss across bit nozzles for all types of drilling fluids. Based on gas fraction of fluids and flow regimes, different nozzle flow models are available.

One of the best models for estimating bit pressure drop for foam has been developed by Okpobiri and Ikoku[20]. The equation for pressure drop across bit nozzles is expressed as

$$\Delta P_b = \frac{\bar{A}}{B}\left(\ln \frac{P_{bh}}{P_{bh} + \Delta P_b} + E v_n^2 \right) \tag{4.45}$$

$$\bar{A} = \frac{m_g \bar{z} R T}{M\left(m_g + m_l\right)} \tag{4.46}$$

$$B = \frac{m_l}{\rho_l\left(m_g + m_l\right)} \tag{4.47}$$

$$E = \frac{1}{2\bar{A}g_c} \tag{4.48}$$

where

ΔP_b = Pressure drop across the bit

P_{bh} = Bottom-hole pressure

v_n = Nozzle velocity

M = Gas molecular weight

m_g = Mass of gas

m_l = Mass of liquid

Note that, compared to the nozzle velocity, the upstream velocity is negligible and the equation above can also be written as

$$P_{bh} = P_{up} \cdot e^{\frac{(B\Delta P_b)}{\bar{A} - Ev_n^2}} \tag{4.49}$$

4.46 References

[1]Sibree, J.O. "The Viscosity of Froth," *Faraday Soc. Trans.*, 1934, 30, pp. 325.

[2]Grove, C.S. "Viscosity of Fire Fighting Foams," *Ind. Eng. Chem.*, 43, No. 5, May 1951, pp. 1120.

[3]Fried, A.N. "The Foam Drive for Increasing the Recovery of Oil," U.S.B.M. Report of Investigation No. 5866, 1961.

[4]Raza, S.H., and Marsden, S.S. "The Streaming Potential and the Rheology of Foam," *SPE Journal*, 7, No. 4, 1967, pp. 359–368.

[5]David, A., Marsden, S.S. and Sullivan, S. "The Rheology of Foam," SPE 2544 presented at the SPE Fall Meeting, Denver, CO, USA, September 28–October 1, 1969.

[6]Einstein, A. "Eine Neue Bestimmung der Molekuldimensionen," *Annalen der Physik*, 1906, 19, No. 5, pp. 289.

[7]Hatschek, E. "Die Viskositat der Dispersoide," *Kollid Z.*, 1910, **7**, pp. 301.

[8]Mitchell, B.J. "Test Data Fill Theory Gap on Using Foam as a Drilling Fluid," *Oil and Gas Journal*, September 1971, pp. 96–100.

[9]Beyer, A.H., Millhone, R.S., and Foote, R.W. "Flow Behavior of Foam as a Well Circulating Fluid," SPE 3986 presented at the SPE Fall Meeting, San Antonio, TX, USA, October 8–11, 1972.

[10]Blauer, R.E., Mitchell, B.J., Kohlhaas, C.A. "Determination of Laminar, Turbulent, and Transitional Foam Flow Losses in Pipes," SPE 4885 presented at the SPE annual California Regional Meeting, San Francisco, CA, USA, April 4–5, 1974.

[11]Reidenbach,V.G., Harris, P.C., Lee, Y.N., and Lord, D.L. "Rheological Study of Foam Fracturing Fluids Using Nitrogen and Carbon Dioxide," *SPE Production and Engineering*, January 1986, pp. 31–41.

[12]Martins, A.L., Lourenco, A.M., de Sa, C.H. "Foam Properties Requirements for Proper Hole Cleaning While Drilling Horizontal Wells in Underbalanced Conditions," SPE 64382 presented at the SPE Asia Pacific Oil & Gas Conference and Exhibition, Brisbane, Australia, October 16–18, 2000.

[13]Sanghani, V. "Rheology of Foam and its Implications in Drilling and Cleanout Operations," MSc thesis, U. of Tulsa, Tulsa, OK, USA, 1982.

[14]Ozbayoglu, M.E., Kuru, E., Miska, S., and Takach, N. "A Comparative Study of Hydraulic Models for Foam Drilling," SPE 65489 presented at the Horizontal Well Technology Conference, Calgary, Canada, November 6–8, 2000.

[15]Guo, B., Sun, K., and Ghalambor, A. "A Closed Form Hydraulics Equation for Predicting Bottomhole Pressure in UBD With Foam," SPE 81640 presented at the IADC/SPE Underbalanced Technology Conference, Houston, TX, USA, March 25–26, 2003.

[16]Moore, P. *Drilling Practices Manual*, Petroleum Publishing Co., Tulsa, OK, USA, 1974, pp. 228–239.

[17]Guo, B., Ghalambor, A. *Gas Volume Requirements for Underbalanced Drilling*, Pennwell Co., Tulsa, OK, USA, 2002.

[18]Krug, J.A, and Mitchell, B.J. "Charts Help Find Volume Pressure Needed for Foam Drilling," *Oil and Gas J.*, February 7, 1972, p. 61–64.

[19]Blauer, R.E., and Kohlhaas, C.A. "Formation Fracturing with Foam," SPE 5003 presented at the SPE Annual Meeting, Houston, TX, USA, October 6–9, 1974.

[20]Okpobiri, G.A., and Ikoku, C.U. "Volumetric Requirements for Foam and Mist Drilling Operations," *SPE Drilling Engineering*, February 1986, pp. 71–88.

[21]Guo, B., Miska, S., and Hareland, G. "A Simple Approach to Determination of Bottom Hole Pressure in Directional Foam Drilling," *ASME Drilling Technology*, February 1988, PD 65, 329.

[22]Lord, D.L. "Mathematical Analysis of Dynamic and Static Foam Behavior," SPE 7927 presented at the SPE Symposium on Low Permeability Gas Reservoirs, Denver, CO, USA, May 20–22, 1979.

[23]Sporker, H.F., Trepess, P., Valko, P., Economides, M.J. "System Design for the Measurement of Downhole Dynamic Rheology for Foam Fracturing Fluids," SPE 22840 presented at the SPE Technical Conference, Dallas, TX, USA, November 6–9, 1991.

[24]Liu, G., and Medley, G.H. "Foam Computer Model Help in Analysis of Underbalanced Drilling," *Oil and Gas J.*, 94, No. 27, 1996, p. 114.

[25]Gardiner, B.S., Dlugogorski, B.Z., and Jameson, G.J. "Rheology of fire Fighting Foams," *Fire Safety J.*, 31, May 1998, pp. 61–75.

[26]Iyoho, A.W. "Drilled-Cuttings Transport by Non-Newtonian Drilling Fluids through Inclined Eccentric Annuli," Ph.D. Dissertation, University of Tulsa, Tulsa, OK, USA, 1980.

[27]Larsen, T.I. "A Study of Critical Fluid Velocity in Cuttings Transport for Inclined Wellbores," M.Sc. Thesis, University of Tulsa, Tulsa, OK, 1990.

[28]Clark, R.K. and Bickham, K.L. "A Mechanistic Model for Cuttings Transport," SPE 28306 presented at the Annual Technical Conference and Exhibition, New Orleans, LA, USA, September 25–28, 1994.

[29]Gavignet, A.A. and Sobey, I.J.: "A Model for the Transport of the Cuttings in Highly Deviated Wells," SPE 15417 presented at the SPE Annual Technical Conference and Exhibition, New Orleans, LA, USA, October 5–8, 1996.

[30]Nguyen, D. and Rahman, S.S. "A Three-Layer Hydraulic Program for Effective Cuttings Transport and Hole Cleaning in Highly Deviated and Horizontal Wells," *SPE Drilling & Completion*, 13, No. 3, September 1998, pp. 182–189.

[31]Saintpere, S., Marcillat, Y., Bruni, F., Toure, A. "Hole Cleaning Capabilities of Drilling Foams Compared to Conventional Fluids," SPE 63049 presented at the SPE Annual Technical Conference and Exhibition, Dallas, TX, USA, October 1–4, 2000.

CHAPTER 5

Air and Gas Drilling (Drilling Dry and with Mist)

Bill Rehm, Drilling Consultant
Arash Haghshenas, Boots & Coots
Abdullah Al-Yami, Texas A&M University

5.1 Introduction

This chapter discusses the advantages and challenges involved in gas drilling. In this discussion the term "gas drilling" will refer to both air and gas unless otherwise defined. It includes an explanation of common field operating procedures and a discussion of the basic theory and mathematics that control the gas drilling system. Finally, there are illustrations of some of the principles and challenges of drilling.

This material does not cover air compression mathematics, the details of air volumes versus humidity, temperature, density altitude, and other basic derivations. Further discussions of those items are covered in *Air and Gas Drilling Manual* (Lyons et al., 2001).

Mist drilling is part of the gas drilling experience and is included in the discussion. Gas drilling is the ultimate underbalanced drilling operation, but it is not properly by formal definition a managed pressure drilling (MPD)/underbalanced drilling (UBD) operation since managing wellbore pressure is not a practical part of gas drilling operations.

Almost all gas drilling takes place on land. Some of the earliest records of gas drilling were in Mexico in the 1920s. Formalized gas drilling in the US. appears to have started in the San Juan Basin of New Mexico in the 1950s when El Paso Natural Gas started using lease gas to improve their drilling rate. The geysers wet steam field in California in the 1970s used air drilling. Drillers in the Appalachian region quickly discovered the value of gas drilling in their hard rock country, and many of the small Eastern US drill rigs have air compressors instead of mud pumps.

Dry gas as a drilling fluid produces the minimum bottom hole pressure allowing maximum drilling rate. With proper design, gas will cool the bit and expel the cuttings. It provides no support to the borehole walls, nor does it act as a barrier against formation flow. In most operations gas drilling has one simple purpose—to increase the drill rate. There are, however, some secondary advantages to gas drilling. These include:

- Drilling in very low pressure formations to avoid lost circulation
- Finding reservoirs that are hidden by liquid or overbalance drilling fluids
- Protecting sensitive reservoirs in tight gas sands, shale, and coal

5.2 Definitions

5.2.1 Air Volume Standards

Gas or air volume measurement used in this book is the API Mechanical Equipment standard for dry air of 14.696 psia at 68°F. This is also very close to the ASME standard. The formal value varies slightly in different disciplines and countries:

SCFM: standard cubic feet per minute

m^3/m: standard cubic meters per minute

Mmscfd: million standard cubic feet per day

m^3/d: standard cubic meters per day

psia: Absolute pressure as apposed to psi, which is gage pressure

5.2.2 Blooie Line

The flow line is renamed the blooie line when air drilling.

5.2.3 Bottom-Hole Pressure

The wellbore pressure gradient is very low with clean gas circulation, roughly equivalent to about 0.003 psi/ft (.068 kPa/m), as compared to normal formation pressure expressed by a column of water to the surface of 0.43 psi/ft (9.73 kPa/m). However, bottom-hole pressure and cuttings transport capability is sensitive to cuttings in the flow stream, friction, and surface back pressure. Figure 5–1 shows an example of the circulating annular pressure profile with backpressures of 14.7 psia (1 atm) and 50 psia (3.4 atm) at a gas injection rate of 2,000 scfm (56.66 m^3) in a 8.5 in. (216 mm) hole. The pressure profile is related to the density of the gas stream and the flowing frictional pressure. When drilling, the pressure also responds to the weight of the cuttings and liquid in the annulus.

The complexity of the problem is due to gas compression. As the pressure increases, the density of gas increases but the volume decrease reduces the velocity and the frictional pressure loss. Drill cuttings or surface backpressures add another variable to compression. The annular pressure profile is the sum of frictional and "density" pressures, two terms which are interrelated and require an iterative process for solution.

5.2.4 Choking

Choking occurs when there is not enough gas velocity present and floating beds start to appear. The well appears to be "choked."

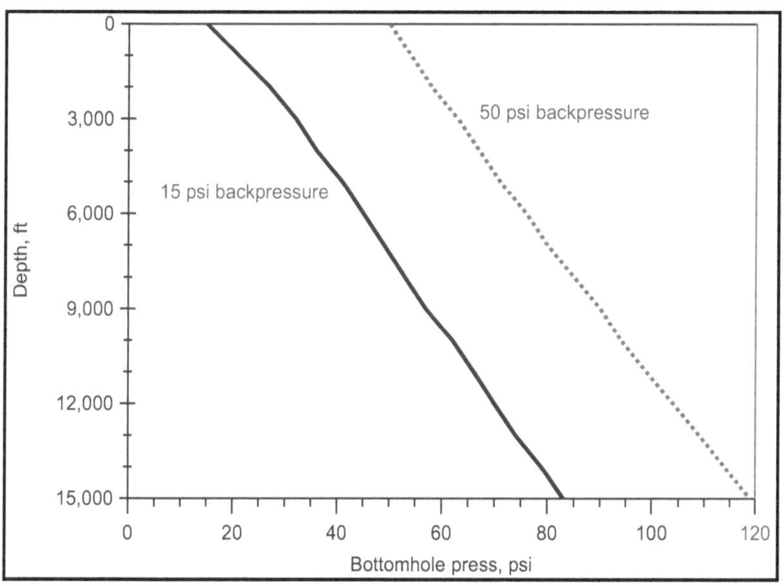

Figure 5–1 *Annular pressure profile*

5.2.5 Floating Bed

When the upward gas velocity abruptly decreases due to an enlarged annular area, the drill cuttings can no longer be carried upward and they form a "floating bed" on top of the higher velocity air.

5.2.6 Free Air

A measure of the actual volume is in ft^3 or m^3; when used with compressors it is a measurement of compressor displacement that ignores mechanical efficiency, the effect of elevation, temperature, and humidity in the volume measurement.

5.2.7 Mist Drilling

Drilling with limited water containing a misting agent that is added to the air stream to help clean cuttings out of the hole. Gas is the continuous phase. Note that this is quite a different condition from "'foam" drilling that is discussed in Chapter 4 where a fluid is in the continuous phase.

Anti-corrosion agents are added to the misting water. This includes a material to keep the pH above 9, generally lime (CaO), or potassium hydroxide (KOH), and one or some of the inhibitors discussed in Chapter 13, Corrosion in Drill Pipe and Casting.

5.2.8 Misting Agent

A misting agent is a detergent that reduces the surface tension of the water and helps form a mist as well as helping to "cut" the mud rings. Most modern misting agents are biodegradable detergents.

5.2.9 Mud Ring

A mud ring is a build up of cuttings in the hole that forms a ring (or hollow cylinder) that restricts the expansion of gas below it. This is usually due to dampness or a small water or condensate flow that wets the drill cuttings and facilitates their packing up against the wall of the hole, reducing the hole diameter.

5.2.10 Slugging or Surging

With too much fluid in the hole, slug flow occurs where a section of the annulus is filled with water, followed by a section of annulus filled with gas. This produces alternating slugs of water and gas at the blooie line. Slug flow can lead to wellbore collapse or stuck pipe.

5.3 Rotary and Hammer Drilling

5.3.1 Rotary Drilling

Rotary air drilling provides very high drill rate. In some rotary operations the drilling rate can be limited in dipping beds by the light bit weights and high rotary speed required to keep the hole straight. Since clearing the cuttings from the hole depends upon velocity, "packed" strings, square drill collars, and heavily stabilized bottom-hole assemblies can restrict the expansion of the gas and cause hole cleaning problems. This leaves pendulum systems and light bit weight as the normal mechanism for keeping the hole straight. Light bit weight in hard rock, even with high rotary speed, restricts the drilling rate. However, stabilized bottom-hole assemblies have been used with success to rotary drill high angle holes in the Texas Panhandle.

Figure 5–2 illustrates general bottom-hole assembly designs to control wellbore direction.

5.3.2 Hammer Drilling

This discussion includes only the "typical" oil field hammer drill. There are a number of other "hammers" used primarily in water well and mining drilling that are significantly different in their air volume, air pressure, and bit requirements.

The "oil field" air hammer uses a free floating piston to hammer against the anvil or driver on the top of the hammer bit. This directs the impulse forces directly into the bit and spares the bottom-hole assembly the majority of the impact force. Light bit weight and slow rotary speed work well with the air hammer. The impact force needs very little bit weight from the drill collars, just enough to keep the hammer on bottom. Normal bit weight will be in the range of 500 to 1,000 lbs per inch of bit diameter (90 to 180 kG/cm of diameter). The required (minimum) bit weight can be observed while drilling by watching the drill rate. The drill rate drops abruptly if the bit is not coupled to the rock.

Many air hammers work at about 1,800 spm. This can be measured from the drillpipe by a simple hand held vibration meter. If the hammer is not striking at the proper rate, it is not operating properly.

The drillpipe needs to be rotated from 20 to 60 rpm to keep the hammer bit from drilling an angular under-gauged hole that will stick

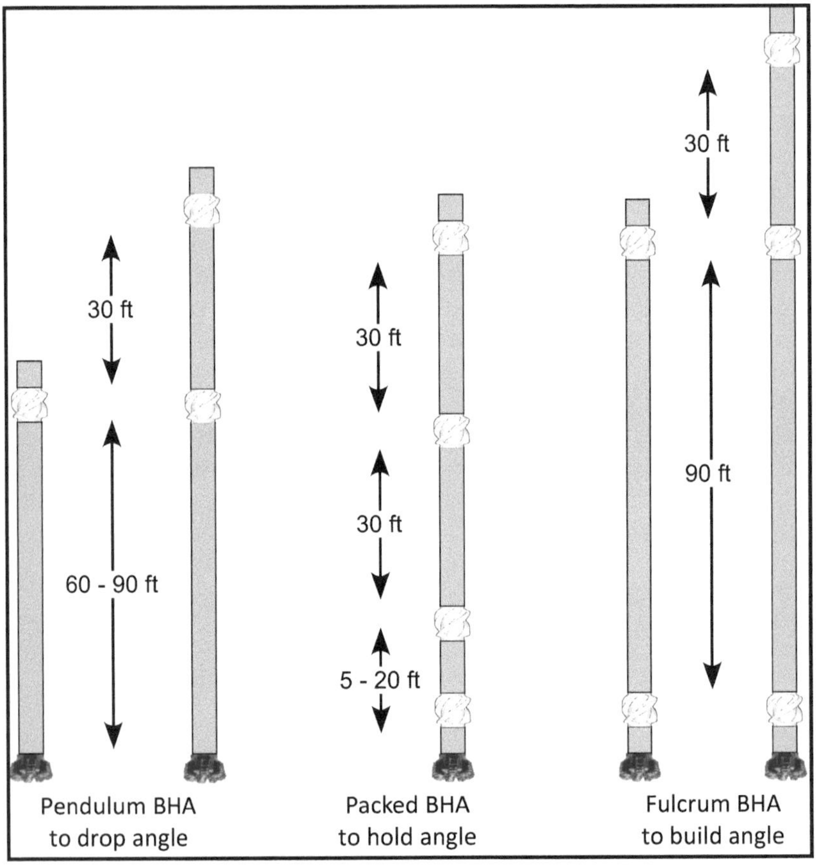

Figure 5–2 *Different BHA configurations to hold, build, and drop angles*

the bit. Higher rotary speed poses little advantage and tends to reduce bit efficiency.

It was once common practice to put the bit on bottom before starting the gas to avoid excessive hammer impact transmitted to the tool joints; the newer hammers have an off bottom bypass that eliminates this problem.

Bit flounder is a common problem with hammer drilling. The bit can out-drill the ability of the gas to clean cuttings from under the bit. With some of the most aggressive bits, the cuttings are too large to be moved by the gas. With a hammer that will handle extra gas volume, drilling rates can be further increased over what would be available with a normal calculated gas flow.

Most modern hammers will handle the water involved in mist drilling, but it degrades performance.

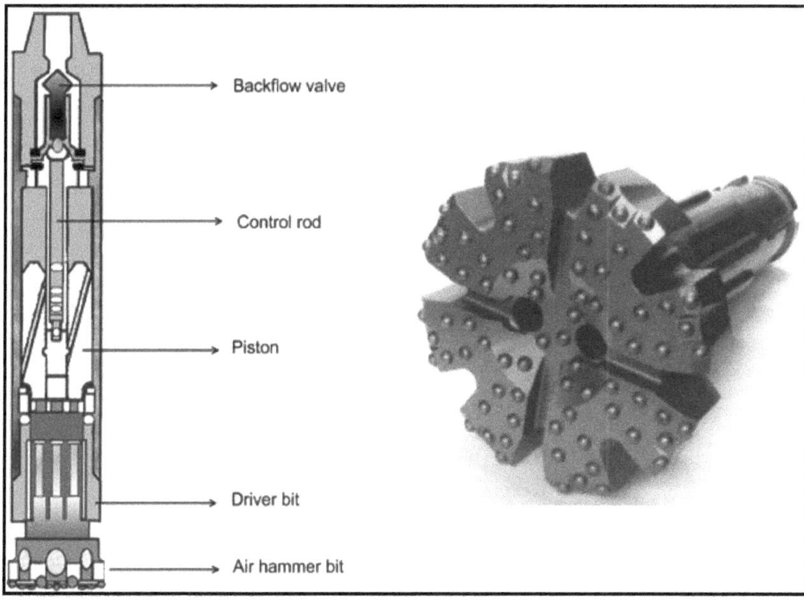

Figure 5–3 *Air hammer (Courtesy of Diamond Air) and air hammer bit (Courtesy of Bit Brokers International)*

The impact of a hammer drill is too great for conventional tricone bits, so a special hammer bit is used. The standard hammer bit has no moving parts. The top of the hammer bit is an anvil or a "driver" that is struck by the free piston (see Figure 5–3). The bottom may be flat, convex, or concave with various gas grooves. Carbide nodes on the bottom of the bit are sometimes reinforced by a diamond coating to provide the hard resistance to wear and transmit the impact into the formation (see Figure 5–4).

Hammer bits tend to drill under gauged holes because of excessive gage wear. It is not unusual to have to reduce gauge size of 1/8 in. (3 mm) with each succeeding bit. It is not practical to ream with a hammer bit, and reamers cannot be used in a gas drilling string because they suppress the expansion of the gas.

HAMMER BITS MUST BE ROTATED! Never drill without rotating the bit at least 20 rpm. High speed rotation (60 rpm) will shorten bit life.

5.3.3 Horizontal Drilling with Air Hammers

Horizontal drilling with gas is a special process because of hole cleaning, directional control, and vibration (with subsequent damage to the survey tool). It is successful in the North East US Appalachian region

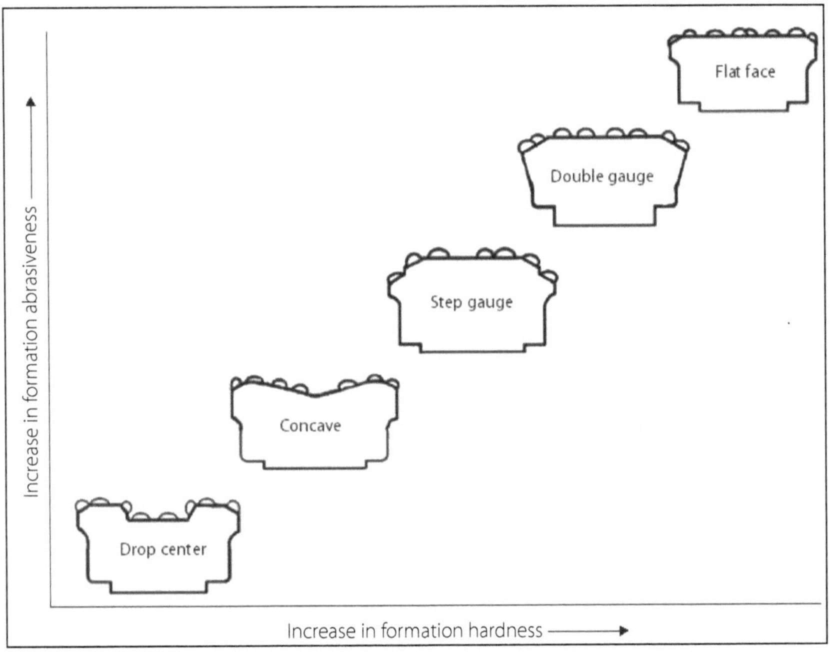

Figure 5–4 *Different designs are available for different formation hardness and abrasiveness (Lyons et al., 2001)*

where the hard Berea sandstone is horizontally drilled using air hammers with rotation and directional control by means of bent housing motors (Pletcher et al., 2010). Horizontal drilling in the coal beds of anthracite or semi-anthracite coal, which are found between harder limestone or sandstone beds, is also successfully practiced in that area.

In the gas horizontal drilling, a near bit stabilizer and an up hole stabilizer are used, which tend to restrict air expansion, and which requires fine (small) cuttings.

5.3.4 Dual Drillpipe

The mining industry makes intensive use of continuous core slim exploration holes and dual drillpipe with reverse circulation to drill. Dual drillpipe has been used in the oil industry, primarily in the Oregon Basin to drill basalt and also in Montana and the Yukon Area of Canada. The dual makeup of drill-pipes and the availability of larger drill-pipes have limited the use of this system in hydrocarbon drilling operations although it is common in slim hole hard rock mining operations. However, Vestavi et al. (2010) have proposed the use of dual drill-pipes for long horizontal wells; larger dual drill-pipes may become more available.

5.4 Advantages of Gas Drilling

In the discussion below, the section on advantages of gas drilling (for oil and gas) is much shorter than the concerns or challenges to gas drilling. This generally reflects the distribution of gas oilfield drilling. Where it works well, which is in the hard rock areas, it is much faster and more economical than using drilling mud. Where there are problems with water flows, wet formations, high pressure fluids, or hydrogen sulfide (H_2S) bearing formations, gas drilling is normally not suitable and there are few gas drilling operations.

5.4.1 Increased Drilling Rate

The primary advantage of gas drilling is greatly increased drill rate. Gas drilling takes advantage of the increased drilling rate that occurs with reduction of bottom-hole differential pressure. In a gas drilled well, there is no damping effect of a liquid drilling fluid and that, along with the reduced wellbore pressure, allows the energy from the drill bit to be directly coupled to the formation. Low or negative differential pressure also tends to make the formation appear more brittle. The result is an explosive breakout of cuttings. Figure 5–5 illustrates the effect of differential bottom-hole pressure on the drilling rate.

The limit of drilling rate is a function of the energy from the bit against the formation, the reaction of the formation to the bit impact, and the ability of the gas system to clean the cuttings from under the bit. It is not unusual for gas drilling rates to be five to ten times that of drilling with a drilling mud. Hammer drilling, as opposed to rotary drilling, usually accounts for the highest drilling rate.

5.4.2 Hidden Gas Zones

The oilfield is full of stories about gas zones that produced "good" flows of gas while being drilled, but the production could never be found or made to flow after the hole was filled with water and the casing run and cemented. While logging, completion equipment, and analysis has been much improved with time, gas drilling still gives up undamaged shows that are overlooked with regular mud drilling operations.

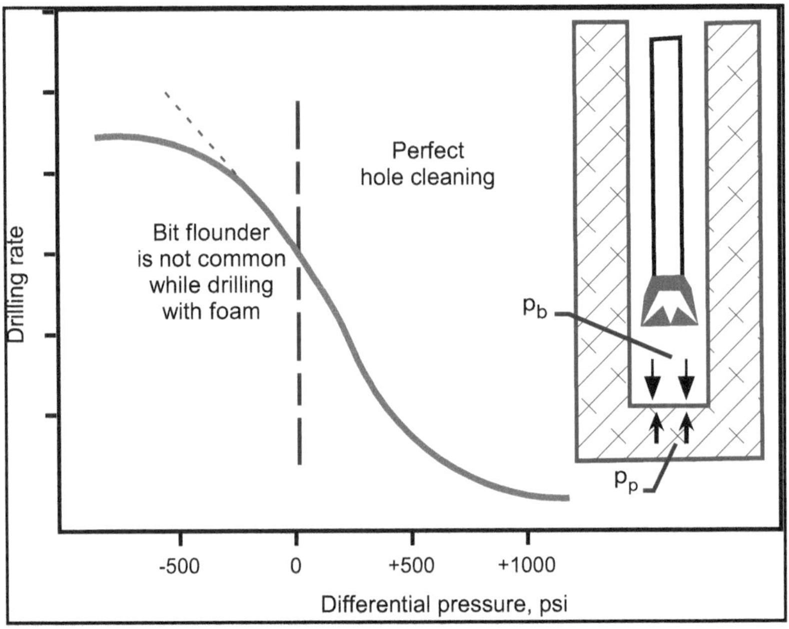

Figure 5–5 *Differential pressure versus drilling rate*

5.4.3 Drilling Bentonite or Water Sensitive Formations

There are several instances where dry gas has been used to drill a very water sensitive formation. In most cases these water sensitive formations would not stand a mist or foam system. On a more common basis, there are a number of shale formations that are well documented to drill very well with a dry gas and collapse or slough after a short period of time when water is encountered or mist is added to the gas stream. One of the best documented of these is the Mancos/ Pierre/Ft. St John shale of the US and Canadian Rocky Mountains.

5.4.4 Formation Test

Gas drilling is a continuous test of producing formations. It provides an excellent initial test of relatively undamaged formations. In most drilling applications there is no provision to actually test volumes or to run an open hole production test. Considering the reservoir damage from filling the hole and cementing casing, it may be worth while to stop short of a target and take open hole production test from a "stray" or unknown zone.

5.4.5 Open Hole Completions

Wells, where the casing is top set above the production zone and the reservoir drilled with gas, can produce from minimally skin damaged zones. In "unconventional gas"—especially low permeability shale gas and coal bed methane—there has been some effort to top set the casing and drill the productive zone with gas to avoid as much reservoir damage as possible. Open hole completions with horizontal drilling with gas and new and improved bottom-hole assemblies have become more common in the Appalachian and Australian coal bed methane fields.

5.5 Limits, Extremes and Challenges to Gas Drilling

The main mechanism for carrying cuttings out of the hole is an upward gas velocity greater than 3,000 ft/min or 50 ft/sec (15.25 m/sec) (Angel, 1958). This is the equivalent of 34 mph or 55 km/h. As the hole becomes deeper, the air in the annulus becomes denser and the velocity lowers, so more gas is needed. Velocity in the wellbore depends upon the ability of the gas to expand from a compressed state at the bit jet to larger volumes and higher velocity in the annulus. Surface restrictions and restrictions or washouts in the wellbore require a greater injected gas volume or a finer cutting size.

5.5.1 Water or Wet Holes

Dampness or water in the hole is the primary cause of gas drilling failure. Dampness occurs frequently and results in muddy cuttings plastered on the side of the hole that form mud rings; mud rings restrict the expansion of the gas below the restriction. Cuttings will not come out of the hole and the fines build up and start to stick to the pipe. This can be alleviated by adding water and detergent to the air stream (mist drilling). However, the water required to wash away the mud rings represses the expansion of the gas and destabilizes shale sections.

Water flows build up pressure as the air column carries the water out of the hole. This in turn requires more air volume. In an 8 ½ in. hole, water flows above 7 gal/min (25 l/min). This, along with the mist volume generally added to help clean the hole, start to cause problems with pressure surges and washouts.

Misting theory and volume requirements are discussed in the operations section of this chapter.

5.5.2 Hole Enlargement

When the cross section of the annular area increases, the velocity of the gas or air stream decreases. The flow patterns change opposite the enlarged zone and the larger cuttings form a floating bed. The hole will not clean up and the cuttings fall to the bottom on connections.

Hole instability caused from stress is common in the mountain regions where much of the gas drilling is done. The hole enlargement related to the principal stress is often not recognized but attributed to "gas" drilling. The introduction of mist or more gas extends gas drilling for a short time, but the problem is in part time related, and eventually the gas drilling has to be abandoned.

Other common concerns (which may also be stress-related), include shale formations. Some shale or mud stone drills with little immediate trouble when the formations stay dry, but shale tends to have weak lamination zones. The low bore-hole pressure as a result of gas drilling, along with bore-hole damage due to drillpipe impact, will cause some shale to cave into the hole and leave an enlarged bore-hole.

In other shale there is enough pore water to both weaken the shale and form a mud in the hole. A damp bore-hole forms a mud ring with the fine bit cuttings that resists removal from the hole. This causes turbulence and restricts the air expansion that might lift the cuttings out of the hole. A very general rule of thumb is that after three days, a wet air drilled hole in the shale will have caving and washouts. This problem may actually be related to geopressured shale where the internal pore pressure is great enough to cause wellbore instability when drilled with a drilling fluid with wellbore pressure less than the pore pressure.

One of the most common cases of hole enlargement is drilling through broken coal in mountain building areas. The coal collapses into the hole until it reaches its angle of repose and leaves a distinct "washout."

5.5.3 Depth Limits

Depth limits with gas drilling are difficult to quantify. Based on the published mathematics, air compression with depth in any oilfield hole should not be a qualifying factor. However, from both experience and the literature, it is difficult to air drill below about 15,000 ft (4,500 m). The general response is that the hole finally resists cleaning, and then a connection part of the new section needs to be "washed down." The clean out becomes increasingly time consuming and difficult with each connection. More air or mist does not solve the problem.

Almost all the dry holes that reach that depth will be in limestone, dolomite or sandstone with a very low permeability.

The mining industry uses a dual drillpipe with reverse circulation. Air is circulated down the annulus between the two drill-pipes and up the inner drillpipe. The annulus between the wellbore and the outer drillpipe has a very small clearance and little circulation. In hard rock mining operations, the wellbore is normally more stable (granite, rhyolite, basalt, etc.), and cuttings removal is not dependent upon the open-hole annulus, so greater air drilling depths have been recorded.

In perfect conditions of no water flows or dampness, and a homogeneous formation, the limit of the air drilling would be wellbore stability. In Table 5–1, the minimum injection rates, bottom-hole pressures, and standpipe pressure are given for a 7 in. hole with 4½ in. drillpipe. The calculations were made using Guo's model (Guo, 2002), which is similar to Angel's calculation, but uses a different friction factor coefficient.

The overburden gradient is assumed to be 1 psi/ft and the differential pressure between the formation and wellbore is calculated. As a simple way of looking at rock failure, if the differential pressure between the formation and the wellbore exceeds the uniaxial strength of the rock, failure (wellbore instability) occurs. Several other factors affect the pressure that wellbore instability occurs, including the layering and type of the formation, tectonic and horizontal stress distribution, pore pressure, inclination of the formation and the wellbore, etc.

The uniaxial strength of sandstone is between 13,000–20,500 psi as measured in laboratories. In reality, presence of cracks (possibly induced during drilling), fractures, and bedding reduces the strength of the formation.

The ultimate depth of the dry wellbore is limited by the strength of the formation. As the formation wall on the wellbore breaks and more and larger cuttings are generated, the hole starts getting larger and larger, intensifying the problem so that more air is unable to solve the problem of fill up on connections.

Figure 5–6 illustrates the differential pressure between the wellbore and the formation. Formation instability (breakage of rock) occurs in the shaded box.

5.5.4 Floating Bed

Where the drill collar string and the drillpipe meet, the annular volume increases and the annular velocity is reduced. This is an area (along with upper hole washouts), where floating beds form. The larger cuttings float on the air column because the velocity below is

Table 5–1 Depth versus Differential Pressure when Dry Air Drilling

Depth ft	Min. Inj. Rate scf/min	BHP psi	SPP psi	Overburden psi	Differential Pressure psi
7,000	1,297	62	131	7,000	6,938
8,000	1,370	70	144	8,000	7,930
9,000	1,440	78	155	9,000	8,922
10,000	1,508	87	167	10,000	9,913
11,000	1,574	97	179	11,000	10,903
12,000	1,638	106	190	12,000	11,894
13,000	1,701	116	201	13,000	12,884
14,000	1,763	127	212	14,000	13,873
15,000	1,824	138	223	15,000	14,862
16,000	1,883	149	234	16,000	15,851
17,000	1,941	161	244	17,000	16,839
18,000	1,998	173	254	18,000	17,827
19,000	2,055	185	265	19,000	18,815
20,000	2,110	198	275	20,000	19,802

great enough to lift them, but the velocity in the larger annular area is not great enough to carry the cuttings upward. The larger cuttings circulate around in the floating bed until they are broken up by the action of the drillpipe and continue up hole, or until the air is shut off for a connection and the cuttings fall back down around the drill collars. The larger cuttings show up as fill after the connection or, in the worst case, they stick to the pipe (see Figure 5–7).

5.5.5 Fishing Operations

Most gas drilled holes need to be mudded up to fish out lost pipe. There are some examples when parted pipe was fished from a hole, but generally parted pipe will drop so fast in an open hole that it corkscrews when it hits the bottom. Items dropped in the hole have

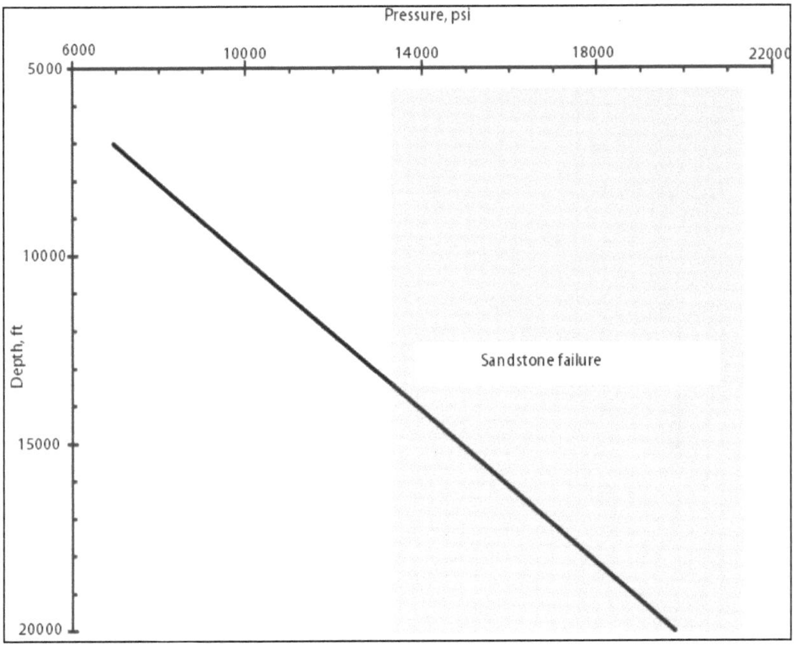

Figure 5–6 *Formation instability for differential pressure between the wellbore and formation*

rarely been recovered without mudding up the hole. Performing any wash over or milling is very difficult in an empty gas drilled hole.

5.5.6 Flashbacks—Fire in the Blooie Line from the Flare

Flashbacks up the blooie line are probably more common than downhole fires, but there has been little study of their frequency. The flashback from the flare line is not normally apparent from the floor; the only indication is surface burns on the rotating control device (RCD) packer. Methods of preventing flashbacks by using a water leg (or water "U" tube) are discussed in Chapter 12, Flaring.

There have been several fires and/or low order explosions in the separator when used to limit dust at the blooie line. When a closed tank, like a separator or frac tank is used, there can be a build up of static electricity if the separator is not completely and properly grounded. The static electricity can cause ignition with the dust and hydrocarbon gas from the hole (Goodman, 2007). This is a particular danger with workover rigs or small drill rigs drilling with air and using small blooie or flow lines.

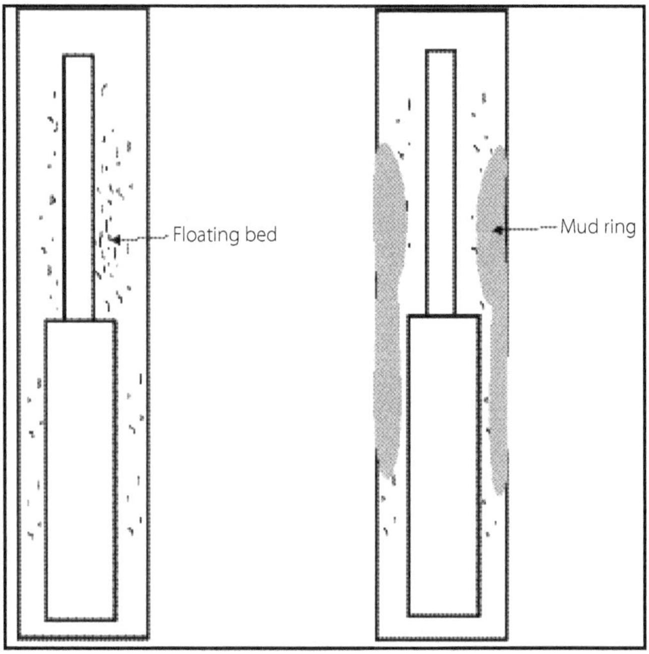

Figure 5–7 *Floating bed and mud ring may occur at the top of drill collars where gas velocity drops as it enters a larger geometry; top of drill collars are usually considered a critical point of hole cleaning requirement to avoid these problems*

5.5.7 Dust from the Blooie Line

Dusting and flares are a major nuisance value and can be a health hazard. The solutions are discussed in further detail in Section 5.6.5, page 320.

5.5.8 Down-hole Fires

The following commentary is based on general experiences in air drilling by one of the editors (Rehm). There is a second opinion or conclusion about drilling with air by Malloy et al., 2007 as well as Grace and Pippin, 1989, and it is expressed in Section 5.5.9. The readers are encouraged to compare the basic references (Malloy et al. as well as Grace and Pippin) and draw their own conclusions. There is relatively little formal documentation of down-hole fires due to air drilling available in the US.

5.5.8.1 Down-hole Fire

Down-hole fires are a curious and somewhat uncommon result of drilling with air. A down-hole fire is almost never evident at the surface until the bit stops drilling. There are no immediate indications like heat, sound, or odor, but there may be a pressure increase on the standpipe gage.

When the bottom-hole assembly is brought to the surface, the bit may be gone, or the bit and several collars may have disappeared. There are also stories of holes in drill collars, elongated collars or pipe.

Seldom is it possible to go back into the hole and drill through the fire area. There appears to be iron slag or bits of steel on the bottom, and the general case is that the well has to be sidetracked.

Down-hole fires almost always occur when *all* the following conditions are met:

- When drilling with air
- When a condensate is encountered
- With elevated bottom-hole temperature
- And slightly elevated bottom-hole pressure

THE USE OF MIST DOES NOT ELIMINATE THE POSSIBILITY OF A DOWN-HOLE FIRE! The point can be made that a limited amount of mist actually adds to the effect of the fire.

Table 5–2 shows ignition temperature and pressure for various hydrocarbons.

The result of these conditions is that in a field where down-hole fires occur, they always occur when drilling with air. Where drilling does not fit the conditions, fires do not occur.

Down-hole fires do not occur:

- When drilling with natural gas or nitrogen
- With foam or gaseated systems
- It is not clear if down-hole fires have occurred when drilling dry methane/ethane zones even with air. In general, field experience indicates this is rare if it occurs
- It is not clear if a down-hole fire could occur as the result of a flashback from the flare, but there appears to be little evidence of it occurring

There is also little formal evidence of down-hole fires when air drilling a coal section. Coal dust fires are not generally well documented,

Table 5–2 Ignition Temperature for Different Hydrocarbons in Air at Atmospheric Condition (Grace and Pippin, 1989)

Hydrocarbon	Pressure, atmosphere	Temperature, °F
Methane	1	1,292
Ethane	1	1,067
Propane	1	986
Ether	1	657
Gas oil	1	676
	5	580
	10	**500**
	15	**450**
	20	**435**
Crude oil	1	716

but apparently fine coal dust and methane in the air stream will ignite under certain conditions. It is not clear, because of lack of records, what part the rank of the coal (bituminous, anthracite) plays in such fires.

The down-hole "fire" appears to be a low order explosion. It occurs when a condensate with a low ignition temperature and low ignition energy flows into a wellbore containing oxygen. The ignition source can be from temperatures generated by the bit on the rock or from collars or tool joints rubbing against the hole, or possibly heat due to compression of the air. The common denominator is the presence of a condensate. The condensates have a lower ignition temperature and ignition energy requirement than methane or the light hydrocarbon gasses.

The common impression is that the fire is like a diesel engine cylinder explosion (Grace and Pippin, 1989) because the driller normally sees a pressure increase at the time of the fire. A flow of condensate will wet the dust and cuttings in the hole and form a mud ring that could increase the bottom-hole pressure. It can be seen from Figure 5–8 that compression under the proper conditions could cause ignition. The fact that the driller may see a several hundred psi increase (±1,000 kPa) in standpipe pressure could also be the result of the fire, not the cause.

Malloy et al., 2007, discussed the compression phenomenon and concluded that it was the potential source of energy for a down-hole

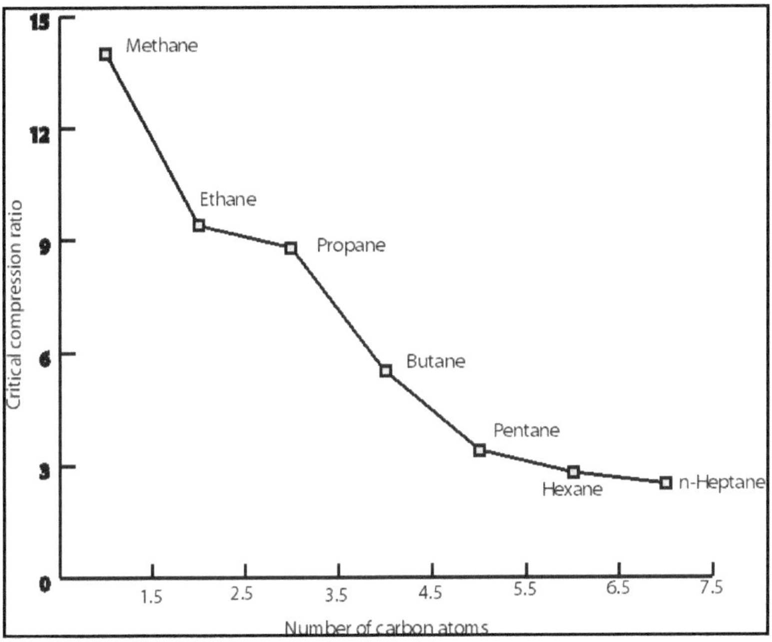

Figure 5–8 *The effect of fuel structure; effect of the number of carbon atoms in the molecule on critical compression ratio of paraffin hydrocarbons; critical compression ratio decreases as the number of carbon atoms in the molecule structure increases (Grace and Pippin, 1989)*

fire. Gokhale et al., 2005, in a discussion of drillpipe failure showed that tool joint temperature could rise above 1,500°F if the pipe was rotated in a tight hole. Other references in the 1960s showed that drillpipe quenching often caused drillpipe failure. Whatever the cause, there appears to be enough energy to initiate an explosion.

Grace and Pippin stated that there was little danger of a fire with methane because the ignition temperature and energy was so high. This conclusion appears to be corroborated by field experience. Fires occur when a light condensate which has a much lower ignition temperature and energy requirement is encountered.

The only sure way to avoid a down-hole fire is to use natural gas or nitrogen and avoid oxygen in the hole. Grace and Pippin felt that misting would decrease the danger of down-hole fire, however, field experience does not bear that out. Misting with the normal amount of mist water does not stop down-hole fires. The introduction of mist may actually enhance the effect of the fire.

What is fairly certain is that the danger of down-hole fires is the greatest when a condensate is encountered. In areas where a fire has occurred, drilling with air into that formation will invariably cause a down-hole fire.

5.5.8.2 Air Drilling and Hydrogen Sulfide Gas

The obvious problem of H_2S gas—it being dangerous to personnel at the surface—is only one of the problems. Traces of H_2S or other sulfur compounds change the ignition and fire characteristics of the hydrocarbon/oxygen mixture. The gas also causes corrosion and steel embitterment. Dry H_2S gas also has a low auto ignition temperature. This is further discussed in Chapter 13, Corrosion in Drillpipe and Casing. The presence of H_2S gas should terminate any air drilling operations.

5.5.9 Drilling with Compressed Air is Hazardous (A Second Opinion about Down-Hole Fires)

Another view of the problem of drilling with compressed air has been discussed by several authors, Malloy et al., 2007, as well as Grace and Pippen, 1989. Malloy et al. proposed that drilling with compressed air is inherently not safe where there is any chance of drilling into a hydrocarbon bearing zone:

> *While the ignition temperature and explosive limits for any gas composition can be measured, the operator has only the most general idea of the actual conditions that exist down-hole. This poses a serious uncertainty when drilling with compressed air in a potential hydrocarbon zone.*

The classic fire triangle (see Figure 5–9) consisting of fuel, ignition energy and oxidizing agent cannot be ignored. The conditions for the completion of this triangle that leads to a down-hole fire do occur.

Refer to Table 5–2, Ignition Temperature for Different Hydrocarbons in Air at Atmospheric Condition (page 314), for an illustration of the concept of down-hole fire potential.

The ignition source can be friction, metal to metal, metal to rock, or possibly adiabatic compression.

There is written and anecdotal information about damage or loss of down-hole equipment with the ensuing NPT. The danger to personnel on the surface is not as clearly documented, but clearly has occurred.

Figure 5–9 *Classic fire triangle*

The authors concluded when there is the potential for penetrating a hydrocarbon bearing zone, compressed air should be changed for natural gas, nitrogen, or a regular drilling fluid.

5.6 Special Rig Equipment for Gas Drilling

5.6.1 Rotating Head

Rig equipment for gas drilling starts with a low pressure rotating control device (RCD). In drilling operations, it is not contemplated that pressure will build up because a minimum annular pressure is required to let the gas expand. A 250 psi (1,700 kPa) rated rotating head is adequate for dry gas and misting operations (see Figure 5–10).

5.6.2 Bit Float and String Float

The drillstring should contain a near bit float to keep back flow or fill from plugging the bit. In addition to the bit float, a string float near the top of the drillstring is installed to avoid having to de-pressurize the entire drillstring to make a connection.

5.6.3 Fire Float and Fire Stop Float

In some areas of the US Rocky Mountains a fire float or a fire stop float has been used in the drillstring just above the bit float. These are

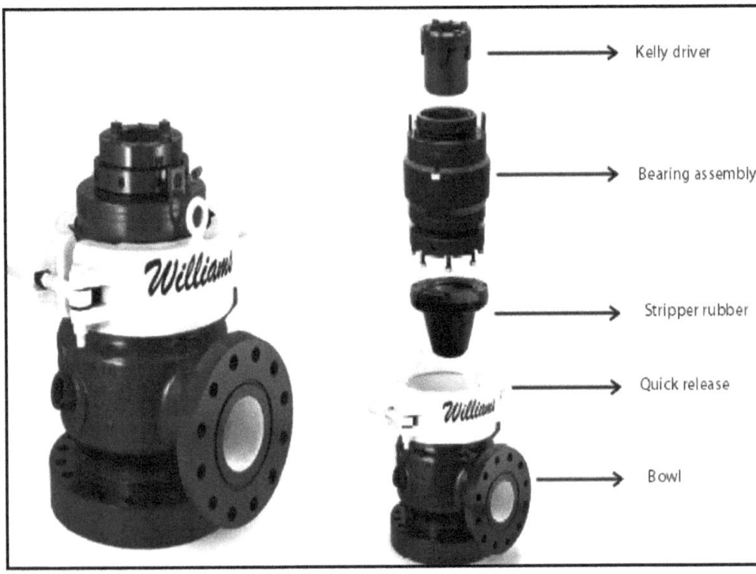

Figure 5–10 *Weatherford/Williams low pressure rotating control device*

used in air drilling in areas where there is uncertainty about the presence of condensates which could cause a down-hole fire. Figure 5–11 illustrates a flapper model. The floats are essentially an upside down flapper or piston valve that is held open by a ring of low melting point metal alloy (see Figure 5–11a). In the case of a down-hole fire, the ring melts and allows closure of the drillpipe. This turns off the source of oxygen for the fire (see Figure 5–11b). It also raises the compressor pressure to open the compressor bypass and so warns of a down-hole fire.

5.6.4 Blooie Line

Blooie lines should be a minimum of 200 to 300 ft. (65–100 m) long and have at least the cross sectional area of the upper hole annulus. The line should be staked down in such a manner that it cannot rise up under the influence of a gas/liquid slug and break off. Blooie lines need to be laid to avoid kinks or bends in the line that would cause the line to twist or break. The low or atmospheric pressure blooie line should never have a downstream shut off valve that could cause it to pressure up and burst. Any shutoff to the blooie line should be in the high pressure feed line or at the well head. In some gas drilling operations, dual or triple lines have been laid to assure that the line will end down wind of the drill rig.

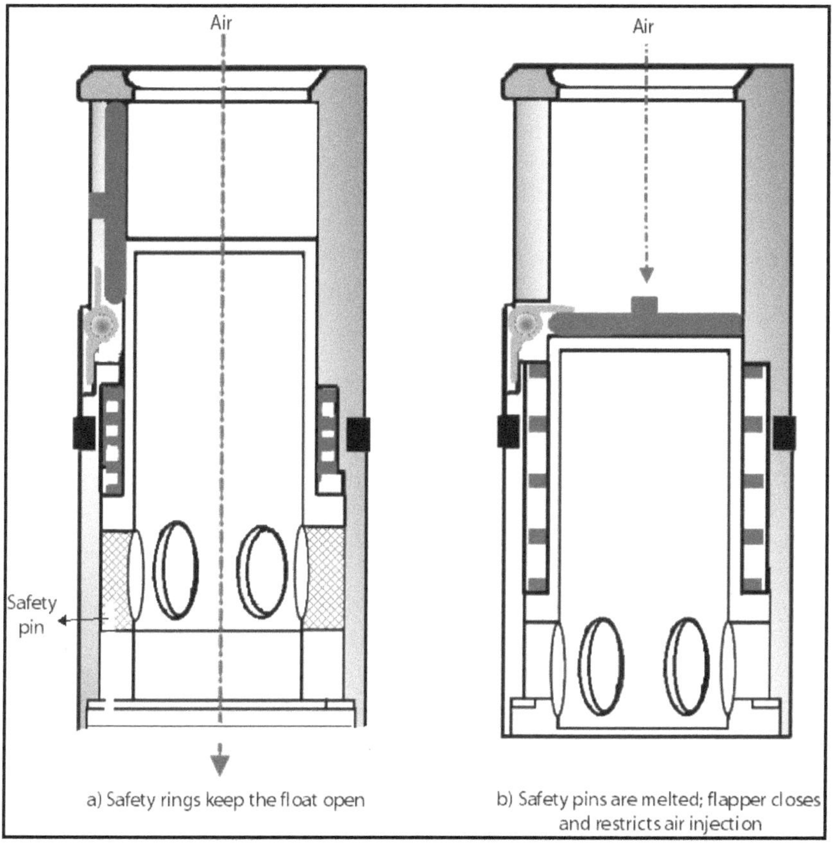

Figure 5–11 *Fire float-low temperature safety pins melt and allow sleeve to close*

A gas bypass line from the compressors or gas line needs to be plumbed into the top of the blooie line about three or four feet (1 m) from the well head at about 45° pointing down away from the wellhead. (Some authorities prefer the bypass line be inserted four or five diameters from the end of the blooie line). Bypassed gas during a connection or trip will pull a slight vacuum on the wellbore and tend to pull any gas from the well head down the blooie line. Another gas bypass should be connected to the standpipe to allow pressure from the standpipe, Kelly, and drillpipe to be released to the blooie line before connections and trips.

5.6.5 Separators, Dedusters, and Mufflers

Separator-deduster-muffler functions can be separate or combined into a single unit. Dust control is managed by a spray in the blooie line where the water mixes with the dust and settles out as a mud. Any large separator/deduster will act as a muffler to contain and dissipate the shock waves from the blooie line. A muffler can be any large vessel that diminishes the sound of the escaping air.

There have been rigs built and commercial systems built to control blooie line discharge. It is critically important in all these systems that separators-dedusters-mufflers be grounded to eliminate static electricity buildup.

Most gas drilling operations discharge from an open blooie line. Unless it is necessary to control dust or sound, or to separate oil from a natural gas drilling discharge, it is safer to avoid using a closed vessel. Blasting the dust or mist and water from the drilling operation into an ungrounded container or steel pit always raises the chance of a static electricity build up, or flare back.

Weatherford U.S., LP has a closed discharge (blooie line) tank system for air/gas drilling, the *Sentry* system, that is claimed to address these problems and makes air drilling a cleaner, safer process by separating formation cuttings, gases and fluids in a closed tank, from which they can be independently routed for proper disposal. The system eliminates the need for open pits traditionally used in air drilling operations, which minimizes the risk of contaminating soil or groundwater and stops the dust to protect personnel and avoid the nuisance value from the fine solid dust particles in the air. Gas or gas and air from the blooie line are sent to a flare line. A solids collection and pumping system transports liquids and solids slurry to a solids processing system for further handling (see Figure 5–12).

Previous closed and semi-closed tank systems have had problems with static electricity, flashback from the flare line, and disposing of the fine dust. Weatherford U.S., LP has approached this problem with a "deluge" water system, careful grounding of the tank, and proper flare line flash back control.

5.6.6 Injected Air or Gas

Gas or air injected to the standpipe manifold should be dry and close to ambient temperature. High temperature air will damage the rotary hose. Damp air makes it hard to dust. All compressors and boosters should have final stage coolers and demisters.

The air hose or pipe that connects the compressors to the standpipe should be pressure tested to at least twice the compressor booster

Figure 5–12 *Sentry system for closed air drilling (Courtesy of Weatherford U.S., LP)*

pressure. The section of pipe or pressure hose from the compressor manifold to the rig standpipe manifold has the greatest potential for causing injury if it fails. Since it would be full of compressed air, if it broke loose from a connection or split, it would whip around with great force. The line should be staked down and the connections should have safety chains.

5.6.7 Mist Pump

The mist pump is an important part of gas drilling operations. The pump is normally diesel or electric driven. A mist pump used with jointed pipe should have horsepower and pressure capability to pump between 0.5 gpm (2 l/min) and 5 gpm (20 l/m) at any pressure up to 2,000 psi (14,000 kPa). Engine systems should have a clutch and transmission so that the pump can work against pressure at any volume. Coiled tubing pressure requirements can be as high as 5,000 psi.

Mist pumps should have two tanks where detergent and chemicals can be mixed on a batch basis. A practical size for each tank would be that it contains enough volume for an hour of misting, at least 6 bbl or 1 M^3. The tanks should have internal volume markings.

Since many of the air drilling operations are in desert areas or areas where there is limited fresh water, misting water should be clean and treated for excessive hardness, and with biocides if necessary. Hard water will leave a residue on the inside of the drillpipe, which under some conditions may actually plug the drillpipe, Van-Vegchel, 2009.

Polluted water can cause drillpipe and casing corrosion. Saline or salt water add to the corrosion problems and limits the effectiveness of the misting detergent.

5.7 Gas Drilling Volume Requirements

The major concern in a gas drilling program is to obtain enough gas to clean the hole. The original work by Angel (1958) was based on quarry drilling and understated the need for dry gas by about 20%. When using this classic source, the understatement on volume needs to be taken into account (see Figure 5–13).

As a VERY rough estimation of air required for a gas drilled hole, double the hole diameter in inches and add 20%. The sum will give a rough number for the minimum air requirements in SCFM (metric hole diameter in mm times .03 equals m^3/min).

In the drilling program, the general rule is to look for the deepest gas drilling interval and compare it to the largest diameter gas drilled interval to obtain enough gas to maintain a minimum of 50 ft/sec, or 3,000 ft/min (1,000 m/min) of annular velocity.

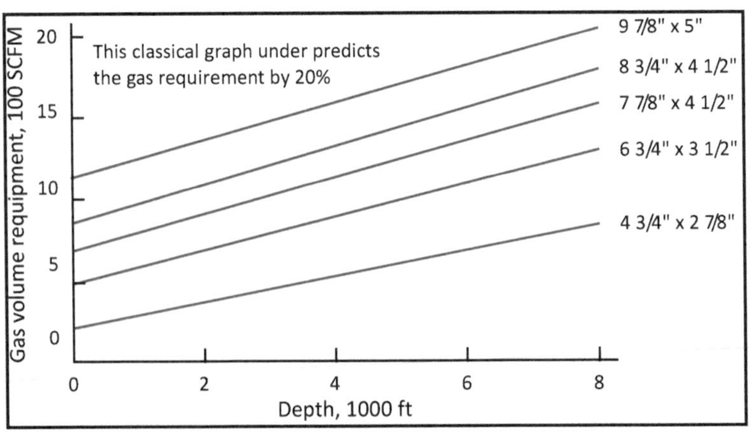

Figure 5–13 *Gas volume requirement for different wellbore geometry (Angel, 1958)*

When hammer drilling or mist drilling, the minimum volume and pressure requirements need to be increased by another 10 to 20%. More gas is almost always better: more gas may clean out floating beds and dry damp formations.

However, other than cost, there is an upper limit to gas volume, especially in deep slim holes. With three or four times the normal gas volume in deep slim holes, the air cannot expand enough and it results in a regime where more air does not help clean the hole. The effect is similar to the friction dominated regime seen in gaseated systems. When misting, the problem of further air and water not cleaning the hole is generally due to slug flow in the deep annulus where gas is no longer the continuous phase. However, note the hole stability discussion in Section 5.5.3, page 308.

Excluding high pressure gas pipelines and cryogenic nitrogen from a tank, air and nitrogen gas volume comes in compressor units. Compressors and boosters are fixed volume machines and operate most efficiently at rated volume capacity. Bypassing air from a compressor to meet limited volume requirement is not a good plan. Air bypasses through an orifice in inverse order of need. When standpipe pressure starts to increase, it generally means more air is required in the hole. With the increase in standpipe pressure, more air is bypassed and less is sent to the hole. If it is necessary for some reason to bypass air from the compressor, be sure to put the air volume meter downstream of the bypass so it will measure air sent to the wellbore.

Compressors and boosters have significant down time, so extra compressors are almost always a good investment.

5.8 Gas Drilling Operations

5.8.1 Unloading the Hole

Most gas drilling operations start out by unloading a hole full of water. The easiest way to unload the hole is to go to the bottom of the casing and start pumping gas. When the compressor booster pressure starts to reach near maxim, pump water with the gas until the pressure goes down, then turn off the water and repeat the process until the well starts to unload. The water increases the average density in the drillpipe and lowers the compressor pressure. Once the hole starts to unload, the compressor pressure will drop.

5.8.2 Drying the Hole

To finish cleaning the water out of the hole, add misting detergent with some lime or potassium hydroxide (KOH) for pH control and use as an inhibitor. Another method to blow "free" water out of the hole is to add a quart (one liter) of misting agent down the drillpipe while circulating dry gas. Another technique is to add the quart of misting agent, close the rams and pressure up the hole to 1,000 psi, then open the rams and flow line and let the hole blow down.

Normally after several hours of blowing with dry air, it is still necessary to drill and generate cuttings to dry the rest of the moisture from the casing and hole. When drilling starts, it is common to drill a little and then circulate and drill again until the well starts to dust. Dust will come in puffs at first and finally in an increasing flow. It may take up to a day of drilling for the drillpipe to "bang" all the damp muddy cuttings out of the casing. Open hole is very hard to dry, probably because of its roughness. It will take drilling and circulating in the open hole to dry it enough to see consistent dust.

5.8.3 General Drilling Operations

There is no fluid in the hole to buffer pipe movement, nor is there any ECD problem from pipe movement. Running in the hole, the bit can hit a ledge and unlatch the pipe from the elevators. Besides the hazard at the surface, pipe dropped in a dry hole drops rapidly and tends to twist when it hits bottom. It is hard to fish. Restrain pipe running speed in a gas drilled hole. Turn the air on at least one joint off bottom so that the bit is not pushed into any fill.

Gas holes tend to make key seats and ledges and unless the drillpipe is in the casing, it is better to reduce pulling speed to avoid jamming into a key seat.

Drill pipe couplings may be damaged or fail when reaming an under-gage air hole because of torq peaks on the couplings. Reaming with a tricone bit also tends to pinch the bearings and wear the bit out of gage. Gage wear is a problem with any bits in abrasive formations. It is always better to pull the bit early to be sure that the hole is in gage.

5.8.4 Connections

On connections, the hole should be circulated until dusting decreases. With a gage hole this will usually take less than ten minutes. As the hole gets increasingly rugged, time to clean the hole may increase to a half hour. Pick up and rotate pipe with the gas on. When the gas is turned off, any cuttings or cavings left in the hole will fall to the bottom and show up as fill after the connection. Hole cleaning time on connections can be significantly improved by the use of a constant circulating sub, Ridley et al., 2011.

There should be a minimum over pull on connections. *Never* pull hard on pipe in a tight gas drilled hole. If there are cuttings around the bit or collars, that will jam the cuttings even tighter, and if it is a key seat it will pull into it tighter. With, tight hole, keep the gas on and move the pipe until it comes free. Any jarring should be down.

5.8.5 Hole Cleaning

At low annular air velocities, a process called "choking" occurs where cuttings are not carried up and out of the hole. There is an increase in compressor pressure, Watson et al., 1997. When annular velocities are increased, the compressor pressure decreases as the cuttings are carried out of the hole. While the industry generally uses 3,000 ft/min (50 ft/sec) as the minimum annular velocity required, it really depends upon the cutting size. On a practical basis, long tooth bits and hammer bits with large inserts generate larger cuttings that can cause "choking" until the cuttings are broken up in the air stream. Smaller cuttings are better from the view point of daily drilling rate and hole cleaning.

In some parts of the Eastern US, where small slim holes are common, 10 ft of fill on a trip would be considered poor hole cleaning. On the other hand, the same 10 ft of fill in a deep large hole in the Western US or Canada might be considered good hole cleaning (Maurer Engineering, 1998).

5.8.6 Well Kicks—Detection

Well kicks with air or gas drilling are primarily a problem of detection. The primary method of detecting a kick is an increase in standpipe pressure, but the increase may be subtle. The general response with both dry gas and a mist is about the same.

The next best detection method is a flow check. Circulate the hole long enough to get the cuttings up hole or to the blooie line, and then bypass the drilling gas. The hole should stop blowing in two or three minutes if there is no gas. In the case of liquids the flow check will not show anything, but when the drill gas is started again, there will be a standpipe pressure increase.

5.8.6.1 Condensate Kick

An influx of a condensate into air drilling generally causes a down-hole fire as discussed in Section 5.5.8 and 5.5.9. The same condensate influx using gas or nitrogen might slightly increase the standpipe pressure, but it would be difficult to detect until it reached the blooie line. With gas as the drilling fluid, condensate would show as a slight increase in the flare and some black smoke. With nitrogen as the drilling fluid, there would probably be no effect except as a possible drip of condensate at the blooie line. If there was a free water knockout at the flare line, there might be no evidence except the subtle increase in standpipe pressure.

5.8.6.2 Gas Kick

Gas influx with air as the drilling fluid might possibly show up as an increase in standpipe pressure, but it would have to be a very large gas influx to effect the standpipe pressure. Generally a gas kick with air will show up as a flare at the blooie line. While it is possible to get a down-hole fire with natural gas, the ignition temperature and energy required makes it an unusual occurrence.

Gas influx with natural gas as the drilling fluid will show an increase in the flare, but it may not be noticed.

Gas influx with nitrogen as the drilling fluid will be hard to detect because the nitrogen will suppress any flare.

5.8.6.3 Water, Oil, and Heavy Oil

Kick with liquids will cause an increase in standpipe pressure. The pressure increase may be the result of the density of the liquid in the air column and/or the development of a mud ring.

Oil influx when air or natural gas drilling will show up as black smoke and change in the blooie line flare.

Oil influx when drilling with nitrogen will probably not show up in a flare because the nitrogen represses combustion. The oil will probably show up as an oil drip in dry drilling or as an emulsion in the misting water.

5.8.7 Well Kicks—Solutions

The solution to a well kick depends in part upon the drilling gas but follow these guidelines:

- The kick can always be killed by pumping water down the drillpipe and around the annulus.

- Bypass the blooie line and go through the high pressure choke line with an open choke. This will provide enough backpressure along with the fluid column to kill or control the well.
- This also gives the option to increase annular pressure if the kick has an unexpected high volume or high pressure. Be prepared to convert to drilling mud as soon as possible to stabilize the wellbore.
- Standard well control procedures i.e., Driller's Method, will not be effective until the drillpipe and annulus are full of water or drilling mud.
- Be concerned about the ECD pressure at the last casing shoe.

5.8.7.1 Drilling with Air

Anytime the drilling fluid is air with its 20% oxygen content, drilling hydrocarbons pose a risk. However, it is quite common to continue to drill with air when encountering small quantities of gas or heavy oils. As a practical matter, many of the small gas stringers are drilled and depleted before any sort of surface action can be taken. The addition of mist will help with oil shales or heavy oil. The ignition temperature and energy of both gas and heavy oil is high enough that there have been minimal problems. Nevertheless, there is some risk with air drilling in gas or heavy oil zones, and it becomes a matter of policy or expediency. The best choice is to convert to foam or a drilling mud.

The addition of mist may help clean the hole, but probably does not help with repressing a down-hole fire.

Any evidence of light oil or condensate should cause an immediate stop to air drilling.

5.8.7.2 Drilling with Cryogenic or Membrane Nitrogen

There is little danger of fire or explosion when drilling with nitrogen. The best solution to nitrogen drilling with a well kick is to continue to drill or at least circulate enough to clean the hole. The problem with nitrogen and gas is that the nitrogen suppresses the flare. Nitrogen and gas or water will require mist or it may not be possible to clean the hole and the hole needs to be displaced with a drilling mud.

5.8.7.3 Drilling with Natural Gas

Drill ahead! Gas, condensate, or light oil will only make a bigger, smokier flare.

Very heavy oil may require mist to clean the hole. The well can be killed at any time using the procedure in Section 5.8.7.

5.8.7.4 Cautions with Air/Gas Drilling Kicks

In any air or gas drilling operation when there is a kick or the potential for a kick, watch the RCD carefully for surface leaks. Use a hand held gas detector if possible. A leaking RCD can fill the substructure or floor with an explosive mixture.

The flare pilot needs to be hot and in good condition.

Keep air or gas going though the blooie line or choke line to prevent a flashback.

5.9 Mist Drilling Operations

Almost all gas drilling operations finish by using a mist system to extend the gas drilling operation. In theory, a small amount of water can be added to the system that ends up as a mist or fog. In practice, enough water needs to be added to the system to:

- Cut the mud cake formed from formation water and cuttings dust
- Carry the larger damp cuttings out of the hole

In an 8 ½ in. hole, a minimum of about 2 gal/min (7 l/min) can be useful but generally requires about 4 gallons/minute (15 liters/min) 6 bbl (1 m^3) of water per hour. The limit to water additions is unstable compressor pressure and reversion to a slug flow that produces alternate heads of water and gas.

Slug flow generally starts at somewhere above 7 gal/min (25 L/min) in an 8 ½ in. hole. Heading generally leads to either hole collapse or an inability to clean the annulus and excessive drag or stuck pipe.

Figure 5–14 illustrates annular/mist flow pattern during mist drilling. High velocity gas pushes liquid phase to the wellbore and drillpipe. Gas moves faster than the thin layers of water and creates shear force and turbulence at the surface of water layers, removing small droplets of water as mist and carries them in the annulus as mist. The steady state pattern is upset by hole roughness and washouts and there may be a number of intermediate flow patterns.

In misting operation, the gas quality is maintained above 95% at the bottom-hole condition. These are very imprecise values because hole problems from excessive mist water also depend on cutting sizes

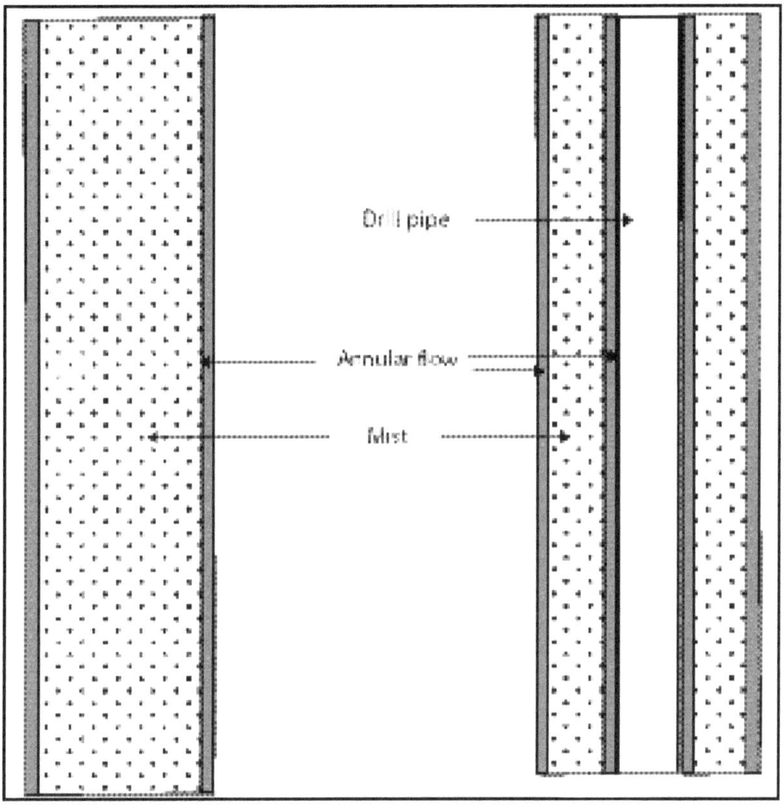

Figure 5–14 *Annular/mist flow pattern exists during mist drilling; thin layers of water form on the wellbore and drillpipe and small drops of water flow as mist in the annulus*

from sloughing and washouts. The velocity of gas and liquid phases in the wellbore are not equal; therefore normal calculation of gas and liquid volume is not an accurate estimate. Gas quality value requires complex calculation using iterative methods to balance the thickness and shear force on the water layer in the bottomhole. The mathematics of annular flow are illustrated in Section 3.9.1, page 133.

When air injection stops, the layer of water falls back down the hole with the cuttings. Later when the air injection is resumed, the difference between gas and liquid velocity shows as a lag time between the time gas and liquid flow appear at the surface.

Mist operations utilize from 25% to 50% more gas than dry gas drilling and require about 25% more pressure. Misting is a sloppy imprecise procedure dealing with a number of unknowns in the wellbore. It is more of an art than a science and the drilling program has to yield to field experience.

There have been some attempts to use foaming agents (and more water) in place of the simpler misting detergent. Foam in the bottom of the hole would have a better cleaning capacity at the bit and around the collars. The problem is that foam has a water continuous phase. Up hole when the gas becomes the continuous phase and goes through a slug flow sequence, there is a loss of lifting capacity and increased potential for stuck pipe. The foam/mist idea is an attractive approach, but at present has had limited success. Misting works reasonably well and foam is a good drilling fluid. Combining the two, however, has some worrisome problems.

5.10 Conclusion

Gas and mist drilling is as much art as engineering. Gas velocity is the critical element, but it is difficult to be sure of what is happening at the bit, around the drill collars, and up hole. The best solution is an experienced attentive driller and careful consideration to standpipe pressure, pipe drag, and the blooie line output.

Drilling with air introduces an uncertainty about down-hole fires. Drilling with nitrogen or natural gas is less uncertain but more expensive.

Mist drilling can cause corrosion problems on the drillpipe and casing. Mist drilling also tends to be limited to several days before hole instability occurs. Despite these problems, the drilling rate advantage is so great that gas drilling will always have a place. In unconventional gas, where fracturing is not required, gas drilling may provide excellent open hole completion.

5.11 Questions

1. What is the primary advantage to gas drilling?
2. What are the main reasons that will terminate a gas drilling operation early? Give three reasons.
3. What are the two main reasons for using a hammer drill?
4. What are the upper limits and lower limits for mist water volumes in an 8 ¾ in. hole? (Either use a number in gpm or l pm, or explain the limits.)

5. How many air compressors are required to drill an 8 ¾ in. air hole 12,000 ft deep with 4½ in. drillpipe at 60 ft/hour? Each compressor has a theoretical output of 1,100 scfm or (31 m^3/m).

6. A 15,000 ft well is planned to be drilled using air drilling technique. 9 5/8 in., 43.5 lb/ft casing is set at 12,000 ft. The ID of casing is 8 ¾ in. The drill string is made of 500 ft of 6 × 2 ¼ in. and 5 in., 25.6 lb/ft drillpipe. The ID of drillpipe is 4.0 in. The maximum drilling rate is expected to be 90 ft/hr using 7.5 in. bit and 3 × 20 nozzles.

   (a) What is minimum gas injection requirement?

   (b) Calculate pressure at casing shoe, top of drill collars, and bottomhole.

   (c) Calculate pressure at the top of nozzles.

   (d) How much pressure is expected at the standpipe?

   (e) How much pressure is expected at the standpipe if misting is done at the rate of 6 bbl/hr and formation water influx is 4 bbl/hr?

5.12 Answers

1. The primary advantage to gas drilling is increased drilling rate.

2. The basic reasons for early termination of air drilling operations are:

   Wet hole, too wet for misting

   Caving formations, especially shale or coal

   Major wash outs in the hole

   Down-hole fire

   Fishing

3. The two main reasons for using a hammer drill are:

   It drills faster than straight rotary because there is more energy expended on the bottom of the hole.

   It drills with light weight so that you can drill a straight hole with a pendulum BHA.

4. The lower limit for mist volume is enough liquid and detergent to cut the mud rings. In an 8 ½ in. hole this is about 3 bbl/hr or 2 gallons/min (8 l/min). The upper limit of mist water injection is near 8 bbl/hr or 5 ½ gal/min (21 l/min).

5. The Angel's tables are about 20% low. 1,900 ft^3/min × 1.20 = 2,280. This would be stretched at 2 compressors and require a lower drill rate, but it probably could be done. The technical question is whether or not to provide backup and adequate air with washouts. Are you willing to pay for a third compressor?

6. (a) 1,500 scfm

   (b) pressure @ CSG = 74 psi, pressure @ top of collars = 123 psi, BHP = 152 psi

   (c) 168 psi

   (d) 101 psi

   (e) In this case, minimum gas requirement is 1,700 scfm, BHP is about 200 psi, and expected standpipe pressure is 160 psi.

5.13 References

Angel, R.R., *Volume Requirements for Air and Gas Drilling,* Gulf Publishing Company, Houston, TX, USA, 1958.

Chen, S., Xinming, N., Nas, S. and Holt, C. "Managed-Pressure Drilling Reduces China Hard-Rock Drilling by Half," SPE 105490 presented at the IADC/SPE Drilling Conference, Amsterdam, The Netherlands, February 20–22, 2007.

Curlee, C.K., Broulliard, S.J., Marshall, M.L., Knode, T.L. and Smith, S.L. "Upstream Onshore Oil and Gas Fatalities: A Review of OSHA's Database and Strategic Direction for Reducing Fatal incidents," SPE 94416 presented at the SPE/EPA/DOE Exploration and Production Environmental Conference, Galveston, TX, USA, March 7–9, 2005.

Gala, D., Viera, P., "Advancements in Drilling Increase Penetration Rates in Deep, Hard, and Abrasive East Texas Formations," IPTC 13229 presented at the IPTC Conference, Doha, Qatar, December 7–9, 2009.

Gokhale, S., Reynolds, N. and Sean E. "Rotating While Packed Off May Cause Unexpected Heat-Induced Drillpipe Tensile Failures," SPE 92429 presented at the IADC/SPE Drilling Conference, Amsterdam, The Netherlands, February 23–25, 2005.

Grace, R.D., Pippin, M. "Down-hole Fires During Air Drilling: Causes and Cures," *World Oil*, May 1989, pp. 42–44.

Guo, B., and Ghalambor, A. *Gas Volume Requirements for Underbalanced Drilling—Deviated Holes,* Pennwell Corporation, Tulsa, OK, USA, 2002.

J.E., Neufeld, M.H. and Sorrells, D.C. "Development of the CottonValley Geopressure Zone in Panola Country, Texas, Using Air/N_2 Drilling and Open Hole Completion Techniques," SPE 14657 presented at the SPE East Texas Regional Meeting, Tyler, TX, USA, April 21–22, 1986.

Lyons, W.C., Guo, B., Seidel, F.A. *Air and Gas Drilling Manual*, McGraw-Hill, 2001.

Malloy, K.P., Medley, G.H., Stone, C.R. "Air Drilling in the Presence of Hydrocarbons: A Time for Pause," SPE 108357 presented at the IADC/SPE Managed Pressure Drilling and Underbalanced Operations Conference and Exhibition, Galveston, TX, USA, March 28–29, 2007.

Maurer Engineering, (Chaffin, Medley, Rehm) *Underbalanced Drilling and Completion Manual*, DEA 101, November 1998

Mehta, S.A., Moore, R.G., Pratt, C.A., Gair, S.D. and Hoyer, C.W.J. "High-Pressure Flammability of Drilling Mud/Condensate/Sour Gas Mixtures in Deoxygenated Air for Use In Underbalanced Drilling Operations," SPE 37067 presented at the SPE International Conference on Horizontal Well Technology, Calgary, Alberta, Canada, November 18–20, 1996.

Pletcher, J., Scarr, A., Smith, J., Swadi, S. and Rogers, C. "Application of Air Percussion Drilling Improves Drilling Efficiency in Horizontal Sandstone Wells," SPE 135308 presented at the SPE Annual Technical Conference and Exhibition, Florence, Italy, September 19–22, 2010.

Rehm, B. *Practical Underbalanced Drilling and Workover*, Petroleum Extension Service, University of Texas, Austin, TX, USA, 2002.

Ridley, K, Portas, B., Robbie J., Pinkstone, H., "Continuous Circulation Reduces NPT," *World Oil*, 232, No. 3, March, 2011.

Van-Vegchel, J., Personal communication, 2009.

Vestavik, O., Syse, H., Hole, O., Aleksandersen, J. "New Approach to Improving the Horizontal-Drilling Reach," SPE 137821 presented at the

SPE Canadian Unconventional Resources and International Petroleum Conference, Calgary, Alberta, Canada, October 19–21, 2010.

Watson, R.W., Adewumi, M.A. and Temple, R.C. "Pneumatic Transport of Solids Using Dry Air" SPE 97-88 presented at the Annual Technical Meeting of The Petroleum Society, Calgary, Alberta, Canada, June 8–11, 1997.

Section 2
Limits to Water Volumes in Mist Drilling

Arash Haghshenas, Boots & Coots

5.14 Introduction

Almost all air/gas drilling operations change to "mist" drilling to enhance cutting transport and prevent the formation of mud rings. Formation fluid influx, or a damp formation, causes wet cuttings and causes cuttings to stick to the formation and form mud rings. Water with chemicals is injected at the surface to enhance cutting transport and cut or dissolve the mud rings.

Previous work (Lyons, Guo and Seidel, 2001) described in detail the maximum water volume limits to mist in the air stream. Their description is the maximum amount of water that would form a mist or fog in the airstream. In field operations it is obvious that more water must be added to the air stream than it can support as mist; this "extra" water is required to cut the mud rings and transport the large cuttings to the surface.

The mechanism for transporting the necessary water appears to be annular flow. Current models provide information on the limits of liquid injection rate for misting. In mist drilling, the minimum amount of water that would be effective appears to be the mist limit; the maximum amount of water would be the upper limit of annular flow where it would break into slug flow.

Injection of water at the surface requires a higher gas injection rate to clean the hole. The presence of water increases the pressure profile in the wellbore and compresses the gas. The compression of gas further increases the pressure profile and reduces the velocity of gas in the wellbore.

5.15 Dry Gas Injection Rate

The minimum dry gas injection rate for cleaning the wellbore is determined by either minimum velocity criterion or minimum kinetic energy. The minimum velocity criterion estimates the slip velocity and particle transport velocity of cuttings in the wellbore. The minimum required velocity in the wellbore is the sum of slip velocity and particle transport velocity.

The minimum kinetic energy criterion is based on minimum kinetic energy of 3 lb-ft/ft^3 of the gas, liquid, cutting mixture. Details of minimum velocity criterion or minimum kinetic energy are discussed by Guo and Ghalambor (2002) and Angel (1958).

Results of minimum velocity criterion or minimum kinetic energy are usually very close. However, the minimum kinetic energy criterion provides a more conservative value (Guo and Ghalambor, 2002). Figure 5–15a and Figure 5–15b show the minimum injection rates for two different hole and drill-string sizes for various drilling rates.

5.16 Standpipe Pressure with Different Water Injection Rates

Available models for gas drilling are able to predict the pressure profile in the wellbore within engineering accuracy. The gas models also provide the bottom-hole pressure while liquid is injected into the wellbore for mist drilling. An extensive number of graphs for minimum gas injection requirements for different hole sizes and depth were generated by Guo and Ghalambor (2002) and Lyons et al. (2009).

5.17 Mist Drilling Requirements

Poor hole cleaning causes several operation problems and liquid injection has been necessary to enhance the hole cleaning process.

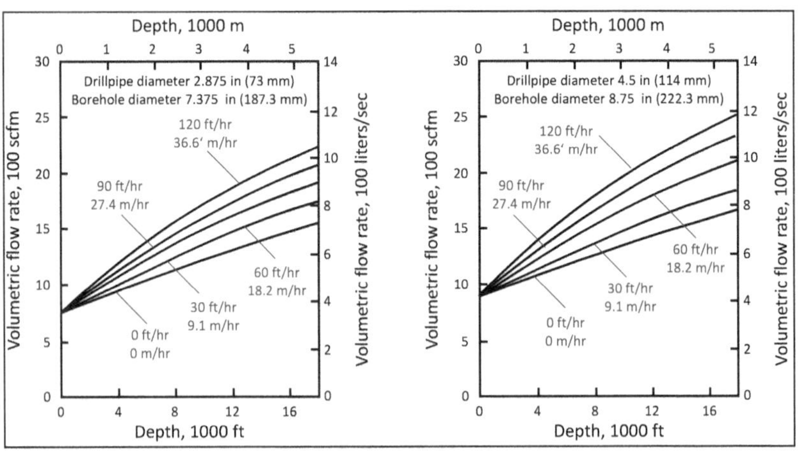

Figure 5–15 *Minimum air injection rate requirement for different wellbore and drill-string sizes (Angel, 1958)*

Insufficient gas injection rate creates a cutting bed in the wellbore or a stationary cluster of cuttings on the top of drill collars which restrict the flow and increases the bottom-hole pressure. Moreover, as the hole cleaning process in gas drilling depends on the velocity of the gas, the higher pressure is related to lower in situ velocity of the gas. As the result, the hole cleaning efficiency deteriorates with time. As the circulation stops, cuttings not removed from the wellbore cause fill and deposit on the bottom-hole assembly and may cause stuck pipe.

Slight influx of formation fluid into the wellbore wets the cuttings and increases the tendency of cuttings to adhere to each other and restrict the flow. The phenomenon of cuttings adhering to each other is known as mud ring formation. The most common place for the formation of mud rings is at the top of the drill collars; this is observed while pulling the drillstring out of the hole as shown in Figure 5–16. To prevent the formation of a mud ring, water is injected to enhance the cutting transport process and to sweep the cuttings at the top of the drill collars. Higher gas injection rate is required to clean the hole from the cuttings and liquids.

Field experiences indicate limits for liquid injection rates. The limits of liquid injection rate can be estimated with a mechanistic modeling approach. The mechanistic model provides better insight into the dynamics of the gas and liquid in the wellbore but complicates the calculations. If liquid injection rate is not the concern of the operation, the simplified mist drilling models of Lyons, Guo and Seidel (2001), Guo and Ghalambor (2002), or Hasan and Kabir (2007) can be applied.

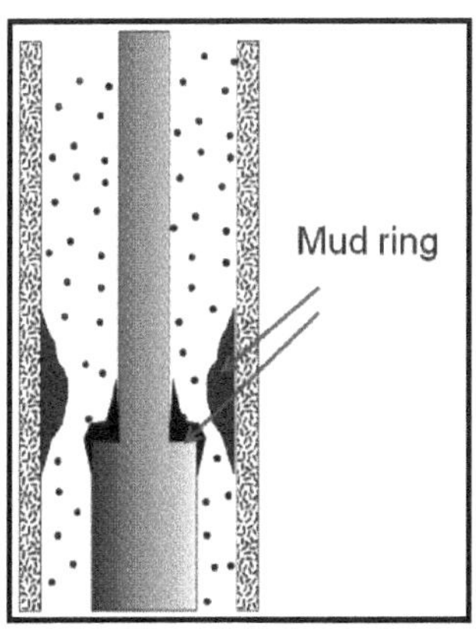

Figure 5–16 *Mud ring forms on the top of the drill collars*

Mechanistic models follow the flow pattern of gas and liquid in the wellbore and estimate the pressure profile based on the prevailing flow regime. (Multiphase flow and flow regimes are described in Chapter 3, Section 4, page 174.) This model considers two limits for liquid injection rate in addition to hole cleaning. The first limit is to ensure presence of a thin layer of water on the drill collars at a turbulent flow regime. It is believed that the presence of a turbulent layer of water sweeps (washes) the cuttings on the top of the drill collars and mitigates the formation of mud rings. The other limit of the liquid injection rate is the limit of slug flow in the wellbore. An excessive amount of liquid in the wellbore creates slug flow with severe pressure fluctuation at the bottomhole. The most probable places for the occurrence of slug flow are bottomhole and any other location where the wellbore geometry changes, such as the top of the drill collars or casing seat as shown in Figure 5–17.

Shoham (2006) describes the annular flow model developed by Alvez for inside tubing. In here, annular flow is a certain type of flow regime which liquid phase moves on the boundaries of the conduit (Chapter 3, Section 4). Field data indicates that the Alvez model predicts the pressure profile in the wellbore with very good accuracy. The Alvez model can be modified to model the gas/mist drilling operation. Figure 5–17 illustrates a schematic of the wellbore during gas/mist drilling.

Three regions exist in the wellbore: two layers of liquid and a core of gas as illustrated in Figure 5–18. Two layers of liquid move up to the surface along the drillstring and wellbore. The core of gas moves up to the surface in-between the two layers of liquid. At low gas velocities, no liquid entrains the gas core. As the gas velocity increases,

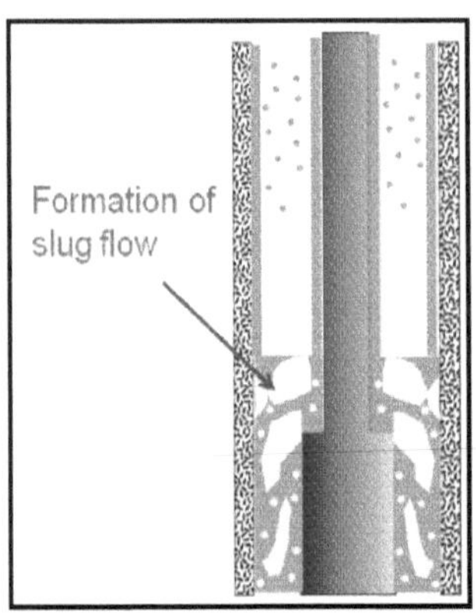

Figure 5–17 *Slug flow is the upper limit of liquid injection rate in the wellbore*

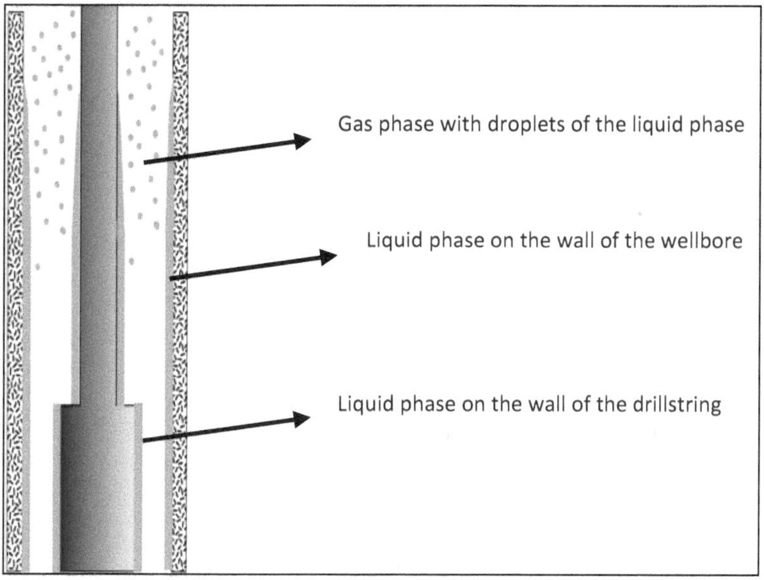

Figure 5–18 *The layers of liquid on the wellbore and drillstring divides the cross sectional area into three sections*

some liquid entrain into the gas core due to the presence of high turbulence at the boundary between the liquid and gas phase.

As the turbulence at the boundary of the liquid and gas core increases, more liquid entrains the gas core and the thickness of the liquid layer decreases. When more than 90% of the liquid has entrained the gas core, homogenous models such as Guo and Ghalambor (2002) and Hasan and Kabir (2007) can be used to determine the pressure profile.

The model is derived by applying momentum balance equations to the layers of liquid and the gas core. At any given location the flow is assumed incompressible and the change of the momentum is neglected. Liquid phase forms the layers and entrains into the gas core. The momentum and mass balance equations for all three regions are combined and summarized in Eq. (5.3). Figure 5–19 illustrates the schematic of annular flow and used notations for modeling in Eq. (5.1) and Eq. (5.2).

$$-\tau_{wl}\frac{S_{lw}}{A_{fw}}+\tau_{iw}\left(\frac{S_{oc}}{A_{fw}}+\frac{S_{oc}}{A_c}\right)+\tau_{lp}\frac{S_{ic}}{A_c}-g\left(\rho_l-\rho_c\right)=0 \tag{5.1}$$

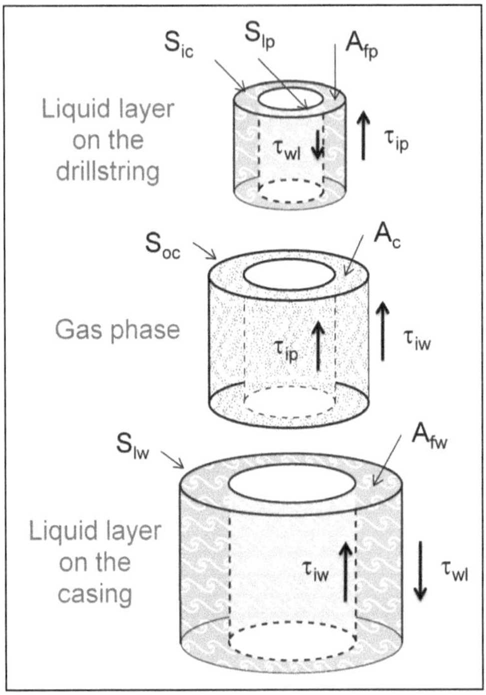

Figure 5–19 *Illustrates schematic of annular flow model*

$$-\tau_{pl}\frac{S_{lp}}{A_{fp}}+\tau_{ip}\left(\frac{S_{ic}}{A_{fp}}+\frac{S_{ic}}{A_c}\right)+\tau_{lw}\frac{S_{oc}}{A_c}-g\left(\rho_l-\rho_c\right)=0 \tag{5.2}$$

$$q=q_w+q_p+q_E \tag{5.3}$$

where

q = The total flow rate of the liquid phase

q_w = The flow rate of the liquid in the liquid layer formed on the wall

q_p = The flow rate of the liquid in the liquid layer formed on the drillstring

q_E = The flow rate of the liquid entrained into the gas core

All three equations should be solved simultaneously to determine the thickness and flow rate of the liquid in each layer. For the details of calculating the parameters in Eq. (5.1) and Eq. (5.2), readers are referred to Shaoham (Chapter 4, 2006) or other books on two-phase flow.

The fraction of liquid entrainment into the gas core can be estimated using the Wallis correlation

$$f_E = 1 - \exp[-0.125(\varphi - 1.5)\,] \tag{5.4}$$

$$\varphi = 10^4 \frac{v_{sg}\mu_g}{\sigma}\left(\frac{\rho_g}{\rho_l}\right)^{0.5} \tag{5.5}$$

$$q_E = q \times f_E \tag{5.6}$$

where

f_E = fraction of liquid phase in the gas phase

v_{sg} = superficial velocity of the gas phase

μ_g = viscosity of the gas phase

ρ_g = density of the gas phase

ρ_l = density of the liquid phase

σ = surface tension of the liquid phase

The value of f_E, which is less than or equal to zero, means that if no liquid entrains the gas phase then the total liquid flow rate is the sum of liquid rates in the layers. The homogeneous flow model is used when the fraction of liquid entrainment is greater than 0.9.

The interfacial shear stresses for τ_{iw} and τ_{ip} are calculated by

$$\tau = f_I \frac{\rho_c (v_c - v_l)^2}{2} \tag{5.7}$$

where

τ = interfacial shear stress

f_I = interfacial friction factor

ρ_c = density of the gas core with liquid entrainment

v_c = velocity of the gas core

v_l = velocity of the liquid in the liquid layer

The interfacial friction factor has been correlated to the friction factor of the gas core using a factor I. Several correlations for I factor is available, however, the simplest and the most practical one is proposed by Wallis.

$$I = 1 + \frac{300}{d} \tag{5.8}$$

$$f_I = I \times f_{sc} \tag{5.9}$$

The superficial friction factor of the gas core, f_{sc}, is the friction factor in a smooth pipe. In annular flow, the gas core is in turbulent flow and the friction factor can be simplified as a function of the Reynolds number.

$$f_{sc} = 0.046 \mathrm{Re}_{sc}^{-0.2} = 0.046 \left(\frac{\rho_c v_{sc}}{\mu_c} \right)^{-0.2} \tag{5.10}$$

Mixture properties could be used for the gas core properties. For simplicity, the gas core parameters can be changed by gas phase properties. Then the superficial friction parameter becomes

$$f_{sg} = 0.046 \mathrm{Re}_{sg}^{-0.2} = 0.046 \left(\frac{\rho_g v_{sg}}{\mu_g} \right)^{-0.2} \tag{5.11}$$

The shear stress between the liquid film and the surfaces (wellbore and drillstring) depends on the flow regimes of the liquid phase. The liquid layer may be at either the turbulent or laminar condition: appropriate friction factor should be calculated. The general shear stress equation is

$$\tau = f \frac{\rho_l v_l^2}{2} \tag{5.12}$$

The film friction factor, f, can be determined for smooth and rough surfaced pipes. For smooth pipes the Blasius equation can be used:

$$f = C \mathrm{Re}^{-n} \tag{5.13}$$

where C is 16 for laminar and 0.046 for turbulent flow. For rough surfaces, like the wall of the wellbore, Fanning friction factor can be used, as given by

$$f = 0.001375\left[1+\left(2\times10^{-6}\frac{\varepsilon}{d}+\frac{10^{6}}{\mathrm{Re}}\right)^{1/3}\right] \tag{5.14}$$

Shear stresses are a function of the Reynolds number. If the liquid flow rate in each layer is known, the Reynolds number can be simplified as shown in Eq. (5.15a–c).

$$\mathrm{Re} = \frac{\rho\, v\, d}{\mu} \tag{5.15a}$$

$$\mathrm{Re} = \frac{2\rho\, Q\delta}{A\, \mu} \tag{5.15b}$$

$$\mathrm{Re} = \frac{4\rho\, Q}{\pi\, \mu(2d_1 - \delta)} \cong \frac{2\rho\, Q}{\pi\, \mu\, d_1} \tag{5.15c}$$

where

d_1 = diameter of the flow

δ = thickness of the liquid layer

Studies show that the thickness of the liquid layer on the wall of the pipe is limited to 6% of the wall diameter. Eq. (5.15c) is simplified by neglecting the liquid thickness when compared to the double of the wall diameter. This simplification introduces a maximum error of 3%. Wen the Reynolds number of the liquid layer on the wall is a direct function of the liquid, flow rate provided by other parameters are constant. As the gas expands on the way to the surface of a well with a constant geometry, more liquid entrains into the gas core which consequently reduces the Reynolds number of the liquid layer as the liquid moves upwards.

Solving Eq. (5.1) to Eq. (5.3) requires iteration on the thickness of the liquid layers and the fraction of flow rate in each layer. The iteration process can be simplified by limiting the thickness of the liquid layers to the 6% of the outer diameter (Shoham, 2006). For a stationary drillstring, the liquid layer on the wellbore has greater surface area (s_{lw}) and roughness (formation face). Therefore, the degree of shear stress is greater on the liquid layer on the wellbore. As a result, liquid phase has a tendency to move in the liquid layer on the drillstring, or rather, the liquid phase passes through less resistant path.

The rotation of the drillstring develops centrifugal force on the liquid phase on the drillstring, which spins off the liquid away from the drillstring. Therefore, the liquid thickness and flow rate on the drillstring will be less than the stationary drillstring scenario. However, the upward velocity of the liquid layer on the drillstring is several times greater than the rotational speed of the drillstring and may not significantly affect the solution.

The annular flow model can be simplified using the hydraulic diameter concept which changes the annulus between the drillstring and the wellbore to a pipe configuration. The diameter of the tube would be the difference between the outer and inner diameter in the wellbore. The simplified tube model can be solved by solving the Eq. (5.16a) and Eq. (5.16b) simultaneously.

$$-\tau_{wl}\frac{S_{lw}}{A_{fw}}+\tau_{iw}\left(\frac{S_{oc}}{A_{fw}}+\frac{S_{oc}}{A_c}\right)+\tau_{lp}\frac{S_{ic}}{A_c}-g(\rho_l-\rho_c)=0 \tag{5.16a}$$

$$q_l = q - q_E \tag{5.16b}$$

Assume a 7 7/8 in. vertical well at a depth of 8,000 ft with 500 ft of drill collars. The diameter of the drillpipe is 4.5 in. The last casing is set at the depth of 3,000 ft. The inner diameter of the casing is assumed to be the same as the hole diameter. The schematic of the well is given in Figure 5–20. For the given operational condition, the vertical velocity of the liquid is about an order of magnitude higher than the rotational speed of the drillstring operated at 60 rpm. Therefore, a significant effect of fluid retention with the drill-string rotation is not expected.

The Reynolds number of the liquid layer on the drillstring is calculated for a gas injection rate of 2,500 scfm and liquid injection rate of 3 and 4 gpm as shown in Figure 5–21. The Reynolds number reduces along the bottom-hole assembly as the liquid layer moves upwards because more liquid entrains the gas core. When the flow reaches the top of bottom-hole assembly, the gas velocity reduces while moving into a larger area above the bottom-hole assembly. The reduced velocity of gas core is less capable of moving the liquid

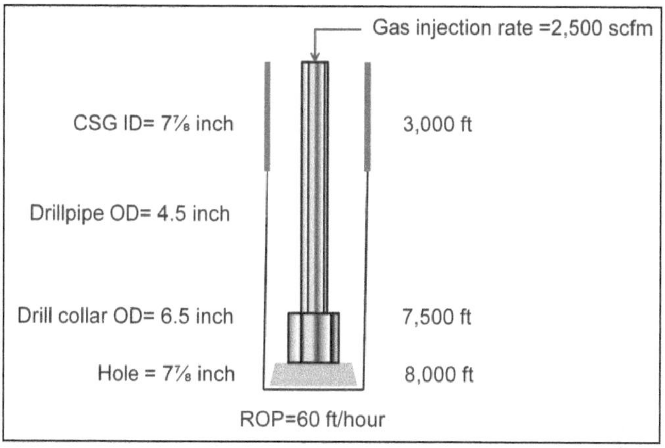

Figure 5–20 *Schematic of vertical well for mist drilling operation*

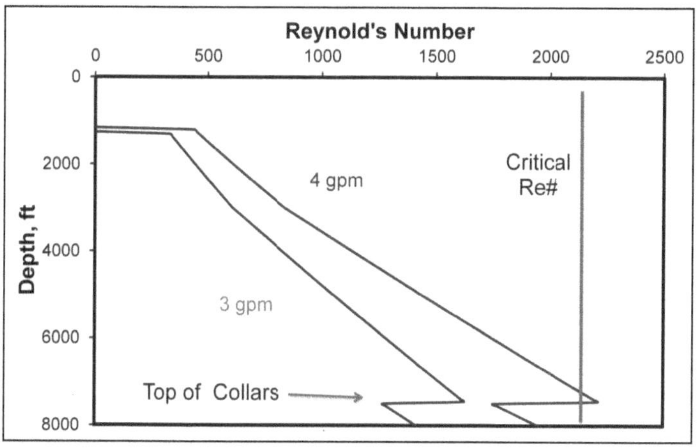

Figure 5–21 *Reynolds number of the liquid layer on the drillstring*

upwards; therefore, some of the liquid is transferred from the gas core to the liquid layers. The top of the drill collars is the critical point where mud rings usually occur. The liquid layer at the top of the drill collar should be in turbulent flow to prevent the formation of a mud ring.

The Reynolds number decreases as the liquid layer moves upwards and more liquid entrains the gas core. The slope of the Reynolds number is related to the pressure profile. When the liquid enters the casing, the trend of the Reynolds number changes due to the change in the pressure profile caused by lower wellbore roughness in the casing compared to the borehole. When more than 90% the liquid phase has entrained the gas core, the liquid layer is ignored and mist flow calculation is used. The Reynolds number is zero during mist flow.

For the liquid injection rate of 3 gpm, the layer of liquid is at laminar condition. However, increasing the liquid injection rate to 4 gpm creates turbulence at the top of the drill collars. Therefore, the liquid injection rate of 4 gpm is considered the lowest limit of the liquid injection rate for cleaning the hole with a gas injection rate of 2,500 scfm.

The upper limit of liquid injection rate is slug flow as shown in Figure 5–22. For the liquid injection rate of 7 gpm, slug flow occurs in the wellbore. As shown in Figure 5–22 slug flow occurs at the top of the drill collars, where the superficial velocities of gas and liquid phase drops and causes the flow pattern change from annular to slug flow. Slug flow increases the bottom-hole pressure, deteriorates the cutting transport efficiency, and causes pressure surges in the drillpipe and bottom-hole pressure.

Figure 5–23 illustrates the limit of gas and liquid injection rate for the well given in Figure 5–20. For the scenarios that no misting is required, the minimum gas injection rate required to clean the hole is 2,200 scfm. However, when misting is required, a discontinuity is indicated in the results. The minimum gas injection for misting is 2,400 scfm and a minimum of 4 gpm is required for proper hole cleaning. The lower limit of liquid injection rate, for mitigating mud ring, remains almost constant with increasing the gas injection rate. The upper limit of liquid injection rate, for avoiding slug flow, increases by increasing the gas injection rate.

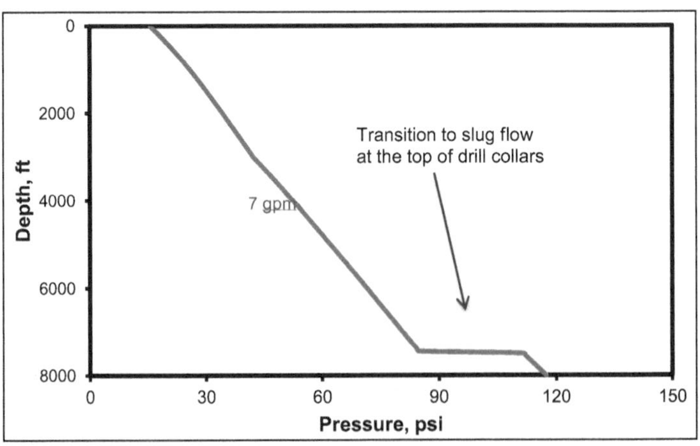

Figure 5–22 *Pressure profile in the wellbore with occurrence of slug flow at the top of drill collars*

The range of liquid injection rate is different for various wellbore geometries. In general, the range of liquid injection rate should be modified for any changes affecting the bottom-hole pressure. Figure 5–24 provides an initial estimate for the range of liquid injection rates. The minimum liquid injection rate requirement for the given hole sizes is 4 gpm (6 bpm). Liquid injection rates less than 4 gpm may deteriorate hole cleaning efficiency. The solutions shown in Figure 5–23 is based on a gauge wellbore with roughness of 0.3 in. Actual field conditions and added chemicals to the injected water may shift the water injection limits.

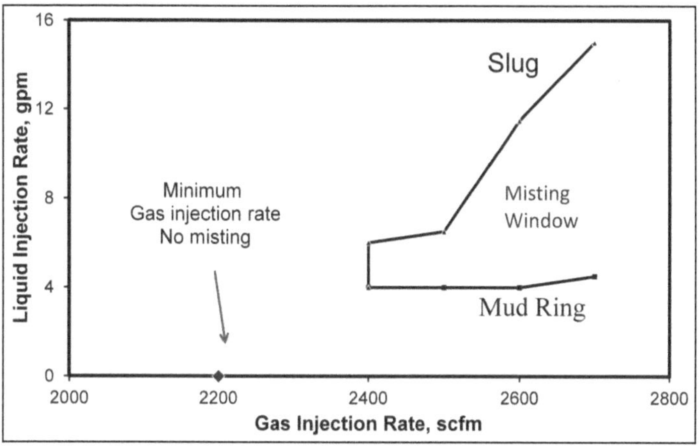

Figure 5–23 *The window of gas/mist drilling for a 7 7/8 in. vertical well*

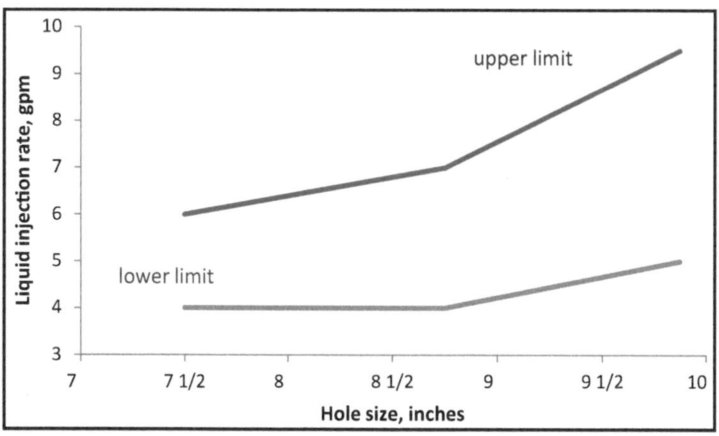

Figure 5–24 *Illustrates the limits of liquid injection rates for different hole sizes*

It is suggested that the outside diameter of the drill collar be 1.5 in. less than the hole diameter. Small clearance between the drill collar and the hole causes excessive pressure drop and very narrow range for liquid injection rates. For hole sizes less than 7 in., an extensive study is required to determine the range of liquid injection rate. The limits of liquid injection rates may affect the drill-string design to avoid slug flow in the drill collar-wellbore annulus.

5.18 References

Angel, R.R. *Volume Requirements for Air or Gas Drilling*, Gulf Publishing Company, Houston, TX, USA, 1958.

Guo, B. and Ghalambor, A. *Gas Volume Requirements for Underbalanced Drilling-Deviated Holes*, Pennwell Corporation, Tulsa, OK, USA, 2002.

Hasan, A.R. and Kabir, C.S., "A Simple Model for Annular Two-Phase Flow in Wellbores," *SPE Production and Operations*, 22, No. 2, 2007, pp. 168–175.

Hasan, A.R. and Kabir, C.S., *Fluid Flow and Heat Transfer in Wellbores*, Society of Petroleum Engineers, Richardson, TX, USA, 2002.

Lyons, W., Guo, B., Graham, R. and Hawley, G. *Air and Gas Drilling Manual*, Elsevier, Boston, MA, USA, 2009.

Lyons, W., Guo, B. and Seidel, F.A., *Air and Gas Drilling Manual*, McGraw-Hill, New York, NY, USA, 2001.

Shoham, O. *Mechanistic Modeling of Gas-Liquid Two-Phase Flow in Pipes*, Society of Petroleum Engineers, Richardson, TX, USA, 2006.

CHAPTER 6

Snubbing and Underbalanced Drilling

Mike Ponville, Boots & Coots

6.1 Introduction

Snubbing refers to the act of controlling pipe movement into or out of the wellbore while maintaining well control. Unlike conventional drilling and completion methods that require the use of kill weight fluids that increase cost and can damage formations, hydraulic work over (HWO) /snubbing units offer the same benefits of traditional rigs while working underbalanced.

A snubbing system can be utilized as a "rig assist" to aid a drilling operation where well pressures at the surface exist. Examples of these operations are handling pipe light tubulars, well control, fishing under pressure, completion installation, and running of TCP (tubing conveyed perforating) guns to name a few.

Snubbing involves the use of a special hydraulic system, and a series of slips and cylinders to overcome the forces on tubulars generated by hydraulic and frictional force pressures in the wellbore that limits the ability of the drilling operation to freely move the tubulars.

Snubbing jobs are performed while the wellbore is in an underbalanced state. As a result, some form of pressure control must be used to keep fluids or gas in the wellbore under control The most common forms of this pressure control equipment include BOP rams, annular preventer, and some form of stripping rubber. By calculating snub forces and tubular properties the correct snubbing unit and BOP stack rig up can be determined.

The jack type "HWO/snubbing" unit has been configured into a "rig assist" mode for quick installation into a conventional design drilling or workover rig for working on a live well. As shown in

Figure 6–1 the blowout preventer (BOP) system is installed into the jack frame and the unit is rigged up inside the rig by a single crane lift.

6.2 Basic Snubbing

6.2.1 The Hydraulic System

Some snubbing systems use a mechanical system for controlling the tubular movement. The safety features of using a hydraulic system are the precise control of force due to liquid incompressibility (hydraulic fluid). The main advantages are:

- Precision tool manipulation: Hydraulic systems are fast and precise in response time. Any change in pressure by manipulation of the hydraulic control valves causes a virtual instantaneous response everywhere within the system. The hydraulic system is used to control the movement of the drillpipe, tubing, or casing through the rams, annular preventer, or stripping rubber as noted in the paragraphs below; it also controls valve movement to control pressures.
- An important safety feature in hydraulics is the pressure relief valves used to protect against overload damages. This protection is accomplished by diverting pump delivery to a bypass mode when loads exceed valve settings. The settings of these valves are in accordance to the capacities of the down-hole tools or other components, or simply the safe working limitations of the tubulars used. This safety feature takes away the possibility of 'human error" in exceeding the limitations of the tubulars. The possibilities of exceeding the tensile strength rating of tubulars, over torque of tool joints, or corkscrewing the work-string are minimized because maximum hydraulic pressure is set with a safety factor included that limits the force delivered by the system.

6.2.2 Stripping Ram to Ram

The use of BOP rams involves the process known as "stripping ram to ram." In this scenario, two BOP rams, a spacer spool, an equalizing loop and a bleed off line are used. This configuration is known as a stripping stack or stripping loop. This system gives the snubbing pro-

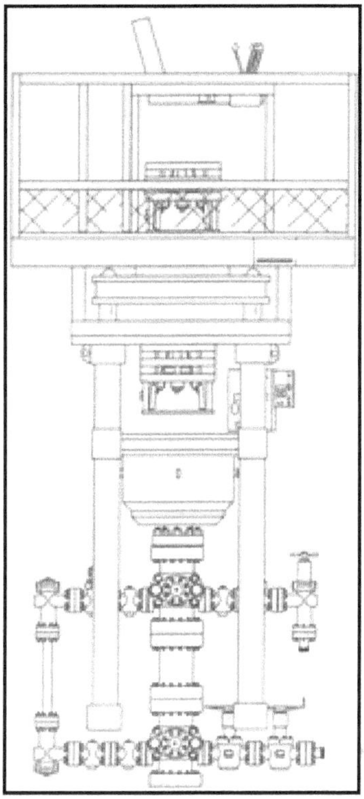

Figure 6–1 *Jack-type workover frame with BOP*

vider the highest form of pressure control. The stripping ram to ram process involves:

1. Closing the lower stripping ram
2. Tripping the pipe into the wellbore so that the tool joint is just above the closed lower ram (# 2)
3. Closing the top stripping ram (#1)
4. Opening the equalize valve
5. Opening the #2 stripping ram
6. Proceed in the wellbore with the tubing to the next tool joint or coupling

This process is repeated until the job is completed.

6.2.3 Stripping with Annular Preventer or Stripping Rubber

Typically, this form of stripping is limited to smaller tubular, lower pressures and working in a sweet well environment. In this case, a single annular preventer or stripping rubber is used to seal the annulus and allow controlled movement of the pipe. When tool joints or couplings are present, the rate of pipe movement is slowed to allow the elastomar system to adjust to the different diameter as the couplings pass though the elastomer element.

6.2.4 Pipe Light

Snubbing is often done under considerable wellbore pressures. The condition when the weight of the pipe is less than the forces generated by the wellbore pressure is known as "pipe light." At this point of the operation, the snubbing unit utilizes the push/pull cylinders and slips to overcome the wellbore forces to snub in the pipe. Once enough pipe weight has been snubbed into the well so that the pipe weight is equal to the wellbore forces, the pipe becomes neutrally buoyant. This condition is known as balance point; every successive joint ran into the well increases the tubular weight in the well.

6.2.5 Pipe Heavy

The condition once enough piping has been ran into the well so that it exceeds the wellbore forces is known as "pipe heavy." Once in pipe heavy mode, the snubbing unit cylinders and slips are used to control the descent of the pipe into the wellbore. The snubbing unit must have the capability to handle the snub loads as well as have enough capacity to handle the pipe weight coming off the bottom in pipe heavy mode.

6.2.6 Fluid Flow

When tubulars are stripped into the hole, the displacement (capacity plus displacement) of the tubular would increase the well pressure, so well fluid needs to be released through the choke system to keep the annular pressure constant. This may require a separator, spare tanks, and a flare system.

When tubulars are pulled from the hole, the annular pressure is kept constant by adding drilling fluid equal to the tubular displacement, or by pumping across the top of the annulus under the rams.

6.3 Snubbing Units

While this discussion is primarily about rig assist snubbing units, it is worth mentioning the general range of units since any type may be employed is some sort of underbalanced operation. The snubbing system can be divided into four main categories of components:

1. Basic snubbing unit
2. Work-string and components
3. Well control equipment
4. Auxiliary equipment

6.3.1 Basic Snubbing Unit

The basic snubbing unit is the mechanical or hydraulic machine used to generate push, pull, and torsional forces on the work-string for accomplishing specific tasks in subject wells. The three most commonly used types of snubbing units are the hydraulic jack (HWO/snubbing), hydraulic long-stroke, and the mechanical "rig assist" (conventional) unit.

6.3.1.1 Rig Assist with Jack Type Unit

The "jack" type unit is the most widely used type of snubbing unit. This is due to its capacity in handling a greater variety of well problems at moderate to high surface pressures. The jack unit can operate with or without a conventional drilling/workover rig for support.

Advantages of this type of snubbing unit when compared to the long-stroke and conventional snubbing units include:

- The snub load capacity is greater than the other units
- The lift capacity is greater than the other units
- Higher rotary torque capacity
- The design is compact
- Handles a wide range of tubulars, 3/4 in. through 13 3/8 in.
- Thrust forces are normally applied to the wellhead, which is beneficial while working on a damaged platform
- The height of the unit can be adjusted to install, additional BOPs for enhanced well control and extended BHA "lubrication" area

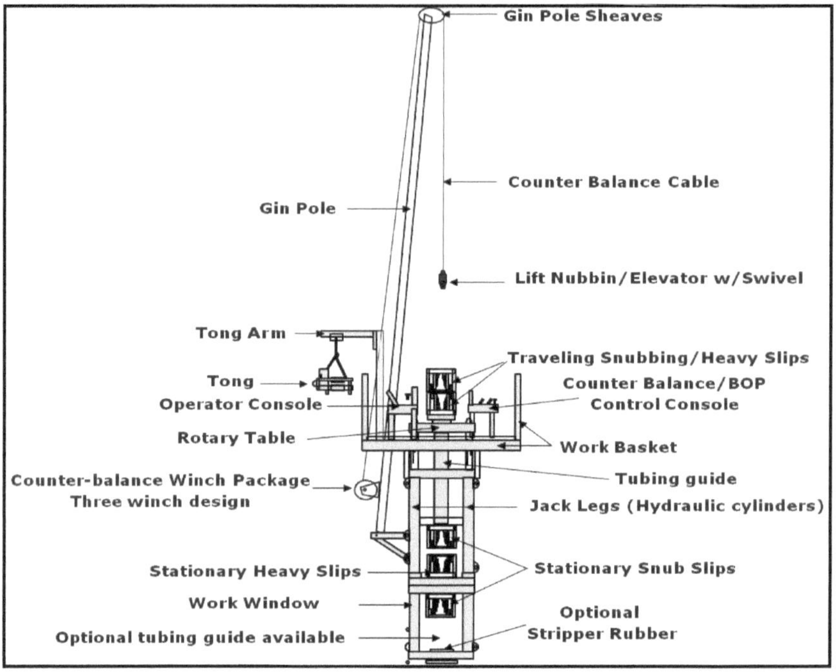

Figure 6–2 *The HWO snubbing unit*

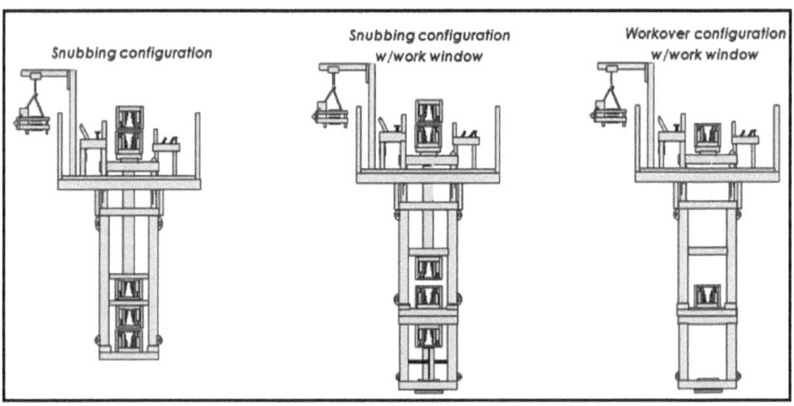

Figure 6–3 *Basic HWO/snubbing unit configurations*

Limitations to jack type snubbing unit are:

- Longer trip times
- Requires a crane or derrick to rig-up
- Runs work-string in single joints

6.3.2 The Work-String and Components

The full spectrum of steel tubulars utilized in standard oilfield drilling and workover operations can be handled by snubbing systems to perform hydraulic workovers. Generally, the contracting operator supplies the tubulars and associated equipment, because the choice of tubular depends on the well and the operation and is completely independent of the snubbing system. Other equipment includes:

- Back pressure valve
- Stabbing valve
- Inside blow out preventer (BOP)
- Circulating swivel
- Circulating hose
- Work-string nipples and plugs
- Bottom-hole assembly (BHA) components

6.4 Well Control

6.4.1 Primary Well Control System

During snubbing operations, well control safety is provided by the primary well control system or stripping arrangement. Typically, the stripping arrangement consists of stripper bowl, stripper rams, and sometimes an annular blow out preventer (BOP). The various elements and configurations follow.

6.4.1.1

The stripper rubber is used as a pack-off (sealing element) during stripping operations using non-upset or tapered upset tubing. A variety of pack-off systems are available throughout the snubbing industry. All pack-off systems are designed to preclude ram to ram stripping on low pressure wells, typically below 3,000 psi (200 bar), expediting completion of workover programs. The stripper rubber is installed into the stripper bowl in the base of the snubbing jack or window.

6.4.1.2

An annular BOP is also used for stripping purposes, but only when a back-up BOP is provided in the secondary well control system.

6.4.1.3

Stripper rams and associated equipment are utilized in snubbing applications to allow for controlled movement of upset and non-upset tubulars in wells with surface pressures. By alternately opening and closing the two upper most pipe rams in the BOP stack, tool joints can be safely and quickly stripped into the well while retaining full control of annular fluids and/or gas content. The minimum required components consist of:

- Two stripper (pipe) rams
- Equalizing loop assembly
- Bleed-off assembly
- Secondary or "safety" well control systems in place below stripper system for annular shut-in during maintenance of the stripping system
- Plugs or NRV (Non Return Valves) inside the tubular (more than one is used)

6.4.2 Secondary Well Control System

The purpose of the secondary well control system is to maintain well control in the event of failure, or maintenance of the primary well control system.

The secondary well control system provides both annular and inside drillpipe well control. The nipple/plug subs in the work-string are considered the inside secondary well control along with the inside BOP(s) available in the work basket. The number of BOP elements and their configurations vary according to the governmental and industry specific requirements. Minimum requirements for the secondary well control system consist of:

- Two pipe ram BOPs
- One blind and one shear (or combined blind/shear) ram BOP
- Choke line and kill line valves below the lowest ram
- Inside BOP and stabbing valve

In addition, there may be slip rams, wireline rams, variable bore rams, and additional choke line and kill line connection valves on the BOP stacks. Additional equipment is dependent upon the type of the snubbing operations.

The components utilized in snubbing operations, as pertains to the secondary well control systems, do not differ appreciably from the same systems in use for drilling or conventional workover applications. A common HWO/snubbing BOP stack consisting of four pipe rams, one blind ram, one shear ram, stripping system, choke and kill lines minimizes the risk of adverse well control conditions by providing operational and emergency contingency flexibility. BOP stack operational ability can be enhanced by the addition of annular, blind, slip, or variable bore ram assemblies and non return valves (NRV) or stabbing valves.

6.4.3 Tertiary Well Control System

The purpose of the tertiary well control system is to maintain well control in the event of failure to, or maintenance of, the primary and secondary well control systems. The tertiary well control system consists of a single shear seal BOP with independent control system.

The main control station is located in the work basket. The dedicated alternate control station is positioned at a safe distance away from the well head in a way that it will not be immobilized by a leak or fire in the snubbing system. The alternate system can be controlled from a backup control unit for the secondary well control system.

6.5 Auxiliary Equipment

6.5.1 Pipe Handling System

The pipe handling system is designed to transport tubulars from the pipe rack lay-down area to the work basket and back; it is also set up to connect and make up the work-string during a trip in and out of the hole. The pipe handling system performs the following functions during HWO/snubbing operations:

- Supplies tubulars to and from the work basket,
- Pipe torque make up systems, both manual and hydraulic,
- Suspension of wash joint, circulating swivel, stabbing valve, and circulating hose during washing and milling operations.

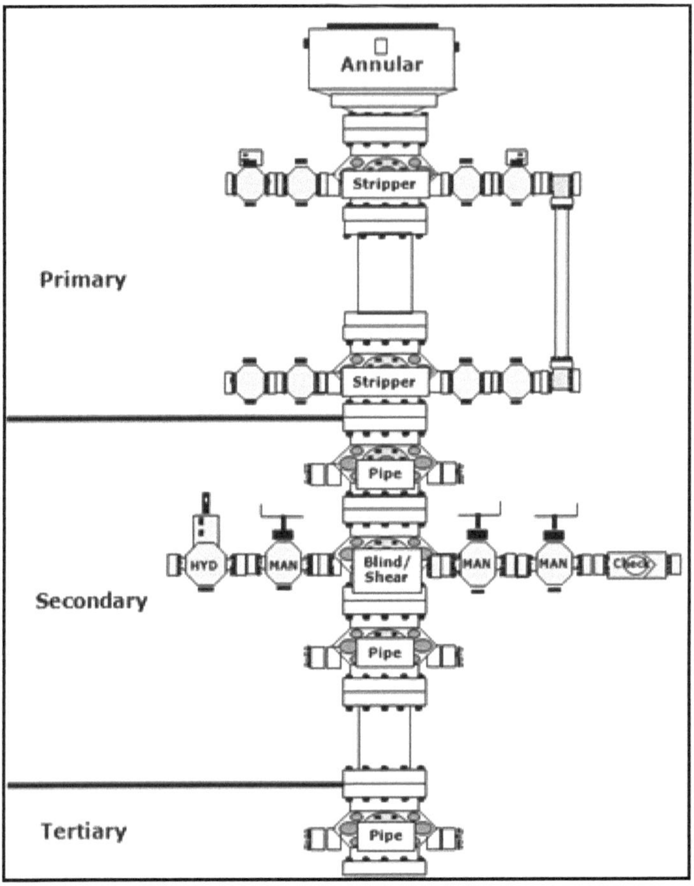

Figure 6–4 *Well control system for stripping*

6.5.2 Work Basket Access

An access ladder provides for safe access to the jack work basket. Additionally, an emergency escape arrangement is installed. The emergency escape system is set up by one of the following:

- Geronimo line
- Fire pole
- Catwalk
- Other ladder or stairway exit paths

6.6 Snubbing Operations

6.6.1 Temporary Securing of the Well

It may occur that well operations need to be suspended. This section covers the minimum requirements for securing the well during live or dead well workovers:

- If practical, pull out of the hole (POOH) and lay down the tubing or run in the hole (RIH) until the pipe is "heavy." A minimum of 5,000 pounds is required on the hook load (consider the effects of a migrating gas bubble).
- Stroke the jack head up 3 ft from the bottom stop and set the stationary slips. Close the traveling slips.
- Install a safety clamp on the work-string in the window area.
- Close and lock both stripper and upper pipe BOPs.
- Bleed off the pressure (if applicable) above the upper safety ram.
- Close all valves in the equalizing loop, bleed-off line, choke line, kill lines, TIW valve, (two safety valves if possible), and install the kelly hose and any other site specific valves, etc.
- Charge the accumulator bank and isolate it from the power pack.
- Ensure the shear, blind, and undersize ram handles are locked open.

6.6.2 Lubrication

Lubrication is the process of inserting or removing bottom-hole assembly (BHA) components from the wellbore on a live well. Well fluids, which may be diverted during the process to a safe area, are contained by internal barriers in the work-string and normal BOP devices. Tools requiring lubrication can be attached to the end of a work-string or at any point above the end.

Lubrication of the BHA with its non-conforming diameters is compulsory within the industry. No other procedures for entering the well should be contemplated without prior consent from the HWO/snubbing contractor and the operating company. The consent should be in writing from both parties. Snubbing operators shall deny any request of snubbing work without lubrication.

Lubrication can be divided into the following techniques: standard, in-line, wireline assist downhole, down-hole barrier lubrication, and shear or connect/disconnect in the BOP stack.

6.6.2.1 Standard Lubrication

Standard lubrication is the most commonly used technique. It uses the area between the upper stripper ram and the Christmas tree swab valve to lubricate the tools as shown in Figure 6–5.

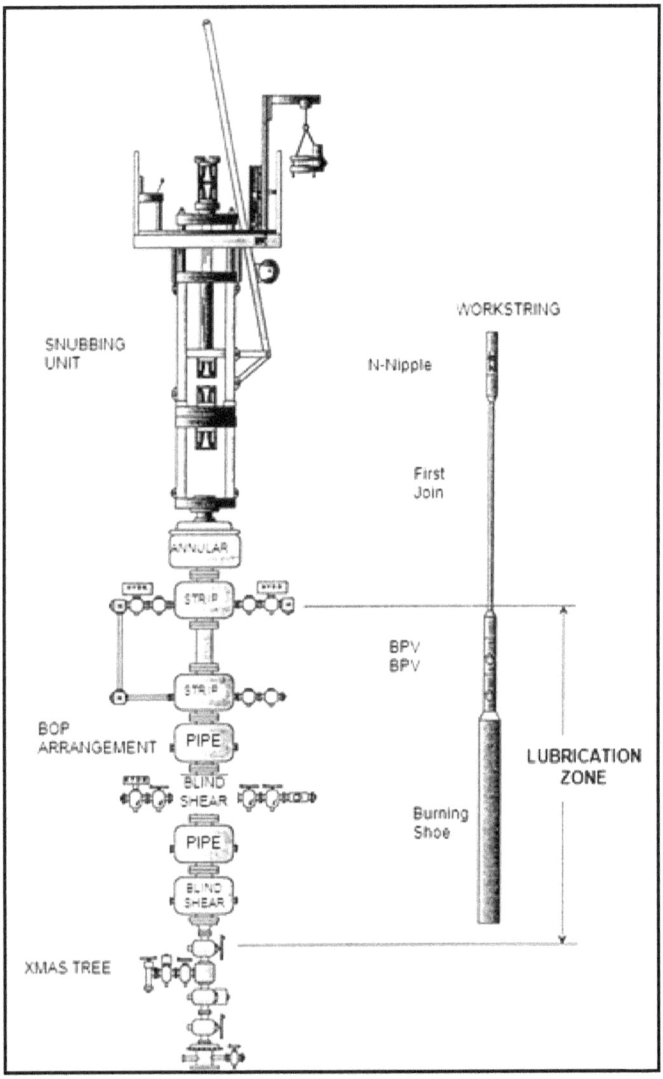

Figure 6–5 *Lubrication zone in the BOP stack*

6.6.2.2 In-Line Lubrication

In-line lubrication is the ability to control the annular pressure around a tool positioned anywhere in the work-string during an operation. The lubrication area is normally equal to the length available between the upper and lower stripper BOPs while stripping drillpipe or tubing (see Figure 6–6).

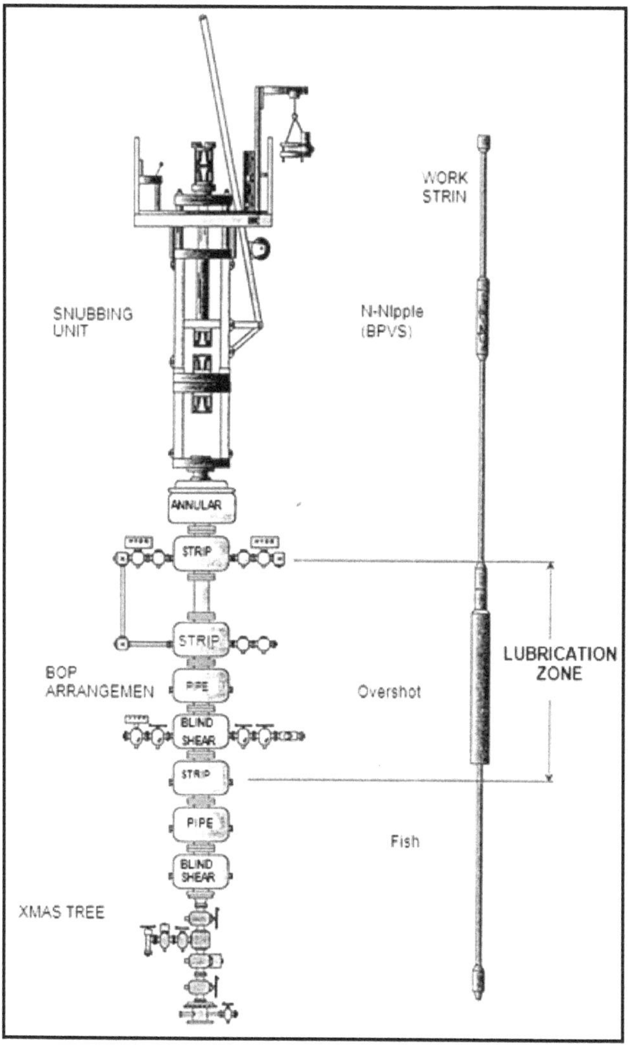

Figure 6–6 *Typical stripping lubrication zone while stripping pipe*

6.6.2.3 Wireline Assist Lubrication

Wireline assist lubrication utilizes the wire line lubricator in conjunction with the HWO/snubbing stack-up to increase the total lubricator length available at the surface (see Figure 6–7).

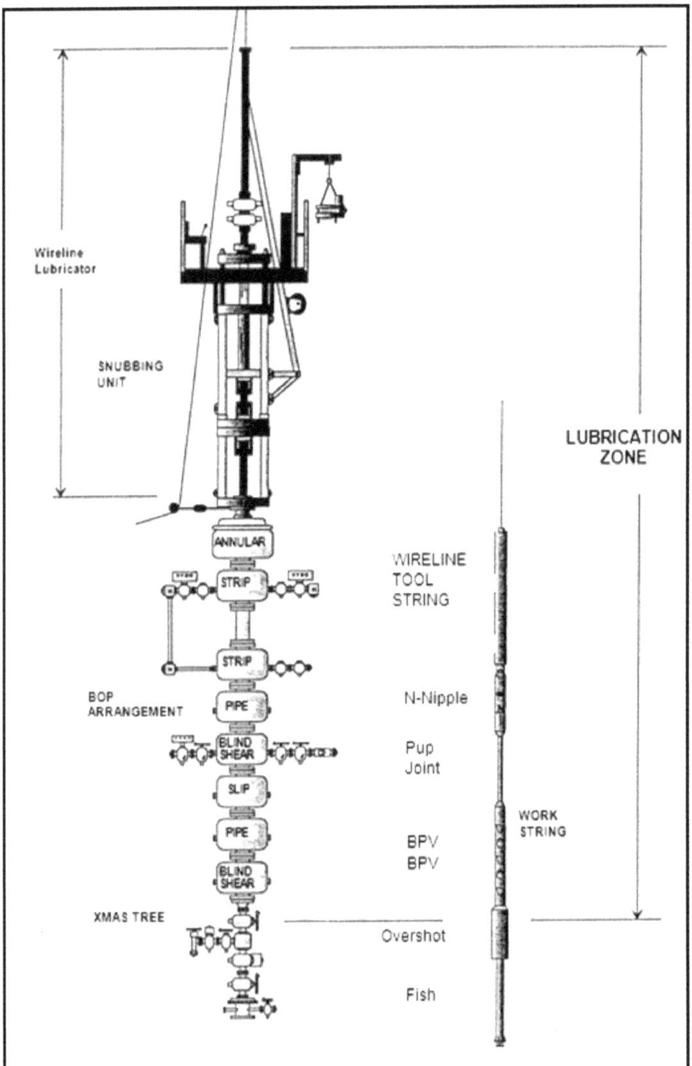

Figure 6–7 *Wireline assist stripping*

6.6.2.4 Down-Hole Barrier Lubrication

This technique requires the use of the down-hole safety valve (DHSV), or other sub-surface barrier, as the lubricator isolation valve; this technique allows the longest BHA length compared to the other lubrication methods. To date (September, 2010), this is 4,100 ft (1,250 m) in the horizontal with 3-3/8 in. OD tubing-conveyed perforating (TCP) guns. This distance will probably be eclipsed by the time this book goes to publication (see Figure 6–8).

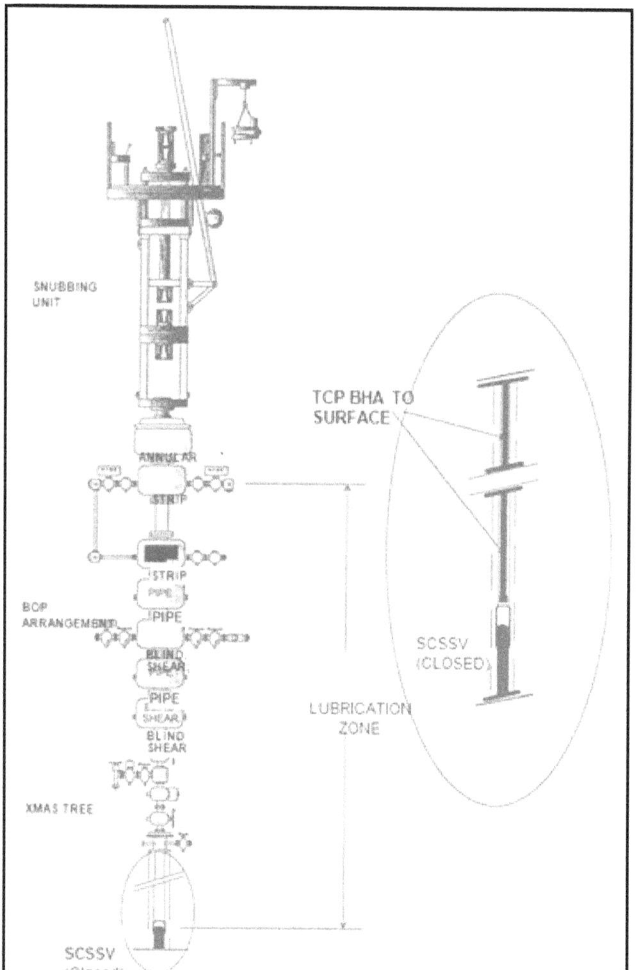

Figure 6–8 *Stripping with a down-hole safety valve (DHSV or SCSSV)*

Some general guidelines for down-hole barrier lubrications are as follows:

- Conduct pre-job safety meeting
- Perform emergency drills related to any (DHSV) leak
- Develop contingency procedures
- Test DHSV
- Displace to freshwater/glycol above DHSV
- Monitor hole volumes with a trip tank
- Only threaded connections allowed in the BHA
- Shear/ram capable of cutting and sealing every BHA element
- Any BHA cutting is hazardous: use specific procedures and deploy an emergency release tool

6.6.3 Shear or Connect/Disconnect in BOP Stack Lubrication

BHAs, or fishes, are simply assembled or disassembled inside the BOP stack at or near the surface on a live well. Sections of wireline, coiled tubing, drillpipe, TCP guns, or other tubulars are retrieved at the surface on live wells by shearing or connecting/disconnecting inside the BOP stack (i.e., a fishing operation for collapsed coiled tubing).

Commercial "deployment" systems are available which allow the assembly and disassembly of a bottom hole assembly (BHA) inside the BOP stack at the surface on live wells. For example, 4,000 ft (1,220 m) long TCP gun assembly can be installed or removed in single gun sections consecutively on a live well. Failure to properly plan for BHA or tool lubrication can create a hazardous situation during live well workover operations. Lubrication requirements must be determined prior to mobilizing for a HWO/snubbing operation in order to prevent potential hazard to crews and loss of rig time (see Figure 6–9).

6.7 Wireline Procedures

6.7.1 Wireline through the HWO/Snubbing Unit

Wireline operations can be performed through the snubbing unit, either inside the workstring or through the BOP stack if the workstring is out of the hole. Note that the gin pole is available for use during slickline services to support the wireline lubricator, but is usually not structurally able to support the weight associated with electric line operations.

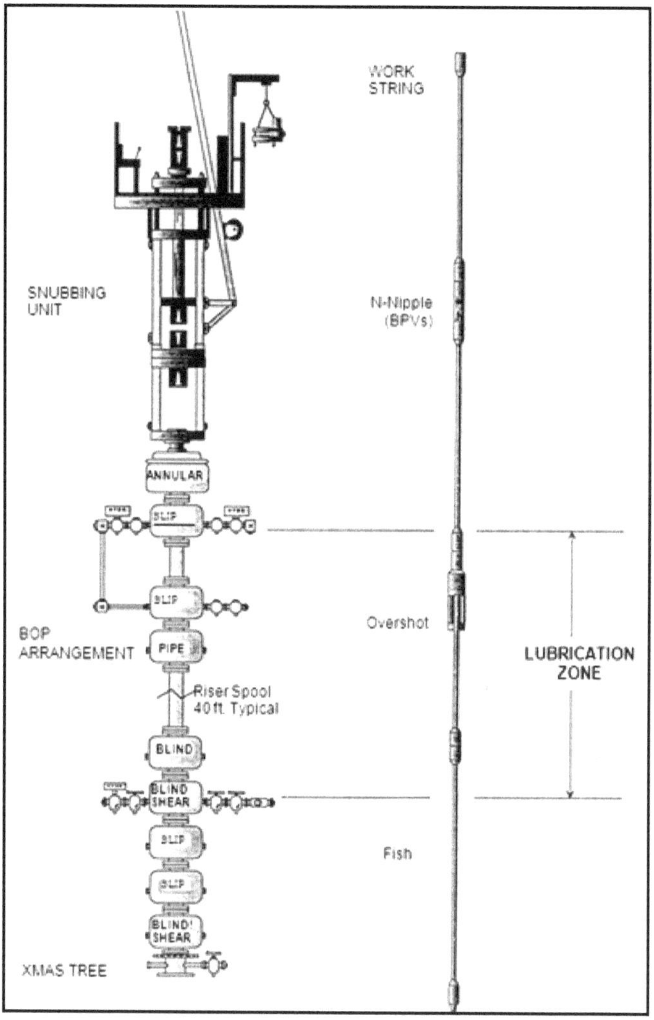

Figure 6–9 *Wireline stripping lubrication zone*

6.7.2 Wireline through the Work-String

If possible, space out the work-string to position the box of the uppermost joint at the work basket level. A minimum of two safety valves (TIW, Gray valve, etc.) should be installed onto the uppermost box connection. Then the wireline crossovers, BOPs, and lubricator components should be rigged up as required. The snubbing BOP stack is utilized by engaging at least one pipe ram or annular to the work-string secured by setting at least one pipe ram on the work-string with the snub slips (if used) being closed on the joint located in the jack stack. If the stripper BOPs are equipped with shear/seal BOPs, they should be capable of shearing the work-string and sealing with wireline inside, however, operator requirements will vary.

6.8 General Stripping Procedures

Stripping allows the insertion and/or removal of jointed tubulars on live wells while controlling well fluids. Stripping methods commonly used in snubbing live well work are:

- Packoff with stripper rubber
- Packoff with annular BOP
- Annular to ram
- Ram to ram

When stripping external upset tubulars using the packoff methods mentioned above, snub force is significantly increased due to the larger diameter tool joint sealing in the packer element.

6.8.1 "Packoff" Stripper Rubber

The stripper rubber may be a rotating control device (RCD), or, with workover and low pressure wells, a type JU stripping rubber or similar item may be used. With this technique:

- Generally a single, well energized, urethane element is used.
- Dual element designs with a pressure control system have been developed but are not commonly used.
- Normally circulation is stopped to change a stripper element. The well must be stable during this operation.
- The recommendation is to change a stripper element in the "pipe heavy" mode only.
- A longer service life is obtained with smooth upset tubulars. Footage of 40,000 ft (12,000 m) with 3 1/2 in. tubing at 2,200 psi (150 bar) WHP is obtainable.
- Typical maximum working pressure is 3,000 psi (200 bar).
- This is usually the most economical method of stripping tubulars.

6.8.2 "Packoff" Annular BOP

With this stripping technique a normal annular type BOP is closed on the workstring and maintains a pressure seal during the stripping process. The "packoff" annular BOP:

- Is normally used when stripping drillpipe at medium to low pressures
- Has been successfully utilized for spiral drill collars
- Requires a surge bottle
- Closing force must be regulated
- Has provided good service life with moderately lower surface pressures (3,000 psi / 200 bar)

A special version is available with a quick change feature. Packer elements can be renewed in approximately 5 minutes, while continuing to circulate.

6.8.3 Annular To Ram

This stripping technique is more common to conventional drilling and is rarely used in HWO/snubbing. However, this technique:

- Is normally used on drilling and conventional workover rigs when stripping in the "pipe heavy" mode.
- Requires an annular BOP and two fixed pipe rams. The annular BOP and the upper pipe ram are used for stripping with the lower pipe ram acting as the master safety pipe ram.

6.8.4 Ram to Ram

In this process, fixed pipe rams are opened and closed around the work-string tool joints while fluids are bled, equalized, or pumped to ensure that rams are actualized in a balanced condition—i.e., rams have equal pressure above and below the sealing element prior to actuation to the open or closed position. Some rules of thumb for the ram to ram process:

- Normally required for "in-line lubrication of tools"
- Utilized for upset pipe
- Recommended method for high pressure and critical tasks

- Normally use fixed pipe rams (variable rams do not provide adequate service life)
- Wear inserts provide longer service life, special compounds and configurations are available for rotating, high temperature, etc.
- Extra care is required when snubbing in hole when the work-string is "pipe light"
- Fixed rams provide positive tool joint positioning when pulling out of the hole and during complex procedures—e.g., extraction of BHA and fish from well

6.9 Pipe Handling

Pipe handling requires special consideration during HWO/snubbing work due to variations in platform size and structural design, rig up elevations, weather restrictions, tubular dimensions and weight, BHA design, and frequent mobilization and demobilization. Work-string tubulars are normally laid out on boards in rows, washed, visually inspected, drifted, and measured as part of the rig up program. The dual winch system is used to pick up or lay down each single joint with a screw-in vented lift sub. Pin end thread protectors of various materials are used as required to prevent connection damage. The following arrangements (in order of frequency of use) are used to transport tubulars to or from the workbasket for make up and break out of the work-string:

- The CBW (counter balance winch) lifts a joint and a cable/sheave trolley is used to secure and tail the pin end.
- The CBW lifts the joint and a roustabout helper tails the pin end with a rope.
- The CBW lifts a joint from the "V" door of a drilling rig and a roustabout helper guides the pin end (rig job).
- The platform crane lifts the tubulars into and out of the workbasket area with screw in lift subs or collar catch elevators. This is typical for the handling of casing joints, wash pipe, gravel screen assemblies, TCP guns, etc.
- The CBW lifts joint from a "lay down machine" of the type used for drilling: this is typical for running casing.

6.10 Acknowledgments

Special thanks also to:

Leonard Goin—VP Operations North and South America

James Wurst—Operations Mgr GOM

Larry Hatcher—HWO Snubbing Technical Instructor

as well as present and past employees of Boots & Coots International Well Control Inc., A Halliburton Company, that helped in the creations of information in this article.

CHAPTER 7

Mud Cap Drilling in Fractured Formations

Dennis Moore, Signa Engineering

7.1 Introduction to Mud Cap Drilling

In mud cap drilling (sometimes called "drilling blind"), a sacrificial fluid, typically water, is pumped down the drill pipe to clean the bit and to flush the drill cuttings into the zone of lost returns. Drilling continues without returns to the surface. A semi-static annular fluid column is used to prevent kicks without having to continuously pump down the annulus. No attempt is made to control the problems of lost circulation while drilling in fractured formations. This technique reduces the time and cost associated with continuous well control issues and loss of drilling fluid.

The basics of mud cap drilling have been thoroughly addressed in an earlier book (Rehm et al., 2008). This chapter reviews the basics of classic mud cap drilling and discusses variations and special applications of the technique, topics that often come up and methodologies that are sometimes used.

A discussion of mud caps used for trips in underbalanced conditions is part of Chapter 1, Section 2, Techniques Common to Underbalanced Drilling.

7.2 Background to Mud Cap Drilling

7.2.1 Beginnings of Mud Cap Drilling

The term "mud cap drilling" was first widely applied in the Austin Chalk fields of South and Central Texas. This fractured carbonate was exploited using horizontal wells. Since the wells were horizontal within the same formation, reservoir pressure was essentially the same throughout the lateral if no depletion from other wells was encountered. Production was from fractures in the carbonate, some of which were very wide.

In the simplest and most common situation, casing was set horizontally into the top of the chalk. When the first fracture was encountered, mud was lost and/or a kick was taken depending upon the pressure in the fracture and the mud density. The pressure in the fracture could then be balanced either statically or dynamically. Plugging these wide fractures with lost circulation material or cement was impossible in many cases.

At first, it was possible to adjust the mud density and choke pressure to maintain circulation and control the influx to the level where it could be managed with the available rotating control device (RCD), adjustable choke, and mud gas separator. As development of the field progressed, the depth, formation pressure, and gas volumes increased, and it became more difficult to maintain proper control. However, as long as only one fracture was open, control was usually practical.

When several fractures were open but widely spaced along the horizontal wellbore, it became impossible to balance all the fractures while maintaining circulation; this was due to the difference in pressures caused by circulating friction (ECD) along the wellbore. This was sometimes further complicated by pressure depletion of some of the fractures by other wells.

In the deeper, higher-pressured areas, high production rates and high surface pressures exceeded the capacity of the RCD, so drilling was stopped and the blow out preventers (BOP) were closed. The well was then circulated below the BOPs until the surface pressure could be reduced to an acceptable level. Since multiple fractures were open, often with different pressures, down-hole cross flow and mud losses occurred. Balancing the well pressures required a lot of time, and large quantities of weighted drilling fluid were lost to the formation.

Mud cap drilling was developed to reduce the loss of time and drilling fluid caused by the higher-pressure, higher production rates, and lost returns. In its simplest form, a "cap" of heavy drilling fluid was pumped, or bullheaded, down the annulus until the well went on a vacuum. Drilling was then continued, pumping fresh water down the drill string with no returns to the surface.

Periodically the well would kick and additional heavy drilling fluid was bullheaded down the annulus until the well was again on a vacuum. Significant volumes of this heavy "kill" mud were often required to maintain control of the well. While no cuttings were returned to the surface, the measurement while drilling (MWD) package with gamma ray and directional information was adequate for drilling purposes.

Virtually the same procedure was used for tripping, with heavy drilling fluid pumped down the annulus as needed, to keep the well under control.

A lot of drilling fluid and water was lost while drilling and tripping, but the total mud and time loss was much less than trying to drill conventionally.

7.2.2 Floating Mud Cap

With a floating mud cap, the annular fluid is heavier than is required to balance formation pressure and the fluid level floats somewhere down-hole at whatever level balances the formation pressure in the loss zone. The floating mud cap tends to be somewhat unstable during drilling and the cap often has to be replenished with additions of fluid, sometimes even requiring a small constant stream of fluid. To help reduce the rate at which gas migrates through the annular fluid and thereby reduce the amount of mud required, the viscosity of this "cap" may be increased. The resulting mud cap then overbalances and controls the well. Since contact with this variable or floating fluid level is not maintained, gas migration to surface may periodically occur but is contained by the RCD. When this occurs, additional heavy mud is pumped into the annulus once again, overbalancing the well and sending the fluid level downhole.

7.2.3 Viscosity

To minimize the migration of formation fluids, the viscosity of a water based mud cap can be high, ideally 600–800 seconds/quart funnel viscosity at the surface. The higher the viscosity of the annular mud, the slower gas will migrate through it. Also, the higher the viscosity of the cap mud, the lower the rate at which "fluid swapping", due to differences in density, will occur. This further helps to stabilize the annular fluid column since fluid swapping often results in changes in the height of the different fluids because of differences in hole size, BHA, etc.

A high viscosity water based mud cap can easily be made using polymers. It is much more difficult to achieve the high viscosities desired when using diesel or 'synthetic' oil based annular fluids. However, gas migration is not nearly as serious a problem when using oil based annular fluids since, at downhole conditions of temperature and pressure, any gas in the wellbore goes into solution instead of migrating.

7.2.4 Pressurized Mud Cap

The next step in mud cap operations is when mud cap techniques were applied to very thick fractured formations and formations with sour gas. The presence of sour gas made it unacceptable to allow any gas to reach the surface as was common when using the previous floating mud cap techniques. While it is possible to balance a single point in the reservoir or to balance several points that have the same formation pressure gradient, it is not possible to balance fractures with very different pressure gradients. This is a common condition in thick gas filled carbonates with fractures at significantly different depths. The pressurized mud cap was developed to continuously monitor the annular pressure at the surface. The technique is variously called "pressurized mud cap", "light annular mud cap," and "closed-hole circulation drilling".

The basic technique is to displace the entire annulus with a full column of mud that is slightly less dense than is required to balance the formation pressure. Drilling is conducted through an RCD with the well shut in against a closed choke or valve, and sufficient surface pressure is applied to balance formation pressure when combined with the hydrostatic pressure of the annular mud. The surface annular pressure is an indicator of what is going on down-hole. Sacrificial drilling fluid, again typically water, is pumped down the drill pipe to clean and cool the bit, and all the fluid and drill cuttings are pumped back into the fracture. By maintaining a static column of fluid in the annulus, mud losses are greatly reduced and constant contact with the loss zone is maintained.

The static surface pressure is the difference between the formation pressure at the top of the uppermost fracture and the hydrostatic pressure exerted by the annular fluid above the loss zone. The pressure that must be balanced is somewhere between the pressure at the top of the formation and the pressure at the uppermost fracture in that formation. Exactly how much higher this balance pressure will be than that of the top fracture is primarily a function of the conductivity of the top fracture and the productivity of the formation "matrix" above that point. Pumping annular pressure is typically higher than static and is determined by the friction pressure required to pump into the fractures.

If gas migration occurs, the annular pressure rises as annular fluid above the top fracture is replaced by gas. To counter gas migration,

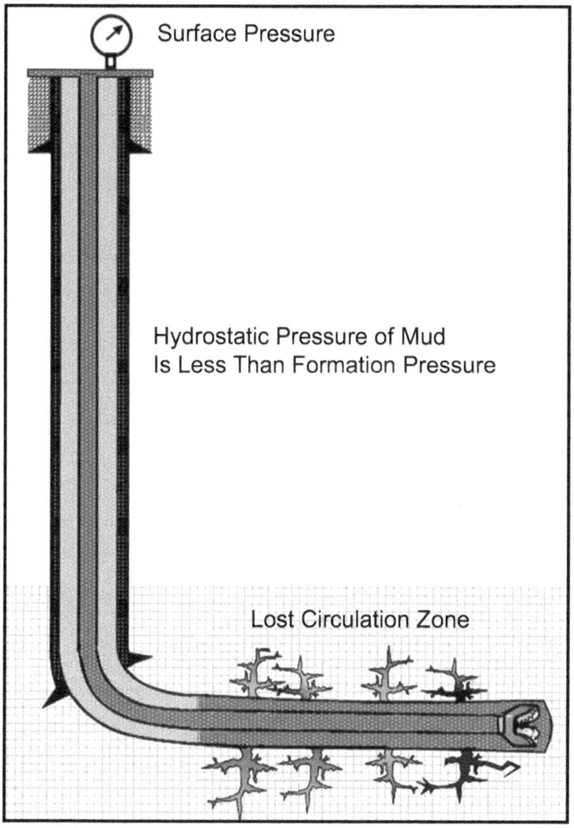

Figure 7–1 *Pressurized mud cap in a horizontal well (Stone, 1995)*

when the annular pressure rises by some pre-determined amount, additional fluid is pumped into the annulus displacing the gas and contaminated fluid back into the fracture until the previous annular pressure is restored. In this way, control of the well can be maintained and gas, especially hydrogen sulfide, will never reach the surface.

Sweep, Bailey and Stone (2003) reported that this technique made it possible to drill very thick, highly-fractured, sour reservoirs that, in some cases, had not been previously completely penetrated.

7.2.5 Trips with Pressurized Mud Cap

To trip out of the hole, the pipe is stripped out under pressure while bullheading sufficient mud into the annulus to replace the pipe pulled and to prevent gas migration. The pressured annulus poses a problem when the last of the pipe and bottom-hole assembly has to be pulled out of the hole. One of the earliest solutions to pressure in the annulus was the use of a partial column of heavier "kill" mud. By pumping a sufficient length of heavier mud into the annulus, it is possible to replace the surface annular pressure with hydrostatic pressure and reduce the surface pressure to zero. The mud used to do this is typically 2–5 ppg heavier than the annular fluid to reduce the volume required. This partial volume of heavier mud is also sometimes referred to as a mud cap, especially when this tripping method is used to balance the well when drilling underbalanced. The critical element of tripping when using the pressurized mud cap technique is to balance the well exactly and not overbalance the well sufficiently to send the fluid level so low that it cannot be readily monitored. When using heavier mud for pulling the bottom-hole assembly out of the hole, workers need to be aware that gas may still migrate up the hole. For this reason, it is essential to only pump the minimum volume of heavier mud required to exactly balance the well while pulling the BHA. Once the surface pressure is reduced to zero, the BHA can be pulled through or re-inserted with an open annular preventer and RCD.

In some cases, when running the drillstring back in the hole, some displacement of the heavy mud at the surface may be observed. Until the bit passes the bottom of the kill mud, this has no effect on BHP. However, once the bit passes the interface between heavy and light mud, as the heavy mud is displaced out of the hole, the height of heavy mud in the hole is reduced and the total hydrostatic annular pressure will go down. For this reason, the RCD seals are installed as soon as possible after the BHA is run so that stripping under pressure can be done as soon as the pipe light point is passed.

7.3 Mud Cap—Geology and Drilling

Drilling fractured formations presents special problems, especially when the interval to be drilled is either very thick (>100 ft) if drilled vertically or very long when drilled horizontally. Lost circulation occurs due to the inability to balance formation pressure with wellbore pressure throughout the fractured interval, either statically or dynamically, with the same drilling fluid density.

The losses and kicks associated with unbalanced pressure in the wellbore result in non-productive time (NPT) and expense, in some cases the drilling operation must be stopped before completely penetrating the objective. Mud cap drilling is one solution to this problem.

7.3.1 Geology—Suitable Formations

The fractured formations in this discussion are typically carbonates, though the principles and procedures described in this chapter can be applied to any formation that is not susceptible to damage by exposure to water. For simplicity, the term carbonate is used since the vast majority of the applications are in carbonate formations.

The fracturing described is most often caused by tectonic activity, weathering, or in some cases, dolomitization. When weathering is involved, wormholes, vugs and caverns can also develop. Openings in the rock caused by these processes are large enough to allow the passage of whole drilling mud, including the solids it contains. Again for simplicity, the term fracture will be used here but the principles and procedures for dealing with the problem can be applied to all these features. This discussion is directed at drilling formations that already contain these fractures, vugs, wormholes, caverns, etc., naturally as opposed to fractures that are artificially induced by drilling with a mud weight that is too high.

7.3.2 Drilling Problems

The following example illustrates what often happens when drilling fractured formations.

7.3.2.1 Drilling with a Static Overbalance

Drilling is progressing with a static overbalance of 0.2–0.5 ppg. The annular circulating friction commonly results in an increase of wellbore pressure, or equivalent circulating density (ECD), of 0.2–0.5 ppg above the static mud weight. This brings the pressure in the wellbore up to 0.4–1.0 ppg higher than the formation pressure. When the bit penetrates the top fracture of the formation, the result is lost circulation. The pumps are shut down and the hole observed. Fluid is pumped into the top of the wellbore in an attempt to keep the hole full. The rate at which drilling fluid is lost is primarily a function of the size of the fractures and the pressure differential between the wellbore and the formation. When the pumps are stopped, the loss rate typically decreases. If it is not possible to keep the hole full with the drilling fluid, a lighter fluid is pumped on top of the annular mud until the hole will stand full. By measuring the amount of lighter fluid pumped into the wellbore, one can calculate pressure at which losses occur which, in a fractured formation, is the same as the formation pressure.

7.3.2.2 Drilling Ahead with Mud Losses

In many cases, the fractures are so small that, even though the mud weight or ECD is higher than formation pressure, the loss rate is low enough to be manageable. Assuming that the first losses are not too severe, it is possible to continue drilling with partial returns.

By taking into account the static and dynamic loss rates, the amount of mud in the active system, the amount of mud readily available in reserve pits, and elsewhere, and the rate at which mud can be mixed on the rig, etc., it is possible to calculate how long drilling can continue under the current conditions. Building a spreadsheet to make this calculation has proven to be an extremely useful tool to allow this calculation to be quickly repeated whenever loss rates, mud inventory or other pertinent data changes. When is it feasible to do so, continuing to drill with some losses can be an effective way to minimize NPT since, if the fractures are small enough that the loss rates are moderate, it is quite common for the fractures to plug with drill cuttings and mud solids as drilling continues. Even if that does not happen, it may be possible to continue drilling the entire interval if more or larger fractures are not encountered.

7.3.2.3 Care Needed When Drilling Ahead With Losses

Caution must be exercised when drilling with partial returns. Since additional formation is being drilled and additional fractures may be exposed, and exposed fractures may be partially or completely plugged, the loss rate can fluctuate continuously. If the loss rate is very high it is often not possible to distinguish between formation fluid being brought to surface and formation plugging. Therefore, everyone involved should be especially alert for warning signs of an influx.

This situation is most important when drilling with synthetic oil based mud (SBM) in deep water with a subsea BOP stack. Gas will go into solution under bottom-hole conditions and can be difficult to detect until it is brought up the hole high enough for the wellbore pressure to fall below the bubble point, and it breaks out of solution resulting in a sudden increase in volume. This depth can often be in the riser above the BOP stack with potentially catastrophic consequences. For that reason, drilling with partial returns and substantial losses is not prudent when using a subsea BOP stack. Drilling with minimal or "seepage" losses is commonly and safely done. The loss rate that is acceptable must be addressed on a case by case basis by the parties involved, taking into account such things as the pump rate and effects of a possible distributed influx.

7.4 Constant Bottom-Hole Pressure

There are other options if the loss rate becomes too high for drilling to continue, either due to lack of adequate mud supply, cost of mud losses, or to the presence of excessive gas in the returns. If the correct equipment is present, it may be possible to use what is commonly called the constant bottom-hole pressure (CBHP) method of managed pressure drilling (MPD) to reduce the pressure differential between the wellbore and the formation sufficiently to reduce the loss rate to an acceptable level. Simply put, annular pressure from a drilling choke manifold is used to replace circulating friction pressure with surface pressure during connections. This ensures that the formation is always exposed to the same pressure instead of a higher pressure when circulating than when the pumps are stopped.

By using a mud weight that is lighter than is required to statically balance the well, it is possible to maintain a wellbore pressure very close to the formation pressure by adding ECD and surface back pressure, thus reducing the loss rate. If the fractures are not too large, this reduced loss rate may be low enough to continue drilling and possibly even finish the hole.

At first glance, this would seem to solve the problem, making it possible to exactly balance the formation pressure with wellbore pressure and eliminate losses and kicks. Unfortunately, the situation is a bit more complicated than that. First of all, the term constant bottom-hole pressure is a bit of a misnomer. Since annular circulating friction is distributed over the entire hole, and surface pressure is applied at a single point when the pumps are shut down, it is only possible to keep the pressure constant at one point in the hole. The operator can select where that point is to be, but it is still only one point. If that point is always the bottom of the hole, then the target pressure is continuously changing as drilling continues. No matter which point in the hole is selected to maintain constant, the pressures at all other points in the hole change every time the pumps stop and start.

To make things even more difficult, the pressure in the formation increases with depth due to the density of the fluid it contains, while pressure in the wellbore increases with depth due to the density of the drilling fluid. The drilling fluid is almost always more dense than the formation so the pressure in the wellbore typically increases with depth much faster than the pressure in the formation. Even though it may be possible to maintain the wellbore pressure exactly the same as the formation pressure at one point in the well, they will not be the same at any other point in the well with a full column of mud.

7.4.1 Pressure Profiles

Figures 7–2 and 7–3 illustrate the pressure profiles in the wellbore and the formation assuming a mud weight and pore pressure at the top of the formation of 11.0 ppg. It is apparent from the graphs that if a constant wellbore pressure is maintained at the top of the formation that exactly balances the formation pressure, then the bottom of the formation will be over 400 psi over balanced. Conversely, if the mud weight and surface pressure are reduced to exactly balance the formation pressure at the bottom of the interval, the top of the formation will be over 400 psi underbalanced.

If the fractures are large, it is easy to see that the first situation will result in losses and the second will allow an influx from the formation.

To insure that the second problem is prevented, the point in the wellbore selected to hold constant should be at the top of the productive interval, not the top of the losses.

By keeping the pressure at the top of the productive interval at or slightly above the formation pressure, it is possible to minimize the pressure differential from the wellbore to the formation and reduce the loss rate. If the pressure is maintained at the minimum safe level and the loss rate is still unacceptable, then some form of a mud cap is the next choice.

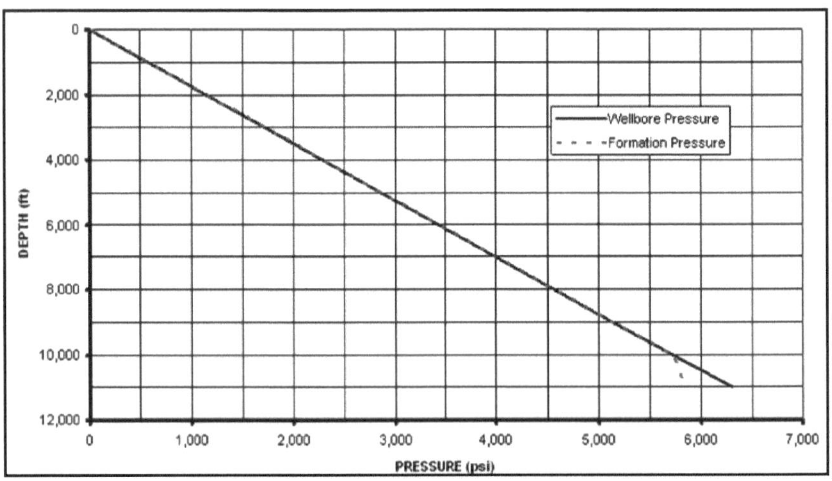

Figure 7–2 *Overall pressure profile in the wellbore*

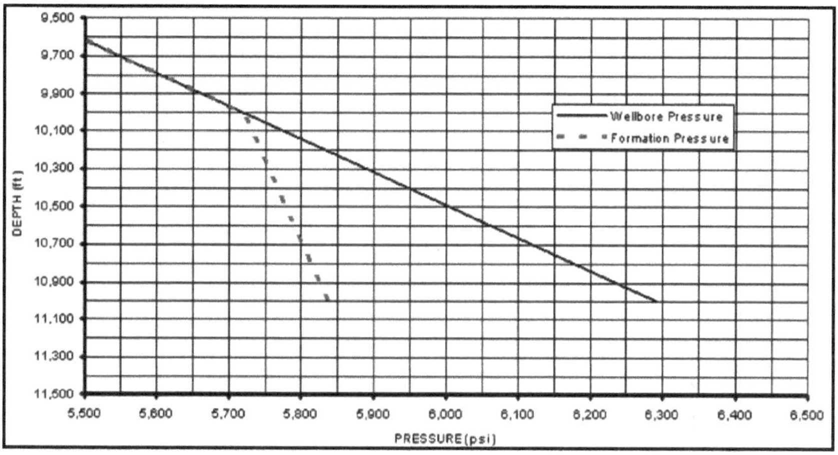

Figure 7–3 *Detail of pressure profile in the wellbore*

7.5 Horizontal Wells

Horizontal wells face very similar problems to those just described for vertical wells. In a horizontal well, the problem is no longer a difference in static gradient between the wellbore and the formation, but it is now the difference in ECD between the heel and toe of the well, Figure 7–1. The result is basically the same since the increase in ECD results in the wellbore pressure exceeding the formation pressure at the toe if the pressure at the heel is balanced, and conversely, being underbalanced at the heel if the formation pressure is balanced by wellbore pressure at the toe.

7.6 Decision Tree for Drilling Fractured Formations

Figure 7–4 presents a very generalized decision tree illustrating the thought process followed when faced with drilling a fractured formation.

7.7 Stabilizing Conditions with Mud Cap Drilling

To reiterate, mud cap drilling (MCD) does not try to stop the losses, but stabilizes well conditions to allow drilling to proceed safely and efficiently without fighting either losses or kicks. It can take the form of either floating mud cap drilling (FMCD) or pressurized mud cap drilling (PMCD). The most widely applicable method is pressurized mud cap drilling (PMCD).

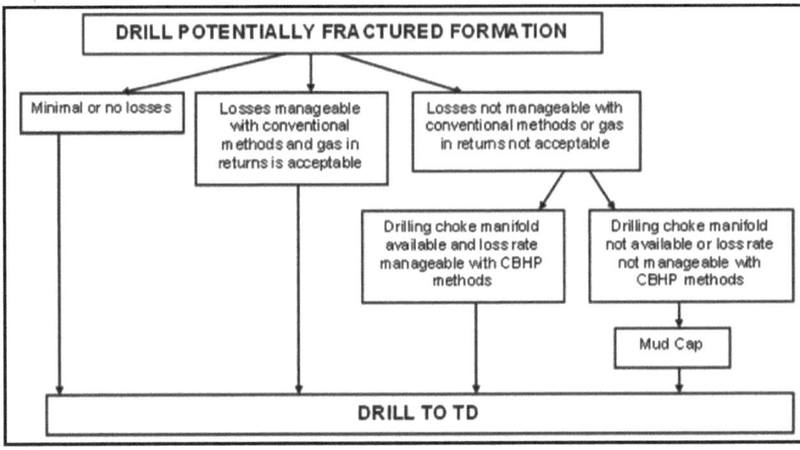

Figure 7–4 *Decision tree for drilling fractured formations*

7.7.1 Pressurized Mud Cap Drilling (PMCD)

Classic PMCD is very simple. A rotating control device (RCD) is used at the surface to allow drilling while maintaining pressure on the annulus. An annular fluid of the desired density is bullheaded down the annulus until the entire annulus has been displaced. The density of this fluid is less than needed to balance down-hole pressure so a surface pressure is below the RCD. The well is shut in, so flow from the formation is prevented.

If formation fluids begin migrating to the surface, the annulus surface pressure will increase. When the surface pressure increases, additional annular fluid is pumped into the top of the annulus to displace all the formation fluid back into the formation. This can be continued as long as necessary to completely penetrate the desired interval. Drilling is continued by pumping a sacrificial fluid (almost always water) down the drillstring to clean and cool the bit and to operate down-hole tools. All drilling fluid, cuttings, contaminated mud and formation fluid are displaced into the loss zone.

7.7.2 PCMD with Partial Losses

Most, if not all, previous discussions have addressed the situation where total losses are encountered. When that happens, classic PMCD works very well and is quite simple to apply. A more difficult and more common occurrence is to encounter loss rates that are too great to continue drilling for extended periods of time but are not total losses. This situation is often accompanied by background gas that is higher than is considered safe to continue drilling. PMCD may offer a solution to this problem.

When fractured formations are encountered and mud losses occur, it may be very difficult to control them with conventional lost circulation material (LCM). Even if they can be controlled with LCM pills or cement, as additional formation is drilled and new fractures are encountered, much time can be lost if it is necessary to repeatedly stop and heal the losses. When drilling thick (or long) sections, or in extremely prolific gas producing formations, gas cut mud can become a serious problem. Since high mud losses are already being encountered, it may not be feasible to increase mud weight to control the gas cutting. When in either (or both) of these situations, it is quite simple to determine whether or not PMCD is a viable option.

7.7.2.1 PMCD Total Losses Applicability Test

When total losses are encountered, or if it is determined that some remedial course of action is required due to high loss rates or high background gas, an injectivity test should be performed. To conduct this test:

1. First, make sure the annulus is filled to the surface. In case of total losses, fill the hole with a lighter fluid (either water or base oil depending on conditions and preference). If a lighter fluid is used, closely measure the volume required so that bottom-hole pressure can be accurately calculated.

2. Pump the drilling fluid currently in the hole and pits down the drillstring at drilling rates and monitor the annular pressure.

3. Once a stable annular pressure is established, shut the pump down and record the time required for the surface annular pressure to bleed to zero.

There are three possible outcomes to the injectivity test:

- The annular pressure continues to rise to a predetermined maximum (for example, the last leak off test) and is still increasing. In this case, PMCD is not a viable option. However, under these conditions, it should be possible to control the losses with LCM or similar conventional means.
- The annular pressure rises very little and classic PMCD techniques can be readily applied.

- The annular pressure rises to some pressure and remains constant while pumping down the drill pipe at a rate equal to the drilling rate. PMCD is still a viable option, but a slightly different approach is required.

7.7.2.2 Determining the Required Annular Mud Weight

Example Conditions:

Current Mud weight = 10.5 ppg

Loss zone TVD = 10,000 ft

Surf Ann Press @ Drlg Pmp Rate = 2,000 psi

Static loss rate = 0

Maximum allowable surface pressure = 2,500 psi

EMW @ loss zone @ drlg pmp rate = Current mud weight + (Surf Ann Press @ Drlg Pmp Rate/(.052 × TVD))

= 10.5 + (2000/.052/10,000)

= 14.4 ppg

The formation will accept mud at drilling rates at an equivalent mud weight of 14.4 ppg. If 2,000 psi is an acceptable annular surface pressure while drilling, then drilling can continue with the mud weight that is in the hole.

Although 2,000 psi may be below the maximum allowable surface pressure and RCDs with this working pressure exist, most people are more comfortable working with a lower surface pressure. It is a simple matter to determine what mud weight is needed to achieve the desired surface pressure. For example, if a surface pressure of 700 psi was deemed acceptable, surface pressure while drilling the mud weight can be increased as calculated below:

Desired mud weight = Current MW + ((Surf Press with Current MW – Desired Surf Press) / (.052 × TVD))

=10.5 + ((2000 – 700) / (.052 × TVD))

= 13 ppg

So if the annulus is displaced with 13 ppg mud, the surface annular pressure while drilling at full pump rate will be 700 psi.

7.7.2.3 Connections with PMCD

Another factor that must be considered is leak off while making a connection. The static leak off pressure is established prior to the injectivity test and the last step in the test is to measure how long it will take to leak off to that pressure. The time it takes for the pressure to fall from a 13 ppg EMW to 0 is how long it will take for the pressure to fall to zero when the pumps are shut down for connections. One of the primary objectives is to never lose contact with what is happening in the wellbore; if this time is long enough for connections to be made without the pressure falling to zero then the annulus can be displaced by bullheading 13 ppg mud from the surface and drilling can continue.

When using this method to select an annular mud weight, it is usually desirable to minimize the surface pressure when drilling. This mud weight should not be high enough to allow the surface pressure to fall below a minimum required to accurately monitor (typically 50–100 psi) during connections. The time for this to occur should always be determined during the injectivity test as described above.

Just as with classic PMCD, it is highly advantageous to increase the viscosity of the new annular fluid to the point of practicality. A funnel viscosity of 600–800 sec/qt is commonly used when using water based mud. This high viscosity serves several purposes:

- Static loss rates are greatly reduced during connections and trips reducing mud requirements.
- Gas migration rates are greatly reduced, thereby reducing the required frequency of bullheading additional fluid down the annulus and the amount of mud required.

It is interesting to note that when mud viscosities in this range are suggested, the most common response is that it is not possible to mix or pump mud that thick on the particular rig involved since, in most cases, it has not been attempted before. In almost every case, it has proven to be possible to achieve viscosity in this range. It should be noted that since this fluid is not circulated conventionally, ECD is not an issue.

7.7.2.4 Trips with PMCD

When tripping using this technique, the static loss rate is usually acceptable, especially if the viscosity of the annular fluid is very high. Before starting out of the hole for a trip, the static loss rate should be established with the mud in the hole. If this loss rate is higher than can be handled, a pill of lighter mud can be bullheaded into the top of the annulus to reduce the loss rate to a rate that can be managed with the mud available. By filling with lighter mud as the pipe is pulled out of the hole, the hydrostatic pressure and the loss rate will continue to be reduced.

If the equivalent mud weight at the formation gets too low, the well will start to flow. A workable approach is to establish upper and lower limits of loss rate that are deemed acceptable and to manage the loss rate by filling the hole with either heavier or lighter mud to increase or decrease the loss rate as desired. If this approach is taken, great care must be taken to keep up with exactly where the lighter and heavier mud is in the well and what the hydrostatic effects of its position are as it moves down the hole. The effects of changes in annulus size and hole angle on the hydrostatic effect of the various mud weights and their effects on loss rate can be anticipated and changes in loss rate due to these effects understood. This method has been successfully used with extremely high rate inclined gas wells and when properly applied, works very well.

It should be noted that since the formation pressure is exactly balanced when the well is shut-in and static, whenever any pipe is pulled out of the hole, an underbalance immediately results accompanied by an influx equal in volume to the pipe removed. Therefore, it is extremely important that a pump is running any time the pipe is moved upward. If the pipe is picked up without a pump running, mud should be immediately bullheaded down the annulus to force it back into the formation before it can start migrating up the hole and become more difficult to get rid of.

7.8 Floating Mud Cap Drilling—Depleted Formations

In floating mud cap drilling, the annular fluid used is heavy enough to overbalance the formation pressure and the fluid level remains below the surface. Additional fluid is pumped down the annulus as required to prevent gas migration or to overcome migration that is occurring and maintain the surface pressure at the desired magnitude, often zero. There are some very good applications for FMCD in specific cases that will be discussed in more detail later. In reality there is very little difference between what happens at the formation in FMCD and in PMCD since both use annular pressure to control the formation. PMCD uses hydrostatic pressure plus surface pressure while FMCD uses hydrostatic pressure alone.

Most of the discussions regarding PMCD address the situation where the formation pressure is high enough to support a full column of fluid to surface and some additional surface pressure. The question that sometimes arises is how to handle these same formation conditions of fracturing or weathering when drilling in depleted formations?

7.8.1 Water Availability

If sufficient water is available and the formations that are open to the wellbore above the loss zone are not water sensitive, it is possible to pump water with no additives down the annulus continuously and keep all formation fluids displaced back into the formation and never face a migration issue. This is the simplest approach and is quite effective. However, if there is not sufficient water available or if there are water sensitive formations open, a floating mud cap can be used. This approach is widely used as long as the potential surface pressure will not be too high when a kick is taken and as long as H_2S is not present. This upper limit is usually determined by the rated working pressure of the surface equipment, specifically the RCD. It should be noted that using high viscosity annular fluid as previously described has proven to dramatically reduce annular mud requirements when using a floating mud cap.

7.8.2 Fluid Level Measurement

Another approach is to use an acoustic fluid level measuring device that "shoots" an annular fluid level, interprets the acoustic readings, and plots the data. This is automatically repeated at very frequent intervals giving both a quantitative measurement of the fluid level and a qualitative comparison of the measurements over time so that the movement of the fluid level can be continuously monitored. This approach gives the operator the ability to tell what is going on downhole with FMCD just as is done with classic PMCD (Schubert and Wright, 1998).

Since the fluid moves up and down the hole when the pumps are started and stopped, determining exactly what is happening in the wellbore is a bit more complicated than when using classic PMCD. However, the principle is the same. The difference is that, in this case, all the pressure the formation sees is hydrostatic while with classic PMCD the pressure on the formation is a combination of hydrostatic and surface pressure. In reality though, the formation cannot tell the difference, so this technique may be considered to be a form of PMCD. When the fluid level rises reflecting the increase in annular pressure due to migration, arriving at the exact difference in hydrostatic pressure must include corrections for:

- How much of the rise is due to injected water
- How much is due to formation fluid
- Changes in hydrostatic due to changes in column height of the various fluids if they encounter a change in cross-sectional area in the annulus

There are an increasing number of situations that would benefit from this approach, and its application is expected to expand.

It is very helpful when using a pressurized mud cap to have an annular pressure gauge that reads in one psi or one kPa increments. For very low pressures, an old drill stem tester's trick can be used where a small hose or tubing is put into a bucket of water (see Figure 7–5). Pressure will cause bubbles, and a vacuum will cause the water to be sucked into the tubing and lower the water level. This can also be used to determine movement of fluid in the annulus when using a floating mud cap.

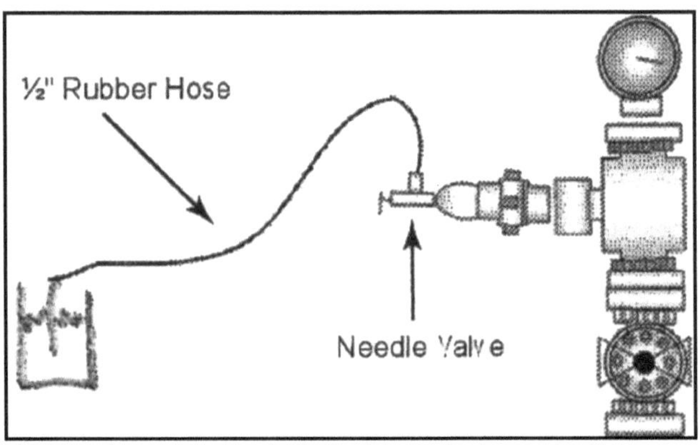

Figure 7–5 *Pressure indicator*

7.8.3 Continuous Annular Injection PMCD

Continuous annular injection of water using FMCD was mentioned earlier in the discussion of drilling depleted formations. Continuous annular injection has also been used successfully as a version of PMCD. Due to the volumes of fluid required, this is most often done with water. In some cases, the annular fluid is viscosified to further reduce gas migration rates. With this method, water is continuously pumped down the annulus at a rate high enough to overcome gas migration, keeping all the gas displaced back into the formation.

7.8.3.1 Trips with Continuous Annular Injection

Tripping with continuous annular injection can be conducted just as is done with other mud cap methods. If done on a well being drilled with surface annular pressure, a high viscosity weighted "cap" can be bullheaded down the annulus to balance the well long enough to remove the BHA, just as when tripping any other mud cap well. Once the BHA is removed, the well is shut in and the pressure monitored. Before opening the well to run back in the hole, additional heavy mud is pumped down the annulus to eliminate any surface pressure.

Another alternative sometimes employed when it is time to make a trip on a well using continuous annular injection and surface annular pressure is to set a plug in the well on the way out. The bit is stripped well up into the casing under pressure while continuing to pump continuously down the annulus. Before pulling out of the hole, a gunk (oil, gel and cement) plug is spotted in the casing to secure the well and allow displacing the hole above it with mud. The remainder of the trip is conducted normally until the bit is at the top of the gunk plug on the way back in the hole. Before drilling out the gunk plug, the mud above it is displaced with whatever annular fluid was being used, the plug is drilled out with water, continuous annular injection is resumed while stripping, and drilling is continued. When a liner is needed, a bit and sacrificial motor is run on the bottom of it to clean out the gunk plug on the way in the hole. On some programs where multiple wells have been drilled using this method, the bugs have been worked out of the process and the people involved have learned how to do it correctly. In many of those cases it has worked very well. There have also been cases, usually involving smaller programs, where a great deal of trouble was encountered related to the gunk trip plug. Some of the difficulties encountered have been plugged drill-strings, cementing drill-strings in the hole, and failure of the gunk plug resulting in pressure communication with the reservoir on trips. The gunk trip plug method also requires substantially more time than the simpler approach. If this method is to be used, very careful contingency planning and training is essential.

7.9 Water Sensitive Formations Exposed

Historically, there has been some reluctance to use mud cap methods when water sensitive formations have been exposed in the wellbore, especially when those formations are shales. There has been fear that the water used as sacrificial fluid will contact the sensitive formations causing a wellbore stability issue, so an inhibitive mud is used, even though there are no returns of any of the drilling fluid to surface. In some cases, even oil-based mud has been used, if that is what is needed to stabilize the exposed shale. Whether or not MCD can be successfully used is largely a function of how far these zones are above the loss zone and whether PMCD or FMCD is being used. If the interface between the water pumped down the drillstring and an inhibitive annular mud cap remains below the sensitive formations, then mud cap methods can be used without sacrificing wellbore stability.

If PMCD is the method being applied, the water sensitive shale can be very close to the loss zone. The interface between the inhibitive annular fluid and the sacrificial water being pumped down the drillstring is never allowed to come up the hole above the top of the losses since the annulus is completely shut in.

When using FMCD, the interface will often move up the hole when pumping down the drill-string, so care must be taken that there is sufficient separation between the two zones to insure that the interface never reaches the sensitive zone.

The use of FMCD with oil based annular mud in severely depleted formations has resulted in substantial savings in mud and NPT due to logistics issues in situations where UBD was not a viable option. Once again, the use of ultra high viscosity annular fluid has been successful in drastically reducing the annular mud required by reducing the frequency that pumping needs to be employed to overcome migration.

7.10 Mud Caps versus Gas Assist UBD

Many times, some type of gas assist UBD is used when drilling depleted intervals to reduce or eliminate mud losses. However, in some cases it is not practical to drill producing intervals underbalanced, especially if there is no good way to handle the hydrocarbon production at the surface. One example of such difficulties that was solved with MCD was due to a lack of storage facilities in the case of oil production in a remote location. Another was the lack of space for additional separation and production handling equipment on small offshore facilities.

7.11 Mud Cap and Hole Cleaning

The question of hole cleaning always arises whenever MCD is being discussed. Many people find it difficult to believe that adequate hole cleaning is possible without returns to surface, especially when only water is being pumped down the drillstring. In practice, hole cleaning is very seldom an issue. Although water does not have great carrying capacity, the cuttings do not have to be moved very far and the pump rates used are generally at the upper end of the operating range for the down-hole tools (MWD, LWD, motor, etc.). In many cases, especially when one or more changes in hole geometry due to drilling liners are present, it is actually easier to clean the hole under mud cap conditions than it is when drilling conventionally since the cuttings do not need to be brought to surface, only moved to the nearest fracture that is taking fluid. By closely monitoring torque and drag it is easy to follow trends and to detect cutting buildups should they occur. Viscous sweeps will usually facilitate hole cleaning, and if necessary, a short wiper trip can be made to remove any buildups that do occur. Many people routinely pump viscous sweeps on a regular frequent schedule as a preventative measure, but they are usually unnecessary. The use of good cuttings transport models has been effective in predicting whether or not hole cleaning will be an issue. If there is still doubt about the ability of water to move solids, remember the video clips of floods transporting cars and houses without the benefit of viscosifiers.

7.12 High Bottom-Hole Temperature

In deeper wells and longer horizontals, high bottom-hole temperature can be a very serious problem as the bottom-hole circulating temperature approaches the limits the down-hole tools will tolerate. MCD results in a much lower tool temperature than is observed with conventional circulation since relatively cool water is continually pumped down the string. With casing set into the top of the target formation, it has sometimes been advantageous to intentionally induce fractures so that mud cap drilling could be used to lower the bottom-hole temperature below the limit of the directional tools, allowing longer holes to be drilled.

7.13 Down-Hole Isolation Valves

PMCD allows drilling with total losses while maintaining contact with and control of what is happening down-hole. When tripping, the well is balanced at the surface by introducing enough heavier mud to replace the shut in surface pressure with hydrostatic pressure. In situations where it is feasible to do so, a down-hole isolation valve (DIV) can be installed, making it possible to shut the well in down-hole, and remove the need for adding the heavier mud, decreasing trip time. A DIV is a full open valve installed in the last casing string run (usually a tieback). Drilling is conducted below this valve. The valve is opened while the bit is below it and closed once the bit is pulled above it, isolating the open hole from the surface. The pressure can then be bled off above the valve and the drillstring tripped normally. When the string is run back in the hole and the bit is just above the valve, the annulus is pressured up at the surface until the pressure above the valve equals the pressure below it. The valve can then be opened, the string run to the bottom and drilling resumed. These valves are generally reliable, especially when using clear fluids. Drilled solids and solids in the mud sometimes interfere with the valves closing properly and/or sealing completely. When the valves work properly, they can be very effective in saving time and as another barrier to the flow on trips. However, all pertinent personnel should be trained and all required preparations made to balance the well on trips should the DIV fail to close properly or seal completely.

7.14 Concentric Annuli

Many other schemes have been tried and called MCD. One notable example was done using a tieback run with circulating ports as are used for concentric gas injection UBD. Various fluids were then pumped down the drillpipe, the drillpipe casing annulus, and the concentric casing annulus while returns were taken up the two annuli. After drilling was completed, the tieback had to be pulled to remove the inject ports and restore casing integrity. Although the wells were finished using these methods, formation fluids found their way to surface up both annuli and much more time and mud were required for these wells than for offset wells drilled using normal PMCD. Overall, these methods have proven to be overly complex and generally unsatisfactory.

7.15 Constant Surface Circulation Approach

A PLC controlled choke system with a continuously running auxiliary pump and so called "automated" controls has also been used to do mud cap drilling. The logic behind this approach is to use a pump pumping into the return line upstream of a flow meter and a manifold containing a PLC controlled choke. The software in the control package for the PLC controlled choke is then used to maintain flow out the same as flow in. While this would theoretically work, the response time for the choke to properly control the well is such that either small influxes will be allowed, more annular fluid will be lost than necessary, especially if it is used any time the pipe is moved or the rate at which the pipe is moved changes. In practice, the approach makes MCD more complicated and less efficient. If this equipment is installed specifically for MCD, it also significantly increases cost since it can be quite expensive.

7.16 Different Pressure Regimes

A question that is frequently asked when planning a MCD well is: What happens if a permeability barrier is crossed and a new formation is encountered having a substantially different pressure regime? The simple answer is: Avoid that. MCD is best applied when drilling a formation having a single gradient, i.e. a formation that is in hydraulic communication from top to bottom. If a permeability barrier is crossed and a different pressure regime encountered, especially if it is also fractured, then down-hole crossflow can occur, and it may not be possible to stop it. Depending on the difference in depth and pressure between the two zones and the conductivity of both zones, it may or may not be possible to isolate them using such methods as barite plugs, gunk plugs, etc. It is therefore very important to prevent this from happening. When drilling using MCD, drilling behavior changes, LWD, and any other available tools should be employed to prevent connecting two dissimilarly pressured formations.

It should be noted that if two dissimilarly pressured formations are inadvertently exposed together while drilling, the risk to personnel and equipment is not materially increased. Mud cap methods only seek to balance one point in the wellbore and that will still happen. Drilling and tripping can continue safely. The difficulty that arises is in isolating the formations from each other. As discussed, this may become very difficult or even impossible so this situation should be avoided if at all possible. If, however, it is accidentally encountered, safety of the personnel and rig are not automatically compromised.

7.17 No RCD Available

Most wells drilled using MCD are drilled using an RCD to allow pressure to be contained on the annulus and to help avoid getting gas to the surface. However, there have been many wells successfully and safely drilled with total losses and without an RCD, usually because the losses were not anticipated on an exploratory well and no RCD was available. Some form of FMCD is obviously needed since no surface pressure can be maintained on the annulus. Continuous annular injection has been used successfully as has the use of an acoustic fluid level device referred to in the discussion of drilling depleted zones. As previously described, using FMCD and the acoustic fluid level device is very similar to using PMCD since it is possible to detect gas migration and to prevent gas from reaching the surface by using only hydrostatic pressure instead of a combination of hydrostatic and surface pressure, as is normally done with PMCD.

7.18 Deepwater and Floating Rigs

Deepwater wells present some special challenges to the use of mud cap drilling. It is of paramount importance to use every means available to avoid allowing gas to enter the riser. As previously discussed, this precludes drilling with partial returns. Installing an RCD on a floating rig is more complicated than on a fixed rig with a surface stack. Historically, the slip joint near the top of the riser has been collapsed and locked to increase the pressure integrity of the riser to around 500 psi. This can make it difficult to connect to the rigs circulating system when conventional circulation is desired.

At the time of this writing, there are several equipment systems under development that incorporate the RCD into the riser below the slip joint making it possible to connect the riser to the rig normally and thus readily allowing circulation through the rigs circulating system. These systems make it possible to use PMCD though the maximum allowable surface pressure is much less than with surface stacks since most risers will not allow over 1,000 psi and often less, around 500–700 psi.

The increased importance of not allowing gas migration to occur coupled with the reduced allowable surface pressure make it somewhat more difficult to switch to and continue to use PMCD. The use of CBHP MPD for situations where total losses have not yet occurred has proven to be especially useful when drilling fractured formations in deepwater.

Deepwater drilling with a subsea BOP stack also provides some useful options that are not available with a surface stack. By filling either the choke line or kill line with a light clear fluid such as water or base oil (depending on the formation pressure, water depth, and mud type being used), and shutting that line in at the surface, a reliable down-hole pressure gauge becomes available. This serves as another way to tell what is going on down-hole. It also makes it possible to do PMCD without an RCD as described previously. With an RCD, the pressure on the annulus is monitored and changes in that pressure are used to detect formation fluid migration, formation plugging, etc. Exactly the same information can be obtained by monitoring the pressure on the line to the seafloor filled with fluid light enough to always maintain a positive surface pressure at the surface. The formation sees the same pressure whether it is all hydrostatic or a combination of hydrostatic and shut-in surface pressure. The only difference is where the bottom of the pressure sensor is located. In one case, it is at the surface, and in the other case, it is on the seafloor.

With a subsea stack all the positive features of a DIV are present, only better. Not only are the BOPs much more reliable, but all of them are multiples so there is always a backup for each closure method. Trips are very simple. The string is stripped up above the stack under pressure while keeping the hole full by pumping down the annulus as the pipe is pulled. Once the bit clears the BOP, the blind rams are closed, the pressure bled off above them and the remainder of the trip conducted normally. As with a DIV, immediately before opening the BOP on the way back in the hole the pressure above the stack is increased to equal or slightly exceed the pressure below it. The blinds are opened and the string stripped the rest of the way in the hole under pressure through the RCD. It is possible to deploy any casing string that is shorter than the water depth with the well completely contained at all times.

7.19 Casing, Cementing and Zonal Isolation

Most wells drilled using a mud cap are completed open hole or with a slotted liner. However, there have been a number of cases where zonal isolation has been successfully achieved using casing, cement and mechanical devices. In thick, fractured, oil bearing carbonates, it has been possible to achieve zonal isolation using stage cementing techniques. In some cases, it has been possible to modify the casing program so that casing is set and cemented just above the top of the loss zone and a liner set through the loss zone into a competent formation. Cement is then pumped down the string, comes up the outside and goes into the loss zone providing cement isolation at the bottom of the zone. A liner top packer then provides isolation at the top. In some cases, it has been possible to allow returns during the cement job, carefully controlling the rate and then immediately setting a liner top packer as soon as the cement is put in place. These are a few examples of ways that cement jobs have been achieved. Each application must be examined on an individual basis to determine if cementing is feasible and, if it can be done, how best to accomplish it. Some of the new completion tools such as recent improvements in the various forms of external casing packers hold great promise to help improve zonal isolation in MCD wells.

7.20 Conclusions

- Mud cap drilling has proven to be a very useful tool in drilling wells with problems related to loss of circulation.
- The techniques are very simple and are easily adapted to a wide variety of conditions.
- With depletion becoming more widespread and wells being drilled in more and more challenging environments, the number of potential applications of MCD will continue to grow, making it a tool that every drilling engineer should be able to apply when needed.

7.21 References

Colbert, J.W., and Medley, G. "Light Annular Mud Cap Drilling—A Well Control Technique for Naturally Fractured Formations," SPE 77352 presented at SPE Annual Technical Conference and Exhibition, San Antonio, TX, USA, September 29–October 2, 2002.

Rehm, B., Schubert, J., Haghshenas, A., Paknejad, A.S., and Hughes, J. *Managed Pressure Drilling,* Gulf Publishing Company, Houston, TX, USA, 2008.

Reyna, E. "Case History of Floating Mud Cap Drilling Techniques—Ardalin Field, Timan Pechora Basin, Russia," SPE 29423 presented at IADC/SPE Drilling Conference, Amsterdam, The Netherlands, February 28–March 2, 1995.

Schubert, J.J., and Wright J.C. "Early Kick Detection Through Liquid Level Monitoring in the Wellbore," SPE 39400 presented at IADC/SPE Drilling Conference, Dallas, TX, USA, March 3–6, 1998.

Stone, R. "The History and Development of Underbalanced Drilling in the USA," presented at the 1st International Underbalanced Drilling Conference and Exhibition, Amsterdam, The Netherlands, October 2-4, 1995.

Sweep, M.N., Bailey, J.M., and Stone, C.R. "Closed Hole Circulation Drilling: Case Study of Drilling a High-Pressure Fractured Reservoir—Tengiz Field, Tengiz, Republic of Kazakhstan," SPE 79850 presented at IADC/SPE Drilling Conference, Amsterdam, The Netherlands, February 19–21, 2003.

Urselmann, R., Cummins, J., Worrall, R.N., and House, G. "Pressured Mud Cap Drilling: Efficient Drilling of High-Pressure Fractured Reservoirs," SPE 52828 presented at IADC/SPE Drilling Conference, Amsterdam, The Netherlands, March 9–11, 1999.

CHAPTER 8

Underbalanced Liner Drilling

Robert Sanford

8.1 Introduction

Since the mid 1990s underbalanced drilling has been coupled with the useful technique of drilling with casing. The combination of these two technologies has given operators around the world an economic means to drill problematic formations with a single string of casing where previously multiple strings of casing were required. While an entire textbook could be devoted to the subject of underbalanced casing drilling, the focus of this chapter will be limited only to the limits and extremes of underbalanced liner drilling (UBLD) applications.

In underbalanced liner drilling applications, a drillstring is used to convey a liner by means of a bit and running tool assembly. The casing drilling bit assembly conveys the liner but transfers weight to the drillstring. The drillstring is designed to meet all the torsional and load requirements of the well design. Drill collars and heavy weight drillpipe are used if necessary to prevent the drillstring from buckling or drilling under compression. This prevents the liner from experiencing the compressive loads of drilling.

The liner bit drills the formation as a typical PDC bit. Drilling fluid is pumped down the drillpipe and then exits the bit. Because the liner is made up onto the bit, the fluid flows up the tight annulus between the open hole and the liner. Returns are taken at surface through the surface pressure control system. This tight annulus also results in the grinding of the cuttings into the formation wall. This is known as the smear effect. Once total depth of the interval has been reached, the liner hanger is set and the liner is cemented into place. Then the liner top packer is set to isolating the annulus.

The drillstring is released from the liner drilling bit assembly and then pulled out of the hole. Once the drillstring has been pulled above the top of the liner hanger assembly, the system is reverse circulated to remove any cement from the annulus. The end result is a cased hole through almost any hole stability problem.

8.2 Well Candidate Selection and Design Considerations

Underbalanced liner drilling is a particularly effective technology. When applied correctly it can save operators millions of dollars. Severe lost circulation, sloughing shale, rubble zones, fractured formations, wellbore ballooning, over-pressured zones, and under-pressured zones are all issues that can be remediated with underbalanced liner drilling.

However, UBLD is not the solution for every well with stability issues. If changing the drilling fluid system or other design parameters, like hole size, solves the stability problem then UBLD is not the solution. UBLD is for holes where stability is unmanageable through conventional means and multiple casing strings are required over a very short interval of a few hundred feet. UBLD is for cases where the goal is to eliminate a casing string that is set solely for hole stability issues.

UBLD is a combination of two useful technologies that can sometimes stand alone to help solve well issues. In many cases, simply drilling with the liner with an at-balanced or over-balanced system might solve the instability issue. The underbalanced application is necessary when there might be damage to a productive formation from fluid invasion, or when a single bit run might not be possible in an at-balance or over-balanced environment. By drilling the well underbalanced the rate of penetration is maximized and the depth of penetration is maximized.

UBLD is like everything else in a drilling operation: when things work like they should the operation goes smoothly, but when there is a mistake or failure with UBLD the result is usually very expensive. Therefore, careful planning and execution are critical to UBLD success.

8.2.1 Basic Planning—The Bit

The first step towards planning a successful UBLD application is selecting the correct liner drilling bit assembly. The casing bit that will drill the desired interval should be designed specifically for each

Figure 8–1 *Weatherford DwL casing drilling bit; the model shown here is an 8 1/2 in. five-blade bit with 13 mm PDC cutters (Weatherford U.S., LP)*

individual UBLD application and should be similar to the conventional PDC bit that would be used to drill the formation conventionally. For example, if a five blade PDC bit with thirteen millimeter cutters would be used under conventional conditions, the liner drilling bit should also be a five blade with the same size cutters.

The bit is one of the limiting factors for a UBLD application. Retrievable bottom-hole assemblies are available and can be incorporated into the system, but they are far from perfected. Almost all UBLD applications are completed with a single bit run. In liner drilling, changing the bit is problematic. Therefore, most liner drilling applications are designed to be completed with one bit. When an underbalanced system is utilized the single bit will likely drill deeper and rate of penetration will be greater than conventional drilling. A bit failure could result in having to pull out of the hole with the entire liner string. If this happens, the entire liner might have to be laid down and inspected.

The liner should only be pulled out of the hole if absolutely necessary, as in the event of a major mechanical failure, or if the bit becomes plugged to the point of exceeding equipment pressure ratings. The casing bit must be designed to drill required formation interval.

8.2.2 Basic Planning—Hydraulic Design

Hydraulic design is another critical design factor. The flow rates must be high enough to clean the hole, but not so high that circulating pressures in the tight annular space exceed the open hole fracture gradient. This is basic equivalent circulating density (ECD) management. In general, the flow rate should allow for an annular velocity of 175 ft per minute (fpm) as long as the fracture gradient is not exceeded. The rate of penetration might have to be limited if an adequate flow rate to clean the hole cannot be achieved without generating an equivalent circulating density greater than the fracture gradient. If the flow rate is too low the hole could easily pack off and the entire string could become stuck. Hole cleaning should be closely monitored, not only by watching the shale shakers, but by monitoring pick up and slack off weights on connections. The annular velocity will decrease drastically above the top of the liner due to changes in well geometry. Therefore, the hydraulics must be modeled to ensure that the flow rate required for the minimum annular velocity above the liner top does not induce an equivalent circulating density opposite the liner in excess of the formation fracture.

If the UBLD application is used because severe or total lost circulation is expected, at least twice the amount of drilling fluid that might be required should be readily available. If a worst case or total lost circulation situation is expected, the amount of drilling fluid required can be calculated by Eq. 8.1. If this volume of mud economically constrains operations, a floating mud cap should be considered for the application.

$$\begin{matrix}\text{Drilling}\\\text{Fluid}\\\text{Volume}\\(\text{bbl})\end{matrix} = \frac{2.86 \times \text{Problem Formation Thickness}(ft) \times \text{Flowrate}(gpm)}{\text{Anticipated Penetration Rate}\left(\frac{ft}{hr}\right)} \tag{8.1}$$

8.2.3 Basic Planning—Torsional Limits

The torsional limit in a liner drilling application is the yield torque of the weakest component of the liner drilling system. The connections for the liner will likely be the weakest component of the system. Special attention must be paid to this aspect of the liner design. If an improper or inadequate connection is used the liner could quickly go from a useful tool to junk in the hole. Offset records need to be closely studied so that a realistic idea of rotating hours can be estimated prior to drilling the well.

The limiting factor of how long the liner can be rotated is the connection fatigue life. The connection fatigue life is estimated by determining the cyclic bending stress in the build interval and then determining the number of revolutions that can be safely applied to the connection.

The cyclic bending stress is first calculated with Eq 8.2.

$$\sigma_b = \pm 211\ \theta\ D / E_t \tag{8.2}$$

where

σ_b = stress created by bending, in psi

θ = build angle in °/100 ft

D = O.D. of the pipe body, in.

E_t = tension efficiency of the connection, in decimal form

An operator specific safety factor is applied to the bending stress value and then converted into a percentage of pipe body or connection yield strength. In most cases the connection will be the weakest component of the string. If cyclic fatigue loading is a concern, as in high dog leg severity wellbores, rotating speed will need to be minimized to prolong the life of each connection.

The tension efficiency (E_t) of the connection can be obtained from the manufacturer of the connection. Every connection has specific Stress-Cycles, or S-N curve. For example, see the S-N curve in Figure 8–2 for TenarisHydril Series 500 connections. It can be seen that after 100,000 cycles, or rotations, the connection bending stress will be about twenty percent of the original yield.

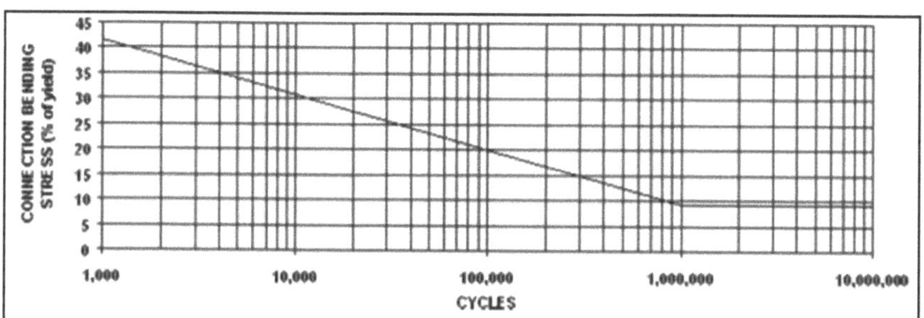

Figure 8–2 *TenarisHydril S-N curve for Series 500 casing*

The connection selected must have a high torsional resistance, and a high tolerance for compressive, bending, and tensile loads. The liner must be designed so that it can withstand tens of thousands of revolution cycles. The connection must be "gas tight," meaning that the leak resistance must be comparable to an API connection. If O-ring type seals are utilized to seal the connections, there will be a temperature limitation that must be taken into consideration. Sometimes special considerations need to be made for wear resistance in abrasive formations. Even premium casing connections were not designed for prolonged rotation through abrasive formations. The torque capacity of such connections allows little tolerance for wear, therefore some casing centralization may be necessary to reduce wall contact and prolong the life of the connection.

Above all else, the connection must be economical. The overall goal of using underbalance liner drilling is to reduce nonproductive time to save money. Premium connections are expensive; so special attention needs to be paid to both the mechanical specifications and the cost. The torque capacity of these liners leaves little capacity for wear, therefore some casing centralization may be necessary to reduce wall contact.

8.3 Advantages of UBLD

8.3.1 Dual Pressure Zones

Drilling through depleted formations can result in expensive non-productive time from lost circulation, especially when there is an over-pressured zone in the same interval. One of the best uses for underbalanced liner drilling is when an over-pressured zone must be drilled in the same interval as an under-pressured zone. UBLD is successful for two main reasons: by managing and minimizing the circulating pressure, the chance of lost circulation is reduced. In addition, the smear effect of the liner against the wall of the formation can help build formation integrity. UBLD significantly reduces the fluid losses in the annulus when compared with conventional drilling, due to the smear effect in the reduced annular clearance created by the liner and related liner tools.

This case study from the Banuwati field offshore of Southeast Sumatra, Indonesia is a good example. Total lost circulation from faulting and reactive shale plagued all previous attempts to drill to the productive zone at about 10,300 ft MD/5,450 TVD. After repeatedly experiencing problems from these trouble zones and multiple sidetracks, the operator took a different approach. At a measured depth of 9,968 ft/5,404 TVD and inclination of 68 degrees, the operator picked up a 2,000 ft long 7 in. 26 ppf VM80 13 CR Vam TOP liner conveyed on an 8.5 in. casing bit assembly. The liner was drilled down 349 ft at 31.7 ft per hour and set without problems. This saved the operator over a million dollars. (Jianhua et al., 2004). The severe lost circulation was mitigated by the smear effect of the liner in the tight annulus. This study showed that high angles and total lost circulation can be overcome with underbalanced liner drilling.

8.3.2 Differential Sticking

UBLD is a good fit when differential sticking is a concern. By reducing the pressure in the annulus through underbalanced drilling, the chance of sticking the liner is reduced. This would be applicable in depleted, permeable formations. Centralization would also reduce the potential for differential sticking, provided that the type of centralizer would not increase the torque or cause problems from reduction of the annular cross section at the centralizer points.

8.3.3 Wellbore Ballooning

UBLD can be used to help mitigate "wellbore ballooning." Formation or wellbore "ballooning" is a challenging obstacle for even the most experienced drilling professionals to recognize and overcome. The exact down-hole mechanics of formation or wellbore "ballooning" are not completely understood. As the well is being drilled, it is thought that circulating pressure, or ECD, exceeds the formation pressure of permeable formations or are near the fracture pressure of tight formations. The formation is then broken down and drilling fluid is pumped away into the formation. The formation is then "charged" with drilling fluid. When the pumps are shut down, the circulating friction pressure is eliminated. Formation pressure hydraulically pushes the drilling fluid out of the formation and back into wellbore. When this happens it can appear on surface that the well is gaining fluid and flowing. If the formation is "ballooned" the return flow rate will steadily decrease and will probably never be greater than immediately after the pumps are turned off. When UBLD is used, the circulating pressure can be managed so that it never exceeds the formation fracture gradient. If the formation pressure is not exceeded, no fluid will enter the formation so ballooning cannot occur. Again, it is extremely important to model and then monitor the hydraulic pressures that will be experienced with the tight annulus that is common for UBLD applications. Even low flow rates can result in extremely high annular pressure losses which result in high ECD.

In the Gulf of Mexico an operator experienced severe lost circulation and wellbore ballooning at 10,983 ft MD with 15.9 ppg synthetic base mud in a 8 1/2 in. hole. In repeated attempts to drill the interval, the wellbore problems were not resolved with conventional means. The hole had a 62 degree inclination with a maximum dogleg severity of 10.3 degrees per 100 ft. After modeling the conditions expected in the well, a 2,510 ft 7 in. 32 ppf HCQ-125 liner with Seal Lock Semi-Flush connections was designed. In 42.75 hours the liner was drilled down 337 ft to 11,320 at an average rate of penetration of 8.4 ft per hour. The parameters were 60 to 90 rpms with 3,000 to 20,000 lbs weight on bit. The surface torque ranged from 11,600 to 16,200 ft-lbs. The liner was drilled in place and cemented without a problem.

8.3.4 Formation Damage

Drilling fluid can damage permeable formations by fluid invasion through fluid plugging of the wellbore pore throats. This increases skin factors and irrevocably damages the formation. UBLD reduces formation damage to the reservoir by reducing and minimizing lost circulation. However, the smear effect creates a near wellbore skin damage problem that at this date has not been clearly defined.

8.3.5 Cementing the Liner

A time and trouble saving benefit is that the well does not have to be killed prior to cementing. In a normal drilling operation, after reaching total depth the well would have to be killed before pulling out of the hole to run a liner. The drillstring might have to be laid down and then the liner would have to be run in the hole and cemented. With UBLD, this entire step is eliminated. Once total depth is reached, the liner can be cemented almost immediately. The drill sting is then pulled out of the hole.

8.3.6 No Obvious Depth Limit

Underbalanced liner drilling (UBLD) is not depth limited and has been used successfully in wells with measured depths in excess of 25,000 ft.

8.3.7 Hole Size Limits

Underbalanced liner drilling can be used in holes smaller than four and a half inches up to sixteen inches or larger.

8.3.8 Casing Centralization and Stabilization

As in conventional drilling, a string of casing which is used to drill with may need to be stabilized to maintain its ability to make a straight hole, be the initial angle vertical or otherwise. Stabilization is required below the neutral point where the casing is in compression and the OD of these stabilizers should be close to the gauge of the hole. Thicker walled casing may also be used here where it is in compression, much like drill collars in a conventional drillstring. Above the neutral point some centralization of the casing is desirable, both for mitigation of connection wear and for the quality of the cement job. The demands on this casing jewelry are different than for conventional casing running, and care must be taken to make the correct selection. The stabilizers may be machined with integral blades and added into the string in the form of subs. For centralization, there are crimped-on centralizers available for specially designed rotating devices. On no account should conventional casing centralizers be used because there is a real possibility of damaging and even cutting the casing through continuous rotation inside centralizers which are not designed for such service. Another alternative has recently become available in the form of spray-metal stabilizers and centralizers, and these are built directly onto the operator's casing to the required dimensions.

8.4 Limits and Challenges with UBLD

8.4.1 Bit Requirements

The limitations of the DwL application are usually related to how long a single bit can last because changing the bit is problematic. Almost all UBLD applications are designed for a single bit run. Pulling out of the hole and laying down a liner is a worst case scenario and should only be done if no other operational alternative is an option.

8.4.2 Liner Availability

As noted above, underbalanced liner drilling can be used in holes smaller than four and a half inches and above sixteen inches. The limiting factor here is usually time. Designing a liner hanger and float equipment that can endure long intervals of continuous circulation and still function as designed is difficult, especially because each UBLD application is different. Most of the liner drilling equipment must be custom made for each application. The service company providing the equipment must have adequate time to design and build all of the necessary components for the job.

8.4.3 Liner Hanger

The liner tools must meet the hydraulic and mechanical requirements for drilling and then properly function to hang off and isolate the liner. This is likely the most challenging aspect of UBLD. The liner tools must withstand the tensile and axial loading along with the compression and torsion strains of drilling.

If possible, the UBLD application should be designed without the use of a hanger. For example, if computer models show that the liner will not helically buckle, the liner string can be set on bottom rather than being hung off with a liner. This will afford more bypass area when circulating and cementing. Another advantage to not using a hanger is that there will be one less hydraulic event to manage.

8.5 Well Control Considerations

Well control considerations for UBLD applications are similar conventional liner running precautions in a managed pressure environment. Along with the appropriate rated rotating head, pipe rams large enough to accommodate both the drillstring and the liner should be used. Floor valves should be tested and located on the rig floors at all times. The likelihood of a failure to one of the liner floats is greatly increased with long periods of circulation while drilling.

As with all underbalanced operations, formation fluids can enter the wellbore on connections if backside pressure is not maintained.

8.6 Drilling Fluid Considerations

The fluid systems used with UBLD have been traditional oil-based, synthetic-based, and water-based mud. Stiff foam and aerated mud have not been attempted. The lubricity of an oil-based mud is ideal when drilling with a liner.

8.7 Special Equipment

One of the most appealing aspects of UBLD is that the only surface equipment needed is equipment that would be required for a normal underbalanced well. All of the special equipment is subsurface. The casing bit has already been discussed in detail, but several other components need to be mentioned. The polished bore receptacle (PBR) can be used to tie back the liner into a string to the surface.

A liner running tool must be chosen to fit the application. This running tool is usually hydraulically actuated. However, circulating pressures must not be greater than the limits of the hydraulic components which can prematurely shear.

The cementing plug used for UBLD operation is a two piece assembly. The plug goes down and wipes the drillpipe and sets in the top of the hanger assembly. The liner wiper is pressured up on and then enters the liner to push the cement and wipe the inside of the liner. Figure 8–3 shows the general procedure for running and cementing a liner in place.

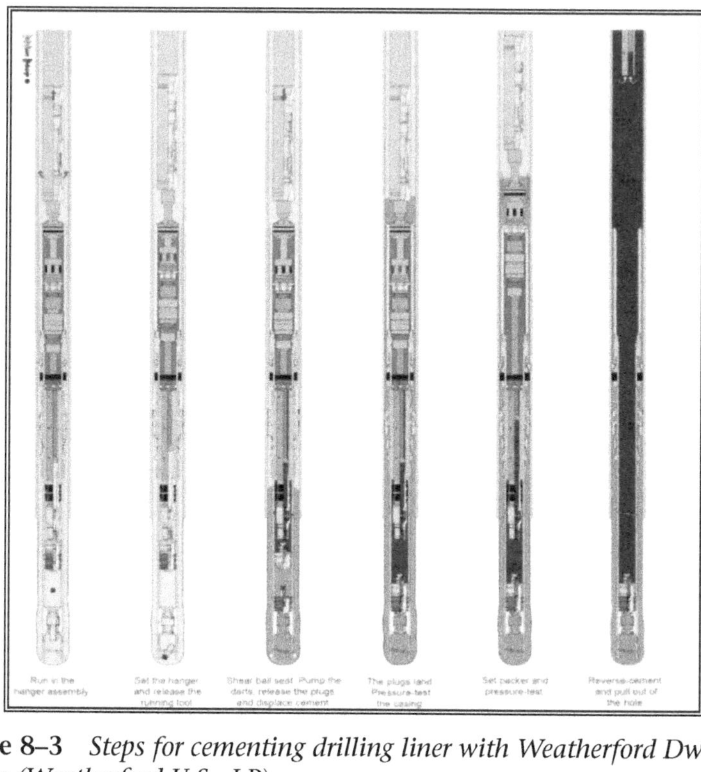

Figure 8–3 *Steps for cementing drilling liner with Weatherford DwL system (Weatherford U.S., LP)*

The last piece of equipment that might be required is the liner top packer. This should be a mechanical set packer. In some applications, the packing element is designed so that hydrostatic pressure will force the element to stay close to the packer mandrel. This prevents the element from prematurely setting. The exact procedure for setting the liner top packer will vary with each application.

8.8 Future Trends

While UBLD in its present state is a useful tool, there are even greater challenges for the technology. The use of an underbalanced liner drilling system with directional capabilities is a reality. There are currently retrievable rotary steerable technologies available for use with UBLD applications with a wide variety of MWD options. The UBLD operation at present is a time consuming operation to plan and execute with the directional aspect adding an increasing layer of difficulty. With experience and tool modifications, it may become a standard operation in place of a specialty tool.

8.9 References

Aliko, E. and Luca, S.D. "Casing-While-Drilling 24-In. Surface Section Enables Operator to Add One Extra Slot to the Existing Drilling Template and Cut Operation Costs in Offshore Congo," SPE 119461 presented at the IADC/SPE Drilling Conference and Exhibition, Amsterdam, The Netherlands, 2009.

Buntoro, A. "Casing Drilling Technology as the Alternative of Drilling Efficiency," SPE 115283 presented at the IADC/SPE Asia Pacific Drilling Technology Conference and Exhibition, Jakarta, Indonesia, 2008.

Cox, R.J., Li, J. and Lupick, G.S. "Horizontal Underbalanced Drilling of Gas Wells with Coiled Tubing," *SPE Drilling & Completion*, 14, No. 1, 1999.

Davies, M., Clark, L., McClain, E. and Thomas J. "A Staged Approach to the Introduction of Casing and Liner Drilling," OTC 17845-MS presented at the Offshore Technology Conference, Houston, TX, USA, May 1–4, 2006.

Fredericks, P.D., Sehsah, O.R., Montilva, J.C. and Vogelsberg, P. "Experience and Results with a New Automated MPD System While Drilling and Cementing Liner in an Onshore Depleted Gas Field," SPE 130319 presented at the IADC/SPE Managed Pressure Drilling and Underbalanced Operations Conference and Exhibition, Kuala Lumpur, Malaysia, February 24–25, 2010.

Gammage, J.H. "Advances in Casing Centralization using Spray Metal Technology," OTC 21979 presented at the Offshore Technology Conference, Houston, TX, USA, May 2–5, 2011.

Gao, X., Li, Y., Zhang, L. and Han, W. "New Development on Expandable Casing Hole Liner and Expandable Open Hole Liner," SPE 131857 presented at the International Oil & Gas Conference and Exhibition in Beijing, China, June 8–10, 2010.

Gordon, D., Billa, R., Weissman, M. and Hou, F. "Underbalanced Drilling with Casing Evolution in the South Texas Vicksburg," *SPE Drilling and Completion*, 20, No. 2, June 2005, pp. 86–93.

Hartsema, K., Aliko, E., Campos, J.F., Clark L., Delgado, F., Folling, P. and Wingate, J. "PDC Casing Drilling Improves HS&E, Cuts Drilling Costs, West Africa," SPE 105595 presented at the IADC/SPE Drilling Conference, Amsterdam, The Netherlands, February 20–22, 2007.

Jianhua, L., Darmawan, A., Chao, Z., Rosenberg, S., Hillis, K. and Utama, B. "Use of Liner Drilling Technology as a Solution to Hole Instability and Loss Intervals: A Case Study Offshore Indonesia," SPE 118806 presented at the IADC/SPE Drilling Conference and Exhibition, Amsterdam, The Netherlands, March 17–19, 2009.

Johnstone, I., Chomley, A., Cernev, G.E. Hoq, M., Atherton, G., Cornel, S.S. and Jacobs, M. "Casing-Drilling Step Improvement: PDC Successfully Drills out Casing Bit and Finishes Hole Section at Lowest Cost," SPE 105395 presented at the IADC/SPE Drilling Conference, Amsterdam, The Netherlands, February 20–22, 2007.

Kristiansen, T.G. "Drilling Wellbore Stability in the Compacting and Subsiding Valhall Field," *SPE Drilling and Completion*, 22, No. 4, December 2007, pp. 277–295.

Kunning, J., Wu, Y., Thomson, I.J, Marshall, L., Daigle, D., Mata, H.J., Pena, R., Hensgens, M. and Eppley, B. "Nonretrievable Rotating Liner Drilling System Successfully Deployed to Overcome Challenging Highly Stressed Rubble Zone in a GOM Ultradeepwater Salt Application," SPE 124854 presented at the SPE Annual Technical Conference and Exhibition, New Orleans, LA, USA, October 4–7, 2009.

Mody, R.K. "Enabling Our Enabling Technologies: What Will It Take? And What Is at Stake?," SPE 113442 presented at the Europec/EAGE Conference and Exhibition, Rome, Italy, June 9–12, 2008.

Mota, J.F., Campo, D.B., Menezes, J., Smith P. and Jackson, T. "Rotary Liner Drilling Application in Deepwater Gulf of Mexico," SPE 99065 presented at the IADC/SPE Drilling Conference, Miami, FL, USA, February 21–23, 2006.

Pickles, R., Brand, P. and Savage, P. "Utilization of Underbalanced Drilling Techniques to Exploit a Low-Pressure Reservoir in Indonesia," SPE 91591 presented at the IADC/SPE Underbalanced Technology Conference and Exhibition, Houston, TX, USA, October 11–12, 2004.

Reid, D., Rosenberg, S., Montgomery, M.M., Sutherland, M. and York, P. "Drilling-Hazard Mitigation-Reducing Nonproductive Time by Application of Common Sense and High-Value-Well Construction Techniques," OTC 18084-MS presented at the Offshore Technology Conference, Houston, TX, USA, May 1–4, 2006.

Rosenberg, S.M., Gala, D.M. and Xu, W.J. "Liner Drilling Technology as a Mitigation to Hole Instability and Loss Intervals: A Case Study in the Gulf of Mexico," SPE 128311 presented at the IADC/SPE Drilling Conference and Exhibition, New Orleans, LA, USA, February 2–4, 2010.

Rosenberg, S.M. and Gala, D.M. "Liner Drilling Technology as a Mitigation to Wellbore Ballooning—A Successful Case Study in the Gulf of Mexico Shelf," SPE 140261 presented at the IADC/SPE Drilling Conference and Exhibition, Amsterdam, The Netherlands, March 1–3, 2011.

Sinor, L.A., Tybero, P., Eide, O. and Wenande, B.C. "Rotary Liner Drilling for Depleted Reservoirs," SPE 39399 presented at the IADC/SPE Drilling Conference, Dallas, TX, USA, March 3–6, 1998.

Srinivasan, A., Frisby, R., Talbot, T.J. and Paraschiv, M. "New Directional Drilling with Liner Systems Allow Logging and Directional Control While Getting Casing across Trouble Zones," SPE 131391 presented at the International Oil & Gas Conference and Exhibition, Beijing, China, June 8–10, 2010.

Steppe, R., Clark, L.I. and Johns, R. "Casing Drilling vs. Liner Drilling: Critical Analysis of an Operation in the Gulf of Mexico," SPE 96810 presented at the SPE Annual Technical Conference and Exhibition, Dallas, TX, USA, October 9–12, 2005.

Terrazas, M., Estrada, M., Lopez, V. and Jardines, A. "Drilling with Liner on Horizontal Oil Wells," SPE 105403 presented at the IADC/SPE Drilling Conference, Amsterdam, The Netherlands, February 20–22, 2007.

Torsvoll, A., Abdollahi, J., Eidem, M., Weltzin, T., Hjelle, A., Rasmussen, S.A., Krueger, S., Schwartze, C., Freyer, C., Huynh, T. and Sorheim, T. "Successful Development and Field Qualification of a 9 5/8 in. and 7 in. Rotary Steerable Drilling Liner System That Enables Simultaneous Directional Drilling and Lining of the Wellbore," SPE 128685 presented at the IADC/SPE Drilling Conference and Exhibition, New Orleans, LA, USA, February 2–4, 2010.

Vogt, C., Makohl, F., Suwarno, P. and Quitzau, B. "Drilling Liner for Depleted Reservoir," SPE 36827 presented at the European Petroleum Conference, Milan, Italy, October 22–24, 1996.

Vonthethoff, M.L., Schoemaker, S. and Telesford, A.L. "Underbalanced Drilling in a Highly Depleted Reservoir, Onshore South Australia:

Technical and Operational Challenges," SPE 121632 presented at the 8th European Formation Damage Conference, Scheveningen, The Netherlands, May 27–29, 2009.

Vrielink, H.J., Bradford, J.S., Basarab, L. and Ubaru, C.C. "Successful Application of Casing-While-Drilling Technology in a Canadian Arctic Permafrost Application," SPE 111806 presented at the IADC/SPE Drilling Conference, Orlando, FL, USA, March 4–6, 2008.

Warren, T. and Lesso, B. "Casing Drilling Directional Wells," OTC 17453-MS presented at the Offshore Technology Conference, Houston, TX, USA, May 2–5, 2005.

Warren, T.M., Schneider, W.P., Johns, R.P. and Zipse, K.D. "Running Casing on Conventional Wells with Casing Drilling Technology," SPE 2004-183 presented at the Canadian International Petroleum Conference, Calgary, Alberta, Canada, June 8–10, 2004.

Whanger, K. and Lowe, E. "Liner Drilling Beats Lost Circulation at 25,000 ft," *Offshore*, 70, No. 6, 2010, pp. 58.

York, P.L., Prichard, D.M., Dodson, J.K., Dodson, T., Rosenberg, S., Gala, D. and Utama, B. "Eliminating Non-Productive Time Associated with Drilling through Trouble Zones," OTC 20220-MS presented at the Offshore Technology Conference, Houston, TX, USA, May 4–7, 2009.

CHAPTER 9

Coiled Tubing and Underbalanced Drilling

Earl Dietrich, Blade Energy Partners

9.1 Introduction

Within the constraints of this book, coiled tubing has a special place that is rapidly changing with technology. The past limits of coil have generally related to the metallurgy of the coiled tubing, the small size of the tools, and the inability of the system to rotate, although it is possible to rotate the bottom-hole assembly. There are many changes taking place in the metallurgy and the material, so some of the limits that are discussed at length will be modified in the near future.

Small tool sizes are now being developed as the result of increased coiled tubing operational experience.

The inability of the tubing to be rotated has led to problems with torque and with hole cleaning, especially in horizontal or high angle wells. Torque problems have lead to the development of the Ag-itator and tractors. Hole cleaning is still a problem, but special modifications of drilling mud have increased hole cleaning ability without increased solids or increasing the annular pressure loss (APL).

Some of the limits to UB coiled tubing drilling are a function of the coiled tubing mechanics. These are discussed in some detail in Table 9–1.

9.2 Preplanning

As with any drilling project/program/well, significant preplanning is required for successful implementation of the job. Historically, jobs with the least amount of preplanning and upfront engineering have had some major unforeseen problems. Sufficient research on similar applications, as well as peer reviews of the proposed work, and outside

Table 9–1 Advantages and Limits of Underbalanced Coiled Tubing Drilling (UBCTD)

Advantages	Disadvantages
Slim hole—smaller bits, faster ROP	Slim hole—fishing difficult
Fast tripping speeds	Removal of BHA is time consuming
Constant BHP simpler with no upset due to connections	BHA more expensive than rotary
Fewer loads/move time less than similarly capable rotary rig	Rig spread more expensive than rotary
Inherently pressure-tight/gas tight system allows easy adaptation to UBD/MPD techniques	Much higher surface pressure for gas injection. Cryogenic N2 equipment required. High pressure fluid pumps required
Excellent history of success in case histories	Not economically or mechanically feasible in many cases
High speed MWD telemetry in e-line	Limited BHA availability
Proven low-cost reservoir access method	BHA cannot handle many mud solids
Ability to utilize service CT units	Limited tubing life
No connections, minimal rig floor work—HSE	Limited ability to handle collars and casing
Crews are experienced in working on a live well	Limited crew availability
CT well service standards already in place with vendors to cover using multiphase fluid/gas mixtures while circulating	Limited vendors and vendor locations
	BHA lost in hole cost is very high
	Sensitive to corrosion

expertise should all be used to minimize potential problems during implementation. Research should not be limited to this module or the references listed in the compilation of this module.

9.2.1 Candidate Selection

CTD is very efficient at drilling consolidated formations of high, normal, or sub-normal pressure. Unconsolidated, water sensitive, mobile formations are extremely challenging for CTD, as the slim hole conditions do not allow for much down-hole clearance between the wellbore and the toolstring.

Using a similar process to UBD candidate selection, candidate wells need to be screened for applicability of UBCT techniques. Not every well is a good candidate, and some wells that have not been considered are likely very good candidates. The selection process starts with data gathering on the subject well and should include the following minimum data (see Table 9–2).

9.2.2 Pressure Categories and BOP Stack Requirements

API 16ST (for coiled tubing) has details on each of the pressure categories and recommended BOP stack configurations to meet the minimum requirements for pressure, risk, and contingencies. The recommended configurations may or may not meet operator regulations, so consideration needs to be given to the optimum stack design.

UBCTD operations tend to fall under API Pressure Category "0" (PC-0) or PC-1.

API Pressure Category "0" is for wells demonstrated as incapable of unassisted flow to the surface, and API PC 1 is for wells where the Maximum Annular Surface Pressure (MASP) is less than 1,500 psig. The required minimum BOP Stack Pressure Rating for both categories is 3,000 psig (20,637 kPa). PC-0 requires at least one barrier and PC-1 requires 2 barriers.

API 16ST categories 2 through 4 cover MASP (Maximum Allowable Surface Pressure) up to 12,500 psig (86,000 kPa).

9.2.3 Coiled Tubing Mechanical Considerations

Unlike drillpipe for rotary drilling, CT strings are a consumable item. The performance is very important due to rapid bend cycle fatigue and increasingly challenging applications. The operating limits of CT strings must be known at all times, *especially for UBCTD operations where multiple wiper trips will occur at the same depths.*

Depending upon the type of CT rig being utilized for UBCTD, the string sizes and options are limited. For Hybrid CTD rigs, the reels can handle much larger OD CT strings, such as 2 7/8 in., or 3 1/2 in. For service type CT rigs, 2 in., 2 3/8 in. or 2 5/8 in. CT can be used.

Table 9–2 Mechanical and Reservoir Considerations to UBCTD

Mechanical Considerations	Example
ID/OD restrictions	Profiles, nipples, changes in ID
Junk in Hole/Other restrictions	Scale, plugs, collapsed tubulars
Casing, Tubing, Caliper History	Tubular size, weight, grade, setting depths
Well Surveys, Inclination	TVD of KOP, inclination
Well history (tubing leaks, cement squeezes)	Could cause restrictions or inadequate isolation
Reservoir Considerations	**Limits**
Indications of lost circulation while drilling motherbore	Heavy LCM concentration
Frac history	Screenout, loss circulation, ball sealers
Reservoir Pressure/Temperature	As well as pressure of any other open zones
Water sensitivity of reservoir	Oil based drilling fluid may be required
Value of existing current production well	Economics of a plug back
Net pay in reservoir	Depth of gas/water, or oil/water contact
Completion options	Open hole/slotted liner/screen
Reservoir Description	Rock type, faulting, fractures, dip, geological changes in zone
Azimuth of target	Can CT make the required turn to reach the target
Reservoir fluid composition	Gas gravity/composition Oil gravity/composition Water composition Normal production WGR, GOR, WOR

Table 9–2 Mechanical and Reservoir Considerations to UBCTD (cont'd)

Other Conditions	Examples
What other formations will CT need to drill through?	Coal, Salt, Shales are troublesome
How much lateral length is required?	Weight transfer in long laterals becomes more difficult
Is the reservoir under-pressured, or severely depleted?	Managed pressure, or underbalanced techniques and equipment may be required

Drilling strings, CT wall thickness starts at about 0.156 in. and may be as high as 0.204 in. Typical material yield strength of 70,000–100,000 psi is standard for CTD. Some wells require a tapered drillstring; this is done in order to provide adequate strength at surface and minimize string weight to allow standardized injector heads to easily pull the string out of hole.

Pre-job engineering normally encompasses using CT modeling software to perform a string design. This design is done using the following variables:

- Open hole size, well path, and true vertical depth
- Maximum overpull allowed
- BHA weight and strength characteristics
- Length and diameter of existing tubulars if drilled through tubing
- Drilling fluid viscosity, weight, and other rheological properties
- Required weight on bit at TD

CTES (National Oilwell Varco) has a detailed manual available for free download that has an excellent chapter on CT string manufacture, engineering, design, and other considerations.

9.3 Coiled Tubing Equipment

9.3.1 CT Injector

The coiled tubing injector head provides the horsepower necessary to run and pull CT into and out of the hole. Hydraulic systems allow the CT operator/CTD driller a high degree of control over string movement, which is important for weight on bit considerations with CTD. The primary functions of the injector head are to apply dynamic axial force to the CT to manage movement of CT in the well, supply adequate traction to avoid CT slippage, apply static force to hold the CT when neutral, and finally to act as the "platform" for weight, and depth measurement sensors. Dual opposing chains are the primary feature of modern injector heads; the chains hold short gripper blocks that are contoured to match the CT OD. Friction is the only force supporting the CT in the injector head (see Figure 9–1).

9.3.2 Gooseneck or CT Guide Arch

The gooseneck is mounted on top of the injector head. The coiled tubing is "guided" from the reel to the chains on the injector head by the gooseneck. Rollers on the top/bottom of the gooseneck allow the CT to slide smoothly.

9.3.3 Reel and Accessories

The CT reel's function is to safely store and protect the string of coiled tubing. The design of the reel may be optimized for land based operations, offshore operations, trailer mount operations, or heli-portable operations. The reel incorporates a swivel assembly in the center that allows fluids to be pumped through the tubing string while the reel drum rotates.

The reel will have hydraulic connections necessary to operate the drive, braking, and level wind systems. CTD reels are also fitted with a mechanical counter as well as additional monitoring equipment for CT fatigue, ovality, and diameter.

Reels with e-line coil have a pressure bulkhead and collector assembly installed. The collector is where the e-line terminates and makes its connection to surface MWD equipment. This e-line is a conductor for electricity to power the CT down-hole tools and MWD equipment. It also allows for data to be sent back to surface from the down-hole tools. The bulkhead is a pressure seal around the wire which allows simultaneous fluid pumping on one side of the reel and a seal to be maintained around the e-line on the other side of the reel.

The effect of tubing capacity, or actual tubing length on UB operations is significant because of the increase in required surface pressure with the increase in tubing length. The problem is not linear because the tubing has a greater pressure drop when wrapped around the reel than when straightened out in the hole.

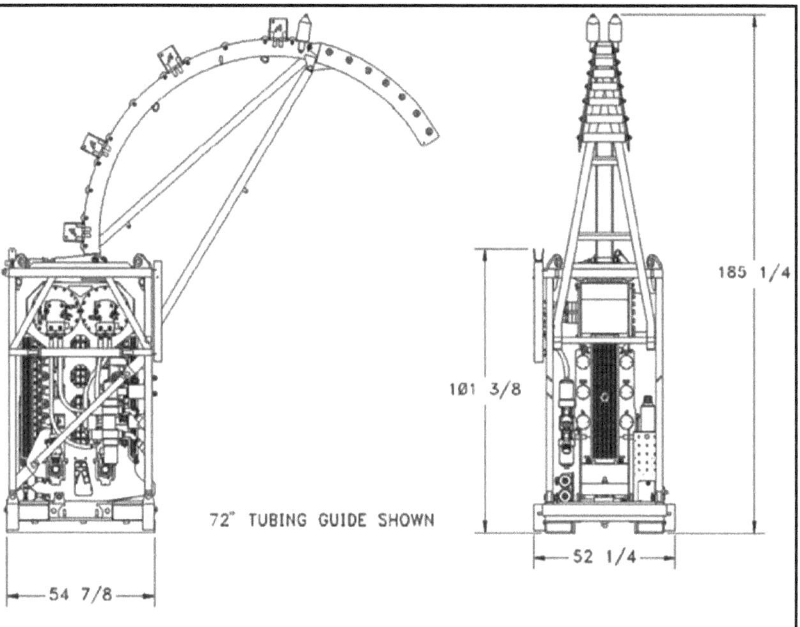

Figure 9–1 *Injector head and goose neck*

The tubing capacity of a reel is calculated using the following formula:

$$L = (A + C)ABK$$

where

L = Tubing capacity, ft

A = Tubing stack height, in.

B = Width between flanges, in.

C = Reel drum core diameter, in.

K = Constant value for different tubing sizes, ft/in^3

9.3.4 Operations Cab

Typical CT service units have a small operations cab, where the CT operator can run the controls for the CT injector, BOP controls, reel controls, and where data is displayed and stored. For CTD operations, a larger operations cab is required, in order to allow the directional

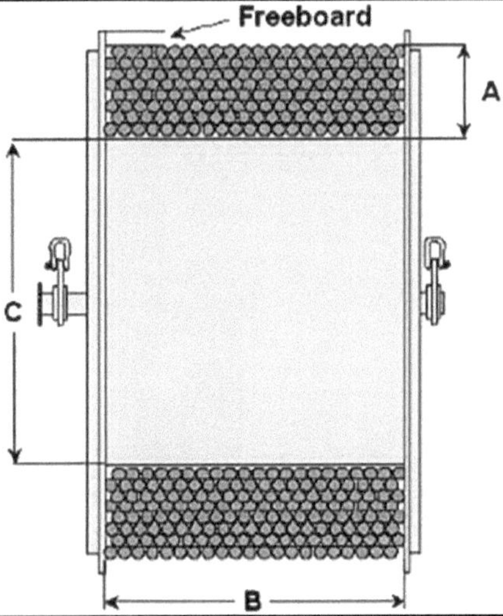

Tubing OD (Inches)	K Value (Ft/in^3)
1	0.262
1 ¼	0.168
1 ½	0.116
1 ¾	0.086
2	0.066
2 3/8	0.046
2 7/8	0.032
3 1/2	0.021

Figure 9–2 *Reel capacity*

driller, as well as the company man to be present and comfortable during prolonged operations. This operations cab will be the center for all data displayed and where all controls on the CT rig are operated.

9.3.5 BOP Stack

The configuration of the well control equipment for a particular well is influenced by many factors such as:

- Worst case scenario conditions that may be experienced
- Operator requirements such as OIMS (Exxon), DWOP (BP), SIEP EP 95-0210. SIEP EP 2002-1500 (Shell)
- Industry recommended practices:

  API RP 53—BOP Systems

  APR RP 16ST—Recommended Practice for Coiled Tubing Well Control Equipment Systems and Operations

  CAODC IRP 01—Critical Sour Drilling

  CAODC IRP 06—Critical Sour UBD Drilling

  CAODC IRP 21—Coiled Tubing

Review of these or equivalent documents along with applicable operator requirements will enable optimum well control equipment design and placement. For wells, where UBD is being planned, the BOP stack needs to address the issue of holding pressure above/below individual rams and allow for equalization of this pressure during pressure deployment operations (see Figure 9–3).

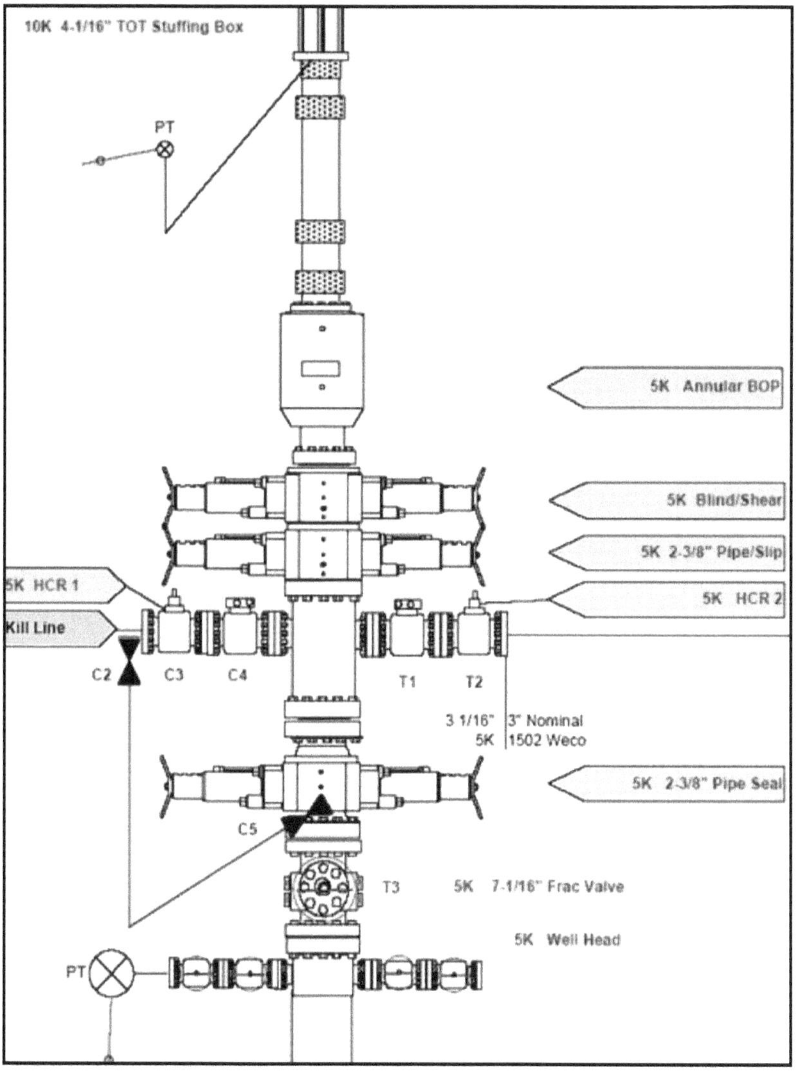

Figure 9–3 *CTD BOP stack*

9.3.6 CT Structure

Hybrid CT rigs have a definite advantage over CT Service units for CTD work due to the conventional rig design incorporating a mast, draw works, and structure. A structure to allow rigfloor work is essential to CTD operations for: deployment of tools, running of completions, and tubulars. Figure 9–4 shows a Hybrid CT rig.

Figure 9–4 *Hybrid CTD rig*

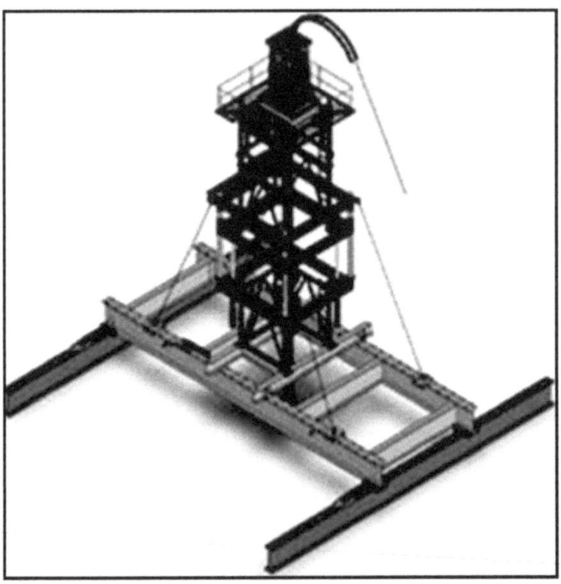

Figure 9–5 *Devin track stack structure*

9.3.7 Pressure Deployment Lubricator

Underbalanced drilling operations may require that the well is kept under-pressure on trips out of the hole. The BHA must be deployed in/out of the hole with pressure at surface to enable this. The pressure deployment lubricator is a specialized extended riser with a winch and grease packoff (see Figure 9–6). Below is a typical procedure involving the pressure deployment lubricator:

1. Once the BHA is at surface, the pipe rams and possibly the annular are closed, and the pressure above is bled off.
2. The riser is disconnected, and the injector is skidded off well center.
3. Once the injector head is out of the way, the pressure deployment lubricator is picked up with a crane, and installed on top of the annular where the riser was broken.
4. The internal winch line is made up to the BHA in order to retrieve it.
5. The pressure is then equalized into the lubricator, the barriers opened, the BHA is lifted into the lubricator with the winch. The grease packoff on the top of the lubricator maintains well control integrity while the lubricator is in operation.
6. Once the BHA is fully out of the hole, the blind ram is closed on the well, the pressure is bled off, and the lubricator is removed to lay down the old BHA and pick up a new one.

9.3.8 Drilling Bottom-Hole Assemblies

Currently there are 2 commercial BHA systems available for CTD, as of 2009.

1. Baker Hughes Inteq (BHI) and Antech make e-line BHA systems.
2. Sperry Sun (Halliburton) makes a mud pulse MWD system for non-eline CT drilling.

National Oilwell Varco (NOV) has many of the components to build a BHA, but does not provide a service.

BJ Services (Baker Hughes) had a competitive BHA system, but it was mothballed in 2008 and is currently unavailable.

Figure 9–6 *Pressure deployment operation*

The BHA system needs to be suitable for the proposed well. The lack of suppliers means that it is likely the BHA will be provided by BHI. The Coiltrak system, from BHI, is extremely robust, has a high temperature capability, and incorporates many features for CTD. The cost is higher for these CTD tools than for standard rotary BHAs due to availability, and "miniaturization" of many Autotrak features into the CTD tool string. Due to the almost "single source" tool availability, the discussion will focus on the Coiltrak capabilities and specifications.

The e-line controlled CoilTrak BHA can be deployed through 3 1/2 in., 4 1/2 in. and larger completions. Tool sizes available are 2 3/8 in. and 3 in. With both systems, hole sizes from 2 3/4 in.–4 3/4 in. are covered; the tools are also designed for high temperature applications as well as underbalanced drilling (also in H2S environment) and geosteering. It is a very modular system that allows a BHA setup to customer needs on location. Even a change from 2 3/8 in. to 3 in. system (or vice versa) can be done locally. All systems are also equipped with a closed loop system for auto-correction. The 3 in. system will get the option to exchange the orienter head with a rib steer motor (designed to drill straight boreholes or 3D well profiles on CT). This rib steer motor module is based on the AutoTrak concept. All other modules of the BHA will stay the same.

9.4 Operation Comments

Normally, two sets of procedures are developed for CTD operations: daily, weekly, per well, and emergency procedures. Normally crews become proficient in the SOPs, but in the event of an emergency, they need to be familiar with these contingency procedures. Frequent crew drills, and/or tabletop exercises are necessary to keep these procedures in the minds of operational supervision staff.

UBCTD pump pressure gas injection pressures are very high. Be sure the transfer line between the nitrogen pump and the standpipe is staked down and chained at joint. It needs to be tested to 150% of highest available pressure.

9.4.1 Pipe Management

Coiled tubing is a consumable drilling item, but unlike drillpipe, it is also relatively weak, easy to damage, and it is sensitive to acid, pressure, temperature, and some other drilling chemicals. Very few systems are available for measuring pipe cycles, fatigue, ovality, and pressure. Without reliable measurement it becomes a manual job to track running footage, pressure, and cycles, and to create a qualitative judgement on fatigue and the life of the pipe. Even with all the instrumentation, failures still occur due to manufacturing issues, operational issues, and unforeseen problems.

Pipe management must be watched on a constant basis. Running feet on the pipe needs to be captured in the daily drilling report each day.

At the end of each well, data should be downloaded from the data acquisition system and sent to operation headquarters for interpretation. A plot with percentage coil life needs to be generated at each well.

9.4.2 General Comments

- Watch solids content and circulating coiled tubing pressure.
- Every sweep should be reported on daily drilling report.
- Do not open EDC in open hole unless forced to disconnect.
- Check the OD/ID of all down-hole equipment for passage.
- Function test MWD equipment near surface when RIH to ensure tool functionality.
- Establish pick up and slack off weights and off bottom pumping pressures.
- Report sweep volume and frequency into daily drilling report along with type and amount (heavy to light) of cuttings brought to surface. This information will be used to form a baseline for most efficient sweep frequency and volume.
- Closely monitor trends of motor differential, surface coil weight, down-hole WOB, and down-hole pressure data.
- Perform a wiper trip every 300 ft unless wellbore conditions indicate a more frequent schedule is necessary.

9.5 Problems and Challenges

9.5.1 Hole cleaning

Removal of cuttings from the open hole section is critical to ensure adequate weight to bit, reduce the chance of stuck pipe, and to optimize hole conditions for running tubulars. In a deviated section, such as the build, the fluid viscosity itself is insufficient to remove the entire cuttings bed. Keeping fluid pumping rates at a mid to high range will help keep liquid annular velocities at an acceptable rate. The cuttings bed will settle more in the 30° to 60° inclination area, as the coiled tubing tends to follow the top of the open hole section, and the cuttings tend to build in the bottom of the open hole section. Agitation is required to stir-up the cuttings, so that the fluid can carry the cuttings to surface. Since coiled tubing does not rotate, and does not have tool joints, the drill bit provides the only agitation during the drilling of open hole. Therefore, wiper trips will be required ft to ensure cuttings are being transported to surface (at least every 150 ft).

9.5.2 Drilling Challenges with UBCTD

Table 9–3 lists some of the drilling challenges with UBCTD.

Table 9–3 Drilling Challenges with UBCTD

Item	Description	Action
ROP	**ROP Reduction**	
	• There may be increased hole drag due to a cuttings bed forming • Lithology changes	• Ensure that the reduced ROP is not due to a mechanical problem. • Pull off bottom. This will allow the jetting velocities to clean the bit. • Ensure that there has not been a lithology change (check a sample). • If there is no change in lithology or mechanical, pull a short wiper trip to check the BASE line and pull out of hole weight. If BASE line weight is normal, run back in hole. If the BASE line weight is high, continue pulling the wiper trip. • A short wiper of between 25–100 ft may correct the hole drag. If the hole drag is still high, a fullwiper trip will be required back to the window area. (ft/min)
Well head **or** **Down-hole Annulus Pressure**	**Pressure Increase** • Well is kicking; possible increased fluid production or a higher-pressure zone has been encountered. • Well may be slugging, if the pressure is increasing and decreasing in a short period of time.	• In both cases the circulation parameters must be checked and adjusted to correct for these changes.

Table 9–3 Drilling Challenges with UBCTD (cont'd)

Item	Description	Action
Well head **or** **Down-hole Annulus Pressure (cont'd)**	**Pressure Decrease** • May be indicating increased fluid production. • If the well head pressure decreases and the down-hole annulus increases, this is an indicator that the cuttings bed is increasing in size and is starting to bridge off the annular space. (Mud Ring)	• The circulation parameters must be checked and adjusted to correct for these changes • Pull a short wiper trip to check the BASE line and pull out of hole weight. If base line weights are normal, and the wellhead and down-hole annulus return to normal, run back in hole. If the BASE line weight is high, continue pulling the wiper trip. • A short wiper of between 25–100 ft may correct the hole drag. If the hole drag is still high, a full wiper trip will be required, back to the window area
Surface Injection Pressure	**Pressure Increase** • Indicates the following: o Higher BHCP o PDM problems (stall) o Surface pumping parameters o Cuttings bed bridging o Bit jet blockage The pressure increase may correspond with an increase in down-hole annulus and decrease in wellhead pressures.	• Stop drilling and analyze the indicators. • The circulation parameters must be checked and adjusted to correct for these changes.
	Pressure Decrease • Indicates the following: o Lower BHCP o PDM problems (bypass) o Surface pumping problems	• Stop drilling and analyze the indicators. • The circulation parameters must be checked and adjusted to correct for these changes

Table 9–3 Drilling Challenges with UBCTD (cont'd)

Item	Description	Action
Down-hole Tubing Pressure	**Pressure Increase** • Indicates the following: o Higher BHCP o PDM problems (stall) o Surface pumping parameters o Cuttings bed bridging o Bit jet blockage The pressure increase may correspond with an increase in surface injection pressure and decrease in wellhead pressures. **Pressure Decrease** • Indicates the following: o Lower BHCP o PDM problems (bypass) o Surface pumping parameters	• Stop drilling and analyze the indicators. • The circulation parameters must be checked and adjusted to correct for these changes.
Drill into Fracture	• A combination of the following will occur: o Possibility of large loss of fluid if BHCP is higher than fracture pressure (i.e., differential sticking possibility) o Change in BHCP and WHCP o Increased weight gauge o Decrease of WOB o Increase of G's o Loss of reactive torque o Increase in solids to surface if the fracture is unconsolidated plus increased removal of the cuttings bed due to increased velocities from gas production o Decrease in BHA differential pressure	• Stop drilling ahead, call for sample. • Begin POOH to window area, or begin to RIH, all depending on well condition and performance. Team decision on next operation. • Evaluate and monitor at all times: o Cuttings samples o All pressures, surface and downhole o Return rates of fluid and gas

9.6 Ag-itator and Tractor Systems

One of the limits to coiled tubing operations that is not necessarily specific to UB operations is the limitation to operating with the tubing in compression as is often done with drillpipe or casing. To avoid placing the tubing in compression, the Ag-itator was developed to break the friction hold on the pipe. The Coiled Tubing Tractors work in a completely different way by pulling the tubing from the BHA. Extended reach wells where achieving adequate weight on bit (WOB) to successfully continue drilling (beyond what is achievable with lubricants) requires use of either a tractor system or a downhole vibration/oscillation tool. Several tractor systems are available, and one system for down-hole oscillation called the Ag-itator is available. Table 9–4 shows the benefits these tools provide.

9.7 Case Histories

The following case histories have been built using available references. Reference materials are listed at the end of this chapter.

9.7.1 Sharjah UBD—CTD "Underbalanced for Life Operation"

CTD was utilized in Sharjah, UAE for a multi-year, multi-well re-entry campaign. A carbonate reservoir with production in decline was the target for the project. In the campaign, 40 re-entries were performed, some through tubing, in a variety of casing sizes. The wells were designed and implemented in an 'underbalance for life' philosophy, where the drilling, tripping, and all other activates were performed on a live well. The only time wells were killed was during well control situations. Casing exits were performed on each well using through-tubing (TTRD) whipstock systems. Several wells had the window milling performed in an underbalanced state in order to maintain productivity from the motherbore. Some of the 40 re-entries were multiple trips to the same well in order to add more laterals in different zones. Produced gas was compressed with on-site gas compression equipment and sent to the gas plant for sales. Implementation of this project required the co-operation of multiple service companies who each supplied "best in class" equipment solutions for the challenging environment. Due to the extended period of the campaign, continuous improvement was seen in all facets of the operation, and the efficiency of the work and success showed the value of an extended campaign (see Figure 9–7).

Table 9–4 Advantages of Ag-itator vs. CT Tractor Systems

CT Tractor	Ag-itator
Enabling the delivery of heavy tool strings to the outer reaches of extended reach wells	Up to 40% reduction in drag friction
Ideal tool for coiled tubing-delivered perforating, wellbore cleanout, acidizing and logging operations	Enables deeper, longer targets to be achieved
Positive gripping mechanism grips a wide range of cased or open-hole wellbore sizes	Provides oscillating force to break friction and improve weight transfer.
Robust construction tough enough for the most demanding operations	Delay in the onset of CT helical lockup
Operates on almost all intervention fluids	Use for perforating, logging to get tools to depth
Debris-tolerant	Use for running small clearance liners and pulling sand screens.

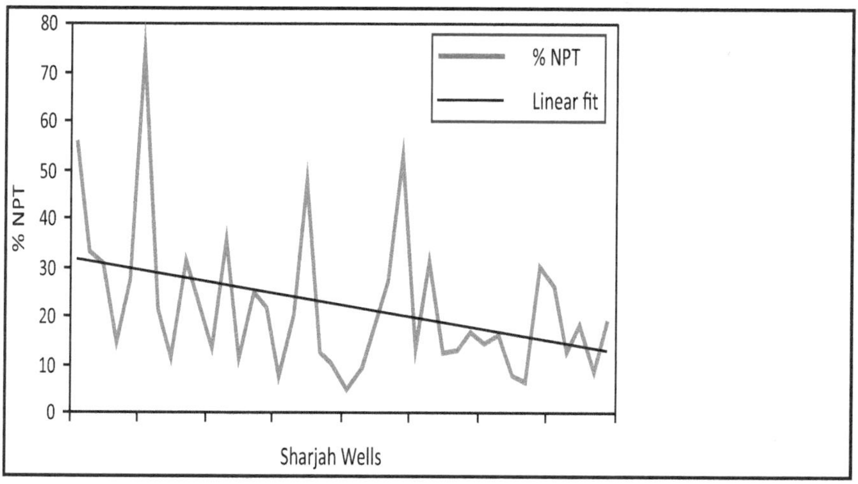

Figure 9–7 *Sharjah reduction*

9.7.2 Sharjah UBD–CTD Retained Personnel and Equipment Limited NPT

Following the first successful campaign in Sharjah, UAE, a second operator utilized all of the equipment, personnel, procedures, and learnings in order to implement a re-entry campaign in a smaller, but equally challenging field. The majority of this case history is taken from SPE 113684 (Mykytiw, C., Killip, D., Brewin, R. and Capps, J., 2008), along with first hand experience. Due to retained personnel, learnings, equipment, and procedures from the previous campaign, only three months were required to do the front end loading for project implementation. The basis of design for the re-entry campaign performed well in advance of project implementation, at which time the completions, mechanical challenges, reservoir uncertainty, and other issues were outlined. As well as the engineering design, both mechanical and hydraulic were performed to determine operability of drilling UBD-CTD. Eleven wells were re-entered, and the production results were far beyond expectations. During the 1-year campaign, additional gas and condensate production of 25 bcf and 301,000 bbls was achieved, at the same time, sales while drilling amounted to 3.29 bcf and 14,372 bbls. Examination of the cost per foot over the campaign shows that the previous Sharjah UBD-CTD campaign had reached its technical limit, and no further improvements were realized on this second project. All of the retained items (as discussed above) clearly enabled unprecedented project success.

9.7.3 Alaska CTD (Conoco/BP/Arco)

Coiled tubing drilling in Alaska is a fully mature technology. The personnel/rigs on the north slope have pioneered, refined, and perfected CTD techniques over the last 20+ years. Lateral learning from the Alaska CTD wells has resulted in project success in several projects outside of Alaska as well. More than 600 wells have been re-entered using CTD technology, and the continuous improvement, technology breakthroughs, and improved equipment designs have allowed the teams to continue to push further, deeper, and into more challenging applications. The majority of the references listed for this module deal with Alaska CTD wells.

9.7.4 Lisburne UBD–CTD

The Lisburne field on the North Slope of Alaska (Prudhoe Bay) is a very large field, with tremendous reserves in place, SPE 108337 (Johnson, M.

et al., 2007). Historically, application of techniques from Prudhoe Bay (Sandstone), have not been effective in Lisburne (Carbonate).

Production of oil from the Lisburne wells has a large gas-oil ratio (GOR), and on the North Slope there is a finite gas production rate per day that is permitted, as the facilities there can only compress and re-inject so much gas per day, and there is not yet a gas pipeline for sales of the gas. Lastly, flaring of the gas is not permitted, therefore, the oil production rate from the Lisburne field is constrained by GOR.

Drilling history in Lisburne has shown the field to more challenging to drill than Prudhoe Bay, and there appeared to be a possible application for UBD techniques in the field, which was expected to increase production by increasing ROP which would allow longer CTD laterals to be drilled. Feasibility work was done on the application of UBD-CTD in the field, and a 2 well UBD-CTD trial began.

Front end engineering for the project included many Alaska-specific concerns such as environmental, SPCC, venting of gas, as well as utilization of existing infrastructure for some of the campaign. Interface with the rig contractor and all vendors was undertaken early in the project to address all the concerns of drilling on a live well with UBD techniques. Noteworthy deviations from previous CTD projects are shown in Table 9–5.

Drilling both pilot wells was done safely and successfully. Some highlights were:

- Excellent HSE performance by the rig team—zero safety incidents
- Total of 9,130 ft drilled in 5 laterals
- Remarkable bit runs of 780 ft, 832 ft, and 977 ft—all of which were Lisburne CTD records
- Most footage in 24 hrs @ 666 ft, and longest CTD lateral at 2,564 ft—both were Lisburne CTD records
- Average ROP of 239 ft/day in well 1,286 ft/day in well 2—double previous ROPS in overbalance wells
- Over 40 safe pressure deployments of the 65 feet long drilling BHA into 700–1,000 psi wellhead pressures
- Five flawless dual string 4 ½ in. scab liner and 7 in. liner window exits
- 14,000 bbl of oil produced from reservoir while drilling laterals
- Each well's GOR was higher than planned. Faults crossed in laterals may be permeable to the gas cap

- The project was completed in 113 days, and 40% over budget. Major learnings were captured and are likely to reduce trouble time significantly in future wells.

9.7.5 North America Gas MPD—CTD

Leveraging the successful Sharjah CTD experience, as well as the ground breaking work done in Alaska with CTD, the technique was chosen for application in a large tight gas field in North Texas. The operator committed to a pilot program to evaluate the use of CTD reentries for existing production wells with an eight to ten well commitment. The pilot experienced drilling success, but had some challenges in getting completions to TD.

Engineering studies were performed following the pilot to evaluate geomechanics, fluid compatibility, natural fracture tendency, ECD effect, and shale stability. The results of the studies indicated that the shale intervals above the reservoir were highly water sensitive, as well as sensitive to inclination when drilling through them. With this information, the second pilot was designed with capability to isolate the shale with a flush joint liner. The CTD equipment spread needed to be capable of running a flush joint liner assembly. Also of concern was the elapsed time between drilling the lateral and placing the completion. A contingency plan that included equipment was done in order to allow the CTD rig to deploy the completion assembly.

The campaign with the improved design allowed the successful drilling and completion of 29 wells, some wells taking multiple

Table 9–5 Project Changes

Lisburne CTD–UBD	Other CTD–UBD Campaigns
Use of gas lift mandrels and field gas for reduction in ECD	Use of Membrane, or cryogenic nitrogen to reduce ECD
Multiple whipstock sets for multiple windows/UBD legs	Open hole sidetracks on different azimuth
Rate Integral Productivity Index work	Reliance on PI, or gas rate to evaluate productivity of lateral
Containment	Alaska has much more stringent regulations on containment for spills and contingency

attempts to mill a window, drill a build section, or place a lateral. A significant difference between this campaign and similar historical campaigns was the use of a small stand-alone drilling rig to abandon the well, mill a window, drill the build section, under-ream the section, and run a flush joint liner. This was an optimization implemented part way through the campaign in order to allow the CTD rig to focus on drilling MPD-CTD laterals, versus doing front-end preparatory work. The contingency plan of having a CT deployed completion quickly turned into the standard completion deployment procedure. The majority of the CTD drilled wells also had the completion deployed with CT.

Operationally, the campaign was successful, with significant learning, improvement, and increased efficiencies. An extended multi-year campaign of CTD in NAG was postponed in 2009 due to a drastic fall in natural gas price and the economic constraints of tight gas reservoirs.

9.8 References

Amor, B.B. et al. "Underbalanced Coiled Tubing Drilling Practices in Horizontal Short Radius Re-Entry Well Applied in Hassi Messaoud-Algeria, Case Study," SPE 106870 presented at the Asia Pacific Oil & Gas Conference and Exhibition, Jakarta, Indonesia, October 30–November 1, 2007.

Alaska Wells Group, *Rigless Operations Manual: Coil Tubing Operations*, Standard Operating Procedures.

API 16ST, *Recommended Practice for Coiled Tubing Well Control Equipment Systems and Operations*, American Petroleum Institute, Washington, DC, USA, 2009.

Baker Hughes Inteq (BHI), *CoilTrak Handbook.*

Baker Hughes Inteq (BHI), *CT Handbook.*

Bawaked, W.K., Beheiri, F.I. and Saudi, M.M. "A New Record of Coiled Tubing Reach in Open Hole Horizontal Wells Using Tractor and Friction Reducer in Saudi Arabia History: A Case History," SPE 117062 presented at the SPE Saudi Arabia section Young Professionals Technical Symposium, Dharan, Saudi Arabia, March 29–30, 2008.

Blount, C.G. et al. "Well-Intervention Challenges to Service Wells That Can Be Drilled," SPE 100172 presented at the SPE/ICoTA Coiled Tubing Conference and Exhibition, The Woodlands, TX, USA, April 4–5, 2006.

BP Sharjah—EOWR.

BP NAG—EOWR.

BP Colombia—EOWR.

BP Lisburne—EOWR.

BP NAG (San Juan)—CTD White Paper.

BP NAG CTD—State of the Business White Paper.

BP Sharjah, Standard/Emergency Procedures.

BP NAG CTD, Standard Operating Procedures.

Brillon, C.L., Shafer, R.S. and Bello, A.A. "Pushing the Envelope with Coiled Tubing Drilling," AADE-07-NTCE-31 presented at the AADE National Technical Conference and Exhibition, Houston, TX, USA, April 10–12, 2007.

Crouse, P.C. and Lunan, W.B. "Coiled Tubing Drilling—Expanding Application Key to Future," SPE 60706 presented at the SPE/ICoTA Coiled Tubing Rountable, Houston, TX, USA, April 5–6, 2000.

da Silva, T.P. et al. "A Process Delivery Template for an Underbalanced Coiled Tubing Drilling Project from Concept to Execution," SPE 107244 presented at the SPE/ICoTA Coiled Tubing and Well Intervention Conference and Exhibition, The Woodlands, TX, USA, March 20–21, 2007.

Denton, S., Dietrich, E., Ortiz, R., Cadena, J. and Ohanian, M. "Cleveland Tight Gas: CTD/MPD Re-Entry Campaign Results," SPE 120846-PP presented at the SPE/ICoTA Coiled Tubing and Well Intervention Conference and Exhibition, The Woodlands, TX, USA, March 31–April 1, 2009.

Dietrich, E., Denton, S., Cadena, J., Ortiz, R. and Ohanian, M. "Coiled Tubing MPD for Tight Gas Field Re-Entry Work," SPE 122272 presented at the IADC/SPE Managed Pressure Drilling and Underbalanced Operations Conference and Exhibition, San Antonio, TX, USA, February 12–13, 2009.

Duque, L.H., Guimaraes, Z., Berry, S.L. and Gouveia, M. "Coiled Tubing and Nitrogen Generation Unit Operations: Corrosion Challenges and Solutions Found in Brazil Offshore Operations," SPE 113719 presented at the SPE/ICoTA Coiled Tubing and Well Intervention Conference and Exhibition, The Woodlands, TX, USA, April 1–2, 2008.

Fraser, R.G. and Ravensbergen, J. "Improving the Performance of Coiled Tubing Underbalanced Horizontal Drilling Operations," SPE 74841 presented at the SPE/ICoTA Coiled Tubing Conference and Exhibition, Houston, TX, USA, April 9–10, 2002.

Faure, A.M., Zijlker, V.A., van Elst, H. and van Melsen, R.J. "Horizontal Drilling with Coiled Tubing: A Look at Potential Application to North Sea Mature Fields in Light of Experience Onshore The Netherlands," SPE 26715 presented at Offshore Europe, Aberdeen, UK, September 7–10, 1993.

Frink, P.J., Leslie, C. and Wooten, J.E. "Transient Simulator-Based Training of Rig-Site Decision Makers and Rig Crews for Through-Tubing Coiled Tubing Underbalanced Multilateral Project," SPE 89532 presented at the SPE/ICoTA Coiled Tubing Conference and Exhibition, Houston, TX, USA, March 23–24, 2004.

Goodrich, G., Smith, B.E. and Larson, E.B. "Coiled Tubing Drilling Practices at Prudhoe Bay," SPE 35128 presented at the IADC/SPE Drilling Conference, New Orleans, LA, USA, March 12–15, 1996.

Johnson, M. et al. "Coiled-Tubing Underbalanced Drilling Applications in the Lisburne Field, Alaska," SPE 108337 presented at the IADC/SPE Managed Pressure Drilling and Underbalanced Operations Conference and Exhibition, Galveston, TX, USA, March 28–29, 2007.

Julian, J.Y. et al. "16Cr Coiled-Tubing Field Trial at Prudhoe Bay, Alaska," SPE 106639 presented at the SPE/ICoTA Coiled Tubing and Well Intervention Conference and Exhibition, The Woodlands, TX, USA, March 20–21, 2007.

Julian, J.Y. et al. "State-of-the-Art Coiled Tubing Technology at Prudhoe Bay, Alaska," IPTC 11533 presented at the International Petroleum Technology Conference, Dubai, UAE, December 4–6, 2007.

LEA, *Intro to CTD.*

Leising, L.J. and Newman, K.R. "Coiled-Tubing Drilling," *SPE Drilling and Completions*, 8, No. 4, December 1993, pp. 227–232.

Leslie, C., Shero, W., Ortiz, R. and Weihe, A. "Challenges of Horizontal Drilling with Coiled Tubing in a Depleted, Hard and Abrasive Sandstone in Algeria," SPE 78309 presented at the European Petroleum Conference, Aberdeen, UK, October 29–31, 2002.

Lheure, S. et al. "Optimizing Performance of Mature Reservoir: An Innovative use of Coiled Tubing Drilling Technology to Tap Unswept Reserves, Alwyn North UKCS," SPE 65138 presented at the SPE European Conference, Paris, France, October 24–25, 2000.

McCarty, T., Stanley, M.J. and Gantt, L.L. "Coiled Tubing Drilling: Continued Performance Improvement in Alaska," *SPE Drilling and Completions*, 17, No. 1, March 2002, pp. 44–48.

Milligan, M.R., Andreychuk, M.T. and Lunan, W.B. "Coiled Tubing Drilling of Horizontal Sidetrack in House Mountain Field, Alberta," *SPE Drilling & Completion*, 15, No. 2, June 2000, pp. 92–96.

Mykytiw, C., Killip, D., Brewin, R., and Capps, J. "Mature-Field Rejuvenation of a Tight Gas Carbonate Reservoir Utilizing Coiled-Tubing Underbalanced-Drilling Techniques," SPE 113684 presented at the SPE/IADC Managed Pressure Drilling and Underbalanced Operations Conference and Exhibition, Abu Dhabi, UAE, January 28–29, 2008.

National Energy Technology Laboratory, *Sound CT Practices*, September 2001.

NOV CTES, *CT Hydraulics Modeling.*

NOV CTES, *CT Manual.*

NOV CTES, *CT Mechanical Calculations.*

Ohlinger, J.J., Gantt, L.L. and McCarty, T.M. "A Comparison of Mud Pulse and E-Line Telemetry in Alaska CTD Operations," SPE 74842 presented at the SPE/ICoTA Coiled Tubing Conference and Exhibition, Houston, TX, USA, April 9–10, 2002.

Portman, L., Nguyen, L. and MacArthur, J. "Cheap, Directional Wells Drilled Underbalanced With Coiled Tubing: An Experience in the Australian Outback," SPE 90783 presented at the SPE Annual Technical Conference and Exhibition, Houston, TX, USA, September 26–29, 2004.

Pruitt, R. et al. "Sajaa Underbalance Coiled Tubing Drilling 'Putting It All Together'," SPE 89644 presented at the SPE/ICoTA Coiled Tubing Conference and Exhibition, Houston, TX, USA, March 23–24, 2004.

Rixse, M. and Johnson, M.O. "High Performance Coil Tubing Drilling in Shallow North Slope Heavy Oil," SPE 74553 presented at the IADC/SPE Drilling Conference, Dallas, TX, USA, February 26–28, 2002.

Samsonsen, B., Jacobsen, B.G., Skagestad, T. and Kerr, S. "Drilling and Completing a High-Angle Well with Coiled Tubing Technology," SPE 48941 presented at the SPE Annual Technical Conference & Exhibition, New Orleans, LA, USA, September 27–30, 1998.

Surewaard, J. et al. "Approach to Underbalanced Well Operations in Petroleum Development Oman," SPE 35069 presented at the IADC/SPE Drilling Conference, New Orleans, LA, USA, March 12–15, 1996.

Suryanarayana, P.V. et al. "Basis of Design for Coiled-Tubing Underbalanced Through-Tubing Drilling in the Sajaa Field," *SPE Drilling and Completion*, 21, No. 2, June 2006, pp. 125–132.

Traonmilin, E.M. et al. "First Field Trial of a Coiled Tubing for Exploration Drilling," SPE 23876 presented at the IADC/SPE Drilling Conference, New Orleans, LA, USA, February 18–21, 1992.

Turner, D.R. et al. "Electric Coiled Tubing Drilling: A Smarter CT Drilling System," SPE 52791 presented at the IADC/SPE Drilling Conference, Amsterdam, The Netherlands, March 9–11, 1999.

Venhaus, D.E. et al. "Overview of the Kuparuk CTD Program and Recent Record-Setting Operations," SPE 100210 presented at the SPE/ICoTA Coiled Tubing Conference and Exhibition, The Woodlands, TX, USA, April 4–5, 2006.

CHAPTER 10

Gases Used in Underbalanced Drilling

Bill Rehm, Drilling Consultant
Abdullah Al-Yami, Texas A&M University

10.1 Introduction

Chapter 10 describes the gases used in underbalanced drilling to reduce the hydrostatic head of the drilling fluid, or in the case of gas drilling, as the primary drilling fluid. This chapter is not intended to be a treatise on the subject of the various gasses, but a general discussion for field operations about uses and problems with the various gases; and when they might be suitable and when the gas, because of its properties, might cause some difficulties. Some of the material in this chapter is forwarded from other chapters in order to complete the picture concerning the particular gas. More detailed technical explanations are to be found in Chapter 12, Flaring, and Chapter 13, Corrosion in Drillpipe and Casing.

The gasses described in the following material are:

- Air
- Natural Gas
- Membrane Nitrogen (N_2 with some oxygen and trace gases)
- Cryogenic Nitrogen (pure N_2)
- Carbon Dioxide (CO_2) and Super Critical Carbon Dioxide

10.1.1 The Ideal Function of the Gas in Underbalanced Drilling

None of the gasses fits all the ideal functions of a gas in a particular system. The choice of a gas is almost always a compromise between properties, availability, and cost. The purpose of this discussion is to present the basic limits of the systems.

10.1.1.1 Gas Drilling

In gas (and air) drilling, the function of the gas as the drilling fluid is to:

- Move the cuttings out from under the bit
- Transport the bit cuttings and cavings to the surface
- Transport any formation fluids or gas safely to the surface
- Safely and simply release the cuttings and formation fluid and gas at the surface
- Prevent or avoid corrosion of the drillpipe and casing
- Cool the drill bit
- Cool the air hammer or air motor
- Be cost effective

In the case of mist drilling, gas acts as the continuous phase in the annular flow of liquid and gas. In this case, the annular flow of the liquid materially adds to the functions.

10.1.1.2 Gaseated and Foam Fluids

In gaseated and foam fluids, the function of the gas is quite different. These functions are:

- Reduce the hydrostatic head of the fluid by displacing part of the fluid out of the hole
- Avoid any adverse reactions with formation fluids and gasses
- Be consistent within the limits of the General Gas Law (so the gas can be modeled)
- Be cost effective

The other functions of a gaseated or foam drilling fluid are *primarily*, but not completely, a result of the liquid continuous phase. These include:

- Help support the wall of the open hole
- Move cuttings out from under the bit
- Transport the bit cuttings and caving to the surface
- Transport any formation fluids or gas safely to the surface
- Safely and simply release the cuttings, formation fluid and gas at the surface

- Operate the hammer or mud motor
- Help limit corrosion or not add to corrosiveness of the system
- Cool and lubricate the bit and mud motor or hammer
- Moderate some of the effect of pipe movement
- Protect the formation from damage
- Be cost effective

10.2 Air as the Underbalanced Drilling Gas

10.2.1 Air Composition

Air is mixture of a number of gasses (see Table 10–1).

10.2.2 Advantages of Air as a Drilling Fluid

The primary advantage of using compressed air is cost and availability. Any other gas, with the possible exception of lease gas used on the lease, is more expensive. Air compressors and boosters for the volumes and pressures normally used in air or gaseated systems are readily and generally locally available.

Table 10–1 Composition of Air, by Volume

Gas	Chemical Symbol	% by Volume
Nitrogen	N_2	78.08%
Oxygen	O_2	20.95%
Argon	Ar	0.93%
Carbon Dioxide	CO_2	0.03%
Neon	Ne	
Methane	CH_4	
Helium	He	
Krypton	Kr	
Hydrogen	H_2	
Xenon	Xe	0.03%

10.2.3 Problems with the Use of Air as a Drilling Fluid

The primary problem with air as the drilling fluid is the presence of 21% oxygen in the system. This is a problem because:

- Oxygen is a primary actor in corrosion.
- Oxygen is one of the three legs that lead to fire and explosion. (Oxygen, fuel, and an ignition source).
- Cost is always a factor and there are significant compression and accompanying fuel costs to air compression, but the overall cost is less than with other gasses. When air compression is above about 2,000 psi (14,000 kPa or 135 Atm), costs become very high.

10.2.3.1 Air and Corrosion

The problems of corrosion with oxygen in the air are discussed in detail in Chapter 13. Air and dampness down-hole provide a continuous source of oxygen and moisture that can corrode the steel of the drillstring and casing. On a cost basis, it is impractical to try to chemically treat the oxygen out of the air column. The result is a high hidden cost for anti-corrosion chemicals.

The oxygen (in air) has limited solubility in water at surface conditions, about 16 ppm at standard conditions. Solubility decreases with temperature but increases with pressure (see Figure 10–1) (Chitty, 1998).

10.2.3.2 Air and Fire

The presence of oxygen in air has the potential to allow fire or explosion. The explosive limits for ignition with hydrocarbons are in theory quite limited, but in actual cases, the proper oxygen content can be approached in a limited area which still can cause an ignition. The ignition temperature for dry natural gas is quite high, but decreases as gas approaches a condensate, so the greatest danger with fire is with the presence of a condensate. Figure 10–2 shows approximate auto-ignition temperatures for various hydrocarbons and Figure 10–3 shows minimum oxygen requirements for flame propagation of different hydrocarbons.

Downhole explosion is a risk when air drilling. A mud ring can seal or plug off the annulus, increasing pressure and temperature until the hydrocarbon-air mixture ignites. Air leaks through a washout in drillpipe may lead to local hot spot that can ignite the mixture of hydrocarbon and air. To minimize the potential of downhole explosions, the oxygen should not exceed 5% by volume. Air

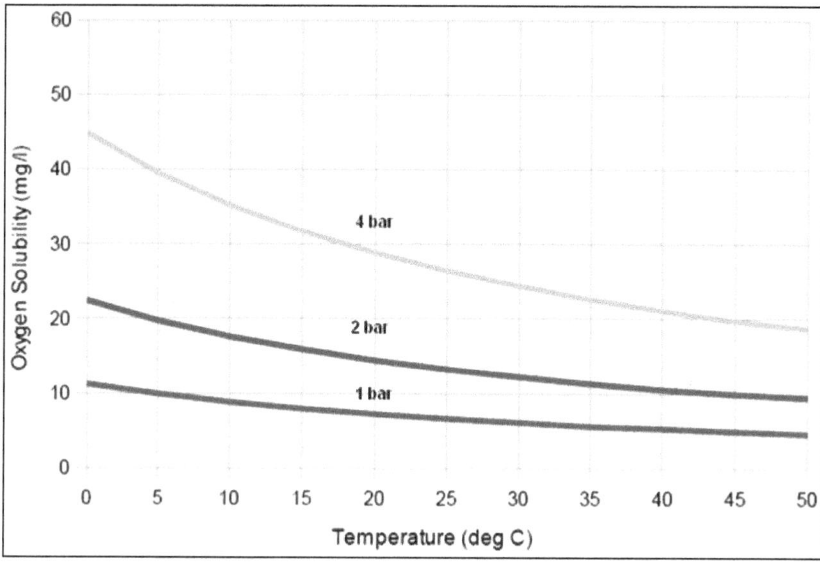

Figure 10–1 *Solubility of oxygen in water (Chitty, 1998)*

Methane 1,300°F
Propane 1,000°F
Condensate 700°F; but at 300 Atm (440psi) auto ignition is 400°F

Figure 10–2 *Approximate ignition temperatures for various hydrocarbons at 1 Atm*

systems also have the potential to cause explosion or fire in the separator systems and may cause flashback in a flare system. For further discussion see Chapter 5, Air and Gas Drilling (Drilling Dry and with Mist).

10.2.3.3 Air and Production Damage

There has been some discussion that air (oxygen) that is forced into the reservoir zone may cause damage to production. This is not well documented.

10.2.4 Comment on Air as a Drilling Fluid

Despite the material on the limits of air, it has proven generally satisfactory over the years and still continues to be used more than any other gas.

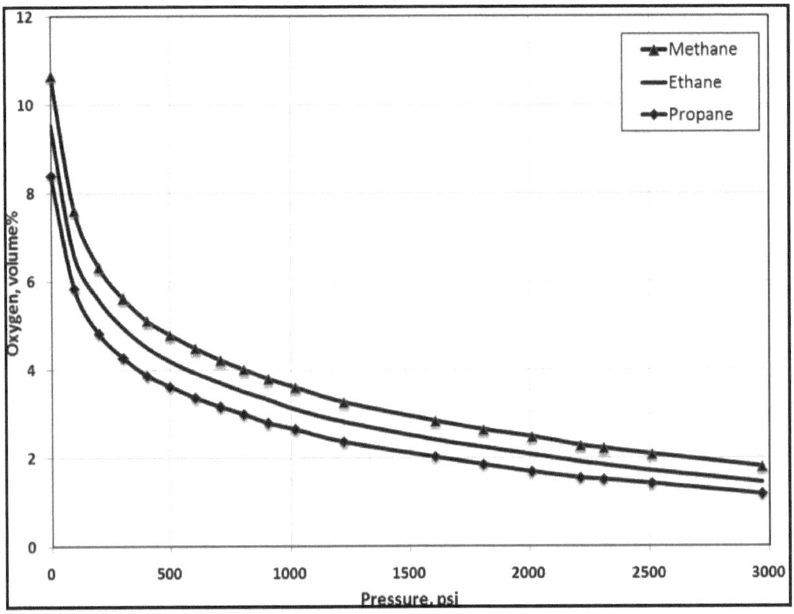

Figure 10–3 *Minimum oxygen requirements for flame propagation of different gases*

10.3 Natural Gas as the Underbalanced Drilling Gas

10.3.1 Natural Gas Composition

Natural gas used with drilling is primarily methane (CH_4) (see Table 10–2). Methane more closely fits the ideal requirements for a gas drilling fluid or a gaseated fluid than does air. Natural gas was one of the most common drilling gasses used until the 1970's when the price of gas started to increase. Other normal constituents of natural gas may be ethane, propane, butane, or one of the higher hydrocarbons. It also may contain some carbon dioxide. Natural gas available for drilling may contain enough of these heavier gasses to have a SpG of 0.8.

10.3.2 Natural Gas Advantages

10.3.2.1 Corrosion

One of the greatest advantages to natural gas is that, without oxygen, it does not cause or add to the corrosion problem. Oxygen corrosion with air drilling or aerated drilling fluids is a major technical and cost problem. This is avoided when using natural gas.

Table 10–2 A General Composition of Natural Gas

Component	Formula	Volume Concentration
Methane	CH_4	70–90%
Ethane	C_2H_6	0–20%
Propane	C_3H_8	
Butane	C_4H_{10}	
Carbon Dioxide	CO_2	0–8%
Oxygen	O_2	0–0.2%
Nitrogen	N_2	0–5%
Hydrogen Sufide	H_2S	0–5%
Rare gases	He, A, Ne, Xe	trace

Corrosion with CO_2 and H_2S can occur in systems using natural gas when there is dampness or water in the wellbore. This is explained in Chapter 13.

10.3.2.2 Down-Hole Fire

The other major advantage is that natural gas without oxygen does not support a fire or explosion. Downhole fires, blooie line, flare line flashbacks, and explosion in the separator are both safety and non-productive time (NPT) problems. When using natural gas *properly,* none of these occur.

Gas around the well head or under the floor of the rig poses a hazard. These areas need to be well ventilated and gas detectors placed in areas where the gas could accumulate. Since natural gas is lighter than air, it tends to collect under the rotary beams or in a similar enclosure. This is a problem in cold weather or arctic operations where the substructure may be closed in. The minimum ignition concentration for methane exposed to air is only 6% as shown in Figure 10–4.

10.3.2.3 Gas from the Pipeline

When pressured natural gas can be obtained from a field pipeline, it simplifies and lowers the cost of compression at the drilling location. Natural gas in such a state may be less expensive than using compressors with added personnel and fuel use.

10.3.3 Natural Gas Problems

10.3.3.1 Natural Gas and Density

Relative to air, natural gas, when it is methane, has about a 0.6 specific gravity. Some of the mixed natural gasses may have SpG of up to 0.8 In gas drilling operations, the basic gas for drilling requirement formula (Angel, 1958) shows that about 20% more methane is required when drilling with methane instead of air. Using other kinetic energy models, the increase in volume required is close to the same (see Figure 10–5 and Figure 10–6).

In gaseated and foam field operations, it is hard to see any increase in volume requirements with natural gas over air use. Models for gaseated or foam systems vary in their results depending upon the basic modeling assumptions, so models may or may not show the necessity for an increased volume.

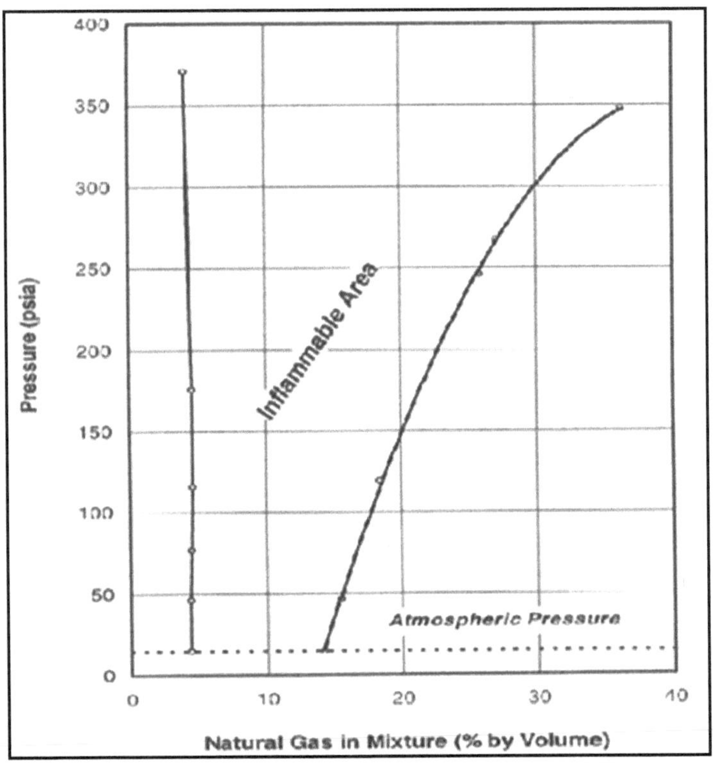

Figure 10–4 *Methane flammability levels*

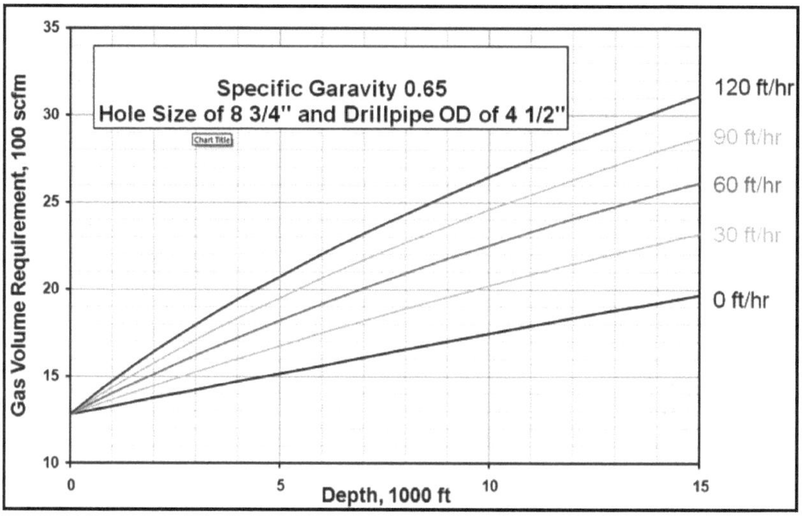

Figure 10–5 *Methane versus air requirements for gas drilling at specific gravity 0.65 (Angel, 1958)*

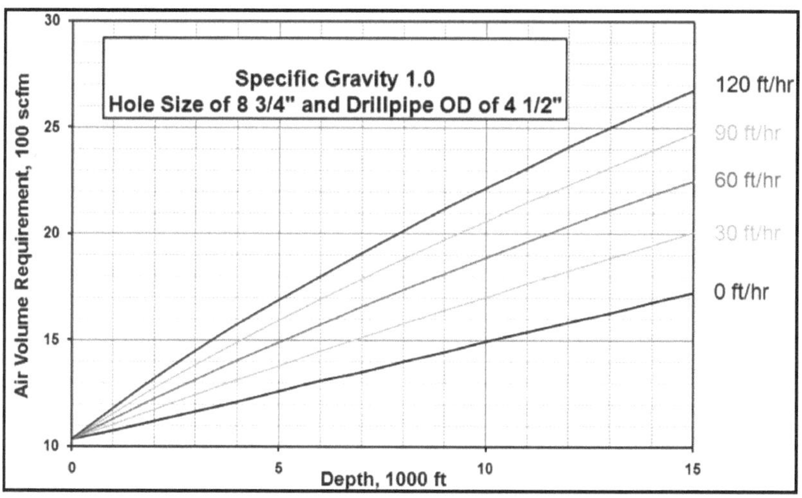

Figure 10–6 *Methane versus air requirements for gas drilling at specific gravity 1.0 (Angel, 1958)*

10.3.4 Natural Gas and Solubility

Natural gas will be dissolved to some extent in any drilling fluid. This can generally be ignored with a water-based fluid, but it is not always clear what the effect will be with oil-based fluids especially the invert emulsions since there is a wide variety of base oil from diesel to synthetic oils. In general, as pressure increases, the solubility of gas in oil increases. As the pressure exceeds the bubble point, free gas cannot coexist with the liquid. At a pressure above the bubble point, only a single phase fluid is presented instead of the targeted two phase drilling operations (with gaseated systems and foam). The reverse occurs when the pressure decreases and gas comes out of solution to form a two phase system or a system with large gas bubbles. Figure 10–7 shows solubility of methane in diesel oil at different temperatures and pressures. This may not represent the effect with any particular invert emulsion.

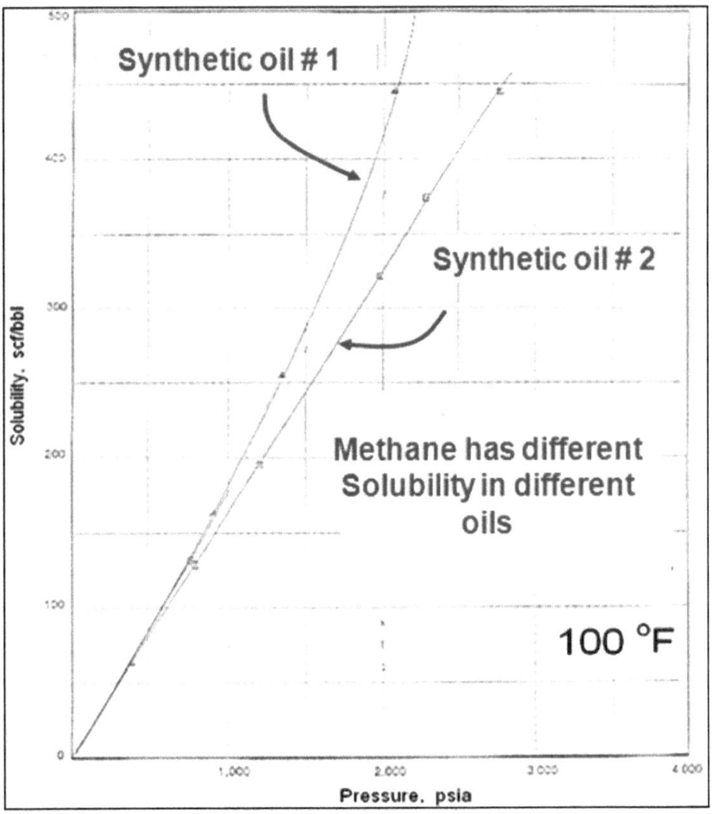

Figure 10–7 *Solubility of methane in different light oils (Watson et al., 2003)*

10.3.5 Comment on Natural Gas as a Drilling Fluid

Natural gas from a pressured lease line is the least expensive drilling fluid gas. In some cases, there may be objections to the irregular use of gas from the line. This has to be locally determined. The safety aspects of using gas are much easier to predict and control than with the use of air.

10.4 Membrane Nitrogen as an Underbalanced Drilling Gas

10.4.1 The Membrane Nitrogen Composition

Membrane nitrogen is produced at the well site by pumping air through a soda straw like membrane that allows the oxygen, carbon dioxide and water vapor molecules to "escape" but retains the nitrogen molecule. This is not an exact cut but depends in part on the pressure across the membrane. The greater the pressure, the more oxygen is forced through the membrane with the nitrogen. Membrane nitrogen as used in drilling operations usually contains 4–6% oxygen plus a small percent of other gasses. Figure 10–8 shows a diagram of a membrane nitrogen system. See Chapter 11, Section 2, page 466, for a more complete discussion on membrane operation and maintenance.

The systems in use at present are about 50% efficient. Twice as much air needs to be pumped into the membrane as nitrogen is produced. Input pressure to the membrane is about 350 psi and output pressure to the drilling compressor/booster is 300 psi. If the input pressure is increased, the membrane will pass more gas, but separation efficiency decreases and the oxygen content of the output increases.

The individual membrane appears something like a soda straw, Figure 10–9. A large number of individual membranes are bundled together in a tank and a number of tanks make up the membrane unit. The individual membranes are very susceptible to plugging, especially from engine exhaust fumes, and a filter system is used ahead of the membrane. The membranes are also temperature sensitive so the filtration system includes a heat exchanger.

10.4.2 Membrane Nitrogen Advantages

10.4.2.1 Fire or Explosion

At 5% oxygen, the wellbore is below the limits of combustion (see Figure 10–10). Nitrogen in a closed separator with only 5% oxygen will also be below any explosive limit from static electricity.

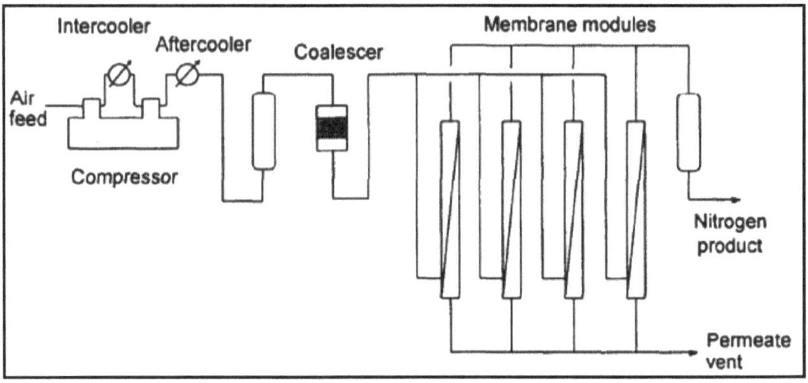

Figure 10–8 *A diagram of a membrane nitrogen system (Gunardson, 1998)*

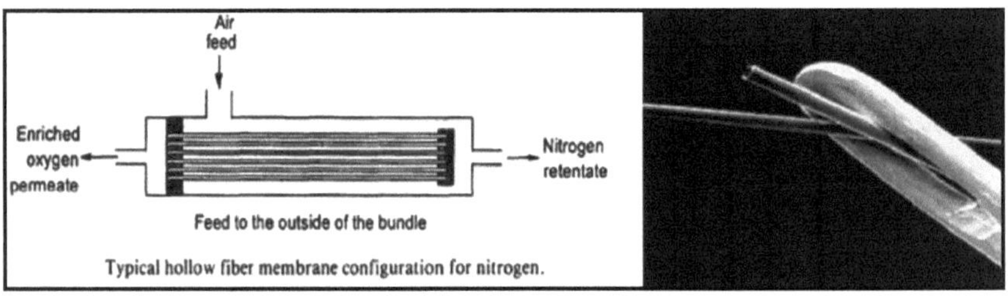

Figure 10–9 *Typical membrane fiber configuration (Gunardson, 1998)*

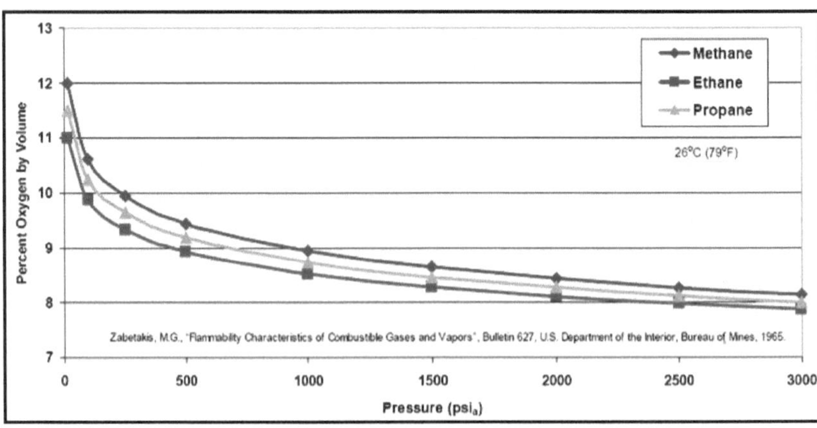

Figure 10–10 *Oxygen content versus ignition potential*

10.4.2.2 Use with Oil

Membrane nitrogen is usable with oil continuous-phase fluids. Unlike natural gas, nitrogen solubility with oil is low. It is the most common gas used with oil gaseated systems and oil foam systems. With oil continuous-phase systems, the potential for any corrosion is very low.

10.4.3 Challenges with Membrane Nitrogen

10.4.3.1 Corrosion

Membrane nitrogen generally contains 4–6% oxygen. As noted above, the saturation for oxygen in water at surface conditions is 14–16ppm. Membrane nitrogen does not solve the corrosion problem. Given the volume of nitrogen used for gas drilling or for gaseated fluids, it is not economically feasible to treat out the oxygen (see Chapter 13).

10.4.3.2 Flaring

Nitrogen in the flare system suppresses the flare and, until a large amount of hydrocarbon gas is available, the flare system will release the produced gas unburned.

10.4.3.3 Cost

Membrane nitrogen at least doubles the cost of air compression. 50% more air compression is required since the systems are only about 50% efficient. In addition there is the nitrogen membrane unit, extra operating personnel, and mobilization.

10.4.4 Comment on Membrane Nitrogen

In spite of the challenges to membrane nitrogen, it fills a need for a relatively inert gas with moderately high cost. It works well with oil systems and reduces any chance of fire or explosion in the hole or on the site.

10.5 Cryogenic Nitrogen

10.5.1 Cryogenic Nitrogen Properties

Cryogenic nitrogen is pure liquid nitrogen at a temperature of –320°F (–160°C). A gallon of liquid nitrogen will produce 93.12 scf of gas or one liter will produce 0.698 standard m^3 of gas. The nitrogen is delivered to the well on land in trucks, semi trailers, or by barges and vessels at sea.

The liquid nitrogen is pumped as a liquid by a special triplex pump or pump and booster, and pressures up to 10,000 psi (70,000 kPa or 680 atm) are standard with higher pressures available. After passing through the pump, the liquid is heated and expanded to the gaseous state (under pressure) for use.

The specific heat of nitrogen is low. The manuals generally quote 134,00 BTU to produce 10,000 scf of nitrogen from the liquid state or 13.4 BTU/ft^3. This level of heat energy is readily available from the truck or pump diesel engine heat exchanger in warm or moderate climates. In arctic or very cold weather usage, a special heater is used to provide extra energy. Figure 10–11 shows a cryogenic pumping schematic.

10.5.2 Cryogenic Nitrogen Advantages

10.5.2.1 No Corrosion from Oxygen Sources

Cryogenic nitrogen has no oxygen and does not support any oxygen based corrosion. It is possible to have downhole corrosion if dampness or water is present along with CO_2 or H_2S. This is explained in Chapter 13.

10.5.2.2 Fire or Explosion

With pure nitrogen and no oxygen, there is no danger of fire or explosion. The nitrogen acts to suppress any potential fire.

10.5.2.3 High Pressures

Since the nitrogen is pumped as a liquid, it can be injected at very high pressures. When using coiled tubing, the injection pressures are so high, that cryogenic nitrogen is the best practical solution as a drilling gas.

10.5.3 Challenges with Cryogenic Nitrogen

10.5.3.1 Cost

The primary limitation of the cryogenic nitrogen system is cost. The basic cost of the nitrogen gas at the plant is not high, but the cost of transportation is very high. Liquid nitrogen needs to be carried in special tanks and pumped and warmed from the transport tank. This equipment is in use from the pickup at the plant until the load is pumped to the drilling operation and the truck or barge returns to the

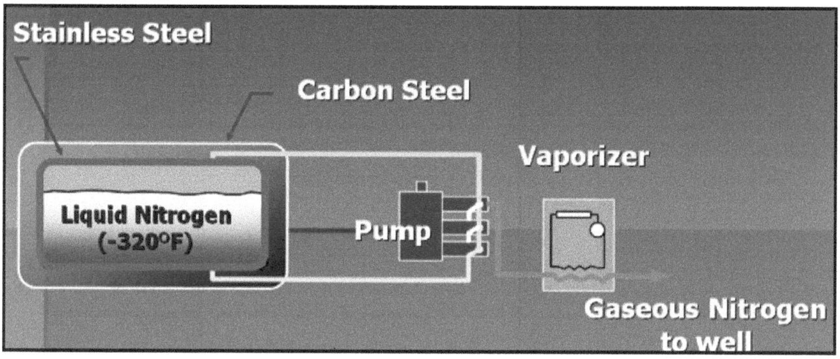

Figure 10–11 *Cryogenic pumping schematic*

plant. The net result is that cryogenic nitrogen costs can be more than five times the cost of compressed air and at least twice that of membrane nitrogen.

10.5.3.2 Boil Off

Boil off of the nitrogen from the liquid state can be a problem in warm areas and where there is a long time from loading to the barge, boat, or truck until use at the rig. In extreme cases, such as Venezuela, boil off has approached 50% of the load. Boil off in Canadian subarctic operations can be negligible even though transport takes a long time.

10.5.3.3 Flares

The presence of nitrogen represses ignition in a flare system, so a significantly greater amount of natural gas must be present for the flare to burn than without nitrogen. Gas from the flare system can be released unburned to the atmosphere.

10.5.4 Comment on Cryogenic Nitrogen

The ability to inject the gas into the drilling system at high pressure makes cryogenic nitrogen the most desirable gas for coiled tubing operations.

Liquid nitrogen in jointed tubing is desirable in areas where corrosion or fire potential exists. The high expense connected with transportation may be less than mobilization costs for a membrane system, especially on a job shorter than a week, or one with limited circulating time. However due to transportation costs, cryogenic nitrogen is very expensive, generally at least three times the cost of membrane nitrogen.

10.6 Carbon Dioxide (CO_2) as a Drilling Gas

The use of carbon dioxide as a drilling gas is a common subject of discussion. Carbon dioxide has some major problems with corrosion and pressure and is not a suitable drilling gas for most gas drilling, gaseated, or foam systems.

10.6.1 Properties of Carbon Dioxide

CO_2 is a gas with some special properties. Much of the use for CO_2 is as dry ice at –109°F or –78.5°C. Dry ice has the particular property of subliming from a solid to a gas without becoming a liquid, although liquid CO_2 does exist. In drilling operations, carbon dioxide may exist in one of several forms primarily depending upon the pH. In water at a pH below about 6 it forms carbonic acid, a strong corrosive acid; from pH6 to 8, it exists as the carbonate ion (HCO^-), and above pH8 it tends to precipitate out as scale, ($CaCO_3$). If dry, it can exist as the gas CO_2 at normal gas drilling pressures. However with traces of moisture it tends to form carbonic acid. This is discussed a greater length in the Chapter 13.

However, a proposal has been made to use super critical carbon dioxide as a liquid for deep drilling The phase diagram (Figure 10–12) of carbon dioxide shows it has supercritical properties in the range of the bottom hole pressure of a deep hole. In the supercritical range, CO_2 can not be identified as gas or liquid but as fluid with viscosity and a very low density.

Gupta et al., 2005, discussed the use of supercritical carbon dioxide as a material capable of operating a drilling motor in a very deep low pressure well where nitrogen would not turn the drilling motor. The proposal is well documented and seems to be a reasonable solution for the deep low pressure well. The proposal outlined some advantages such as:

- The CO_2 has high liquid density in the tubing which would allow it to turn a down-hole motor and generate adequate torque.
- The CO_2 in the annulus is in gas phase which allows a very low bottom-hole pressure (see Figure 10–13).

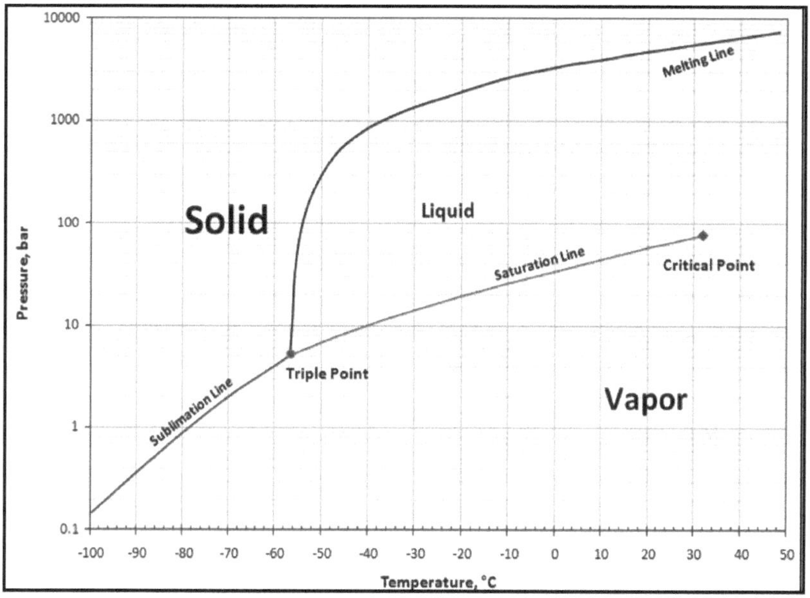

Figure 10–12 *Phase diagram of CO_2 (Gupta et al., 2005)*

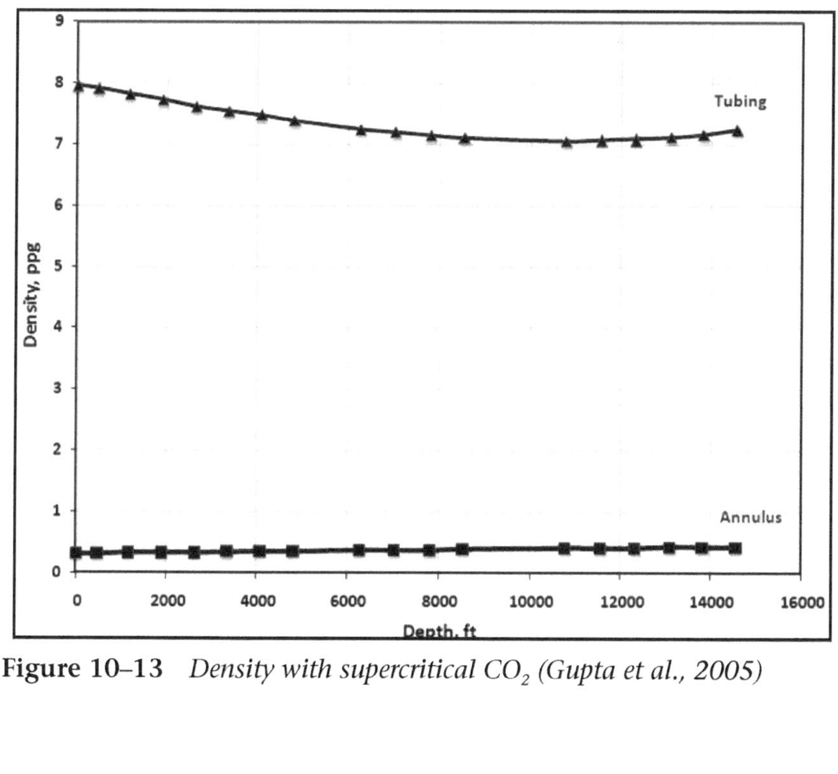

Figure 10–13 *Density with supercritical CO_2 (Gupta et al., 2005)*

10.6.2 Problems with Carbon Dioxide as a Drilling Fluid

10.6.2.1 Corrosion

Dry carbon dioxide is not corrosive until above 750°F.

CO_2 breaks down in the presence of water or water vapor into carbonic acid.

$$CO_2 + H_2O \rightarrow H_2CO_3$$

Carbonic acid lowers the well fluids' pH and increases their corrosiveness. It also reacts with iron to form an iron carbonate (scale) plus hydrogen (Stone et al., 1989):

$$Fe + H_2CO_3 \rightarrow FeCO_3\downarrow + H_2\uparrow$$

Carbonic acid reacts with cement to form calcium carbonate and other reaction materials, the result is a leaching of the cementing material out of the cement (Nelson et al., 1990).

10.6.2.2 Toxic

Carbon dioxide gas is colorless and odorless. It is considered to be toxic above 5% by volume and can cause headaches and nausea at as low a concentration as 1%.

10.6.2.3 Solubility

CO_2 is very soluble in water at standard conditions. Its dissolves on a 0.9:1 by volume ratio with water. Solubility increases with pressure but releases with lowered pressure (such as soda pop or beer). Solubility of gasses in diesel oil is shown in Figure 10–14.

10.6.3 Comment on Carbon Dioxide

In summary, it would require a special set of conditions to justify its use in the drilling operation.

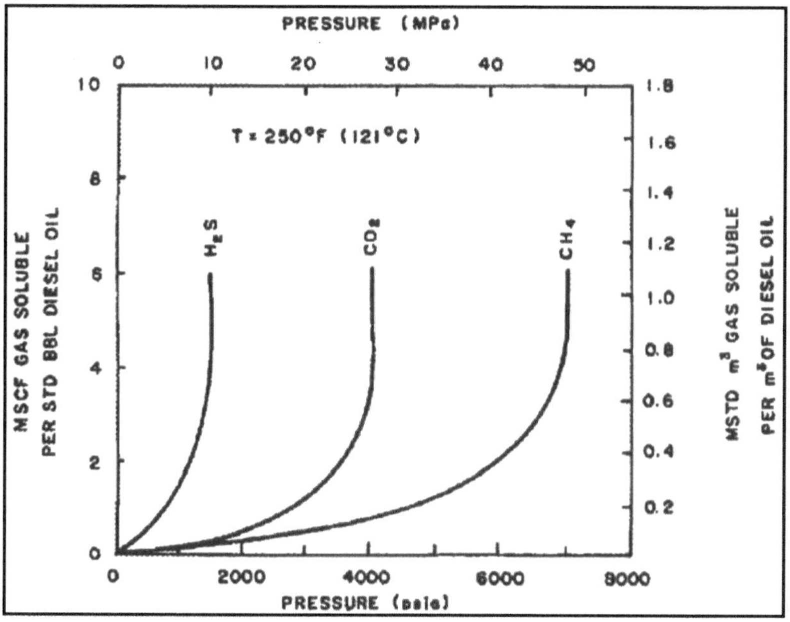

Figure 10–14 *Solubility of gasses in diesel (Watson et al., 2003)*

10.7 Questions

1. How can you decide between using cryogenic and membrane nitrogen?
2. What is the biggest concern of using natural gas in foam UBD?
3. When do down-hole fires usually occur?
4. How does the gas separation fiber work?
5. What issues need to be considered for the proposed CO_2 gas in UBD operations?
6. What will the lower density of natural gas than air result in?

10.8 Answers

1. Job duration tends to favor membrane nitrogen after 4 days.

   Small volume requirement favors cryogenic nitrogen.

High pressure requirement favors cryogenic nitrogen.

Indifferent purity requirements favor membrane.

2. High hydrocarbon repress foam.

3. The problem of down-hole fires normally only occur with air drilling where the air is more than 90% of the fluid/air volume.

4. Oxygen and water vapor are smaller molecules than the nitrogen and they pass though the membrane more quickly. However, some oxygen and water vapor never get time to pass through the membrane and are retained and come out the other end. This is primarily dependent upon the rate that the gas is passed through the membrane.

5. Scale problems

   Hydration problems

   Corrosion problems

   Completion problems

   Environmental concern

6. Lower density of natural gas than air results in:

   - Lower BHP
   - Lower drag forces
   - Higher required circulation rates

10.9 References

Angel, R.R. *Volume Requirements for Air and Gas Drilling*, Gulf Publishing Company, Houston, TX, USA, 1958.

Ceylon, I., Santra, A. and Cullick, A. "Carbon Dioxide, Geochemica, and Rate-of-Dissolution Simulation for Deep Storage Environments," SPE 141031 presented the SPE International Symposium on Oilfield Chemistry, The Woodlands, TX, USA, April 11–13, 2011.

Chitty, G.H. "Corrosion Issues with Underbalanced Drilling in H2S Reservoirs," SPE 46039 presented at the SPE/ICoTA Coiled Tubing Roundtable, Houston, TX, April 15–16, 1998.

Gunardson, H. *Industrial Gases in Petrochemical Processing,* Marcel Dekker, Inc., New York, NY, USA, 1998.

Gupta, A.P., Gupta, A. and Langlinais, J. "Feasibility of Supercritical Carbon Dioxide as a Drilling Fluid for Deep Underbalanced Drilling Operations," SPE 96992 presented at the SPE Annual Technical Conference and Exhibition, Dallas, TX, USA, October 9–12, 2005.

Lyons, W.C. *Air and Gas Drilling Manual,* 3rd Ed., Elsevier, Boston, MA, USA, 2008.

Malloy, K.P., Medley, G.H. and Stone, C.R. "Air Drilling in the Presence of Hydrocarbons: A Time for Pause," SPE 108357 presented at the IADC/SPE Managed Pressure Drilling and Underbalanced Operations Conference and Exhibition, Galveston, TX, USA, March 28–29, 2007.

Masoudi, R. and Tohidi, B. "Gas Hydrate Production Technology for Natural Gas Storage and Transportation and CO_2 Sequestration," SPE 93492 presented at the SPE Middle East Oil & Gas Show and Conference, Kingdom of Bahrain, March 12–15, 2005.

Nelson, E.B. et al. "Well Cementing," TSL-4135/ICN-01557200, Schlumberger Educational Services, 1990.

Shahbazi, K., Mehta, S.A., Moore, R.G., Ursenbach, M.G. and Fraassen, K.C.V. "Investigation of Explosion Occurrence in Underbalanced Drilling," SPE 106770 presented at the Production and Operation Symposium, Oklahoma City, OK, USA, March 31–April 3, 2007.

Stone, P.C., Steinberg, B.G. and Goodson, J.E. "Completion Design for Waterfloods and CO_2 Floods," *SPE Production Engineering,* 4, No. 4, November 1989, pp. 365–370.

Watson, D., Brittenham, T. and Moore, P. *Advanced Well Control,* Society Petroleum Engineers, Richardson, TX, USA, 2003.

Yuan, M., Mosley, I. and Hyer, N. "Mineral Scale Control in a CO_2 Flooded Oilfield," SPE 65029 presented at the SPE International Symposium on Oilfield Chemistry, Houston, TX, USA, February 13–16, 2001.

http://www.chemicalogic.com/co2tab/downloads.htm

CHAPTER 11

Equipment and Equipment Integration

Bill Rehm, Drilling Consultant

11.1 Introduction

Specialty equipment has been discussed in detail in a previous book in this series: *Managed Pressure Drilling*. In this book, *Underbalanced Drilling: Limits and Extremes*, there is a somewhat different problem with specialty equipment and its operation. Early in the development of UBD, Ralmalho, in his paper on "Changing the Face and Feel of Underbalanced Drilling", discussed the problem of equipment integration. Through the years, there have been complaints that it was difficult to administer all the specialty equipment from different vendors. The technical literature contains further descriptions of the problems of equipment integration and field management. Informal discussions in the technical meetings often centered on the problem of specialty equipment cost, administration, and well site supervision.

This chapter contains a description of several systems used by service companies in an attempt to make the operation more responsive and efficient. The following material is not intended to be a comprehensive listing, but rather a number of selected examples of the various service companies' approaches to equipment and equipment supervision, as well as some of the philosophy or marketing behind how the company wants to present its approach.

This chapter is frankly commercial. A question was proposed to the service companies: What equipment and technologies would the company propose to use on an (unspecified) underbalanced operation? Within that charge, the answer is obviously going to be commercial and there is no reason to attempt to hide it. General equipment that is common to oil field practice or covered in detail in *Managed Pressure Drilling* is only mentioned in the following sections. However, special equipment systems, programs, or technologies that appear to be unique to that organization are noted and often illustrated.

11.1.1 Normal Equipment Common to Underbalanced Operations

The standard set of equipment used with UBD operations and not covered in detail in this chapter consists of:

- Rotating control device
- Drill pipe non return valves (NRV)
- Choke and manifold system
- Separator system
- Surface valves and piping
- Flare system

The system may also include:

- Air compressors and boosters
- Nitrogen generators
- Special instrumentation
- Special chemical injection equipment

11.2 Planning and Supervision

The list in Section 11.1.1 (above) shows some of the problems and complexity in this one drilling operation. It is probably unrealistic to expect the Operators Representative (company man) to keep up with his operational program and paperwork, and in addition monitor and control the specialty operations. Most of the complaints about operational control are based on the problem of supervision and authority over a complex set of systems and conditions. Planning and supervision are the real keys to effective control of the rig and all the specialty equipment. Ideally, the planning should start with the basic well program. The solution or question that each of the following examples propose is: At what level is the specialty equipment and supervision integrated into the plan? Is it part of the basic reservoir discussion, part of the drilling plan, or for a limited use in a single part of the hole?

Once the requirements for the UBD equipment system are clear, then what is the level of supervision and how is it to be organized?

What level of training is required and are there any certifications to be obtained?

There is no one definite answer.

The purpose of the following discussions is to present a limited number of views on how all of this should be organized. The references at the end of the chapter contain some other approaches, and many of the service companies and operators have their own manuals on this subject.

Section 2
Halliburton - GeoBalance
Underbalanced Drilling Services

Isabel C. Poletzky, GeoBalance Services, Sperry Drilling

11.3 Project Management Approach

Halliburton Project Management (HPM) is a set of processes, systems, and techniques for effective planning and control of resources necessary to safely complete a project. These processes, systems, and techniques should focus not only on the resources, but should also include the control of hazards associated with underbalanced drilling (UBD) operations. Suitable project management will enhance the safety of the underbalanced program while identifying customer needs; health, safety and environment (HSE) considerations and controlling possible UBD hazards, but it will also reduce the overall cost of the project (see Figure 11–1).

Figure 11–1 *Example of an underbalanced drilling project (Henry Pinkstone)*

There are three main steps in any well executed project: Planning, Execution, Review and Close Out. The more meticulous the planning phase, the more time and money is saved in the execution. The review and close out steps are equally important so that the lessons learned can be applied to subsequent wells. This phase must be completed as quickly as possible to reap the benefits.

The five main phases of Halliburton Project Management are listed below:

- Planning
  - Phase I—Candidate Selection
  - Phase II—Technical Feasibility Study
  - Phase III—Front-End Engineering and Design (FEED)
- Execution
  - Phase IV—Execution, Field Work
- Review and Close Out
  - Phase V—Review, Reports, Lessons Learned

11.3.1 Phase I—Candidate Selection

Proper candidate selection is critical to the success of UBD projects, especially when these projects focus on the reservoir. It is essential that the main project objectives are identified at the beginning of the project to ensure that they can be achieved. The candidate selection process consists of analyzing and evaluating some of the main reservoir and wellbore characteristics: the drilling, geological, reservoir, geomechanical and petrophysical information help to determine whether a particular well and/or reservoir is a potential candidate by evaluating some of the main reservoir and wellbore characteristics.

This phase is intended to determine the viability of underbalanced drilling. The objectives of this phase are to:

- Identify the main drivers to drill underbalanced: reservoir (formation damage), wellbore stability and/or drilling problems, such as lost circulation, differential sticking, rate of penetration (ROP)
- Define the objectives of the project
- Select candidate wells
- Determine if UBD technology will meet the objectives

The project objectives will be defined with regard to client's drivers, so that no conflicts are built into the objectives that would ensure failure of the project. An example would be drilling underbalanced to mitigate formation damage while at the same time wanting to kill the well prior to tripping.

It is also important to delineate the candidate wells to form a rough determination as to whether or not the use of UBD technology will meet the stated objectives. Contra-indicators and benefits will be evaluated for each candidate qualitatively, as well as performing preliminary flow modeling to determine if UBD is possible, and then a ranking for comparison of the different candidates will be developed.

Preliminary wellbore hydraulics modelling is performed to determine the operational feasibility; in other words, if underbalanced (UB) conditions are possible and can be maintained throughout the entire hole section while maintaining adequate hole cleaning and satisfying the downhole motor constraints. If there are multiple candidates, then characteristics of each of these would be compared and ultimately ranked according to the key variables that would have the highest potential for success.

UBD technology is not the solution to all problems, and project success always depends on determining the correct wells where the technology is applicable. If UBD technology is not a solution to the given problem, then this will be the end of the study. However, if UBD technology provides clear benefits to the customer for the outlined candidate wells, the project is taken into Phase II. The deliverable of Phase I is a recommendation to continue or stop the project, along with a short written report summarizing reasoning.

11.3.2 Phase II—Technical Feasibility Study

This phase will provide a more detailed technical and economic evaluation including: reservoir modeling, wellbore hydraulics, equipment selection, examination of contra-indicators, analysis of economics and risks, and comparison to other methods when applicable. A borehole stability study could also be recommended and completed by a third party during this phase.

The potential impact of benefits and limitations of applying UBD will be quantitatively determined for the selected candidate, including:

- Drilling issues, such as loss-circulation potential, poor ROP, differential sticking
- Pressure depletion
- Wellbore instability

- Temperature limits for downhole tools and compatibility of elastomers
- Fluid sensitivity issues
- Well construction, depth/location constraints
- Safety issues, such as hydrogen sulfide (H_2S)

The deliverable of this phase is to recommend which reservoir or candidate wells are suitable for UBD, along with possible well construction and accompanied by a written study.

11.3.3 Phase III—Front-End Engineering and Design (FEED)

The planning that goes into an underbalanced drilling operation is more extensive than the planning of a conventionally drilled well. Some of the tools used in the planning phase of underbalanced operations include: hazard identification (HAZID), hazard and operability (HAZOP), detailed operational procedures, and drawings such as equipment layout drawings (ELDs), process-flow diagrams (PFDs), valve-numbering diagrams (VNDs) and hazardous-area drawings (HADs). These processes contribute considerably to reducing the time and cost of engineering the underbalanced drilling program.

The PFDs are produced to more clearly identify the equipment required and used extensively in the engineering phase (see Figure 11–2). These drawings are critical for the HAZOP and are also very useful in writing procedures. ELDs are produced to ensure that all equipment will fit on the location prior to mobilization. These drawings must be in scale to ensure that everything will fit. Escape routes and safety equipment, such as fire extinguishers and alarm lights, must be indicated here. HADs are produced to ensure that the equipment placement on the layout drawing is in compliance with zoning standards. VNDs are produced to simplify the procedures and to provide a safeguard against the wrong valve being opened during complex operations. Other helpful drawings are: dimensional blowout preventor stack, injection manifold, choke manifold, platform structural and riser system drawings.

A safe approach to underbalanced project management should follow an industry accepted management system model. Halliburton Management System complies with ISO 9001 standards. This system takes in the available information, links it to the customer, and connects back to the management system for initiation of improvement. A high-quality management system is designed to meet operations,

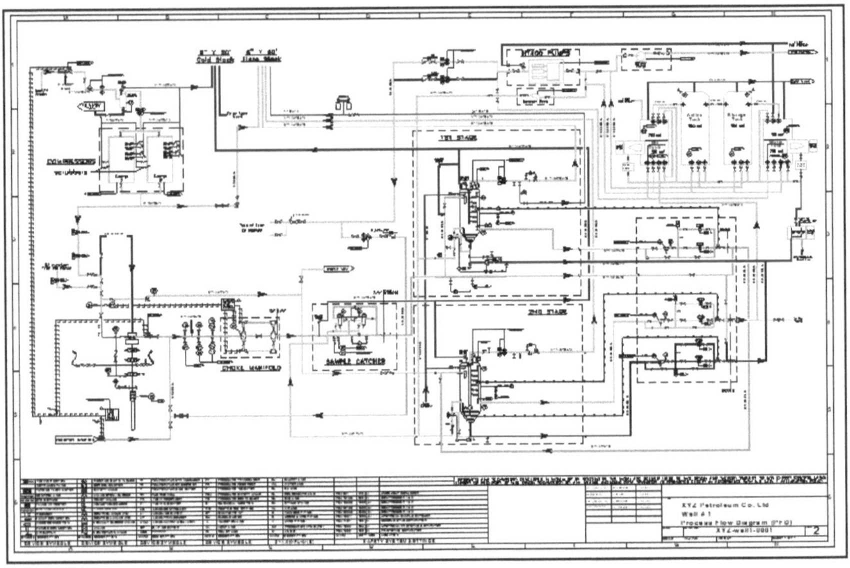

Figure 11–2 *Example of a process flow diagram (PFD)*

quality, and HSE management system needs. Consequently, a high performing management system is customer-focused and driven.

In order to ensure a safe and efficient operation, all project personnel involved in UBD operations have to be familiar with the process, the equipment, and the procedures employed. Therefore, it is critical that HSE issues are considered from the very early phases of the project-planning cycle, and that management of these issues are imbedded in the project management scheme. It is recommended that the risks inherent to UBD are evaluated and controlled using industry-accepted hazards and effects management process (HEMP) techniques such as HAZID and HAZOP studies. In addition to these studies, site-specific HSE guidelines and emergency-response plans should be developed and fully implemented for all UBD projects. The purpose of the techniques and plans described above is to ensure that: a) there is an understanding of required HSE compliance issues; b) there is a system in place to manage HSE throughout the project; c) critical activities are effectively analyzed and controlled; and d) procedures and documentation are in place.

HAZID and HAZOP are part of the overall HSE management system, and they apply specifically to managing hazards and their effects. For both the HAZID and HAZOP process, it is very important to establish a Formal Action Item Closeout process.

The main purpose of a HAZID process is to identify the hazards associated with the specific operation; it is not about trying to solve the problem. A multi-disciplined team consisting of representatives from the operator's drilling, engineering, and production departments, and all services involved in the operation, are part of the HAZID process; all action items are recorded and assigned to specific individuals.

The HAZOP process is conducted following a comprehensive rig visit and after the detailed design phase have been completed. The main purpose of the HAZOP process is to identify the hazards' effects and operability problems for the project, and the process is conducted by a multi-disciplined team. If design, process, or procedural changes are made after the HAZOP session, another (mini) HAZOP session should be employed to ensure that the changes have been appropriately assessed for hazards and effects.

Procedures are another very important part of the planning and execution of the project. These are broken down into three categories:

- **Operational**—Describe tasks that are normally executed in order to drill and complete the well
- **Contingencies**—Non-emergency procedures planned for non-emergency failures
- **Emergency**—Procedures put in place with the primary focus being the health and safety of personnel and the surrounding public

Training is broken down into three stages. The first two are conducted in the planning phase just prior to the execution phase:

- **Stage 1** is designed for office-based personnel with emphasis on basic UBD techniques as they pertain to the specific project. This will advise office staff of the different operations undertaken to complete the well and what to expect during different phases of the operation.
- **Stage 2** is designed to be project specific for the field personnel and is more concentrated and detailed. This training is conducted with all crews one tour prior to execution. Time will be needed for the crews to absorb the information and for their questions to be answered.
- **Stage 3** includes live training exercises conducted at the start of the execution phase.

Some other considerations during this phase are:

- Legislation reviews
- Process selection, including cost reviews
  - Jointed pipe or coiled-tubing drilling (CTD)
  - Standpipe, concentric-casing injection or flow drill
  - Underbalanced fluid selection and flow modeling: the operational window identifies the minimum and maximum flow rates of fluid (gas and liquid) by depth
  - Equipment requirements

11.3.4 Phase IV—Execution, Field Work

Project management and supervision of the project will ensure both safe operations and that underbalanced objectives are met. The execution phase must also include site-specific procedures, proper training, and emergency preparedness preparations in order to ensure that operations can be conducted safely.

A standard crew size per shift for an UBD operation without nitrogen injection would include: one UBD supervisor, one UBD engineer, one data-acquisition engineer, two separation equipment operators, and one rotating-control device (RCD) operator; if a membrane nitrogen system is required for the operation, two more operators for the nitrogen equipment should be considered.

Once integrated, multi-disciplined teams are operational for an UBD project, a process must be established that ensures safety, protection of the environment, execution credibility, and consistency of delivery. This process must become part of the company's management system and should be linked to all aspects which could affect or influence the project. The UBD process should clearly define the scope of the project and then answer who does what, when, where, and how. Critical path processes should be clearly identified during the planning phase. Standards of delivery should be established beforehand, and competency levels of critical personnel involved in the UBD project should be defined and strictly adhered to. Flexibility for operations in any part of the world is necessary to adapt to remote locations, but this flexibility should never compromise the original four elements of the UBD process: safety, protection of the environment, execution credibility, and consistency of delivery.

Particular attention should be placed on safety and protection of the environment. With all the current interest in this application, it is necessary to pay appropriate attention to safety and the environment. The UBD process must include an internal or external audit element to ensure that HSE goals are met. As previously stated, it must be part of the total process and should afford operating entities the ability to respond as necessary with a fully integrated solution. This element of the process should be designed to: (a) assist in development of the Conventional Plan, (b) establish specific operational procedures, (c) audit Conventional Plans, (d) develop the Contingency Plan, and (e) identify equipment and personnel to execute the Contingency Plan. The process includes a Strategic Event Plan (STEP)—a site-specific Contingency Plan that gives a detailed decision tree and actions to take for a host of possible events.

Specific training has been developed and implemented for the supervisors, engineers, operators, and rig crews to ensure that all personnel have a heightened awareness and thorough understanding of underbalanced drilling operations. The training at the rig is provided by the UBD supervisors and focuses on the rig crews as they are typically not familiar with the specific procedures and operations when commencing the first UBD well. While procedures should be discussed at an earlier time, they are no good unless they are implemented. These procedures should describe, step by step, each task to be performed in drilling and completing the well. There could easily be 50–75 procedures in a typical offshore project; consequently, it is beneficial to have a good document control system in place.

Regarding live training exercises, practice making connections and tripping with a live well should be done prior to drilling out the shoe. This training should include all anticipated operations, including emergency drills diverting flow to the secondary flow line to facilitate a possible rotating control device elastomer failure. Crews will also have to become very familiar with the operation, limitations, and maintenance of the rotating control device. This is the mechanical barrier, which has replaced the column of fluid, so operations will have to be rehearsed and procedures well defined and understood. When the crews are comfortable drilling with an energized wellbore, the risk of an incident is greatly reduced. Effective communication during operations is undoubtedly one of the greatest single things to preventing incidents.

Teamwork is a "modern word" that is talked about constantly and maybe ignored too often. This technology, however, requires real teamwork from many diverse disciplines. To be successful, the group must develop and function as a team, not just be a team in name only. Even something as common as making a connection requires the cooperation of at least three different operators to be successful. These people are also separated by distance and will not be in line-of-sight of each other, so to make the operation work, teamwork and communication is essential.

One of the most important considerations during the execution phase is the onsite coordination during the drilling operations to ensure full compliance with HSE regulations and policies. This onsite coordination includes:

- Reporting/communications
- Training
- Supervision
- Equipment set up and operations
- Data acquisition system (DAS) and quality control
- Troubleshooting
- Site preparation

11.3.5 Phase V—Review, Reports, Lessons Learned

After completion, a review of the entire project should be conducted with all parties as soon as possible. This is important so that lessons learned can be communicated and benefit ongoing and future operations. A formal End of Well Report should also be issued in a timely manner and include recommendations for reducing time and money and increasing safely.

The review and close out phase must include a mechanism to capture and communicate the lessons learned to the project team for future and ongoing UBD operations.

11.4 Equipment Requirements

Key components of a surface equipment package for underbalanced drilling can include:

- **Upstream Equipment**—Air compression and membrane nitrogen equipment, liquid nitrogen, gas compression and rotating control devices.
- **Downstream Equipment**—Rig assist snubbing, downhole isolation valve, three-phase or four-phase separation system, solids control equipment and data acquisition system
- **Downhole Equipment**—Pressure while drilling (PWD), non-return valves (NRV) and telemetry systems

Note that these equipment systems typically come with auxiliary flow lines, emergency shutdown (ESD) systems, pumps, metering devices, data acquisition systems, holding tanks, mud pits, flare stacks, etc.

Several equipment setups can be derived from a combination of these key equipment components—from simple to the most complex:

> **Setup 1**—represents the basic setup and uses the minimal amount of equipment, most times requiring only a rotating control device using the flow lines of the drilling rig. The combination of drilling fluid, any reservoir oil, and water is re-circulated back into the injection fluid system and pumped into the wellbore. Although often practiced in some regions of the world, this setup typically does not satisfy safety requirements if there is a risk of traversing a productive interval and receiving hydrocarbons at surface.
>
> **Setup 2**—uses an RCD plus a choke and open or closed surface-separation system.
>
> **Setup 3**—uses Setup 2 equipment, but with an upstream gas generation and/or compression system. Setup 3 will be employed when lighter drilling fluid is needed to induce reservoir influx.
>
> **Setup 4**—uses an RCD, a choke manifold system, a downstream fluid-separation package, and a geologic sampler for continuous reservoir description. The downstream separator could either be a three-phase or four-phase fluid-separation system.
>
> **Setup 5**—is required when a gasified or foam drilling fluid is needed to establish underbalanced drilling conditions. This setup employs equipment required for Setup 4 plus an upstream gas-generation and compression system. This system is typically the full UBD equipment package that is used for drilling.

Many equipment setups are possible, and best practices should always drive the design of the system to appropriately and safely handle the well's potential. This is one key point that at times has been ignored in some instances to the detriment of the safety of the operation. A typical setup for primarily addressing drilling problems would include a RCD, along with a choke and an open or closed surface-separation system at minimum. If the reservoir is extremely depleted, then an upstream gas-generation and/or compression system would be required. If the objective of the UBD operation is reservoir-focused, additional metering, a geologic sampler for continuous reservoir description, and a three-phase or four-phase fluid separation system typically would be used with an adequate data-acquisition system to capture all the surface, rig, and bottomhole data.

Some of the major equipment components are briefly described below.

11.4.1 Drill String Requirements and BHA Components

Drill string for the UBD applications does not substantially differ from the conventional drilling, except that the drillstring has to be of a premium rating. It also has to be inspected, and connections should be gas-tight. Tool joints will have to be tapered at an 18° bottleneck angle to provide a smoother transition through the RCD. There are some additions to the typical overbalanced assembly that are always required in the underbalanced assembly regardless of local traditions:

- **Underbalanced String Floats**—These are only required in the case of standpipe injection to economize on gas volume and rig time.
- **Heavy Weight Drillpipe (HWDP)**—Some companies have recently began eliminating the HWDP and running the entire assembly with drillpipe. This can work well with PDC bits but can be the cause of excessive buckling when using a rock bit.
- **Pressure While Drilling (PWD)/Logging While Drilling (LWD)/Measurement While Drilling (MWD)**—With the exception of the PWD (which should be considered an integral part of any underbalanced assembly), the selection of the telemetry system will be the hardest decision and must be engineered on a specific project basis.
- **Dual Float Sub**—A minimum of two "non-ported" drillpipe floats will be run in the bottom stand of the underbalanced assembly. It is acceptable to run more and to back these up with a retrievable or pump-down system, but there must be at least two. Quite often, these are placed directly above the mud motor so they do not interfere with the instrumentation and telemetry packages. Also, due to the length of the LWD/PWD/MWD package and the need to have these in the bottom stand, it is often necessary to place these subs above the motor.

Drilling a directional or horizontal well safely and efficiently can be successfully addressed with a downhole positive displacement motor (PDM). The downhole motor provides for effective directional, horizontal, and performance drilling while reducing drillpipe, collar and casing wear and minimizing related problems. The PDM is a natural complement to the LWD/MWD systems and surveying systems, allowing for accurate directional drilling. Constraints of a PDM, such as minimum and maximum equivalent liquid rates, are among the most critical that are addressed in the process of a flow modeling for the planning of successful UBD operations.

11.4.2 Rotating Control Device

The rotating control device (RCD) is a key piece of well control equipment and is required to enable closing the flow path to enable a controlled back-pressure application on the annulus through the choke system. The RCD generally meets industry specifications and requirements for its intended service. However, because the unit is used in actual continuous drilling and stripping operations, the RCD can not be designated as the primary well control device. That role is relegated to the blowout preventer (BOP). There are two different types of RCDs in the industry:

- **Passive Rotating Control Device**—This RCD has a sealing element created by friction, and it is wellbore pressure assisted. The passive RCD relies on an interference fit between the sealing element and the drillstring. The sealing element resides in a bowl shaped to encourage greater seal strength as weight is applied, but it is not hydraulically activated, although the pressure from the wellbore provides additional pressure sealing (see Figure 11–3).
- **Active Rotating Control Device**—This RCD has a sealing element that has a hydraulic force applied to the sealing element that creates the seal. The active RCD works like an inverted balloon, with a rubber element pushed against the drillpipe by pressure applied from behind. This arrangement requires an external power source to keep the pressure applied.

11.4.3 Four-Phase Separation System

During UBD operations, the drilling returns consist of drilling mud or fluid, drill cuttings (solids), and reservoir fluids such as water, gas and oil. The separation system is required to separate the reservoir fluids

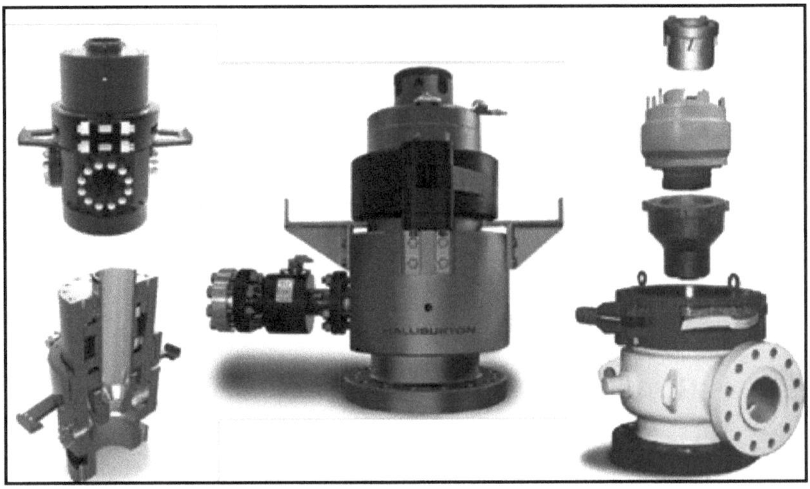

Figure 11–3 *Passive rotating control devices*

from the drilling mud and cuttings prior to returning these streams to the usual rig-provided handling systems. The separated gas and oil phases, if not used for re-injection purposes, must be stored or disposed (flared) on location. The surface separation package used in conjunction with UBD operations must be able to safely control, separate, measure, and redirect both the drilling-related and produced hydrocarbon-related fluids encountered.

The four-phase closed loop separation package is a modular, skid-mounted system that is designed to be used in UBD operations on both land and offshore locations. The separation system consists of several skid-mounted assemblies comprising the choke manifold, geologic sampler, first-stage separator, second-stage separator, pump and piping skid.

11.4.3.1 ESD Valve

The entire well return stream exits the wellbore at the BOP and enters the separation system at the emergency shut-down (ESD) valve. This fail-closed valve is the primary safety mechanism in the event of a process upset to protect downstream equipment from catastrophic failure. The ESD valve is maintained in the open position with hydraulic pressure from the ESD control panel. The hydraulic panel is controlled via the ESD Programmable Logic Controller (PLC) system or standard pneumatic ESD system. Electronic systems enable multiple checks or voting to ensure that an unsafe condition is actually occurring. Pneumatic systems do not lend themselves to the higher

Figure 11–4 *Four-phase separation system*

logic forms as easily as PLC systems that are now the standard in process plants and factories. From the ESD valve, the return stream flows through high pressure piping to the under balanced choke manifold.

11.4.3.2 Secondary Flow Line

A secondary flow line is provided inclusive of the secondary block valve and ESD valve which is connected from the BOP stack below the RCD to the UBD choke manifold. The secondary flow line provides the ability to maintain underbalanced well productivity while the RCD is being maintained or seal elements are replaced.

11.4.3.3 Choke Manifold

Primary well flow control is provided by the 5,000 psi API 16C drilling choke manifold with either a single or double block-and-bleed inlet valve to dual Power Choke drilling choke units that are hydraulically actuated. The block valve arrangement allows flow through either or both of the adjustable chokes (3 in. maximum opening), or through the 5 1/8 in. bore center bypass (belly line). In addition, either choke may be isolated via double block-and-bleed valve arrangements for inspection and maintenance.

The chokes are controlled by a remote hydraulic operating panel. This remote panel allows the operator to control both chokes at a safe location and is complete with redundant hydraulic pumps for control of this critical function.

11.4.3.4 Geologic Sampler

From the choke manifold, the total well return stream flows through six-inch piping to the geologic sampler skid. The geologic skid is complete with two-side stream sample chambers that allow continuous or alternate samples to be taken. Within the sample chambers the fluid stream is subjected to strong centrifugal forces, separating a cuttings sample for analysis. The well fluids are recombined downstream of the sample chambers and sent to the first-stage separator.

The geologic sampler incorporates a 5,000 psi to 600 psi pressure break relief valve set at 1,000 psi, providing mechanical over-pressure protection in addition to an array of ESD PLC alarms and sensors.

11.4.3.5 First-Stage Separator

The first-stage separator is a vertical 84 in. × 10.3 ft S/S ASME B&PV code, 250 psi maximum allowable working pressure (MAWP) unit complete with a 90-degree cone bottom for solids control. An advanced centrifugal inlet provides the first separation and minimizes any potential foaming. Separated gas exits the separator via an 8 in. line and is measured with a Daniel Senior orifice meter. The gas is piped to skid edge for flow to a sales line, recompression, or disposal flare.

The solids are removed with an effective sand-jetting system and exit the separator through the nozzle at the cone bottom. The flush fluid injection is measured with a turbine meter and the solids slurry exit is measured with a mass flow meter. The overall vessel process is unaffected as the flow rate of the solids slurry stream exiting the vessel is controlled to match the flush fluid injection flow rate.

The liquid stream exits the separator under liquid level control to the second-stage separator or to the downstream disposal and/or storage systems in the three-phase separation system.

The first-stage separator incorporates both a 600 psi to 150 psi pressure break relief valve, and a separator relief valve set at 285 psi and 250 psi respectively, providing mechanical over-pressure protection in addition to the array of ESD PLC alarms and sensors.

The liquid stream from the first-stage separator flows via the pump and piping skid to the second-stage separator for further separation. The total combined liquids stream exiting the first-stage separator is measured at the pump and pumping skid en-route to the second-stage separator.

To aid in low-pressure operation, a 4 × 6 × 13 in. inter-stage centrifugal pump is provided should flowing conditions warrant its use.

The pump and piping skid also contains two further 4 × 6 × 13 in. export pumps with associated measurement, liquid interface, and liquid level control systems for the second-stage liquid returns.

11.4.3.6 Second-Stage Separator

The second-stage unit is a vertical 84 in. × 10.3 ft S/S ASME B&PV code, 100 psi MAWP unit complete with a 90-degree cone bottom for solids control. An advanced centrifugal inlet provides the first separation and minimizes any potential foaming. Separated gas exits the separator via a 6 in. line and is measured by a Daniel Senior orifice meter. The gas is piped to skid edge for flow to a sales line, recompression, or disposal flare.

The second-stage separator light liquid stream exits the separator under liquid level control and is measured via the pump and piping skid. To aid in low-pressure operation, a 4 × 6 × 13 in. centrifugal pump is provided should conditions warrant its use. Dual liquid meters are available to measure this flow stream and to provide a greater range of measurement. As the application requires, this degassed light liquid stream may be sent back to the rig system as the drilling fluid, to tanks for storage, or to disposal via crude oil burners.

The second-stage separator heavy liquid stream exits the separator at the bottom and is measured via the pump and piping skid. To aid in low-pressure operation, a 4 × 6 × 13 in. centrifugal pump is provided should conditions warrant its use. Dual-liquid meters are available to measure this flow stream and to provide a greater range of measurement. As the application requires, this stream may be sent back to the rig system as the drilling fluid, to tanks for storage or further processing, or to disposal.

The second-stage separator incorporates a relief valve set at 100, 200 and 250 psi in different separation systems just to protect the downstream equipment from catastrophic failure, providing mechanical over-pressure protection in addition to the array of ESD PLC alarms and sensors.

A third stage of separation or storage can be added in the form of inexpensive atmospheric tanks for the liquid hydrocarbons or the drilling fluids.

11.4.3.7 System Protection

Overall system protection per API 14C is provided by an integrated safety system that monitors critical system and safety parameters such as pressures, levels, gas detection, H_2S, and fire. The system is staged to alarm if these parameters go outside normal expected operating levels, and it is also set to shut the ESD valve and de-energize the pumps should the measured parameters begin to approach equipment maximum allowances. Some systems are pneumatically controlled and some are electronic PLC controlled, depending on the required functionality for the job.

11.4.3.8 Make Up Gas and Flush Liquids

Make up gas and flush liquids connections are provided at various points in the system for flushing of solids in the lines and to allow the system to operate at a desired pressure.

11.4.4 Three-Phase Separation System

The advanced three-phase, closed-loop separation and well control system is designed specifically for underbalanced drilling applications where on-site oil/water separation is not required. This service is typical in dry gas wells, non-emulsified oil wells, water injection, and areas where the drilling fluid matches the reservoir fluid. The benefit of such a system is a simplified process requiring a smaller area and facilitating quicker rig-ups yet retaining minimum requirements for reservoir characterization and exhibiting all the performance requirements for gas separation when compared to the more advanced four-phase system.

11.4.5 Membrane Nitrogen System

The advanced technology membrane nitrogen generation plant comprises of the latest design enhancements in feed air compressors, membrane nitrogen generation, and nitrogen compression.

11.4.5.1 Membrane Air Separation Principles

The heart of any membrane generation unit is the air separation or membrane module. Air separation is based on the principle of selective permeation, whereby each gas constituent has a characteristic permeation rate that is a function of its ability to dissolve and diffuse through a membrane. The module in which the nitrogen and oxygen separation takes place is a cylindrical bundle of hollow fiber membranes. Each bundle contains several million fibers, each about the size of a human hair. Pressurized air enters one end of the fibers and flows to the opposite end on the module through the fiber bores.

Gas separation takes place as the pressurized air contacts the membranes. "Fast" gases such as oxygen, carbon dioxide, and water vapor quickly permeate through the fiber walls and exit as an enriched gas at the vent port on the side of the module case.

Nitrogen, a slower gas, does not permeate through the fiber as quickly under flowing conditions. It flows down the bore of the fibers and exits at the product manifold at the end of the high-pressure shell.

11.4.5.2 System Characteristics

It is completely automated and designed for unattended operation. Pushbutton controls are standard, and product purity is automatically monitored and controlled, even vented, if product purity is not maintained.

The nitrogen membrane system generates a minimum 95% nitrogen purity. The control system continually and rapidly controls purity and monitors all major system parameters. A high accuracy pressure and temperature-compensated flow meter provides instantaneous and totalized flow to the job site data acquisition system. All information is conveniently displayed on a master display panel, indicating all parameter levels and alerting the operator to system alarms and shutdown options. Process control is provided by a PLC and is easily modified in the field.

The nitrogen membrane systems are manufactured to be transported as 40 ft shipping containers and are therefore easy to transport and install, making them easily adaptable to a wide variety of process applications and locations. They are unmatched for portable applications or for installations requiring small footprints like offshore platforms.

The only moving part within the nitrogen generation membrane skid is the main backpressure control valve. The only replacement items are filter elements. All drain valves are automatically dumped and filter pressure drops are monitored to alert the operator for the need to replace loaded filter elements.

11.4.5.3 Membrane Process Description

Compressed air is fed from the feed air compressors to the air receiver, which collects and removes the bulk of the oily condensate that is entrained in the compressed air supply. The air receiver is equipped with a condensate drain line, which automatically dumps excess fluid. A moisture separator further conditions air leaving the air receiver. The primary function of the moisture separator is to remove large quantities of oily water condensate from the feed air supply.

The air is further conditioned by coarse and fine coalescing filters to remove virtually all remaining oil aerosol still present in the feed air despite the moisture separator. The coalescing filters are also equipped with a condensate drain line. Another packed bed of activated carbon pellets adsorb heavy hydrocarbon vapors that may be entrained in the feed air supply.

A PLC-controlled feed valve provides on/off control of the feed air supplied to the nitrogen membrane system. The feed valve is classified as a "fail closed" valve, which enables the nitrogen membrane system to isolate the feed air supply in the event of an instrument air supply failure.

A process heater is provided which optimizes the membrane module performance by ensuring that the feed air stream is heated and maintained at a constant temperature. Another particulate filter removes all remaining particulate matter, thus assuring that only clean, dry air reaches the membrane portion of the unit.

The actual air separation process takes place in the membrane section of the skid. For the nitrogen membrane system, eighteen high efficiency nitrogen membranes are configured in parallel, so each module output adds proportionately to the capacity of the system.

The nitrogen membrane system is equipped with a purity assurance valve. The product valve allows on-spec product gas to flow into the injection manifold. The product vent valve will vent off-spec product gas with too high or too low oxygen impurity. An onboard oxygen analyzer continuously monitors oxygen content in the nitrogen product gas and provides set point signals to the PLC.

A backflow/over pressure check valve is located downstream of the purity assurance product valve to prevent backflow of product gas should pressure on the outlet side exceed the nitrogen membrane system's product pressure.

The nitrogen membrane system makes use of an adjustable dead-band pressure to sense output product demand. When product is not withdrawn and backpressure on the membrane builds up, the system goes into standby, and the feed air valve will close. When pressure in the nitrogen manifold falls below the dead-band value, the nitrogen membrane system will automatically restart.

11.4.5.4 Feed Air Compressor

The feed-air screw compressors gather the air through an air filter at normal atmospheric pressure at ambient temperature (up to 131°F or

55°C). The air is compressed from atmospheric pressure to 350-psig (24-bar) using the Sullair two-stage compressors. A vertical cooler then cools the air to 120°F (49°C) before it is delivered through a common manifold to the nitrogen membrane unit. Each screw compressor is driven by a Caterpillar 3412E DITTA engine.

The feed-air screw compressors are constructed with the compressor and engine mounted inside their own sea can. On one end of each container is the process cooler, and at the other end is a radiator for cooling the glycol and lubrication oil used in the package. Coolers are hydraulically driven, which allows for varying fan speeds to match operating conditions. Both the radiator and cooler are mounted so they can be removed if required.

The units are built with removable doors on the ends and sides. The doors create a protective enclosure for the units to limit shipping damage while providing security when the units are idle. The doors on either end of the sea can use a drawbridge design, so they can be lowered at an angle to deflect the heated air upward and away from the packages and operators. The side doors on the sea cans are designed with louvers. The louvers are capable of providing all air requirements for the packages with the doors closed, while allowing the units to operate in severe weather. Compressed air is then delivered through stainless-steel flexible hoses to the common manifold for delivery to the membrane bundles in the nitrogen membrane unit.

11.4.6 Gas Injection Manifold

A combined nitrogen injection and standpipe bleed-down manifold is provided as a component item of the membrane nitrogen system. The manifold includes all piping, valves and metering to accept generated nitrogen from the nitrogen boosters, direct the maximum nitrogen output at up to 5,000 psi to the standpipe, accept bleed-down returns from the standpipe (making and breaking tool joints, etc.) and discharge under choke control to the separation and vent system.

11.5 Real Time Reservoir Evaluation (RTRE)

Similar to a conventional well test, the real time reservoir evaluation service (RTRE) monitors the rates, pressures, temperatures, depths and fluid properties, and utilizes this data to characterize the permeability, reservoir extent (if sufficient time), reservoir pressure, productivity index (PI), and fracture properties of the zones drilled. From comparison with conventional well testing, it has been found that the radius

Figure 11–5 *Membrane nitrogen system*

of investigation is similar to other well tests and depends on the formation properties and the time spent producing from that particular zone. If underbalanced conditions are maintained throughout the operation, it also provides an opportunity to review the undamaged reservoir potential, which in some cases, reveals productive intervals missed when drilled conventionally and a quantification of their contribution. If, per chance, the system goes overbalanced, then skin damage can also be determined using the analytical and numerical simulators developed for well testing while drilling.

Reservoir characterization while drilling methods and software development began in earnest in 1998 with the development of a model for the moving boundary condition seen while drilling, implementation and testing of the model. From the initial analytical model for a vertical well in a circular reservoir, the development grew to handling of any reservoir or wellbore geometry. Currently, both an analytical model and numerical simulator are used for reservoir characterization. Both allow the handling of a range of complexities from simple well and reservoir geometries with single-phase flow all the way through complex multilaterals in a complex fractured reservoir with multiphase flow in the reservoir.

The RTRE service begins several months prior to the operation with the engineering design and job preparation phase. At this time, data and information is requested from geologists, reservoir and

11.8 Conclusions

- The use of project management tools and techniques are essential for the planning, execution, and final review of a UBD project. Every phase of the project requires full implementation of these tools and techniques. Underbalanced drilling operations can be conducted safely if the planning phase includes hazards and effects management process techniques. The execution phase must also include site specific procedures, proper training and emergency preparedness preparations. The review and closeout phase must include a mechanism to capture and communicate the lessons learned to the project team for future and ongoing UBD operations.
- More equipment, people, and significantly different operations are involved in the underbalanced drilling project which translates into potentially higher risk. However, with the proper planning, hazard assessment and risk mitigation, a UBD operation can be a safe and profitable operation for both the operator and service companies involved.
- Different UBD setups will determine the amount of data that can be obtained from the operation, and it is important early on in a project to determine the objectives of the operation and the type of analysis required.
- To optimize the underbalanced drilling operation and maximize the knowledge gained from reservoir characterization while drilling underbalanced, input from all the specialists involved and integration of the other acquired data are necessary to cross check results.

11.9 References

Ansah, J., Shayegi, S. and Gil, I. "Optimizing Reservoir Characterization During Underbalanced Drilling: Tools, Analysis, Methods and Results," AADE-07-NTCE-42 presented at the AADE National Technical Conference and Exhibition, Houston, TX, USA, April 10–12, 2007.

Ansah, J., Shayegi, S. and Ibrahim, E. "Maximizing Reservoir Potential using Enhanced Analytical Techniques with Underbalanced Drilling," SPE 90196 presented at the SPE Annual Technical Conference and Exhibition, Houston, TX,USA, September 26–29, 2004.

Finley, D., Ansah, J., Shayegi, S., and Gil, I. "Reservoir Knowledge and Drilling—Benefits Comparison for Underbalanced and Managed Pressure Drilling Operations," SPE 104465 presented at the IADC/SPE Indian

Drilling Technology Conference and Exhibition, Mumbai, India, October 16–18, 2006.

Finley, D., Ansah, J., Shayegi, S., Gil, I. and Lovorn, R. "Reservoir Knowledge and Drilling—Benefits Comparison for Underbalanced and Managed Pressure Drilling Operations," Indonesian Petroleum Association, Thirtieth Annual Convention and Exhibition, May 2007.

Hunt, J.L. and Rester, S. "Reservoir Characterization During Underbalanced Drilling: A New Model," SPE 59743 presented at the SPE/CERI Gas Technology Symposium, Calgary, Alberta, Canada, April 3–5, 2000.

Pinkstone, H., Timms, A., McMillan, S., Doll, R. and de Vries, H. "Underbalanced Drilling of Fractured Carbonates in Northern Thailand Overcomes Conventional Drilling Problems Leading to a Major Gas Discovery," SPE 90185 presented at the SPE Annual Technical Conference and Exhibition, Houston, TX, USA, September 26–29, 2004.

Saeed, S. "Underbalanced Data Acquisition: A Real-Time Paradigm," SPE 81630 presented at the IADC/SPE Underbalanced Technology Conference and Exhibition, Houston, TX, USA, March 25–26, 2003.

Sarssam, M., Peterson, R., Ward, M., Elliott, D. and McMillan, S. "Underbalanced Drilling for Production Enhancement in the Rasau Oil Field, Brunei," SPE 85319 presented at the IADC/SPE Middle East Drilling Technology Conference and Exhibition, Abu Dhabi, UAE, October 20–22, 2003.

Schmeigel, K. "UBD Techniques Optimize Performance," *Harts E&P*, October 2005.

Williams, M., Lewis, D. and Bernard C. J. "A Safe Approach to Drilling Underbalanced Starts with Project Management," SPE 85294 presented at the SPE/IADC Middle East Drilling Technology Conference and Exhibition, Abu Dhabi, UAE, October 20–22, 2003.

Section 3
Schlumberger's Approach to Underbalanced Drilling, Engineering, Equipment and Services

Mike Tangedahl and the M-I SWACO Pressure Control Technical Experts

11.10 M-I SWACO, a Schlumberger Company—Pressure Control

The Schlumberger M-I SWACO and Dynamic Pressure Management segments provide a wide range of field-proven underbalanced drilling (UBD) services and equipment, including: engineered solutions, rotating control devices (RCD), nitrogen gas, flow equipment, fluid separation services, and fit-for-purpose pressure control equipment.

Underbalanced drilling applications and requirements can vary significantly. Depending on the operator's specific project conditions, drivers, and required drilling parameters, Schlumberger can be involved in planning and engineering activities far in advance of project commencement.

At the other end of the spectrum, the operator may only request pressure control equipment configured for the proposed operation. Multiple service options are described below, as well as the reasoning behind why certain services are best in specific situations.

Our integrated UBD services are designed to help the operator characterize the reservoir, avoid reservoir damage, maximize drill rate, and minimize non-productive drilling time. All of these benefits directly or indirectly improve the drillability of the reservoir and improve final hole quality. In the end, the solution should make sense to the operator on safety, technical and economic parameters for it to be applied successfully.

11.11 Identifying the UBD Candidate Well

The primary reason for choosing UBD as the drilling method is to prevent formation damage in the reservoir. In addition, UBD operations help drillers avoid inducing fractures that lead to lost circulation throughout the wellbore, as in cases where a depleted zone must be drilled in order to reach the production target.

A number of decision trees have been developed by various UBD experts to help guide operators in defining UBD opportunities.

For example, the tool generated by the International Association of Drilling Contractors (IADC) begins with the question of whether there is a previous history of UBD operations in the field. If the answer is yes, then the operator will want to review and analyze as much of the field data from those operations as possible. Full access to that data represents an ideal situation.

If the answer is no, then the well planning/engineering team goes forward with its own investigation based on cost and safety considerations. Several factors can come into play, including the presence of hydrocarbons, the potential for H_2S intrusion, slow drilling in hard formations, depleted reservoir conditions, reservoir protection needs, and the risk of lost circulation or differentially stuck pipe. It may also happen that the operator decides to drill the well conventionally, and during the course of drilling realizes that there is a non-existent drilling window that will force a decision in favor of drilling with losses and some flow (i.e., underbalanced).

Table 11–1 shows some of the main criteria for making decisions about UBD candidate wells.

Table 11–1 Indicators and Considerations of UBD

Indicators	Considerations
Depleted reservoir	Typically exhibit lost circulation and differential sticking problems. A consolidated formation makes an excellent UBD candidate.
Naturally fractured and vugular formation	Usually exhibit huge losses, which can increase the chance fowler control problems or lead to differential or mechanical sticking, making this type of formation a good candidate for UBD.
Hard rock formations	Usually consolidated and can sustain UBD. UBD will provide an improvement in ROP and bit life in hard rock.
Highly permeable formations	Typically exhibit lost circulation and differential sticking problems. A consolidated formation makes an excellent UBD candidate.
Formations susceptible to formation damage due to fluid invasion	Fluid invasion can be minimized or even eliminated with UBD.

The UBD engineering process has to start with data gathering and candidate screening. Candidate screening may result in a decision which is not favorable to UBD. Examples could be the presence of high levels of H_2S, or highly unstable formations. On the other hand, depleted reservoirs and tight gas reservoirs could pose a good candidate for underbalanced drilling.

The Schlumberger team can help define each UBD opportunity and safely make the most of the operational and well productivity advantages that UBD provides. A feasibility study is often the first step, and the plan goes forward based on the facts identified in the study.

11.12 Planning Phases

The engineering process is usually defined in four stages:

Phase 1—Planning involves preliminary data gathering, candidate screening, and feasibility studies. This results in a quick look at the project and allows for budgetary proposals including basic equipment and personnel set up.

Phase 2—Planning involves detailed hydraulic modeling, a write up of the well plan to be added to the client drilling plan, detailed equipment set up drawings, and process and instrumentation diagram (P&ID), operating procedures and personnel selection.

Phase 3—Execution may or may not involve engineering at site. As a minimum it will require UBD supervision at the rig location. As a standard drilling practice, Schlumberger will follow up with the client on a daily basis to take care of all safety and service quality issues and modify the well plan in pseudo real time to match client's drilling plan and well data.

Phase 4—Close-out will involve issuance of an end of well report, closure of any service quality issues outstanding, and archiving data gathered during the well.

11.13 Project Planning

After the proposed well has been validated as a good candidate, the Schlumberger team works closely with the operator to develop a workable, cost-effective plan that aligns with the project objectives and schedule. The key goal is to effectively integrate well, rig, and

UBD systems. Procedures are reviewed and optimized where possible, and training requirements are also identified for wellsite personnel.

At this stage, the Schlumberger team clarifies expectations, which allows them to determine the tasks needed to support those expectations. They work out the specific well control methods, contingency plans and best practices needed to achieve production objectives. This may entail drilling the well on paper (DWOP), and it includes the delivery of documentation related to target zones, pressure regimes and pressure-while-drilling (PWD) tool applications. The UBD team can facilitate HAZOP/HAZID exercises and report-outs if the operator does not have an established HAZOP protocol for UBD operations.

The typical work flow leading up to execution of services is shown in Figure 11–6.

11.14 Well Engineering

Once armed with a detailed understanding of the proposed UBD candidate that is shared by the operator, Schlumberger moves into modeling multiple hydraulics scenarios that account for pore pressure/fracture gradient (PP/FG) values, pressure drops across the drillstring, pressure control system components, and rig design and capacities. The key outcome of the well engineering process is the operating window. This window describes the minimum operating parameters such as flow rate, surface pressures, gas injection rates and drawdown applicable for the project. The extent and sizing of equipment supply is a direct derivative of this operating window development, and has a direct impact on the commercial proposal.

Integrating the specialized equipment into the existing rig systems is a major undertaking in UBD well design and engineering. The Schlumberger team will analyze available space, load ratings, power supply, wellhead and/or riser specifications, kick detection systems and kick response capabilities.

This is a general description of what is involved—the operator and the Schlumberger team will explore these and other aspects of the proposed operation in depth so they can identify the optimal practices and configurations for each specific well and rig.

11.14.1 Operations Planning

Underbalanced drilling systems should be tested prior to installation and integration at the rig site; this is part of the Schlumberger team work scope. After mobilization to the rig, the systems should be tested

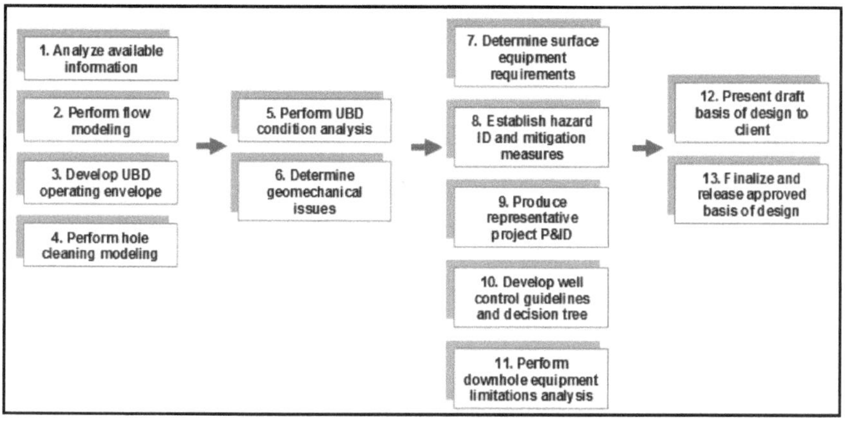

Figure 11–6 *Flow diagram for UBD planning*

again to ensure the integration is safe, effective and reliable. If needed, the Schlumberger group will furnish operations and safety manuals that are specific to the proposed well or field.

The Schlumberger service includes documenting the best locations for UBD equipment components, piping and power cable paths, tie-in points and data monitors.

Data acquisition planning is a critical piece in the UBD operation. Working with the operator, rig personnel and other service companies, the Schlumberger team will help define specific data requirements (rig site and remote), inputs for the UBD service, the various data sources and providers, data communication methods and as mentioned above, the location of data monitors at the rig site.

11.15 Typical UBD Equipment

The following are typical UBD equipment:

- Rotating control device
- One or more chokes and a manifold
- Open or closed surface separation system at a minimum

Gas-injection UBD in depleted reservoirs include:

- Compression system
- Gas generation and injection

Reservoir-focused UBD operations include:

- CARBONTRACKER gas flow meter
- Geologic sample catcher
- Four-phase fluid separation
- Data acquisition (DAQ)

11.15.1 Rotating Control Device (RCD)

The RCD is installed on top of the BOP directly onto the annular or an adapter spool. It safely diverts well returns, such as pressurized gas, fluids, and cuttings to a surface separation system during UBD applications.

The RCD is a key component of Schlumberger's managed pressure drilling (MPD) and UBD service offerings, and can also be provided as a stand-alone service as required. It includes a sealing element made of urethane and utilizes wellhead pressure to form a positive seal on the drillstring. Depending on the application, well returns are diverted through the side outlet of the RCD body to a choke manifold/surface recovery system or directly to the rig shakers.

11.15.2 Choke System

Choke systems are based on the principle of creating a restriction in the flow path of return flow from the well. This creates a back pressure on the well, which is translated downhole as a change in bottom-hole pressure. By virtue of this capability, the choke can be "opened" or "closed" to allow a variation in the resulting back pressure, changing the bottom-hole or annular pressure profile in real time. This is the basic principle applied in UBD operations.

The AUTOCHOKE has proven to be ideally suited for UBD operations. It regulates pressure automatically under all conditions with a patented shuttle system.

Figure 11–7 shows the AUTOCHOKE.

Choke operators no longer have to handle the controls during transition operations, such as mud pump start-up or shut-down, or when mud and gas are flowing alternately through the choke. The ability to automatically regulate casing pressure greatly simplifies the process of managing pressure. It should be noted, however, that the connection procedures involving multiphase operations that need drillstring bleed offs, equalization of pressures between RCD and

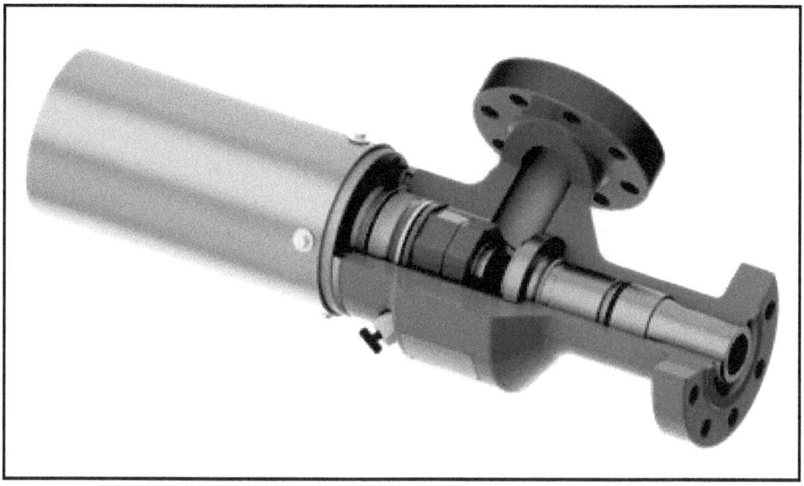

Figure 11–7 *M-I SWACO AUTOCHOKE*

stack, and stabilization of gas injection rates will still need skilled personnel at the rig site.

The AUTOCHOKE also makes stripping pipe simpler and safer by maintaining casing pressure. As the drillpipe is lowered into the hole, an equal volume of fluid is automatically displaced through the choke allowing the bottom-hole pressure to remain constant.

The standard AUTOCHOKE and console are rated for 10,000 psi and 15,000 psi service.

11.15.2.1 The Low Pressure AUTOCHOKE Console (LPAC)

The low pressure AUTOCHOKE console (LPAC) is uniquely adapted to UBD operations because they typically require less than 500 psi backpressure. Achieving precise backpressure control with a 10,000 psi choke is challenging. The LPAC meets that challenge and is widely used in UBD applications. One advantage of the LPAC system is the precise digital adjustments it can make to the programmable logic controller (PLC) remote operation; this allows real time adjustment of set point pressures on the choke, thereby making the back pressure control process safer, faster, and more precise compared to standard joystick or wheel controlled chokes.

11.15.3 Choke Manifold

The choke manifolds used for UBD operations can vary depending on the sensitivity of the well and pressure rating at the surface.

Figure 11–8 *The Schlumberger M-I SWACO LPAC control console features a touch screen pad for precise, rapid remote control from the driller's cabin*

Typical UBD manifolds consist of two choke-legs to provide redundancy and a gut line which allows bypassing the chokes as required. The manifold sizes can vary from 3–6 in. for typical land operations.

The primary concern in manifold sizing would be the maximum flow expected through the manifold as that translates in a pressure drop; higher flow rates may require bigger flow paths.

11.15.4 Mud-Gas Separator (MGS)

Entrained and produced gas must be separated and vented or flared. The Mud Gas Separator (MGS) is used to remove the free gas from the drilling fluid and is typically the first piece of equipment the drilling mud encounters when it reaches the surface.

Depending on the design, the mud gas separator vessel may be pressurized to a specific rating or held at atmospheric pressure. In addition a MGS may be designed to separate two, three, or four phases, meaning that they can separate solids and multiple types of fluid comingled in the return flow stream. During separation the drilling fluid returns have a typical retention time of 1–2 minutes in the MGS.

11.15.5 Degasser

Degassers are designed to remove virtually all entrained gases from drilling mud, including H_2S and corrosive oxygen. Degassers tend to be of two types: vacuum and centrifugal.

Vacuum degassers are more efficient degassers, but have lower throughput capability (see Figure 11–9). They are better suited to lower flow rate mud systems with high gas cuts or systems that are very sensitive to entrained gas. Vacuum degassers require a separate mud pump to operate the eductor nozzle, which can drive up the initial cost of installation.

Centrifugal degassers employ centrifugal forces to separate the gas from the fluid by exerting centrifugal force to the mud, multiplying the force acting on the gas bubbles to increase buoyancy and release. The increase in buoyancy accelerates the bubble-rise velocity. As the bubbles rise toward the surface, they escape the mud and are further broken down by flow turbulence. A light vacuum is pulled by a regenerative

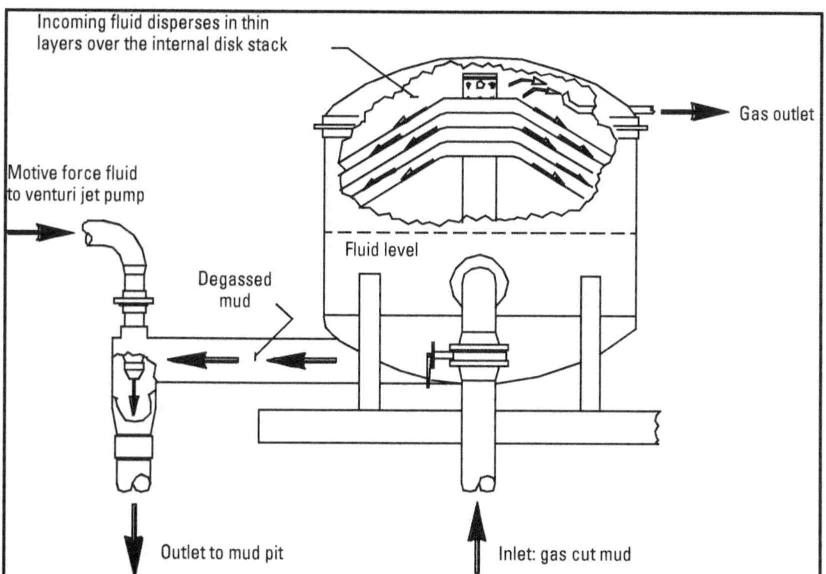

Figure 11–9 *Compact vacuum degasser*

blower to vent the gas. The freed gas and the gas-free mud are then separately discharged as desired.

11.15.6 CARBONTRACKER Flareline Gas-Flow Metering System

The gas-flow meter is particularly useful in tight-gas reservoirs while drilling underbalanced. A meter like the M-I SWACO CARBONTRACKER measures gas-flow velocity, temperature and pressure to determine the actual and standard volumetric flow, as well as mass flow rates. It is installed in the flow line between the MGS and the flare stack and is accurate at flow velocities as low as 0.1 ft/sec (0.03 m/sec). It creates no obstructions in the flare line and meets drilling well control and industry code requirements (see Figure 11–10).

During UBD operations it can help characterize tight gas reservoirs, determine the location of productive gas zones and production rates, and provide an indication of the true production potential of a reservoir. That allows operators to make accurate, more cost-effective decisions because it allows them to avoid relying solely on expensive log data to predict production zones. In addition, it can also provide quantitative data for carbon emissions reporting that the global environmental protection community will require.

11.16 Pressure Fluids Management System (PFMS)

The pressure fluids management system (PFMS) is an example of the total integration of both the surface pressure systems and the drilling fluid management system. PFMS controls surface pressures for UBD and handles, cleans, and returns the drilling fluid to the mud pumps while removing the drilled solids to a solids disposal system. PFMS is designed to take fluid returns from the rigs stack or rotating control valve while maintaining bottom-hole pressure.

Two separate sections ensure thorough processing:

Pressure control utilizes:

- A choke manifold
- Dual patented AUTOCHOKE units for precise control of well pressure, with improved wear characteristics for severe-service jobs and no plugging
- A field-proven M-I SWACO MUD/GAS SEPARATOR for removing high volumes of free gas

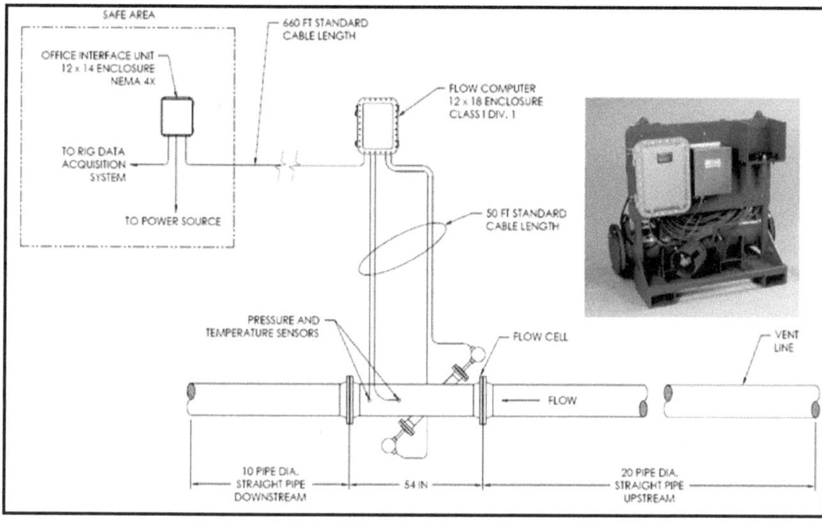

Figure 11–10 *M-I SWACO CARBONTRACKER flareline gas metering system*

- A 154-bbl, three-compartment fluid-collection tank outfitted with pumps, agitators, gun lines and transfer lines
- A vacuum D-GASSER for final removal of any gas still entrained in the fluid after solids-control processing

Solids control utilizes:

- A patented, dual-motion MONGOOSE PT shaker outfitted with patented, pre-tensioned MAGNUM XR mesh screens
- A proven-technology 4T4 D-SILTER for removing finer particles

PFMS requires two operators and a supervisor who controls the system and reports to the operator's representative.

The PFMS has a modular design and can be transported to land and offshore rigs and uploaded in just six lifts.

11.17 Membrane Nitrogen Generation Units

Schlumberger offers three different types of nitrogen units: a fully integrated mobile trailer mounted unit, a diesel skid mounted unit, and an electric drive containerized unit designed for arctic and desert operating environments.

Figure 11–11 *Pressure fluids management system—land version*

Main benefits of membrane nitrogen generation units:

- Mobility and versatility
- Use in remote locations both onshore and offshore
- Continuous supply of $N_{2,}$
- Adjustable range of N_2 purity up to 99%
- Quick set up
- Built for both desert and arctic conditions
- Electrical N_2 unit—very low noise levels

Trailer-mounted nitrogen generation units:

- Fully integrated units—ideal for remote locations
- Highly mobile—quick rig in/rig out
- Each unit comes complete with inlet and outlet metering

Diesel and electric nitrogen generation units

- Consists of:
  - 3 × feed air compressors
  - 1 × nitrogen production unit
  - 1 × booster
  - 1–2 generators

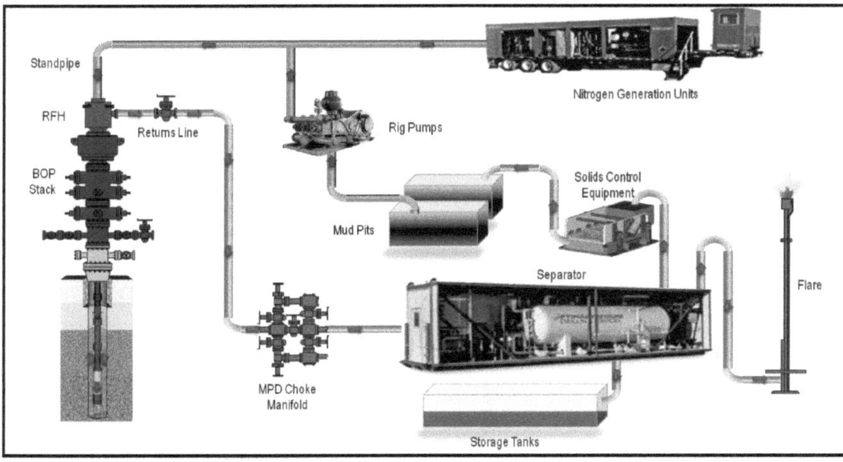

Figure 11–12 *MPD/UBD package layout*

- Available for desert and arctic conditions
- Can be modified for use on offshore applications

11.18 Total Gas Containment (TOGA) System

The total gas containment (TOGA) system is designed to contain, separate, and remove the free and entrained gases from the drilling fluid circulated from the wellbore. Total containment and removal of the gases from the fluid is accomplished through a unique combination of the MGS and the degasser.

Fluid, which may contain gas and other formation liquids or gas-cut mud, enters the mud-gas separator where the free gas breaks out to be sent to the flare line. The fluid level is controlled by the combination of the float valves in the MGS and the degasser.

The liquid phase enters the degasser where the entrained gases are freed from the fluid and sent to the flare line. The degasser is plumbed to circulate the active system if the MGS float valve should be closed.

The TOGA system can be used in UBD operations where small H_2S gas intrusions enter the wellbore. It allows the operator to continue drilling when encountering gases which will be flared.

Figure 11–13 is an example of the process flow diagram (PFD) for UBD application utilizing a total gas containment system (TOGA).

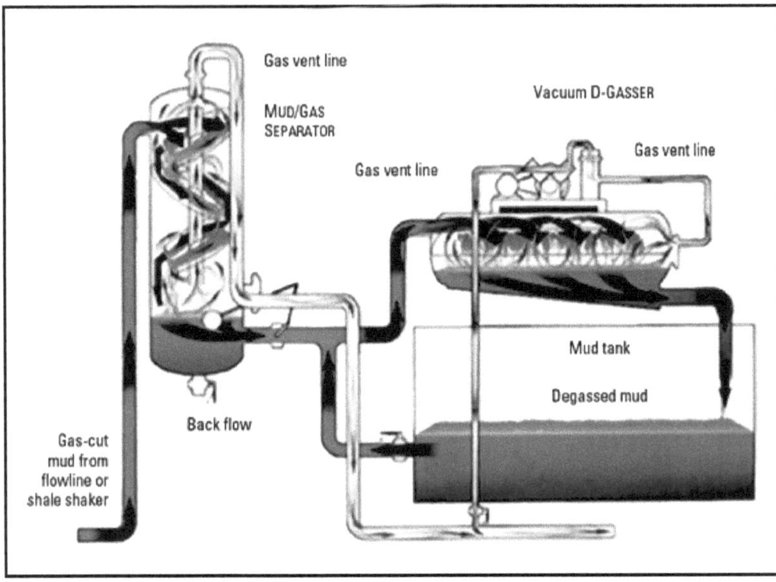

Figure 11–13 *Underbalanced drilling TOGA system*

11.19 Low-Pressure and High-Pressure Well Types

Each well presents specific characteristics, and the equipment selected to support the UBD operation reflects the well type. UBD wells can be broadly categorized into two types, but there are many variations within each category. New or improved UBD technologies can increase the economic attractiveness of proposed candidates. The drilling team has to decide when and why a specific reservoir will benefit from more complex or costly UBD techniques.

11.19.1 Low-Pressure Well

A low-pressure well can be drilled with three key pieces of equipment. Often there is no need for dedicated personnel to remain at the wellsite during drilling to manage the equipment and operations, but it is not uncommon for an RCD operator to be part of the wellsite team on any UBD operation.

These wells usually share these features: no uncertainty about formation pressure, depleted reservoir in a mature field, and no flaring restrictions. Gas and fluid measurement devices may be optional.

The basic line of equipment includes the following:

- Low-pressure rotating control device (RCD)
- Choke and choke manifold (there are several options available, from manually operated to remote controlled)
- Mud/Gas separator (can be a vertical three-phase or horizontal four-phase separator based on well flow characteristics)

A flareline gas meter can also be installed to provide real-time data on gas volume and composition.

Figure 11–14 illustrates the process flow diagram (PFD) for a typical low-pressure UBD well.

11.19.2 High-Pressure Well

A high-pressure well will require more sophisticated pressure control tools, as there may be uncertainty about the maximum pressure to be encountered. These are typically reservoirs where good production is expected and accurate gas measurement is essential. One or two skilled UBD operators will generally be assigned to each shift during drilling operations.

High pressure wells generally require additional equipment to help improve well control response and gas management. Early kick detection becomes very critical, so specialized equipment is installed to augment existing rig systems.

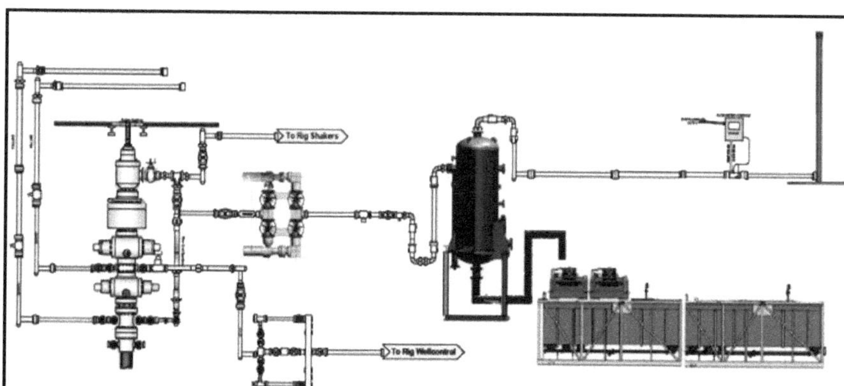

Figure 11–14 *Process flow diagram (PFD) for typical low-pressure UBD well*

The typical equipment package on this type of well would include the following:

- High-pressure RCD
- Choke manifold with a remotely operated control console
- Mud/Gas separator (can be a vertical 3-phase or horizontal 4-phase separator based on well flow characteristics)
- A flareline gas meter

In addition to the equipment listed above, a PWD tool is often included in the bottom-hole assembly to monitor real-time pressures.

Figure 11–15 shows the process flow diagram (PFD) for high pressure application using the TOGA system.

Several equipment configurations can be derived from a combination of these key equipment components—from the simplest to the most complex.

11.20 Final Word about UBD Planning and Equipment

The main purpose of UBD technology is to prevent damage to the reservoir by maintaining underbalanced conditions throughout the drilling operation or to reduce non-productive time caused by lost circulation or stuck pipe.

This discussion of UBD well planning, engineering and equipment presents the baseline options available to operators who are developing fields using UBD technologies.

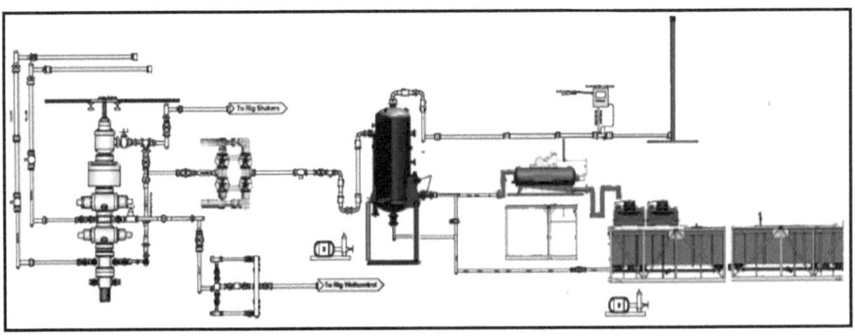

Figure 11–15 *Process flow diagram (PFD) for high pressure application using the TOGA system*

The detailed engineering provided by Schlumberger allows the operator to optimize each well design and take advantage of every opportunity for continuous improvement over a series of wells. As each well is drilled, it "talks." The real-time data acquisition tools deployed on UBD operations (e.g., CARBONTRACKER gas meter, down-hole pressure data tools) help rig site personnel adapt to conditions on the fly.

The anticipated and observed conditions on each well rightly determine the specific practices and equipment configurations required to ensure safe and efficient operations.

11.21 Questions

1. What are the characteristics of a "good" UBD candidate well?
2. What are the four main phases of the engineering process for a UBD well or project?
3. What are the key equipment components used to provide UBD services?
4. What are the reasons for choosing to drill underbalanced?
5. What is a typical work flow in the execution of UBD services?

11.22 References

M-I SWACO CARBONTRACKER Brochures, technical manuals and technical presentations.

M-I SWACO Degassers Brochures, technical manuals and technical presentations.

M-I SWACO Drilling Chokes Brochures, technical manuals and technical presentations.

M-I SWACO Low-Pressure AUTOCHOKE Brochures, technical manuals and technical presentations.

M-I SWACO Mud Gas Separators Brochures, technical manuals and technical presentations.

M-I SWACO PFMS Brochures, technical manuals and technical presentations.

M-I SWACO Pressure Control Manual.

M-I SWACO TOGA Brochures, technical manuals and technical presentations.

M-I SWACO UBD/MPD technical presentations.

Section 4
SunTerra Integrated Underbalanced Non-damaging Reservoir System

Dale Cunningham, SunTerra Oil and Gas LP

11.23 Introduction to the SunTerra System

The purpose of the integrated system is to provide the means to control the underbalance against the reservoir during drilling, trips, and completion, and to provide a drilling fluid that will cause minimum damage to the reservoir under all conditions of drilling, logging, and completion. While SunTerra undertakes reservoir evaluation as the preliminary step, this discussion is limited to the mechanical equipment and its integration into the whole underbalanced project.

This system gives the UBD drilling supervisor complete control over critical activities including the completion procedure. The SunTerra philosophy for equipment design is as follows:

- Small footprint (lessens environmental impact)
- Automation—four workers versus twelve workers per 12 hour shift (lowers personnel risk)
- Only two workers in zone 1 (lowers personnel risk)
- No workers on the drill floor (lowers personnel risk)
- Closed Circuit TV cameras monitoring all critical areas (operational safety)
- Ubits software is used to forecast drilling events which allows for ongoing drilling and HSE planning (operational safety)
- Electronic PLC control of all critical drilling functions allows for precise operations of equipment (operational safety)
- Significant noise reduction in rig operational areas versus conventional rigs (lowers personnel exposure risk)
- Non-conventional drilling fluids used (lessens environmental impact)
- Package integration ensures equipment compatibility (operational safety)

The SunTerra system has two main components, the four-phase separation system (which includes the chokes and manifold) and the solids control system.

11.24 Four Phase Separation System

Integration and automation are key advantages to any four-phase separation system for underbalanced drilling (see Figure 11–16). To lower AFE costs, this system was built with a state of the art wireless SCADA system that collects and calculates all data and operates all key components with built in operational safe guards. This automation of the operational systems reduces manpower requirements therefore reducing operating costs. The SCADA system gives onsite technicians operating the unit a complete overview of all operating parameters while drilling and is controlled through touch screen panels. At any time the technician operating the system can change any operating parameter instantly, adjusting the system to changing influx from the well. The SCADA system is WITS compatible for onsite data sharing with the drilling rig's EDR. Only two men per 12 hour shift are required to run the system which reduces the health and safety risks associated with drilling.

Integration of key components of the system is another very important factor in lowering costs and improving the efficiency for any UBD project. The four-phase system (Figure 11–16), has been integrated onto two skids which significantly reduce trucking costs. This integration also significantly decreases rigging up and rigging down time. The separation vessel has a 260 psi operating pressure, the vessel has a capacity of 170 bbls of fluid and provides the means to separate and circulate all the wellbore returns (gases, liquids, and solids). It is skid mounted, built for sour service and all weather applications. There is a 6 in. gas inlet which allows for a gas capacity of 100+ MMSCF/day. The liquid capacity is 40,000 BBL/day.

Produced gas while drilling is flared with a portable 60 ft × 6 in. anchorless flare stack that is tied into the separator outlet via 6 in. main line, 2–4 in. vent lines, 2 in. pilot line c/w Wes-Tech style nozzle, 1 in. pilot line and hydraulic lift with safety valve.

11.25 Choke and Manifold System

The choke and manifold system are integral to the SunTerra system. For ease of operation and maintenance, and to decrease the overall footprint, the chokes are situated on the same skid as the separator and are in an enclosed environment next to the separator. The system

Figure 11–16 *Four phase separation system control*

Figure 11–17 *Interior of system container*

has two 2–4 in. Willis Cage and Sleeve (C&S) – 5000# sour service automated drilling chokes (see Figure 11–18). There are multiple flow paths which allows for any situation that may arise to be handled safely during the drilling operation, including maintenance without shutting down drilling. Choke operations are controlled by the operator from set-points that are put into the control system software. Once imputed, the system will operate automatically to maintain those set-points.

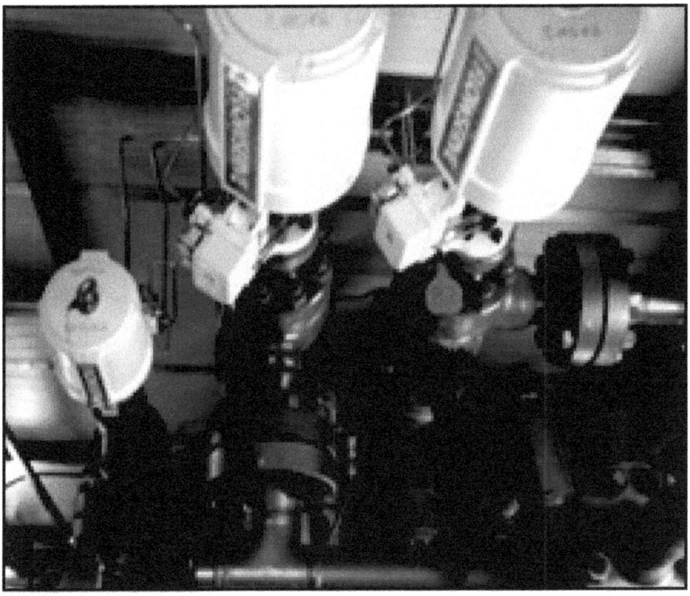

Figure 11–18 *5000# drilling chokes*

The manifold system is fully redundant and features 6 in. 5000# gate valves and a fail last positioner for the 6 in. inlet valve.

11.26 Gas, Fluids Measurement, and Geological Samples

Accurate gas and liquid measurements while drilling are critical, for plotting PIWD, fracture identification, reservoir characterization and end of well reporting. Gas rates are measured with two Rosemount 8800C Vortex gas meters (Figure 11–19) that are tied into the data acquisition and operations software, giving real-time feeds of data to the operations group onsite and office staff via satellite communications. Liquid levels are measured within the vessel with a guided wave radar ensuring accurate liquid level measurements in each compartment of the vessel. There are transparent sight-glasses for back-up.

For accurate fluid flow and identification of produced fluids Coriolis Micro Motion meters on each liquid leg coming off the vessel are tied into the data acquisition system giving instantaneous mass flow, volume flow, and density measurement for determining and reporting of all produced fluids.

Figure 11–19 *Rosemount Vortex gas meters*

11.26.1 Sample Catching

Geological sample catching (Figure 11–20) helps to determine the current formation; typically when drilling underbalanced, samples from the shaker are not adequate for this determination. A high pressure cyclone sample catcher collects drill cuttings for the geologist.

11.27 Specialty Drilling Fluids

SunTerra has worked with and considered many fluid types during the last 10+ years of drilling underbalanced wells. The preferred drilling fluid is a proprietary non-damaging mineral oil. The properties of the mineral oil allow for underbalanced conditions to be present in a normally pressured reservoir with little or no N_2 requirements. The selection of the mineral oil and its specific composition allow for hole stability and proper cuttings transport. Other important factors taken into account are:

- What is the lowest and the highest pore pressure and depth of the target production zone?
- What is the type of formation and the formation damage mechanism?

Figure 11–20 *Sample catcher*

- Reservoir characteristics of the production zone
- Bore-hole stability (geophysics/geomechanical)
- What is the potential type and anticipated rate of production?

11.28 Solids Control System and Transportation of Drilling Fluid

One of the most important design and philosophical criteria is to ensure damage is not created by fines introduction into the reservoir from drill cuttings. Preservation of the reservoir in its virgin state is paramount. A specialty drilling fluid is cleaned to where there are no cuttings present larger than a 5 micron size. SunTerra utilizes a system that was jointly designed by Kemtron and SunTerra that is specific for cleaning of the specialty mineral oil drilling fluid. The system is built into two 56 ft trailers with a capacity of 7,500 gallons. Each trailer is a dual axle rockover style with a roof and side shelters to protect the mineral oil from water contamination during rain or snow fall and to provide a safe and comfortable working environment. This also lowers the risk of possible accidents which may occur during rain and/or winter conditions. The roof and side shelters are retractable during transport to and from well sites (see Figure 11–21).

Transfer pumps on the four-phase separator skid remove drill cuttings and all produced fluids from the pressure vessel.

The cuttings and fluids are then shipped by pump to the first-solids control trailer (shaker tank) over two linear motion ktl-48 shaker units. After going over the shakers, the fluid is pumped to the centrifuges on the second solids control trailer where there are three 14 in. × 48 in. high speed kt-1448 mechanical drive centrifuge units. The centrifuges are connected in a series to enable proper cleaning of the fluid to meet SunTerra's criteria of less than a 5 micron particle size remaining in the drilling fluid. The fluid is then pumped from the second solids control trailer to three 500 bbl storage frac tanks that are connected in a series for onsite storage and further settling of any solids. The fluid flows from each frac tank to the next as each one fills up, the clean fluid from the third and last tank is then pumped back downhole, again ensuring every possible action has been taken to protect the reservoir while drilling.

To ensure that there is no possible way that any contaminants can be introduced into the drilling fluid prior to starting the drilling process, the fluid is transported in two specially designed 7,500 gallon mineral oil tanker haulers.

11.29 Compression and Nitrogen Systems

For some projects and reservoirs, an underbalanced condition must be created. In these cases, due to its inert qualities, nitrogen is generated onsite and is pumped downhole. This is required for depleted or sub-normally pressured reservoirs that are being drilled. A typical compression and N_2 system is a trailer mounted with a fully automated, integrated system capable of up to 2,000 scfm/min. The compression and nitrogen sections of the system are already fully connected, eliminating rig up and rig down time. This type of onsite nitrogen injection at a constant rate and pressure with less than 2% oxygen is preferred. Trailer mounted integrated and automated N_2 systems are preferred for their ease of mobilization, quick rig up and rig

Figure 11–21 *Covered pits*

down time, and low cost of mobilization. Typically, only two operators are required for a 24 hour operation (see Figure 11–22).

11.30 Engineering

Although this chapter is dedicated to equipment, there must be some consideration given to how we arrive at equipment design and purpose. In any project that SunTerra undertakes, engineering and preplanning cannot be overlooked. Underbalanced drilling flow simulation is an integral element in the preliminary engineering and design stages for an underbalanced operation. Several factors must be considered in the determination and analysis of the circulating pressure and hole cleaning criteria. Ideally, the resultant pressures and reservoir inflows in the friction dominated portion of the curve should be designed such that the maximum possible reservoir inflow rate will not exceed the rated capacity of the rotating control device and the surface separation system.

The objective of flow simulation is to assist in this analysis. Therefore, the main goals of modeling are to:

- Investigate the possibility of achieving a stable underbalanced or near-balanced condition downhole
- Ensure adequate annular velocities for hole cleaning

Figure 11–22 *Nitrogen unit*

- Identify operating performance of the downhole motor
- Set the available operating envelope for underbalanced operations
- Establish expected operational parameters such as injection rates, injection pressure, bottom-hole and surface annular pressures, reservoir inflow rates, etc.
- Identify the technical operating specifications and parameters for the separation equipment

SunTerra utilizes Wellflo Multi-Phase Flow Modeling software. The general modeling assumptions that we input into the software program are as follows:

- All calculations based on steady state flow behavior
- *Important*: limited composition of membrane nitrogen (95% N_2, 5% 0_2) used to determine critical properties, not simply specifying the specific gravity
- Minimum hole cleaning velocity based on liquid velocity at 200 ft/min for all cases
- Pressure drop assumed on bottom motor

- Pressure drop across remaining BHA assumed to be 50 psi
- Standpipe pressures modeled assume an on bottom pressure drop, thus higher pressure
- Linear temperature profile assumed, with constant temperature over horizontal
- Eccentricity of the pipe assumed to be 100%
- Equivalent diameter of pipe determined to compensate for tool joints of the drillpipe
- Drill pipe roughness assumed to be average bare steel (0.0018 in.)
- Bore-hole roughness assumed to be that of rough concrete (0.1 in.)

An underbalanced operating envelope (Figure 11–23) will be developed, once the flow simulations are completed, taking into account all relevant well information. The "window" allows a better understanding of the propensity for a successful underbalanced drilling operation. For example, if all the crucial underbalanced elements are achieved (UBD pressure, hole cleaning, and motor performance) and the operating envelope is large, the preliminary engineering of the UBD potential appears promising. However, if the window is critically small or non-existent, then perhaps more research, equipment utilization or casing design may be required.

11.31 General Operational Comments

11.31.1 Organization and Training

Since the goal is integration, automation and small footprint, a key philosophy on underbalanced drilling operations is to have all onsite field personnel fully trained in all aspects of all equipment functions. This allows for any one person to handle any situation that may arise during drilling. Cross-training lowers the number of personnel that are required onsite, which in turn lowers HSE risk and overall cost of operating.

11.31.2 Connections

A very important and often overlooked operational consideration is maintaining underbalanced conditions not only during drilling but while making connections. Going overbalanced during any of the drilling operation can jeopardize the success of the project and

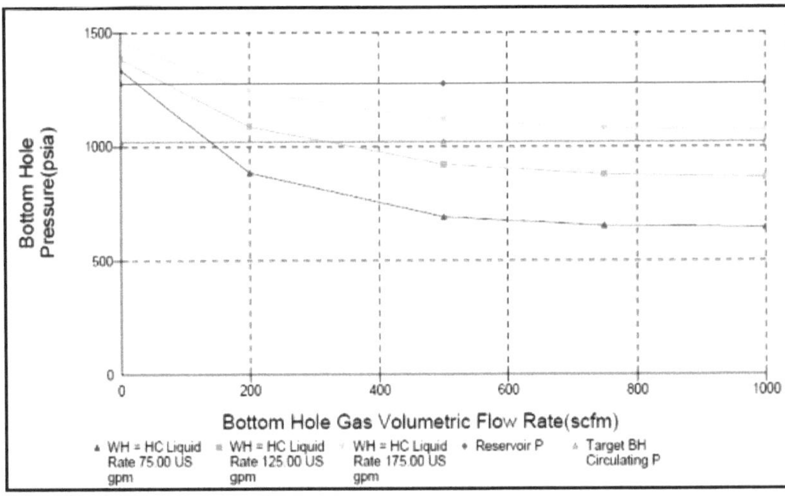

Figure 11–23 *Underbalanced operating envelope gas volumetric flow*

damage the reservoir. Therefore, it is important to consider the interruptions in the steady state of the circulating system such as that observed during drill-string connections. Constant nitrogen injection is critical to maintaining the underbalanced state while drilling, but N_2 is also required during connections and taking the proper operational steps while making a connection.

Figure 11–24 is an example of a pressure profile observed during drilling of a gas well.

From above it may be observed that during a connection, the pressure will initially drop due to the loss of the friction gradient caused from circulation. As time passes, the fluids will gradually segregate in the annulus and form a higher gradient slug at the bottom. As the connection is complete and circulation is initiated, the bottom-hole pressure will increase due to fluid acceleration overcoming inertia and the subsequent lifting of the liquid slugs. Similar pressure cycling will occur during all periods of pump-off and will be more severe while employing a high quality two-phase mixture.

11.32 Conclusion

The vision is to utilize non-damaging reservoir drilling techniques as an oil and gas well completion technology; this includes full integration and automation of equipment and cross training of all field personnel, which translates into a safe and cost effective operation. Having fewer workers onsite means less onsite housing needs to be

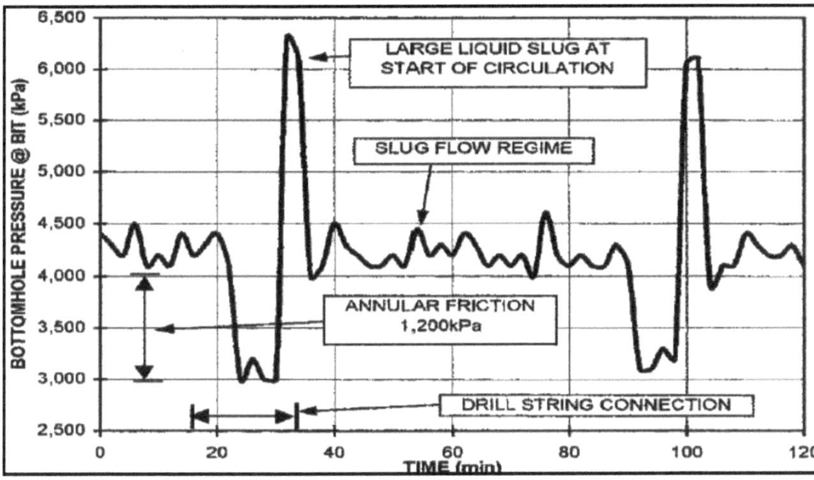

Figure 11–24 *Pressure increase on connection*

provided, which in turn means smaller lease sizes can be built, less water is required for personnel bathing, drinking, and clothes washing. Less onsite power generation is required for electricity for housing, which in turn means a smaller power plant is required and less diesel fuel is used to fuel the power plant again lowering operating costs. These are only some examples of how cross training and integration can be of great benefit to a successful underbalanced project. During each project there will be areas of operational improvement, these experiences should be compiled and reported.

Lessons learned and then reported accurately are integral to any successful project. Having a sophisticated data acquisition and measurement system is not of value unless the data is complied into an end of well report, and the report must be valuable, concise, a record of the well, a means to improve on the next operation, a means to advance the technology, and a training tool.

We start to create a final report while drilling. Charts and tables are built for sections on drilling operations, keeping a running list of challenges and events, while adding details in point form as they occur. An action list is initiated after drilling, charts, and tables are finished. Engineering discussion points should be picked, HSE statistics compiled, and available write-ups edited for inclusion. Finally, an analysis/discussion for submission should be written for peers for review.

There are many more aspects to our systems and philosophy; each contributes to a successful, safe and cost effective operation, and when combined, they are a winning solution.

Section 5
Weatherford Approach to Underbalanced Operations

Brian Grayson, Secure Drilling Services

11.33 Introduction

Underbalanced drilling (UBD) is an important technique for drilling aging fields with depleted reservoirs, narrow pressure windows and/or drilling issues. UBD improvements resulting from these operations stem from mitigating common drilling hazards that are caused by lost circulation, kick/loss scenarios, sour-gas influx or simply slow drilling through hard formation. UBD also increases reservoir productivity and maximizes net present value (NPV) by reducing formation damage and enhancing reservoir characterization

In underbalanced drilling in a reservoir, the well is designed to allow the reservoir to flow to surface while drilling. In underbalanced drilling, the wellbore pressure is maintained below the reservoir pressure at all times, and the resulting inflow from the reservoir is carefully controlled during the entire drilling process.

$$\text{Underbalanced drilling } P_{\text{reservoir}} > P_{\text{bottom-hole}} = P_{\text{hydrostatic}} + P_{\text{friction}} + P_{\text{choke}}$$

The well is still controlled by controlling the wellbore pressure, but this pressure is maintained to always be below the reservoir pressure. Primary well control is no longer an overbalanced barrier of a column of fluid, but is replaced by flow control using a combination of hydrostatic pressure, friction pressure, and surface choke pressure. The BOP stack remains as the secondary well control barrier. It must be pointed out that a UBD well operates on a single barrier.

The bottom-hole circulation pressure is a combination of hydrostatic pressure, circulation friction losses, and surface pressure applied at the choke.

The hydrostatic pressure is considered a positive pressure and is a result of the fluid density and the density contribution of any drilled cuttings and a small contribution of any gas in the well.

The friction pressure is a dynamic pressure (it changes with pumps on or off) and results from circulating friction of the fluid used.

The choke pressure arises from annular back pressure applied at surface.

These three pressures are controlled at all times and ensure that flow control is maintained while drilling underbalanced.

The bottom-hole hydrostatic head avoids the build-up of filter cake on the reservoir formation and avoids the invasion of mud and drilling solids into the formation. This helps to improve productivity of the wellbore and reduces any pressure related drilling problems.

11.33.1 Why Drill Underbalanced?

The reasons for underbalanced drilling can be broken down into three main categories:

- Minimizing pressure related drilling problems
- Reducing formation damage and enhancing productivity
- Reservoir characterization while drilling

The first reason for underbalanced drilling is often to reduce losses and to avoid pressure related drilling problems such as differentially stuck pipe and penetration rate improvements. This became known as drilling enabling UBD; it is widely used as a justification for underbalanced drilling, but it is now being replaced by managed pressure drilling techniques.

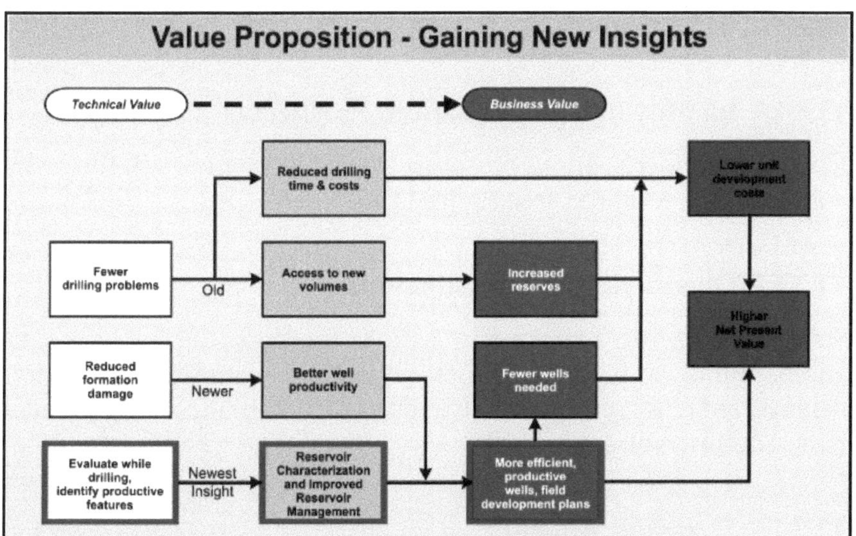

Figure 11–25 *Reasons for underbalanced drilling*

Another reason is to improve reservoir productivity by eliminating reservoir damage caused by drilling fluids and fines and filtrate migration into the reservoir formations. For this application, reduction of skin factor is the main justification for UBD.

Operators are now considering UBD to characterize the reservoirs while drilling. Productive features in the reservoir can be identified while drilling, well trajectories, and well lengths can be optimized to increase reservoir productivity and to identify potentially productive horizons in the reservoir.

11.33.2 Drilling Underbalanced

Before mobilizing or selecting equipment, it is essential that the correct reservoir candidate is selected, as well as the correct well and the correct way to drill underbalanced.

One of the complexities of underbalanced drilling is ensuring that all the issues associated with drilling and flowing a well simultaneously are understood (see Figure 11–26).

11.34 Phase One—Suitability

SURE (suitable underbalanced reservoir evaluation) in Phase 1 is a method of identifying, quantifying, and economically analyzing an underbalanced prospect.

11.34.1 Identifying Underbalanced Prospects

Identifying underbalanced prospects for suitability through the reservoir screening tool, RST (see Figure 11–27).

11.34.2 Quantifying Potential Results

SURE provides the analysis that defines the potential for underbalanced drilling and the processes to effectively mitigate formation damage, reducing non-productive time (NPT) and improving productivity and reservoir knowledge.

11.34.3 Economic Analysis

NPV is estimated on both a well-to-well and field-to-field basis.

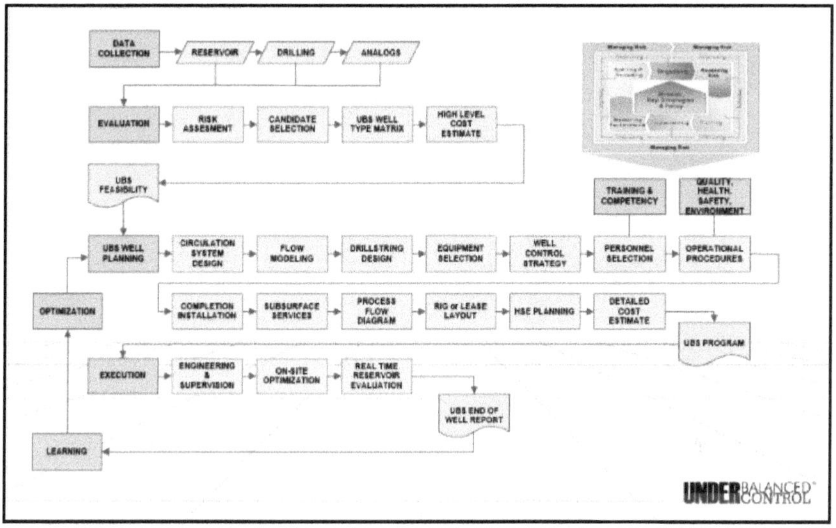

Figure 11–26 *Underbalanced drilling road map*

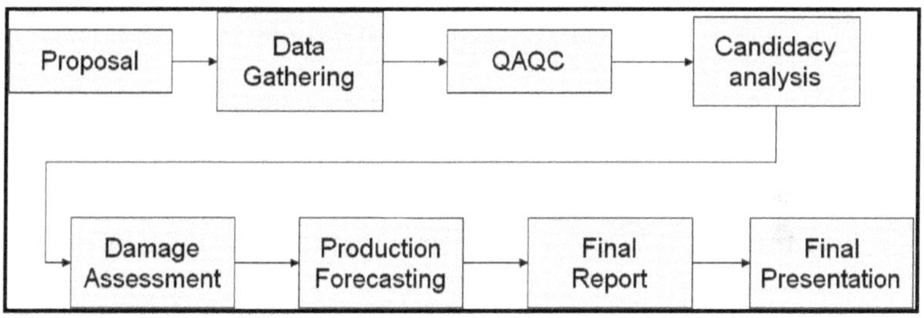

Figure 11–27 *SURE ANALYSIS (Weatherford U.S., LP)*

11.35 Phase Two—In Depth Analysis

In Phase 2, Secure Drilling services include adaptive drilling processes that enable a more precise control of wellbore pressures by using engineered equipment and procedures to provide a plan as well as a reduction in avoidable costs (see Figure 11–28).

Phase 2 is an in-depth assessment of drilling processes that includes:

1. Formation damage evaluation viareservoir damage analysis (RDA)
2. Drilling performance evaluation including NPT via risk-based drilling and cost simulation software

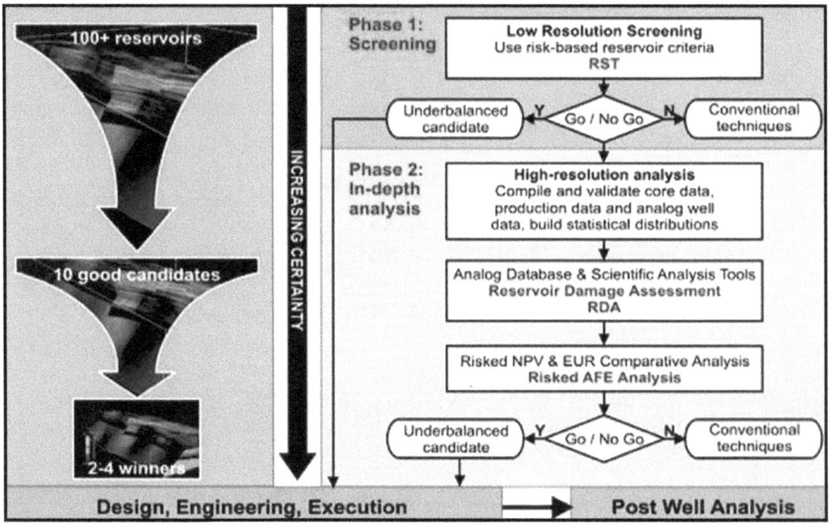

Figure 11–28 *Phase 2, in-depth analysis (Weatherford U.S., LP)*

3. Production forecasting, via RDA software or third-party software

4. Economic evaluation, via RDA software or third-party software

5. To round out the planning, UBD should also be used for real-time reservoir characterization. Continuous flow-rate data and bottom-hole pressure data obtained during UBD provide valuable reservoir information

Mathematical models and associated interpretation methods are used to estimate the permeability and pressure profile along the wellbore. Well testing theories (variable rate) are used to interpret the production associated with the drawdown maintained throughout UBD. Reservoir characterization from UBD operations can be accomplished through two techniques:

Qualitative—provides a rough estimation of reservoir features, such as fractures/high-permeability zones and/or overbalanced pulses, by productivity index while drilling (PIWD) (see Figure 11–29).

Quantitative—offers an in-depth analysis involving more reservoir information by using transient well and reservoir

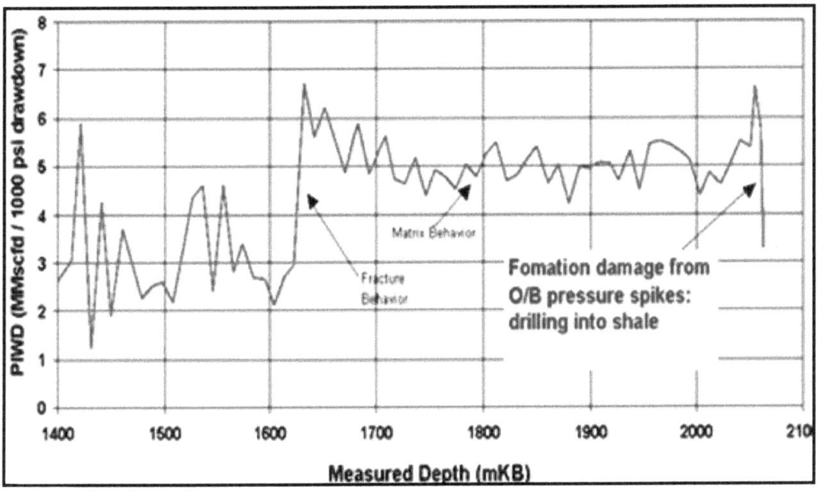

Figure 11–29 *Reservoir characteristics while drilling (Weatherford U.S., LP)*

models. When UBD is used for reservoir characterization, you can obtain:

- production interval selection,
- optimal well length,
- permeability and pressure distribution profile, and
- elimination or reduction of well testing costs.

11.36 Supervision and Monitoring of the Process

UBD ranges from reasonably simple operations, such as gas drilling or single phase liquid systems, to much more complex operations with foam and gaseated systems. Complexity generally increases with an increase in mechanical equipment and may also increase if reservoir tests are to be taken during the drilling process. The job of the onsite well supervisor can be brought into reasonable balance, allowing him to make operating decisions and keep up with the "paperwork" requirements with addition of an underbalanced operations supervisor who is available to control underbalanced operations. The UB supervisor, (like the toolpusher who controls the rig personnel and operations), controls the underbalanced operational procedures and underbalanced technicians.

A full blown underbalanced system may require up to 22 extra personnel (Nas, 2006) to successfully operate and maintain all the extra rental equipment. The notation below is for a 24 hour continuous operation allowing for two shifts with 11 personnel on each shift.

	Supervision and Monitoring	
	Supervisor (2)	
	UBD Engineer (2)	
Separation Supervisor (2)		Compressor Supervisor (2)
Separation Operators (4)		Compressor Operator (2)
DDV/RCD Operator (2)		Nitrogen Specialist (2)
Data Operator (2)		Mechanic (2)

For various forms of UBD, the job requirements may change and some fewer people may be required, but the point is that the operations and logistics of such an operation require specialized knowledge and the ability to control the minute to minute operations.

On remote jobs requiring complex sets of equipment, an equipment supervisor may be required on site or onshore to deal with logistics and maintenance.

11.37 Physical Resources to Deal with the Process

All of the following will require variations on a basic equipment package of RCD (rotating control device), chokes and manifold, separator, and piping. Often UB procedures or methods are changed during the course of drilling, and the initial equipment setup needs to be flexible enough to change. For example, from gaseated to pressurized mud cap drilling and back or from gas drilling to foam. The complexity is further increased with marine applications due to limited space, deck load, and special safety concerns.

11.37.1 Single-Phase Fluid

Reservoir pressure gradients greater than about 3.9 psi/ft (24 kPa/m) could be drilled with a single-phase fluid, either oil or water. If a suitable fluid is available, single-phase will greatly simplify the equipment and procedures.

11.37.2 Gaseated Fluid

A gaseated (two-phase) fluid is more difficult to control and requires a source of gas, generally nitrogen or natural gas. It has the advantage of a much lower bottom-hole pressure gradient than a single-phase fluid and can use almost any suitable liquid as the continuous phase. In many reservoir applications, nitrogen and oil provide suitable reservoir protection (see Figure 11–30).

11.37.3 Foam

Foam, either the common water base system or the more exotic oil base foam, provides a very stable (surge free) environment that has the lowest density of all the liquid continuous two-phase systems.

11.37.4 Gas

Gas drilling, which may use air, natural gas or nitrogen, is the ultimate underbalanced system. Where gas can be used as the drilling fluid, the drilling rates are significantly higher than with any other drilling fluid and reservoir identification and analysis are simple and inexpensive.

11.37.5 Mud Cap or Pressurized Mud Cap Drilling

In zones of extreme lost circulation where water is available, pressurized mud cap drilling is often the simplest solution to total lost returns.

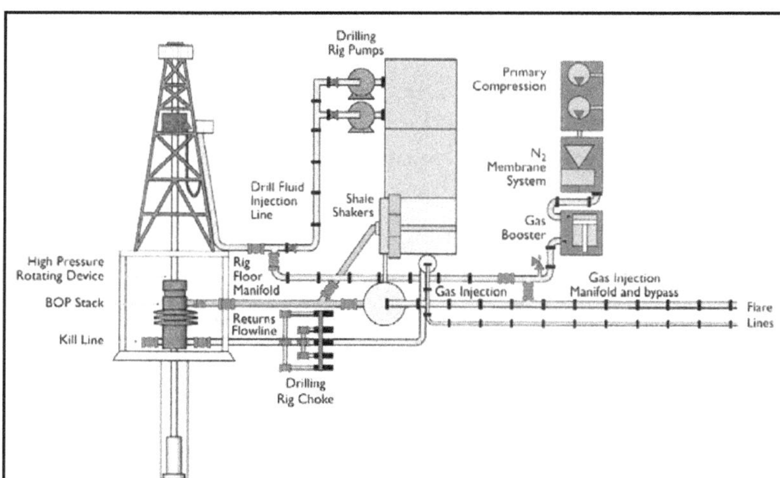

Figure 11–30 *Air drilling/gaseated drilling setup (Weatherford U.S., LP)*

11.38 Special Mechanical Wellbore Approaches

There are a number of ways to configure the wellbore to improve the potential for success in UBD. These techniques are discussed at length in other chapters of this book and in other manuals.

11.38.1 Parasite String

The parasite string consists of one or two strings of tubing or coiled tubing run with (and alongside of) the surface pipe and tied into it at the bottom to allow gas or fluid injection into the upper part of the hole without having to pump it down through the drillstring. This requires a larger surface hole and a modified well head, and tubing to casing attachments. Gas, air, or nitrogen equipment will be required to gasify the mud.

11.38.2 Dual Casing String for Annular Injection

Annular injection through a concentric casing string is commonly used in onshore and offshore projects. The liner is then tied back to surface using a modified tubing hanger to suspend the tie back string.

Gas is injected in the casing liner annulus to facilitate the draw-down required during the drilling operation. The tie back string is then pulled prior to installation of the final completion. The alternative is for an older well to have a completion in place incorporating gas lift mandrel pockets. These can be set up to provide correct bottom-hole pressures during the drilling operation.

The main advantages of using the annulus to introduce gas into the system is that the gas injection can be continued during connections, creating a more stable bottom-hole pressure. This also has the advantage that a single-phase fluid is pumped down the drillstring, allowing for conventional MWD tools to be used in their preferred environment.

11.39 Special Equipment

Special underbalanced drilling equipment has been described in great detail in several other references as well as in Weatherford literature. This paragraph is a listing of some specialized Weatherford equipment with a short comment on its value and use.

Figure 11–31 *Land based RCD systems (Weatherford U.S., LP)*

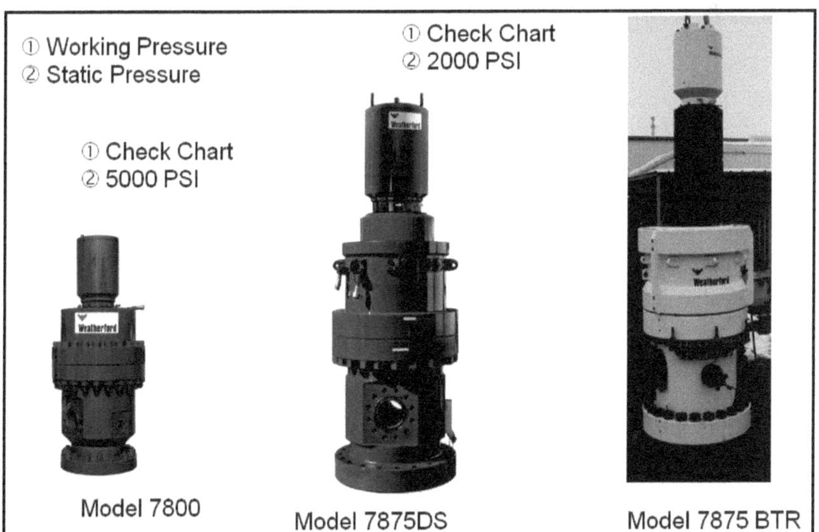

Figure 11–32 *Marine based RCD systems (Weatherford U.S., LP)*

11.39.1 Rotating Control Device (RCD)

The rotating control device (RCD) is the key piece of equipment for all underbalanced operations. The principle use of the RCD is to provide an effective annular seal around the drillpipe during drilling and tripping operations and forces the return fluid down the flow line or choke line. The annular seal must be effective over a wide range of pressures and for a variety of equipment sizes and operational procedures. The RCD system achieves this by packing off around the drillpipe. The RCD comprises of a pressure-containing housing where packer elements are supported between roller bearings and isolated by mechanical seals.

11.39.2 Retrievable Non Return Valve (NRV)

All UBD operations require a float or NRV in the drillpipe for making connections. In scenarios where a two-phase fluid design is used, the use of additional NRV can reduce the bleed off time when making connections, thus improving overall efficiency. The use of a Weatherford Wireline Retrievable NRV also provides a means of replacing the NRV should a problem occur as well as allowing access to the drillstring (see Figure 11–33).

11.39.3 Down-Hole Deployment Valve (DDV)

The DDV is a key piece of equipment used for the drilling and completion of a UBD well (see Figure 11–34). It is a down-hole isolation valve that when closed isolates the formation pressure to facilitate the removal and installation of bottom-hole assemblies while eliminating pressure at the surface. It effectively turns an interval of the casing into a sub-surface lubricator which results in increased safety, reduced tripping times and installation options while maintaining a constant bottom-hole pressure.

11.39.4 Nitrogen Generation

Nitrogen generation units are available for both land and marine use. Weatherford nitrogen generators are locally available in many oil fields and are the standard equipment for most drilling operations.

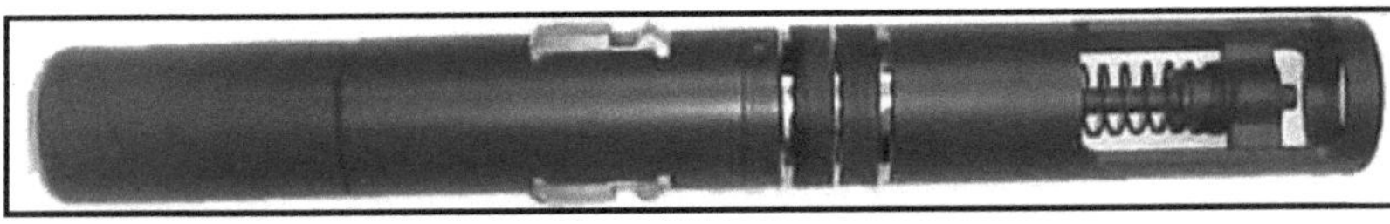

Figure 11–33 *Wireline retrievable non return valve (NRV) (Weatherford U.S., LP)*

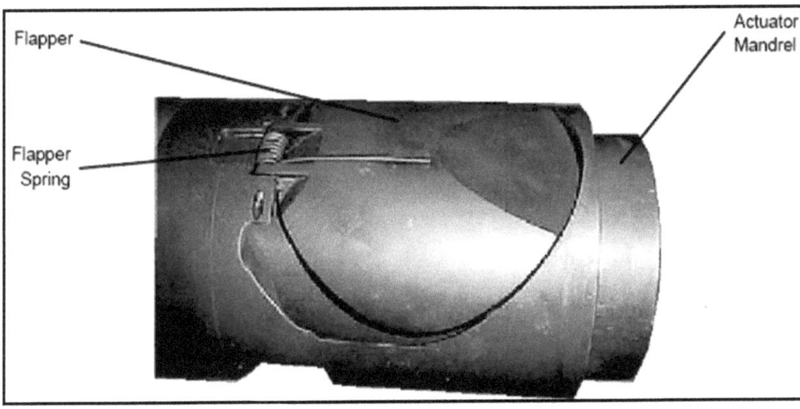

Figure 11–34 *Drilling deployment valve (DDV) (Weatherford U.S., LP)*

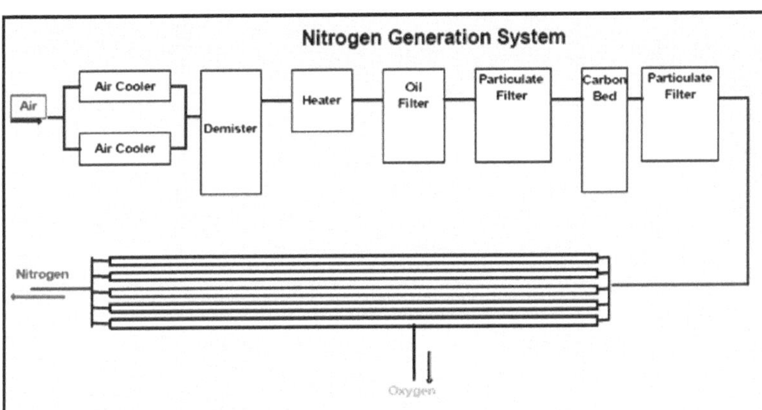

Figure 11–35 *Schematic of nitrogen generation vessel (Weatherford U.S., LP)*

11.39.5 Drilling Gas Analysis

The GC-TRACER advances normal mud logging gas detection into a new era of repeatable information about reservoir gas contents and volumes. The GC-TRACER is a fast cycle membrane chromatograph-type gas analyzer. The basic principle of permeation of gas though a membrane is common to oil field chromatographs but is done on a much faster cycle. The problem of gas in the drilling fluid was resolved by the concept that the amount of gas dissolved in a liquid is directly proportional to the partial pressure of gas above the liquid surface.

The result of this and other advanced technologies is that the gas analysis can be tied to depth to estimate fluid density, rock permeability, biodegradation, maturity grade, fractionating grade, gas leakage, and thermal BTU (Ortez, 2010).

The use of advanced drill gas analysis provides an important new tool for reservoir analysis during the early stages of drilling.

11.40 Data Acquisition While Drilling

Data acquisition packages are a fundamental part of underbalanced drilling systems and fulfill several key roles (see Figure 11–36). A specifically well designed data acquisition package should:

- Enable monitoring of all surface equipment being used at the well site, allowing the entire process to be monitored from a single operator station. This enables well-informed decisions to be made about the entire process
- Feature an alarming system to alert the operator to changes in process conditions that are outside of the expected operating parameters, and alert to potential safety events
- Be able to incorporate from other well site service providers data that enables the operator to make informed decisions regarding down-hole conditions, for example bit depth from the rig and PWD data from the MWD provider

Once data is acquired by the data acquisition package, additional tools are required for working with the data, specifically:

- Report generation software that can generate "daily" and "end of job" reports that provide an overview of events and the process throughout the day or job
- Feature tools for extracting data in a format that is easily imported into other software packages for reservoir analysis

The design of an underbalanced drilling data acquisition package should be:

- Certified for installation in hazardous environments to ensure the safety of all personnel at location
- Designed in modular fashion to accommodate different equipment required for different underbalanced operations
- Designed to withstand the rigors of transport to and from well site
- Designed to be easily installed once it arrives at well site in a minimal amount of time
- Acquire data at a high enough sampling rate that data lends itself to analysis (ideally one data point every ten seconds or less)

At a bare minimum, an underbalanced data acquisition package needs to monitor the following parameters in order to enable reservoir analysis:

- gas injection rate/volume
- liquid injection rate/volume
- injection pressure
- wellhead/choke upstream pressure
- vessel/choke downstream pressure
- gas out rate/volume
- liquid out rate/volume

A well designed data acquisition system is a crucial component of any underbalanced drilling package enabling greater efficiency, understanding and safety of the process.

11.41 Final Report

The final report form fits in the overall analysis according to Figure 11–27. The final report needs to be timely, and this is especially important if there are to be other wells drilled in the series. While final reports can be expected to contain the detailed data sets, it is particularly important to provide a short, clear summary of the plan versus actual operations with a straight-forward explanation of reasons for deviations from the plan and the results of the deviation.

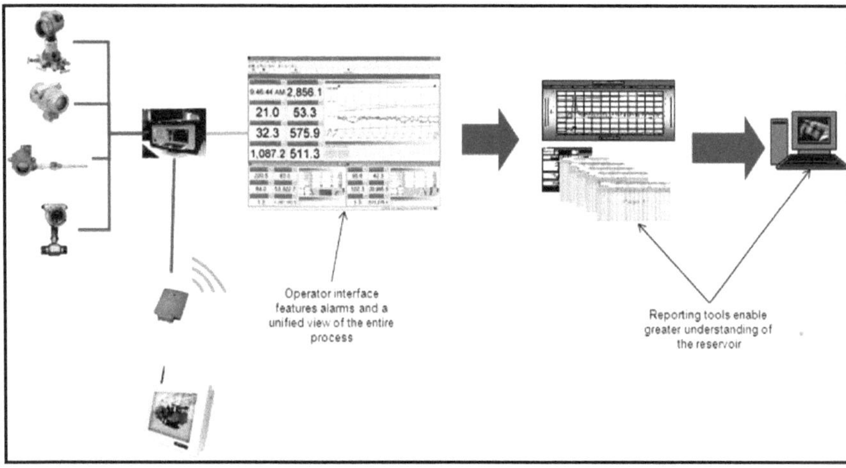

Figure 11–36 *Data acquisition systems enable greater efficiency, understanding and safety (Weatherford U.S., LP)*

11.42 Conclusion

The very use of underbalanced techniques pre-supposes some particular problem or event that needs to be controlled. Weatherford provides planning suitable for the project, whether simple or complex, that can encompass planning, operations, supervision, and reservoir damage control and analysis. Underbalanced drilling is more than just a collection of equipment to modify the mud density during a drilling operation.

11.43 References

Bennion, D. B., Lunan, B. and Saponja, J. "Underbalanced Drilling and Completion Operations to Minimize Formation Damage—Reservoir Screening Criteria for Optimum Application," *The Journal of Canadian Petroleum Technology*, 37, No. 9, September 1998.

Boutalbi, S., van der Werken, T. and Kimery, D. "Reservoir Screening for Horizontal Underbalanced Drilling Candidacy," IPTC 10966, presented at the International Petroleum Technology Conference, Doha, Qatar, November 21–23, 2005.

Nas, S. *Introduction to Underbalanced Drilling*, Weatherford Internal Publication, Ref APR-WUBS-WFT-001, 2006.

Ortez, D. "GC-TRACER MPD Overview and Application in Deepwater and Shale Plays," SPE Drilling Symposium Lecture at the Petroleum Club, Houston, TX, USA, April 22, 2010.

CHAPTER 12

Flaring

Olavo Cunha Leite, Flare Industries LLC

12.1 Editor's General Comment

The following material and the mathematical section (Section 2, page 566) are a discussion of the basic principles of flaring. This discussion is primarily for the onshore operation. Offshore flares are commercially designed and their operation is set forth in rig specific operations manual.

Flares on the land based drill rig will vary from a gas drilling horizontal blooie line with a pilot flame consisting of a five gallon pail burning diesel oil-soaked rags to complex commercial horizontal or vertical flares. Few engineers and rig supervisors in the drilling industry have any technical knowledge of flaring. For this reason, Chapter 12 is a general discussion followed by Section 2, a series of mathematical solutions.

12.2 Introduction

Flaring has become more complicated and safer than just lighting up the waste gas. Operators and regulators are concerned about efficiency. People become more involved on the safety matters and also more concerned about emissions of pollutants, besides smoke, noise, glare, and odor. OSHA and the EPA have become more active resulting in tighter regulations on both safety and emissions control. A flare system basically is safety equipment and if properly designed, it works also as an emission control system with more than 98% combustion efficiency. When burning heavier hydrocarbons, smokeless combustion can be achieved using steam (a normal plant technique) or air assisted (induced and forced draft) flare tips.

12.3 Safety

The primary function of a flare is to safely dispose flammable, toxic, or corrosive vapors by converting them into less objectionable products by combustion. Toxic limits are the greatest concentration of a poisonous substance that can be tolerated in the air for a specific length of time without danger.

Atmospheric discharge of hydrocarbons or other flammables should be designed to avoid the formation of flammable mixtures, exposure of personnel to toxic or corrosive vapors at grade level or on elevated structures. Either elevated flares or ground flares can efficiently accomplish the discharges to atmosphere, when properly designed, based on the characteristics of the flare gas, heat radiation, noise levels, smoke and atmospheric dispersion. Flare stack height and location should also be considered, based on gas dispersion if the flame is extinguished.

12.3.1 Heat Radiation

Heat radiation affects humans and equipment when exposed during a period of time. With a heat radiation intensity of 2,000 BTU/hr/ft^2 (6.3 KW/hr/m^2), the pain threshold is reached in about eight seconds and blistering occurs in twenty seconds. Under emergency conditions, the total exposure period is about eight to ten seconds. That is three to five seconds reaction time plus five seconds for escaping to a safe area. Appropriate clothing extends the exposure time.

Equipment exposure to heat should be evaluated to prevent damages on heat sensitive materials, on flammable vapor areas, and on electrical equipment. Usually, the maximum heat intensity allowed on equipments is 3,000 BTU/hr/ft^2 (9.46 KW/hr/m^2). The effect of heat radiation on the equipment results in an increase of temperature with the exposure time. The equilibrium temperature is reached with an incidence more than 40 minutes.

12.3.2 Dispersion

Dispersion analysis can cover hydrocarbon combustion under both flame out and flaring conditions.

Under flame out conditions an ambient temperature level of 15°C (27°F) above the maximum expected ambient temperature should be used for the plume rise calculations. Under flaring conditions, ambient temperature is negligible when comparing with the flame (products of combustion) temperature. Typically flaring conditions produce negligible values for the ground level concentrations.

Table 12–1 Radiation Intensity

British Thermal Units per Hour per Square Foot	Kilowatts per Square Meter	Pain Threshold (seconds)
550	1.74	60
740	2.33	40
920	2.90	30
1,500	4.73	16
2,200	6.94	9
3,000	9.46	6
3,700	11.67	4
6,300	19.87	2

Maximum flow rate and heat release from the project specifications are used and data points for ground level concentrations at a distance from the flare for vented and flared gas are calculated. These maximum concentrations are reflective of the total gas through-put.

12.3.3 Modeling Dispersion

US EPA developed and recommended software "SCREEN 3" can be used for basic modeling. It uses a range of wind speeds for calculations and selects the worst case wind velocity for each point to maximize the ground level concentration. A few comments on this software:

- **Carbon Dioxide (CO_2)**: The CO_2 emissions are based on a 100% conversion of all hydrocarbons.
- **Unburned Hydrocarbons (UHC)**: A combustion efficiency of 98% is assumed under flaring conditions, consequently 2% of the waste gas is not converted to flue gas. Unburned hydrocarbons in the resulting flue gas will become a minor percentage of the total flue gas.
- **Nitrogen Oxides (NOx)**: The expected emissions of nitrogen oxides from combustion at the flare tip are 0.068 lb/MMBTU as per US EPA AP-42.
- **Sulfur Compounds (SOx)**: In flame out cases for any sulfur compounds, such as hydrogen sulfide (H_2S) present in the gas stream, ground level concentrations will be the same fraction of the total concentration as on the original gas stream. Under

flaring conditions, if sulfur compounds such as H_2S are present, the SOx ground level concentrations will be the same fraction as on the products of combustion. 100% conversion to SOx is assumed during flaring conditions to provide for the worst case scenario.

- **Carbon Monoxide (CO)**: The expected emissions of carbon monoxide from combustion at the flare tip are 0.37 lb/MMBTU as per US EPA AP-42.

12.4 Types of Flares

Elevated flares are primarily used to safely dispose large quantities of combustible gases. The flared gases are injected into the atmosphere using a tip designed to provide a stable flame in high cross winds with a very high turndown. (A "turndown" is a reduction in fuel or gas volume).

Ground flares are used when concealing the flame is required; otherwise the elevated flare is the common choice because it handles larger flow releases more economically.

12.5 Types of Flared Gases

The flared gases can be divided into three types:

- Low heating value gases
- Medium heating value gases, like some mixed well gases
- High heating value gases, like typical drilling gases and refinery gases

Table 12–2 Maximum Grade Level Concentration (Cm)

Threshold limit values (E.P.A.)		Typical odor threshold values	
GAS	PPM (Vol.)	GAS	PPM (Vol.)
Carbon Monoxide	9	Carbon Monoxide	50
Hydrogen Sulfide	0.33	Hydrogen Sulfide	0.0047
Nitrogen Dioxide	0.05		
Sulfur Dioxide	0.5	Sulfur Dioxide	0.47

12.6 Smokeless Flaring

The method of air entrainment changes the smoke behavior of the flare combustion. Flares without smoke suppression systems are known as non-smokeless, and are used to burn non-sooty gases at normal flow (This is typical of underbalanced drilling flares). Smokeless flares use smoke suppression systems like steam injection, forced draft air fan, high pressure gas injection, and other devices to reduce the smoking tendency of heavier fuels by improving air entrainment and mixing. Steam injection is the most used technique with plant operations. It is not used in land drilling operations and will not be further discussed.

The flame of a conventional flare is a diffusion type that occurs on ignition of a fuel jet issuing into air (see Figure 12–1). The other type is the aerated flame which occurs when fuel and air are premixed before combustion. Common examples of an aerated flame would be an acetylene torch or a Bunsen burner.

Figure 12–1 *Drilling rig with diffusion type flare*

12.7 Limits and Cautions

When using a flare, attention should be given to the following:

- Safety and reliable operation from minimum to maximum flows
- Prevention of air ingress into the stack
- Smokeless combustion at normal operating conditions
- Good separation of liquid droplet to prevent liquid flame carryover (burning droplets can hit the ground)
- Properly designed stack height to prevent excessive heat radiation and concentration of hazardous elements below the threshold at ground level

Fuel properties affect the flare operation as follows:

- Gases must be within the flammability limits to burn.
- Gases must be at auto-ignition temperature and proper mixture to burn.
- Gases must have heating value enough to generate flame stability and low emissions.
- The gases carbon/hydrogen weight ratio is related with the propensity to smoke (see Table 12–3).

Nitrogen injection for gaseated, foam, and gas drilling operations cools the flame and causes a problem with the fuel properties. Flares will not burn or are unstable.

Table 12–3 Flame Colors at Different Temperatures

Color	Celsius	Fahrenheit
Dull Red	600–800°C	1,100–1,500°F
Bright Cherry Red	899–1,000°C	1,500–1,800°F
Orange	1,100–1,200°C	1,800–2,200°F
Bright Yellow	1,200–1,400°C	2,200–2,550°F
White	1,400–1,600°C	2,550–2,900°F

12.8 Combustion Principles

Combustion may be defined as the rapid chemical reaction of oxygen with combustible elements of a fuel, resulting in heat release. Hydrocarbons are chemical compounds of carbon and hydrogen, and their combustion results in carbon dioxide and water vapor. Carbon, hydrogen and sulfur are the pure elements. The oxygen comes from the air, which is 21% oxygen, 78% nitrogen and other inert gases, by volume, (or 23% and 76% respectively, by weight). Complete combustion is the combination of fuel with oxygen without fuel left over requiring time, turbulence, and temperature high enough to ignite all the combustible elements.

Ignition temperature is reached when more heat is generated by the chemical reaction than is lost to surroundings and combustion becomes self-sustaining. The temperature of a flame depends on the type of fuel, starvation or excess air, and initial temperatures of both fuel and air. Maximum temperature is reached when a fuel is burning at stoichiometric conditions.

Flammability limits are the lean and rich mixture of a fuel with air or oxygen beyond which practical combustion is impossible. Typically a fully mixed (unstaged) flame is unstable when the combustion temperature is below 2,200°F (1,200°C). Assuming a good mixing, even without inert gases, and using methane as fuel, the theoretical flame temperature correspondent to the lower limit of flammability of 5.4% methane/air is 2,066°F (1,130°C). Adding inert gases (wellsite injected nitrogen) will further cool the flame temperature and will reduce the flame propagation velocity.

12.8.1 Stoichiometric Conditions

Stoichiometric air or theoretical air is the exact amount of air required to provide the right amount of oxygen for complete combustion. The amount of air required for stoichiometric combustion is fairly constant on the air/gas weight ratio, with an approximate value of 16. This results also on a fairly constant net value of each 96 BTU of gas net heat release requiring 1 SCF of air, as shown on Table 12–4, or as general estimate under stoichiometric conditions it takes more than 10 volume of air to completely burn one volume of a natural gas normally flared during a drilling operation. On a weight basis, the air to gas ratio becomes about 16 (see Table 12–4).

The heating or calorific value of a fuel can be determined experimentally in a calorimeter or from its chemical analysis. The high or gross heating value (HHV) is the total heat released of a perfect mixture of fuel

Table 12–4 Heat (BTU) /Stoichiometric Air Required (SAR)

	Formula	HHV, BTU/LB	LHV, BTU/LB	WGT AIR/GAS	Gross BTU/SAR, SCF	NET BTU/SAR, SCF
Paraffins						
Methane	CH_4	23,879	21,520	17.27	105.7	95.3
Ethane	C_2H_6	22,320	20,432	16.12	105.9	96.9
Propane	C_3H_8	21,661	19,944	15.70	105.5	97.1
Butane	C_4H_{10}	21,308	19,680	15.49	105.2	97.1
Pentane	C_5H_{12}	21,091	19,517	15.35	105.0	97.2
Hexane	C_6H_{14}	20,940	19,403	15.27	104.8	97.1
Olefins						
Ethylene	C_2H_4	21,644	20,295	14.81	111.8	104.8
Propane	C_3H_6	21,041	19,691	14.81	108.7	101.7
Butane	C_4H_8	20,840	19,496	14.81	107.7	100.7
Pentene	C_5H_{10}	20,712	19,363	14.81	107.0	100.0
Diolefins						
Propadiene	C_3H_4	20,710	19,755	13.80	114.7	109.4
Butadiene	C_4H_6	20,496	19,436	14.06	111.3	105.7
Pentadiene	C_5H_8	20,416	19,295	14.21	109.8	103.8
Aromatics						
Benzene	C_6H_6	18,210	17,480	13.30	104.7	100.6
Toluene	C_7H_8	18,440	17,620	13.53	104.2	99.6
Xylene	C_8H_{10}	18,650	17,760	13.70	104.1	99.1
Misc.						
Acetylene	C_2H_2	21,500	20,776	13.30	123.6	119.4
Naphthalene	$C_{10}H_8$	17,298	16,708	12.96	102.0	98.6
Ammonia	NH_3	9,668	8,001	6.10	121.2	100.3
Carbon Monoxide	CO	4,347	4,347	2.47	134.6	134.6
Hydrogen Sulfide	H_2S	7,100	6,545	6.10	89.0	82.0

and air originally at 60°F and then cooled to the same temperature. The low or net heating value (LHV) is equal to the high heat value minus the heat released by condensation of the water vapor in the products of combustion (flue gas), i.e., it assumes all products to remain gaseous.

$$LHV = HHV - Q_F, \text{ BTU/lb} \tag{12.1}$$

where

$Q_F = 1{,}040\ w$

w = condensed H_2O/fuel weight ratio

BTU= British Thermal Units/pound
(1 BTU/lb = 0.55573 kcal/Kg)

For hydrocarbon mixtures, the low or net heating values can be given based on the average molecular weight by the following fitting equations:

$$LHV = 51.8M_w + 87, \text{ BTU/SCF} \tag{12.2}$$

$$(1\text{BTU/SCF} = 8.90 \text{ kcal/m}^3)$$

We can simplify the calculation of the stoichiometric air required (SAR) of a mixture with several components using the following expression:

$$SAR = Q/96 \tag{12.3}$$

where

SAR = air flow, SCFH

Q (Net Heat release) BTU/hr = Gas Flow (SCFH) × LHV(BTU/SCF)

SCF = Standard Cubic Foot, 60°F at 14.7 psia

Water vapor content in the air should be taken into account resulting in a reduction of oxygen. A humidity factor as well as a temperature factor should be applied to correct air requirements. Saturated air at 100°F contains 6.45% water vapor, reducing the oxygen volume percentage from 21% to 19.7%. Also, absolute pressure correction should be used at significant elevations above sea level (500 ft or more).

The i is to convert organic compounds to carbon dioxide and water. The products of combustion (flue gas) have a composition containing low concentrations of CO and unburned H/C in addition to CO_2, H_2O, O_2 and N_2, besides the Particulate. If the waste contains sulfur compounds, the flue gas will also have sulfur dioxide (SO_2).

12.8.2 Complete Combustion

If enough oxygen is supplied, the mixture is lean and the flame is oxidizing, resulting in a clear and short flame. If excess fuel occurs, the mixture is rich and the flame is reducing, resulting in a long and smoky flame, consequently incomplete combustion. Nitrogen in the air does not take part in the chemical reaction because it is an inert gas, but absorbs some of the heat, resulting in lower flame temperatures. If there is a shortage of oxygen, the final product may contain carbon monoxide (CO), hydrogen (H), hydrocarbons (HC) and free carbon (C).

12.8.3 Flare Combustion Efficiency

Complete combustion is the combination of fuel with oxygen without fuel left over requiring time, turbulence and temperature high enough to ignite all the combustible elements. Ignition temperature is reached when more heat is generated by the chemical reaction than is lost to surroundings, and combustion becomes self-sustaining. The temperature of a flame depends on the type of fuel, starvation or excess air, and initial temperatures of both fuel and air. Maximum temperature is reached when a fuel is burning at stoichiometric conditions, meaning no excess air.

In practice, combustion is never complete, resulting in combustion by-products like carbon monoxide. Combustion efficiency defines the mole percentage of combustion emissions that are completely oxidized to CO_2:

$$\% \text{ Combustion Eff.} = 100 \times CO_2/(CO_2 + CO + THC + \text{soot}) \quad (12.4)$$

where

CO_2 is carbon dioxide percent by volume

CO is of carbon monoxide percent by volume

THC is Total (unburned) hydrocarbons percent by volume

Table 12–5 Hydrogen/Carbon Weight Ratios vs. Black Smoke Production

Gas	Formula	H/C wgt ratio	% Carbon as black smoke
Paraffins			
Methane	CH_4	0.333	—
Ethane	C_2H_6	0.250	5
Propane	C_3H_8	0.222	12
Butane	C_4H_{10}	0.208	16
Pentane	C_5H_{12}	0.200	18
Hexane	C_6H_{14}	0.194	21
Olefins			
Ethylene	C_2H_4	0.167	32
Propane	C_3H_6	0.167	32
Butane	C_4H_8	0.167	32
Pentene	C_5H_{10}	0.167	32
Diolefins			
Propadiene	C_3H_4	0.111	55
Butadiene	C_4H_6	0.125	47
Pentadiene	C_5H_8	0.133	43
Aromatics			
Benzene	C_6H_6	0.083	> 55
Toluene	C_7H_8	0.095	> 55
Xylene	C_8H_{10}	0.104	> 55
Acetylene	C_2H_2	0.083	> 55

Tests on diffusion flaring with natural gas (methane) have indicated above 99% efficiency without any assist means. Flame temperature was around 1,800°F with a correspondent combustion excess air about 160%. Heavier hydrocarbons will have a combustion efficiency of 98%+.

Other conclusions and observations have resulted from testing, such as:

- Smoking flares are very efficient on the destruction of hydrocarbon gases.
- In many cases, the production of carbon monoxide was negligible.
- Flare with unstable flame can have low efficiency. Flame instability occurs when the jet velocity exceeds the flame velocity.
- Combustion efficiency is high for flares with high velocities operating in the region of flame stability.
- Flame lift-off from the flare is not an indicator of flame instability.
- Flame stability is affected by flare tip design and properties of the flared gas, such as, LHV, flammability and flame speed.

Based on the above testing, EPA ruled on continuous non-emergency flaring, dictating low heating value (LHV) and exit velocity limitations to ensure a combustion efficiency 98% or greater. The minimum allowable LHV of the flared gas is 200 BTU/SCF. If the gas does not have that minimum, it needs to be enriched with support fuel up to that value.

Non-smokeless flares must be designed and operated with an exit gas velocity as follows:

$$Ve < 60 \text{ ft/sec, if } 200 < LHV < 300 \text{ BTU/SCF}$$

$$Ve < 26.6 \text{ antilog}(H_V/849), \text{ ft/sec, if } 300 < LHV < 1{,}000 \text{ BTU/SCF}$$

$$Ve < 400 \text{ ft/sec, if } LHV > 1{,}000 \text{ BTU/SCF}$$

In case of air assisted flares, the minimum allowable LHV of the flared gas is 300 BTU/SCF, and the maximum exit velocity should not exceed

$$Ve < (329.2 - LHV)/11.3, \text{ ft/sec}$$

12.8.4 Flare Tip Diameter Sizing

Unlike the sonic flare tips, the standard non-smokeless and assisted smokeless flare tips should be sized to provide low sub-sonic exit velocities for all operating cases.

The sonic velocity, C, can be given by the following simplified equation:

$$C = 223 \times (nTr/M_W)^{0.5},\ \text{ft/sec} \tag{12.5}$$

where

Tr = gas abs. temp., R

$n = C_p / C_v$ (ratio of the specific heats)

M_W = Mol. Wgt.

Use exit velocity Ve = 0.2 C for normal continuous flow and Ve = 0.5 C for a peak short term emergency flow. Determine diameter of flare to meet all conditions including combustion efficiency requirements and use next standard pipe size.

12.8.5 Flare Stack Height Sizing

Typically the flare stack height is determined by either the two methods given on the API 521: the "Simple Method" and the modified Brzustowski method. The determination of the flare stack height is a function of the allowable heat radiation intensity at ground level or a designated location as well as the location of the flame center. There are other variables like the wind speed, gas flow rate, composition and exit velocity that will affect the determination of the flame center coordinates. From this point on, it becomes a geometry calculation (see Eq. (12.9) through Eq. (12.19) in Section 2 of this chapter).

12.9 Flare Header Design

In any flare header design, the sizing of piping based on ideal gas flow under isothermal conditions will normally be adequate. Piping should be designed to avoid the formation of liquid traps. If the liquid can not be drained to a remote knock out (K.O.) drum, a local K.O. drum should be added, only for liquid collection purpose.

12.10 Elevated Flare Components

The major components of an elevated flare system are the flare stack, flare tip, pilot and ignition system as well as gas seal, liquid seal, and knock out drum (see Figure 12–2).

12.10.1 Flare Stacks

Flare tips may be mounted on guyed supported stacks, derrick supported stacks or self supporting stacks. Flare stack height and location should also be considered, based on gas dispersion assuming an extinguished flame.

12.10.1.1 Guyed Supported Stacks

The guyed supported stack requires a space with a radius close to half of the stack height to connect the guy anchors. These systems have been supplied with overall heights up to 550 ft. Guyed stacks generally are the less expensive type, but this design requires a large amount of real estate. It cannot be used with flares handling gases with much difference from ambient temperature because the thermal expansion/contraction differences between the stack and any guy-wire highly changes the guy-wire tension, leading to very high stresses or causes structural instability.

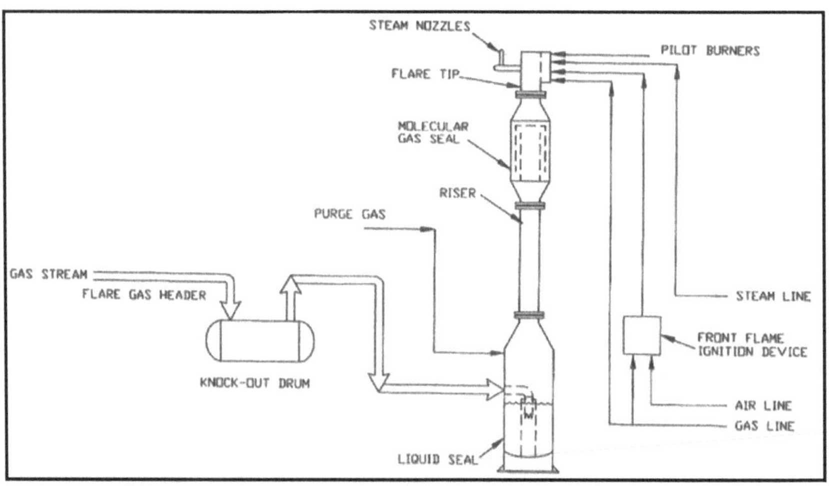

Figure 12–2 *Elevated flare components*

12.10.1.2 Derrick Supported Stacks

The derrick supported stack is ideal for very tall stacks with reduced ground clearance. Due to the structure, these systems are the most expensive including the erection costs. They allow different expansion rates between stack, piping and derrick. Derrick flares have been built to heights of 350 ft.

12.10.1.3 Self Supporting Stacks

The self supporting stack is the most economical and easiest to erect for short flare stacks, requiring less space for installation. The bottom sections are larger than the top, giving a practical height limit of 200 ft. Generally, internals are added at the base section, incorporating a vertical knock out drum or a liquid seal drum.

Generally, the stacks are designed and fabricated in accordance with ANSI, UBC, AISC, and ASME codes. The stacks are shipped with weld-prepared or flanged ends and the majority of components outfitted. If applicable, caged ladders, intermediate rest platforms, and a 360 degree top platform are supplied and designed to meet OSHA requirements.

12.10.2 Non-Smokeless Flare Tip

The non-smokeless flare tip should be provided with flame retention device and heavy duty pilots, allowing the flare to operate at high exit velocities without flame lift off. Flare stabilization is achieved by specially designed flame retention devices built into the flare tip and by the correct positioning of reliable pilots (see Figure 12–3). Addition of aerodynamically designed wind deflectors eliminates the local wind vortices responsible for sucking the flame down the leeward side of flares, extending the operational life of the flare tip, T/C cables, ladder, etc.

12.10.3 Flares Employing the Coanda Effect

The application of the Coanda effect in the design of flare tips has resulted in more efficient combustion with lower radiation levels and shorter flame lengths. Henri Coanda, early this century, noticed the tendency of a fluid jet discharging from a nozzle or slot to adhere to an adjacent surface, entraining the surrounding fluid. This fluid dynamic mechanism was applied to flares resulting in several designs of burner tips.

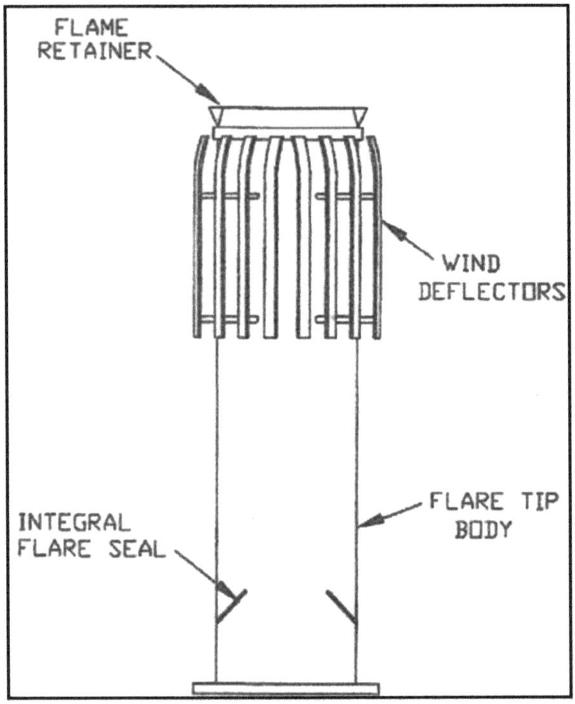

Figure 12–3 *Non-smokeless flare*

On external Coanda profiles, the pressure energy within the high pressure gas is utilized to induce several times its own volume of air. The waste gas adheres to the Coanda profile, producing a smokeless low radiation flame. The flame is initiated equally around the flare resulting in excellent wind stability because at least half of the flame base is sheltered from the winds. The flame propagates from the outside, always keeping a layer of gas protecting the profile body against extremely high temperatures and allowing the flare tip material to be stainless steel alloys without refractory protection (see Figure 12–4).

Coanda flares operate under sonic conditions at elevated pressures with a minimum of 10 PSIG to achieve smokeless combustion. The liquid carry-over up to 25% Wt/g ratios can be burned without fall-out. The flow rate through this flare is a function of the gas pressure assuming a constant annular slot area and constant gas properties. Some of these Coanda sonic flares feature a variable slot throat to increase the smokeless turndown.

The flame is highly aerated, radiating extremely low heat on an average of 0.1 for the fraction of heat radiated. The flame is shorter

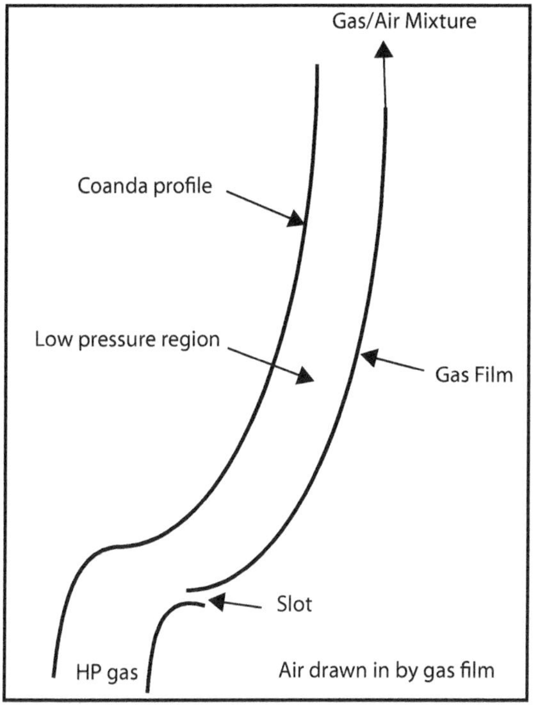

Figure 12–4 *Coanda principle*

than on conventional flares, with a high directional stability, resistant to cross winds.

12.10.4 Air Assisted Flare

Air assisted flare tips have been widely used for smokeless combustion on applications in which heavy saturates and unsaturates are flared. The primary air is supplied by a low pressure fan providing turbulence to mix the gas and air, and also ensuring entrainment of secondary air to achieve a smokeless combustion. A gas/air mixing head provides mixing and turbulence, resulting in a stiff vertical and about 50% shorter flame. A significant lower fraction of heat radiated (F = 0.13) will result in a reduction of the stack height requirement. The estimated center of the flame is at half of flame length and without horizontal deflection at ambient winds.

Good mixing is required prior to combustion; otherwise, smoke formation will occur. Combustion zone temperature is lowered by dilution and turbulence, prolonging the oxidation process and minimizing H/C decomposition. Unsaturates dissociate easier, requiring more primary air for smokeless combustion, at least 30 percent of the stoichiometric quantity, compared with 20 percent minimum for the heavy saturates.

This flare tip should be provided with flame retention device, allowing the flare to operate at high exit velocities without flame lift off. Wind deflectors are not required due to the vertical flame with little influence from ambient wind (see Figure 12–5).

Generally, these flares are furnished with a variable frequency drive (VFD) or with a two speed fan and a pressure switch in the flare header to save energy when it is flaring, at a fraction of the smokeless rate. The flare will smoke when flaring is greater than smokeless design rates, typically under upset or emergency conditions.

Air assisted flares are ideal for applications in remote areas where steam is not available and can also be used in offshore locations. There are several advantages to use this type of flare, including lower maintenance costs, extended life of the flare tip and elimination of steam lines and controls.

12.10.5 Flare Pilots

The flare tips should come complete with reliable heavy duty pilots and designed for continuous operation and flame stability, regardless of wind conditions. Each pilot is fed by its own separate natural draught aspirator, positioned at the flare tip base for maximum performance and reliability. Fuel gas consumption is about 65,000 Btu/hr of fuel gas per pilot. If propane or butane is used as a pilot fuel, consumption would be about the same required heat output. Pilot burners are made out of high nickel alloy to ensure a long operational life.

The pilots can be ignited by direct spark or more commonly by flame front generators. Pilot flames can be detected by optical UV/IR, flame ionization, and acoustical systems as well as by the most commonly used thermocouple system. Although less reliable, the thermocouples have the advantage of detecting the pilot flame without being directly exposed to the flare flame.

12.10.6 Flame Failure Panel

Each pilot typically carries a separate type "K" thermocouple which is attached to the pilot line and enters the pilot nozzle base. The pilot flame failure panel monitors the signals from the thermocouple installed in each pilot on a continuous basis by sensing the circuitry.

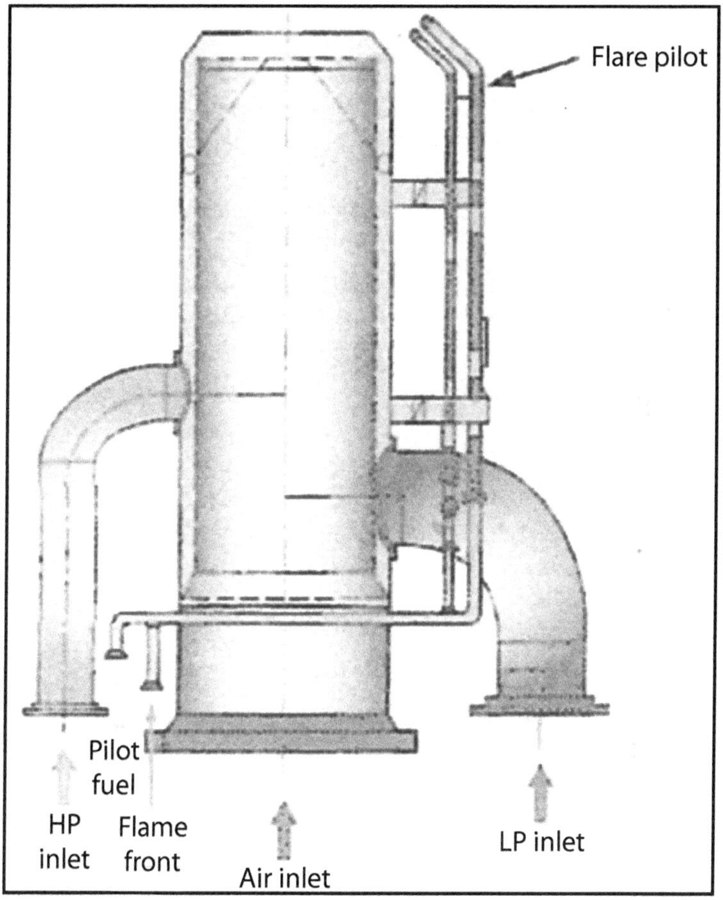

Figure 12–5 *Air assisted flare tip*

Should any pilots fail, a local panel alarm lamp will illuminate. Usually, there is a provision for connection of a remote alarm.

12.10.7 Ignition Systems

12.10.7.1 Manual Ignition

Ignition is achieved by manually operating a pushbutton to energize the transformer and spark plug fitted to ignition tube. During ignition only, approximately 150 SCFH of ignition gas, 1,500 SCFH of air and 0.3 KVA (about 2 Amps) electric power are required.

12.10.7.2 Automatic Ignition

In case of any pilot flame failure, signal from thermocouple energizes the relay, which in turn traps the on/off timer and solenoid valves on fuel gas and air supply lines. Typically an ignition transformer will be energized at a rate that can allow enough time to purge the ignition line between attempts. This sequence will continue until all pilots are lit, closing the solenoid valves, stopping ignition, and turning on the green lights. If any pilot does not light, time cycle ends, and in one to five minutes, the system sends an alarm signal.

12.10.8 Operations of Pilots and Igniters

The flare tip is fitted with continuous pilot burners to ensure ignition, regardless of the wind conditions. The pilots are ignited by a remote front flame generator. Fuel and air are fed via needle valves, non-return valves and restriction orifices to a mixing igniter tube where a spark ignites the mixtures. The mixing igniter tube is connected to the pilots by the ignition line where the front flame travels and lights each of the pilot burners.

The pilot flames are blue and stable and sometimes difficult to see if pilot burners are lit during daylight. Each pilot can be provided with thermocouples, type "K", to monitor pilot flames, activating an alarm to warn off pilot flame failure.

To operate the ignition system safety:

1. Completely purge the flare system with natural gas or nitrogen.

2. Open the air and gas needle valves and set pressure of both at 10 PSIG.

3. Open pilot fuel gas valves.

4. Purge lines for two to three minutes.

5. Spark to light the mixture. The flame front will be seen as a blue-yellow flash in the igniter sight port.

6. If pilot does not light, purge and spark again.

7. If pilot does not light, adjust the air needle valve and repeat steps 4 and 5.

8. When all pilots are lit, close air and gas needle valves, keeping pilot gas open. It is good practice to mark the pressure setting because they depend on fuel gas gravity and heat content.

9. After all pilots are lit, open the flare gas valve.

Although these systems do not require extensive maintenance the following procedures should be performed:

1. Drain condensate from all ignition, air and pilot lines.

2. Check and reset igniter electrode.

3. Check fuel gas and air supply pressures.

4. Clean igniter tube.

5. Clean air aspirator jets on pilot assemblies.

6. Check thermocouples and wiring.

12.10.9 Gas Seals

Gas seals are used on elevated flares to prevent the entry of air which can develop an explosive mixture with the gas. The purge gas maintains an H/C rich atmosphere in the flare stack, and the gas seal reduces the amount of gas consumed for this purpose. There are two main types: molecular and integral seals (a.k.a. dynamic, velocity or fluidic seals).

12.10.9.1 Integral Gas Seals

The integral velocity type seal (see Figure 12.6) is located within the flare tip body and provides low flow resistance in one direction and high resistance in the other. This design has a low pressure drop, yet is sufficient, compact and light, being installed in flare tips without increasing structural loads. To maintain the seal, under the same conditions, it requires a flow rate of natural gas with 0.03 to 0.04 ft/sec velocities to keep the oxygen concentration below the seal at 6%. If nitrogen is used as purge gas, the volumetric flow will be 75% of the purge rate with methane.

12.10.9.2 Molecular Gas Seals

The molecular type seal forces the purge gas to make two "U" turns forming a seal due the different molecular weights and requires a

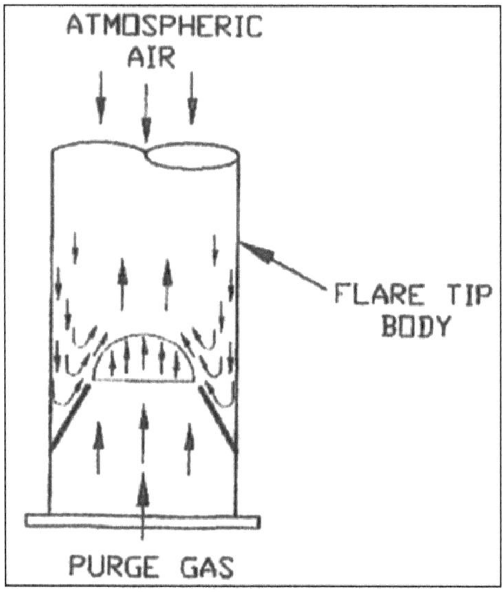

Figure 12–6 *Integral fluidic seal*

purge flow of natural gas with a velocity of less than 0.01 ft/sec to keep the oxygen concentration below the seal less than 1% with winds up to 20 MPH. It is placed just below the flare tip, and it works on principle of buoyancy of the purge gas, creating a zone having pressure greater than atmospheric. If the gas in the stack is lighter than air, the pressure at the bottom of the stack can be lower than atmospheric. The purge gas flow must counteract this situation. When the size of the flare tip is 42 in. or larger, the use of the molecular type seal is recommended (see Figure 12–7).

Purge rates are normally quoted to prevent the ingress of air. However, if purge rates are very low, then the flame begins to burn back into the tip. If this is allowed to continue for long periods, then the life of a tip is severely reduced. Purge rates should be sufficiently high to prevent burn back occurring, resulting in exit velocities typically about 0.35 to 0.5 ft/sec.

When the flare system is filled with high temperature gases, and flaring is interrupted, the gas will cool down, shrinking, giving place to the same volume of air, unless a purge rate is introduced to compensate the shrinkage volume of hot gas. This purge rate is also a function of the time for cooling, generally 15 to 20 minutes. Sometimes, with high temperature flow conditions, shrinking rate becomes the governing factor, until the system cools down. Also, a buoyant condition of gases lighter than air exist, purge volume needs to be

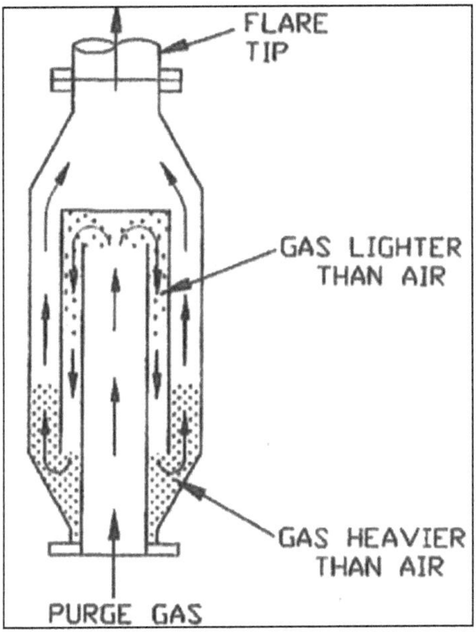

Figure 12–7 *Molecular seal*

added to replace at least equal volume of buoyant light gas, avoiding the entrance of air in the system.

However, in drilling operations the annular gas at high temperature is cooled by expansion in the annulus and through the choke and separator to near atmospheric conditions so it cannot be assumed that there will always be a high temperature shrinkage problem. The whole problem on a remote temporary flare system is complex enough that check measurements need to be taken during operations.

12.10.10 Liquid Seal Drums and Screen Flame Arresters

Screen flame arresters are subject to plugging and flame out with low flow or pulsation.

The most effective method to prevent flame propagation into the flare system is the installation of a liquid seal drum which can also solve some other problems (see Figure 12–8).

Pulsation can cause the flame to go out with low flow. To prevent this effect, a positive back pressure can be generated by the use of a liquid seal dip-leg, see Eq. (12.28).

The liquid seal can also be used as a back pressure device to maintain positive pressure in the flare header

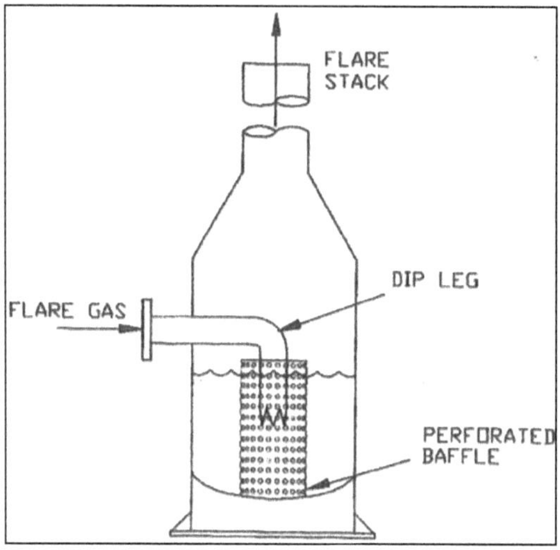

Figure 12–8 *Liquid seal drum*

In normal operation, the gas bubbles up through the "V" notches at the dip leg bottom. The back pressure in the vent header must be higher than dip leg pressure in order for the waste gas to flow through the seal. The liquid seal drum is designed to act as a final or secondary knock out drum for separating liquids from gases. Also, incorporated are special antisloshing baffles, providing plenty of viscous damping, besides the use of notches on the end of the dip pipe, increasing the flow area to minimize surge. Constant skimming of hydrocarbons and seal water level should be maintained by a minimum continuous flow of water. Generally seal drums are designed built with a design pressure of 50 PSI to resist explosions and a corrosion allowance of 1/16 in. minimum, when carbon steel is used.

The main danger in the use of liquid seals is the possibility of freezing, blocking the flare systems.

12.10.11 Knock Out Drums

Knock out drums are used to drop out and collect the liquid before the vapor is sent to the flare. They can be either horizontal or vertical and in a variety of configurations and arrangements.

The flare can handle small sized liquid droplets, making it only necessary for the drum to separate droplets above the 300 to 600 microns range (typically above 400 microns). Generally, the drums are designed and built with the same design pressure as the flare header or with 50 PSI to resist explosions, see Eq. (12.20). A minimum corrosion

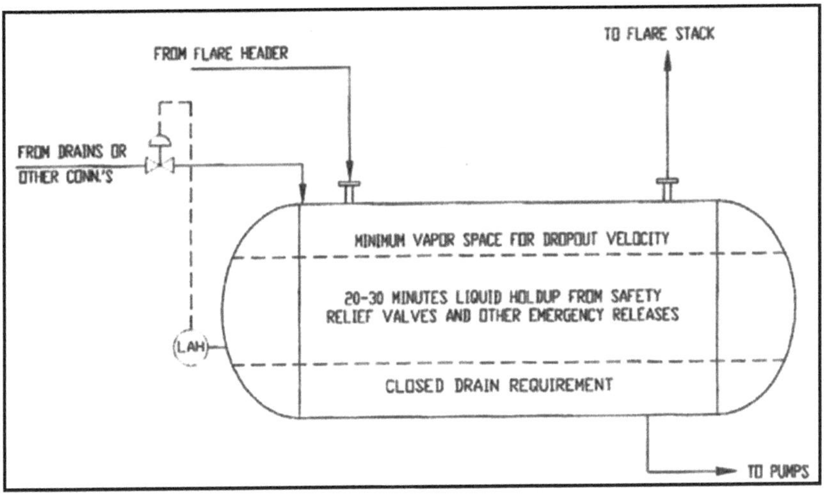

Figure 12–9 *Horizontal knock out drum*

allowance of 1/16 in. (1.6 mm) should be used with carbon steel vessels. Typically, a storage capacity plus a liquid hold up capacity of 20 – 30 minutes release should be provided in any drum. There should be a dual drain system in case of plugging.

12.10.11.1 Horizontal Drums

Horizontal drums must be of sufficient diameter to affect the desired liquid-vapor separation, see Eq. (12.24). On a horizontal drum, a split entry or exit is used and reduces the size of the drum for large flows (see Figure 12–9).

12.10.11.2 Vertical Drums

The vertical drums come with tangential inlet nozzle and also with a cylindrical baffle, giving a swirl effect and improving their effort, see Eq. (12.27). They can be incorporated at the base of the stack. An off size knock out drum should be provided close to the flare when the flare line serves more than one unit or the distance to flare exceeds 600 ft (see Figure 12–10).

12.11 Ground Flares

12.11.1 General

Ground flares are used to conceal the flame and also to reduce combustion noise, generally sources of complaints from the neighborhood. A

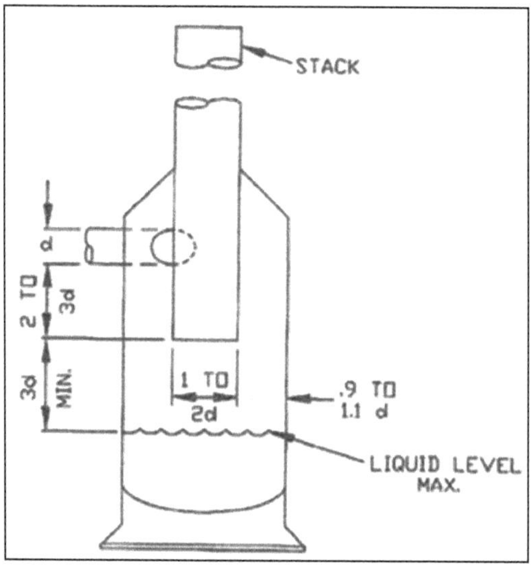

Figure 12–10 *Vertical knock out drum*

combustion noise reduction up to 15 dB is achieved on the ground flare vs. a flare stack.

Ground flares use single or multiple burners placed inside a steel refractory lined open enclosure, either in a round or rectangular open area. Ground flares usually consist of several manifold-mounted flare burners placed near the bottom of the enclosure shell. Some designs use a larger center flare burner that is air assisted to achieve smokeless combustion. Combustion efficiency is 98% and higher, but analysis of the combustion products has shown the need for an average of 165% excess combustion air (see Figure 12–11).

The inside open area is based on the total heat release at the design flow rate. It is common to design the enclosure according with the following rule of thumb:

$$\text{Heat Flux} = 1 \text{ to } 1.5 \text{ MM BTU/hr/ft}^2 \text{ of enclosure open area}$$

("Heat Flux" is heat intensity calculated as heat rate per area. Heat loading would be per volume).

12.11.2 Multi-Stage Multi-Burner Design

The staged ground flare divides the large mass flow through the multi-flare burners, achieving better mixing with the atmospheric air, resulting in a short and smokeless flame. Using multiple burners,

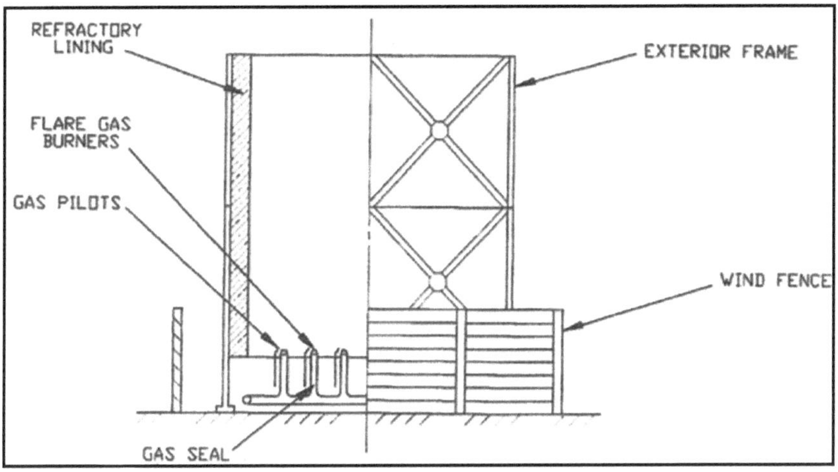

Figure 12–11 *Ground flare system*

staging becomes a useful feature to maximize the turndown of the smokeless condition. The multi-flare tips discharge vertically from the sub headers (manifolds), which are connected just outside the enclosure to a large header. The minimum number of flare tips, N, is based on the maximum heat release, allowed flame length and spacing of the flare tips.

Staging the burner can be accomplished by the use of liquid seal dip-legs or mounting pressure switch/control valve assemblies. This system requires available high pressures to achieve good smokeless turn down capacity, without the assist of steam, water or forced draft air.

A multi-stage multi-burner is generally accomplished by the use of pressure switch/block valve assemblies. The first stage is always open without automatic block valve. As the flow increases, the pressure switch will activate the solenoid to open the second stage control valve. At this point, all first and second stage burners become operational. As the back pressure builds up, the pressure switch will energize the solenoid to open the valve on the third stage. Now all the burners of the three stages are open. The sequence is similar for additional stages.

Each stage block valve should have a by-pass line with a rupture disc or safety pin for safety. Generally, the stage block valve is a butterfly valve with an actuator mounted for fail-open position. Since the flare is at ground level, a shut off valve should be provided on the main header in case of the failure of the pilot(s) ensuring that unburned gases do not accumulate at ground level and create the potential for an explosion.

The height of a multi-burner ground flare enclosure is a function of the flame length. The burner flare tips are generally 5–8 ft off the ground, generating low dispersion. This arrangement can present a pollution problem, especially if some sulfur is present in the waste gas that increases the grade level concentration of pollutants. At least two pilots should be provided between the first and last two burners of each sub-header stage. Pilots need to be designed for continuous service, and incorporating a type "K" thermocouple to send a signal activating the pilot light failure alarm. If all pilots of any stage fail, a signal is sent immediately to close the upstream valve mounted on the main ground flare header, shutting down the ground flare system.

12.11.3 Air Assisted Ground Flare

The air assisted ground flare system is based on a center air assisted flare burner which ensures proper mixing of gas and air to give complete and smokeless combustion in a short vertical flame with low heat radiation. Both high and low pressure gases can be handled in this combustion system. The well gas is connected to the flare burner outer body through a side inlet. A fraction of the air required for combustion is introduced by a forced draft fan connected to the center body of the flare burner, ensuring a very stable flame and creating a highly turbulent mixing zone.

To reduce power consumption, the air blower can be provided with a variable frequency drive (VFD) or with a two speed motor, using only 1/8 of the connected horsepower at low speed. Either pressure or flow transmitters in the flare header can be used to change fan speed to high or low, according to relieved flow rates. The remaining air to complete the combustion is introduced by natural draft through the existing gap on the enclosure bottom. For this type of ground flare, the shell diameter and height are both a function of total heat released.

The ground flare major shortcoming is the difficulty controlling large variations of well gas flow. It may require a conventional elevated flare to work together, on demand, for the release of larger flow rates.

12.12 Pulsation

Pulsation can cause the flame to go out with low flow. To prevent this effect, a positive back pressure can be generated by the use of a liquid seal dip-leg. This auxiliary equipment will also prevent the flashback danger. Properly designed seal drums should be used to provide protection against the possible flashback (see Section 12.10.10, page 559).

12.13 Flare Combustion Noise and Spectrum

This material is not covered in this discussion but references are included in the References.

12.14 References

Bednar, H. *Pressure Vessel Design Handbook,* Van Nostrand, New York, NY, USA, 1981.

Mack, W.C. "Selecting Steel Tubing for High-Temperature Service," *Chemical Engineering,* June 1976.

Schillmoller, C.M. "Solving High-Temperature Problems in Oil Refineries," *Chemical Engineering,* January 1986.

Section 2
Flare Systems

Olavo Cunha Leite, Flare Industries LLC

12.15 Mathematical Expressions

12.15.1 Radiant Heat Flux

Heat radiation intensity, K, Btu/hr/ft^2, is given by,

$$K = tFQ/4\Pi D^2 \tag{12.6}$$

where

Q = Total heat release
t = transmissivity (atmospheric absorption)
F = fraction of heat radiated
D = Distance from the flame mid point to the object
$\Pi = 3.1416$

Table 12–6 Radiation Intensity

BTU/hr/ft^2	kW/hr/m^2	Threshold Pain Seconds
550	1.74	60
740	2.33	40
920	2.90	30
1,500	4.73	16
2,200	6.94	9
3,000	9.46	6
3,700	11.67	4
6,300	19.87	2

12.15.2 Total Heat Release

Total heat release, Q, is given by

$$Q = \text{Flowrate} \times \text{LHV, BTU/hr} \quad (12.7)$$

12.15.3 Low Heat Value (LHV)

$$\text{LHV} = \text{HHV} - Q_F, \text{ BTU/lb} \quad (12.8)$$

where

$Q_F = 1{,}040\ w$

w = condensed H_2O/fuel weight ratio

For unknown mixtures of hydrocarbons, the low heat value (LHV) can be given by the equation:

$$\text{LHV} = 51.8M_W + 87, \text{ BTU/SCF} \quad (12.9)$$

12.15.4 Flame Length

Flame length, Lf, is given by the following fitting curve:

$$Lf = Q^{0.46}/114, \text{ ft} \quad (12.10)$$

12.15.5 Fraction of Heat Radiated, F

The fraction of heat radiated is dependent on several variables including gas composition, exit diameter and velocity, assist air/steam to waste gas mixing, etc. Experimental data indicates that F increases toward a limit. If liquid droplets are present, the F value should be somewhat increased.

Gas	F*
Hydrogen	0.15
Natural gas	0.19
Butane	0.29

* These values for near ideal combustion conditions of diffusion flames, resulting in conservative stack heights.

There are some expressions designed to fit the experimental data:

$$F = 0.048\ M_W^{0.5}$$

and

$$F = 0.19\ (LHV/910)^{0.5}$$

Efficient combustion is not expected at peak flaring rates, and a correction coefficient of 2/3 is commonly used (see API RP 521, 1st Edition, 1969). The maximum value of F for any gas when operating under smokeless conditions is 0.13.

12.15.6 Distance between Flame Center to an Object

The distance, D, between flame center to the object is given by:

$$D^2 = (H + Y_C)^2 + (R - X_C)^2 \tag{12.11}$$

where

R = Horizontal distance from stack to object, ft

H = Stack height, ft

Y_C = Vertical coordinate of the flame center, ft

X_C = Horizontal coordinate of the flame center, ft

Both the flame center coordinates can be determined from these fitting equations as an alternate to the API RP 521 graphs:

$$Y_C = 0.5\ Lf/[1+15*(Vw/Ve)]$$

$$X_C = 0.5\ Lf\ (Vw/Ve)/[0.02+(Vw/Ve)]$$

where

Vw = Wind velocity, MPH, typically 20 mph

Ve = Gas exit velocity, ft/sec

$$Ve = Qe/A,\ \text{ft/sec} \tag{12.12}$$

where A is the flare tip exit area and Qe is the actual volumetric flow

$$Qe = (W/3600)(379.1/M_w)(T+460)/520,\ \text{ACFS} \tag{12.13}$$

where

T = Flow Temp., deg. F

W = Flow rate, lb/hr

12.15.7 Transmissivity

Transmissivity, t, is the fraction of heat intensity transmitted that corrects the effect of atmospheric absorption for distances, D, between 100 to 500 ft

$$t = 0.79\,(100/r)^{0.0625}\,(100/D)^{0.0625} \quad (12.14)$$

where

r = relative humidity, %

Alternatively, the stack height, H, can be calculated for a given maximum required radiation level, K, at a distance R from the base of the stack.

$$H = [(tFQ)/4\Pi K) - (R - X_C)2]0.5 - Y_C,\ \text{ft} \quad (12.15)$$

If Kb = Max radiant heat intensity at the base.

$$D = H + Yc \quad (12.16)$$

and

$$H = [(tFQ)/(r\Pi K_b)]^{0.5} = Y_C \quad (12.17)$$

12.16 Hot Spot Temperatures

When there is no wind, a conservative assumption, cooling is by free convection. The equation describing this steady heat balance, where K = Radiant heat flux, BTU/hr/ft^2, is given by

$$K = 0.1713(T_s/100)^4 + (0.21/E_s)(T_s - T_A)^{4/3} \quad (12.18)$$

where

T_S = Hot spot temp, R

T_A = Ambiant air temp, R

Table 12–7 Material Emissivity

Material	E_S, Emissivity
Black body	1.0
Rough C.S.	0.94 to 0.97
S.S.	0.5 to 0.7
Alum. Paint	0.27 to 0.67
White Paint	0.77 to 0.95

Eq. (12.18) can be used for surfaces exposed to the flare heat radiation, in a calm weather condition. See Figures 12–12 and 12–13 for hot spot temperature under zero wind condition as well as with 15–20 mph wind conditions.

12.17 Grade Level Concentration of Vented Gas

The equations below assume short time operation, steady wind direction, neutral atmosphere and the waste gases at atmospheric temperature. When hotter, the gases tend to rise, reducing the required stack height. This effect, named plume height, results from the exit velocity, and it can be used as a safety factor if not applied on the stack height. This way, the actual height is equal to the effective height of the stack.

Stack Height, H, on the conservative side, is given by the following empirical expression:

$$H = (44.4\ Q\ D_Z/V_W CmD_Y)^{0.5},\ \text{ft} \tag{12.19}$$

where

Cm = Maximum allowable concentration at grade, PPM (Vol.)

V_W = Wind velocity at grade, MPH

Q = Flow of toxic component, CFH

D_Y = Horizontal diffusion coef.

D_Z = Vertical diffusion coef.

If

$$30 < H < 75\ \text{ft}$$
$$N = 0.25$$

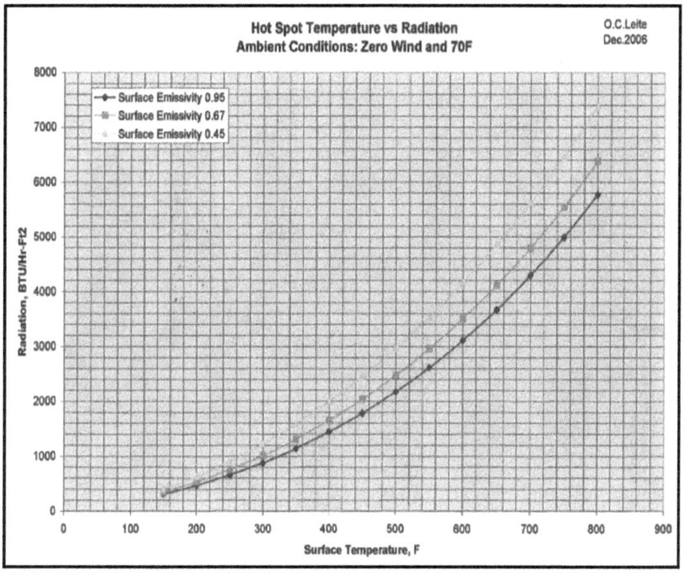

Figure 12–12 *Hot spot temperature versus radiation—no wind*

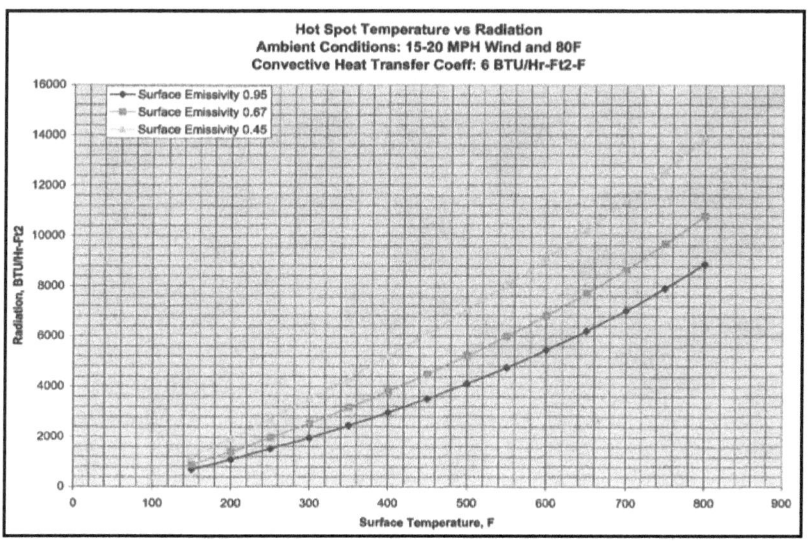

Figure 12–13 *Hot spot temperature versus radiation—15 to 20mph wind*

then

$$D_Y = 0.24 - 0.067\ (H - 30)/30$$
$$D_Z = 0.14$$

with N = Environmental. Factor = 0.25 for neutral.

If

$$H > 75 \text{ ft}$$
$$N = 0.25$$

then

$$D_Y = D_Z = 0.14 - 0.02\ (H - 75)/75$$
$$D_Z/D_Y = 1$$

12.17.1 Horizontal K.O. Drum Design

The liquid drop-out velocity is given by

$$V_Y = K[(W_L - W_G)/W_G]^{0.5}, \text{ ft/sec} \tag{12.20}$$

where

w_G = gas spec. wgt, lb/at operating conditions, (ACF)

w_L = liquid spec. wgt., lb/ft^3

The value of the constant "K" (gas load factor) is primarily a function of liquid particle size and drag coefficient. For a 400 microns particle, API RP 521 uses $K = 0.236\ /C^{0.5}$, although a widely used value is K = 0.417. For critical service, the value K = 0.21 is widely used, regardless the droplet size resulting in drums very close to the ones using API design. "C" is the drag coefficient.

The velocity of vapor, V_X, to give the droplets enough time to settle out, before the vapor leaves the drum, is determined by:

$$V_X = Q/\Pi D^2/4)f_A, \text{ ft/sec} \tag{12.21}$$

where

Q = gas flow rate, ft^3/sec

D = Drum dia., ft

f_A = fraction of drum, cross sectional area (C.S.A)

The gas residence time is

$$t = L_N / V_X \tag{12.22}$$

where L_N = distance inlet to outlet nozzles, ft.

The time required for the droplet to settle out must be at least equal to the residence time.

$$t = f_Y D / V_Y = L_N / V_X \tag{12.23}$$

where f_Y = fraction of height (dia.) used for vapor.

Solving for V_Y:

$$V_Y = f_Y D V_X / L_N \tag{12.24}$$

Replacing V_X by its expression and solving for D:

$$D = (4f_Y QD/pf_A V_Y L_N)^{0.5}, \text{ ft} \tag{12.25}$$

L_N/D ratio, for this type of application, is generally close to 2.5. Consequently,

$$D = 0.714(f_Y Q/f_A V_Y)^{0.5} \tag{12.26}$$

Using $f_Y = f_A = 0.5$ is a good start for this trial-and-error sizing. When the vessel diameter exceeds 10–12 ft, a split entry should be considered, to reduce the vessel diameter by a factor equivalent to reduce the flow rate to 50–60%. The two inlet nozzles should be placed apart at least 1.25 × Dia. of drum from outlet nozzle and the vessel should be sufficiently long (L/D > 5). Additionally, if there is a secondary drum at the base of the flare stack, i.e., water seal or secondary K.O. drum, the K.O. drum diameter can be further reduced.

12.17.2 Vertical K.O. Drum Design

The drop-out velocity is given by the same equation as for the horizontal design although with a different value "K." API 521 uses the same "K" value for both horizontal and vertical design. When using an inlet stream baffled to direct the flow downwards, an equivalent K = 0.21 is commonly used to calculate the liquid drop-out velocity. Both values are based on a 400 microns droplet size. The diameter, D, of the vertical K.O. drum is given by:

$$D = (4Qg/\Pi V_Y)^{0.5}, \text{ ft} \tag{12.27}$$

12.17.3 Liquid Seal Design

The maximum distance, L, that the dip leg is submerged is equivalent to the maximum exit back pressure allowable in the vent

$$L = 12^3\, P/W_L, \text{ in.} \tag{12.28}$$

where

P = Back pressure, PSIG

L = Distance, in.

w_L = liquid specif. wgt., lb/ft_3

Typical seal depths are 9–24 in. for elevated flares and 6–9 in. for ground flares, measured from top of "V" notches to liquid level. The diameter of the vertical drum, should be at least twice the inlet or dip leg diameter. This way, the free area above liquid level will be at least three times the inlet cross sectional area, to prevent surges of gas flow to the flare.

12.17.4 Final K.O. Drum

In order to act as a final or secondary K.O. drum, the liquid seal diameter should also meet the required diameter to give enough disengaging space.

The dropout velocity, V, is given by:

$$V = K[(W_L - W_G)/W_G]^{0.5}, \text{ ft/sec} \tag{12.29}$$

where

w_G = gas spec. wgt., lb/ft^3

For the final/secondary cases, K is double of standard design, becoming K = 0.9 for normal service and K = 0.45 for critical service. Also, a constant vapor velocity of 12–15 ft/sec for all conditions is widely used. The vapor velocity through the dip leg should be based on K = 3.6. Occasionally, it is required to increased dip leg diameter to reduce gas velocity and allow enough room for the slots. The height of the vapor space should be at least one vessel diameter, D, or 3 ft as a minimum to provide enough disengaging space.

CHAPTER 13

Corrosion in Drillpipe and Casing

Bill Rehm, Drilling Consultant
Abdullah Al-Yami, Texas A&M University
Katherine Dimataris, Lamberti USA

13.1 Introduction

A simple definition of corrosion is "the degradation of a substance by its environment." The purpose of this chapter is to identify some of the practical elements of corrosion encountered with underbalanced drilling. The chapter includes descriptions of the various types of corrosion and immediate operating solutions for corrosion problems. Some basic chemistry is used to explain the processes. Chloride embrittlement, stress cracking, formate and bromide salt systems, and biocides are not covered in this underbalanced drilling session. For an introduction to those problems and a deeper discussion of corrosion, consider "Corrosion Control of Drilling Tools Through Chemical Treatments-Effectiveness and Challenges" (Asrar, N., 2010) or source material from corrosion specialists. The more complex chemical reactions can be sourced from the references. For basic terms used in this chapter, see Table 13–1.

This chapter is not a basic reference on corrosion but is intended to be a field guide for the drilling engineer or rig supervisor. Serious corrosion problems or the potential for serious problems need to be handled by a specialist in that field.

Corrosion is a matter of concern to the drilling contractor because of pitting or loss of steel in the drillpipe. The operators have a similar, but greater, concern about the condition of the casing and in particular whether the corrosion is strictly a matter of the drilling process or is to a large degree a function of the formation and reservoir water.

Table 13–1 Terms Used in this chapter

→ Chemical reaction goes in this direction
↔ Chemical reaction is reversible
e^- represents a free electron for further reaction
H↑ represents a gas escaping from the reaction
FeS↓ represents a material precipitated from the reaction
Fe iron or any other material in its common state
Fe^{+2} iron or any other material as an ion with a valence of +2
Cl^{-1} chloride or any other material as an ion with a valence of –1

Some factors that affect corrosion are (Kippie, 2007):

- Oxygen content
- Wellbore temperature
- Acidic gases: H_2S; CO_2
- Drilling fluid composition
- pH
- Bacteria in the makeup water
- High salinity
- High stress
- High pressure

13.2 How Corrosion Occurs

Corrosion is only occasionally and locally experienced as a *significant* problem in general drilling operations because most drilling mud provides a non-corrosive environment. Basic corrosion reactions are shown in Figure 13–1.

In underbalanced drilling:

- The use of air as an injected gas sets up the potential for oxygen-iron corrosion.
- Membrane nitrogen with up to 6% oxygen eliminates the chance for fire or explosion, but does not solve the corrosion problem.

- Natural gas or cryogenic nitrogen eliminates the chance for oxygen-iron corrosion but does not always eliminate the chance of corrosion from other down-hole acidic gases. The most common acidic gases are hydrogen sulfide (H_2S) and carbon dioxide (CO_2).
- Aluminum drillpipe is not subject to oxygen corrosion but is affected by pH and high salinity.
- High flow rates that occur with gaseated mud can strip the protective coat from pipe and accelerate corrosion.

Corrosion first appears on the drillpipe during a trip as red rust or a black film and in most cases does not go beyond this point. If there is little or no corrosion in production operations in the field, any corrosion that occurs while drilling will be forced by some drilling fluid or drilling operational procedure such as packer fluids, aerated mud, or floating mud cap operations. Where corrosion is prevalent in production operations it will appear to some degree during all drilling. Different types of corrosion are shown in Figure 13–2.

13.2.1 Iron and Steel

Below are some simple equations for iron corrosion with regards to iron (and steel) drillpipe. Corrosion starts with an electrical corrosion

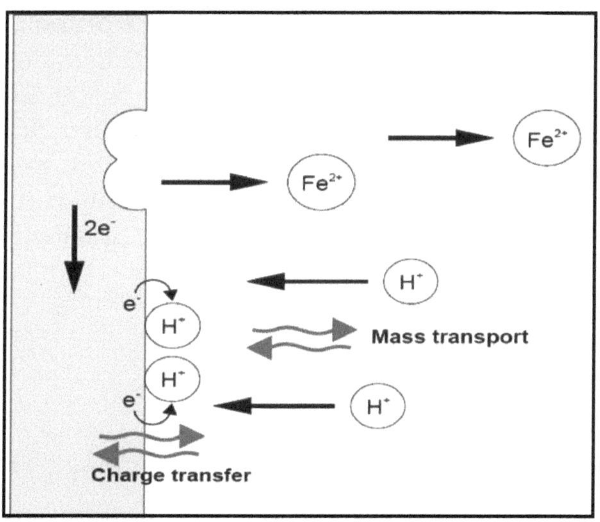

Figure 13–1 *Graphical representation of the corrosion processes (Roberge, 2000)*

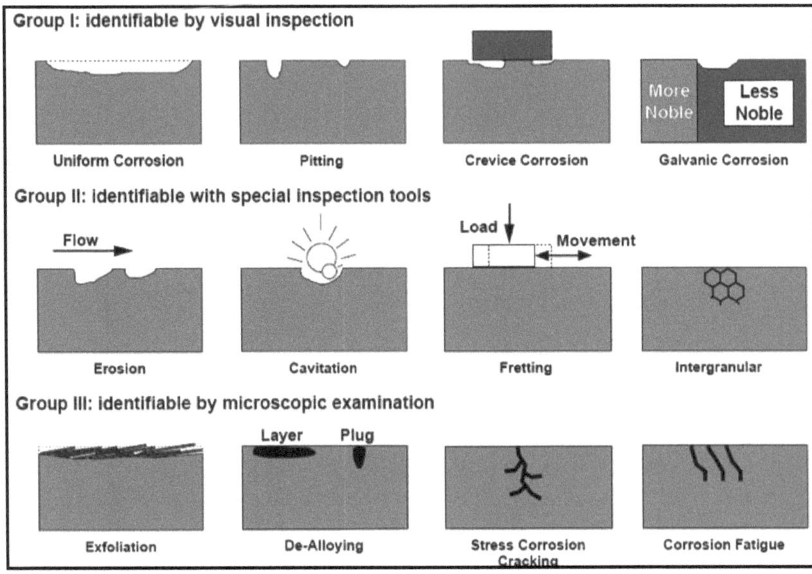

Figure 13–2 *Different types of corrosion (Roberge, 2000)*

cell. Iron corrodes in oxygenated water, see Eq. (13.1), to form hematite, also called red rust.

$$4Fe + 30^2 \rightarrow 2Fe_2O_3 \tag{13.1}$$

Eq. (13.1) is an overall equation. The anode reaction is an oxidation, or loss of electrons, while the cathode reaction is a reduction or gain of electrons. The term OIL RIG (*O*xidation *I*s *L*oss, *R*eduction *I*s *G*ain) is commonly used to remember the difference. Broken down into the anodic (oxidation) and cathodic (reduction) reactions, the details of this are as follows. For the initial oxidation reaction, see Eq. (13.2) and Eq. (13.3), iron degrades into 2 different oxidation states (Fe^{2+} and Fe^{3+}).

$$Fe \rightarrow Fe^{2+} + 2e^- \tag{13.2}$$

$$Fe^{2+} \rightarrow Fe^{3+} + e^- \tag{13.3}$$

At the cathode with low pH solutions, the proton (H^+) is in excess, producing hydrogen gas, see Eq. (13.4).

$$2H^+ + 2e^- \rightarrow H_2\uparrow \tag{13.4}$$

Combining Eq. (13.4) with Eq. (13.3) produces hematite (Fe_2O_3) and hydrogen gas, see Eq. (13.5).

$$2Fe^{3+} + 30_2 \rightarrow Fe_2O_3 \tag{13.5}$$

At the cathode with neutral or high pH solutions, water accepts the electron formed in Eq. (13.2) and Eq. (13.3), to break apart the strong water molecule, see Eq. (13.6).

$$2H_20 + 2e^- \rightarrow 2OH^- + H_2\uparrow) \tag{13.6}$$

Combining Eq. (13.6) with either Eq. (13.2) or Eq. (13.3) produces hydrogen gas and iron hydroxide, see Eq. (13.7) and Eq. (13.8), either iron(II) hydroxide or iron(III) hydroxide, again based on the oxidation state.

$$Fe^{2+} + 2(OH)^- \rightarrow Fe(OH) \tag{13.7}$$

$$Fe^{3+} + 3(OH)^- \rightarrow Fe(OH)_3 \tag{13.8}$$

Iron hydroxide ($Fe(OH)_2$) appears as barnacles or black boils, but can further decompose with time and a pH above 9, to form magnetite, see Eq. (13.9).

$$3Fe(OH)_2 \rightarrow Fe_3O_4 + 8H_20 + H_2\uparrow \tag{13.9}$$

The formation of magnetite, see Eq. (13.9), is a long term reaction (Schikorr reaction) and will not be noticed on new drillpipe, but is a problem for older or reconditioned drillpipe (see Section 13.3.2.2, page 584).

13.2.2 Aluminum

There are a number of alloys for aluminum drillpipe. The alloys vary in their resistance to high chlorides, temperature, and pH. The following is based on Alcoa 2414-T6. Other alloys may be less sensitive in a particular situation, particularly relating to chloride content of the drilling fluid, and care should be given to the type of alloy and the circumstances of use.

Aluminum drillpipe is not normally affected by the same materials that cause problems with steel pipe. This includes oxygen, hydrogen sulfide and carbon dioxide products. Any oxygen available combines with the aluminum for an aluminum oxide coating on the pipe which resists further corrosion.

Upon visual inspection, the drillpipe will have a dull aluminum colored coating, which is the oxide. Pitting can occur so the coating can be rubbed off with a similar soft abrasive to see if any pitting has occurred.

13.3 Identifying the Corrosion Types

Corrosion on the exterior or couplings of drillpipe or tubing is usually obvious on close inspection and often can be identified as to type. When checking for corrosion, always look at any tong marks and at the tool joint/drillpipe interface as well as on the body of the pipe. It is prudent to take before and after photos of the drillpipe if there is any reason to believe significant corrosion will occur.

pH has a strong influence on corrosion. pH is defined as, "the negative log of the hydrogen ion concentration" which goes from 1 (very acidic) to 14 (very alkaline) in normal field operations. (This scale is drastically expanded with some special chemical usage.) The important part of the definition from a practical point is that it is a logarithmic scale. When pH values are taken with paper strips, try to be on the upper end of the required value. A mud value shown as a neutral 7 on a wide range paper strip may still be acidic enough to allow corrosion. *Another technical point: the activity of the H^+ ion increases with temperature and so a neutral pH at surface temperature may actually be corrosive downhole. Corrosion rate also increases with temperature.*

13.3.1 Red Rust

The most common early identification of oxygen corrosion is red rust on the drillpipe. This is an indication that down-hole conditions are conducive to oxygen corrosion. Rust is hematite (Fe_2O_3). It forms rapidly over a few hours and the coating will be red and soft. Scrape the pipe to see if it is only a thin surface film or whether pits are starting to form. Specially compare the top of the drillstring to the middle and bottom joints. This will start to give a clue to the source of the problem. Check the pH of the mud from the flowline. Uniform rust is probably the result of a pH below 7 and/or oxygen/salt-water attack at the surface. In the marine environment rusty pipe in the derrick may only have a little surface corrosion from salt water and air (see Figure 13–3).

Figure 13–3 *Rust on drillpipe*

Figure 13–4 *Oxygen corrosion pits on drillpipe*

13.3.1.1 Pitting and Red Rust

Scrape the pipe to get a clean surface and have a good look at potential pitting. Initial pits from *oxygen* corrosion may have sharp edges but are often shallow, almost symmetrical round pits. Later effects are long wide pits, or sections of pipe where whole areas are depressed as a result of extensive loss of steel. Compare the top, middle, and bottom of the drillstring to see where the worst pitting occurs. Air or oxygen injected in the mud at the surface will rust the entire string. Local areas of rust may indicate a water flow. Pitting normally occurs when the pH is below 7 (see Figure 13–4).

Pitted pipe is bad news because there is a loss of steel. It is very common to use pitted drillpipe in shallow wells (500–2,000 ft) in eastern US air drilling. The strength of the pipe or tubing is well in excess of string weight or rig capabilities. Pitted pipe in wells where drilling or fishing loads may approach pipe strength needs to be replaced.

13.3.2 Black Coating on the Pipe

A black coating may be the immediate result of hydrogen sulfide (H_2S) or carbon dioxide (CO_2) in the drilling fluid. It may also be a form of oxygen corrosion (magnetite, Fe_3O_4) on tubing that has been in the hole for a long time. On the other hand, a black stain on sections of the pipe may only be the result of drilling pyrite, some similar mineral, some minor influx, or even a biological reaction. If the stain is less than 1 mm in thickness, it may not be a significant problem. Follow the same procedure of scraping the pipe to get a good look at any pitting, check the hardness of the coating, and compare the top, bottom, and middle pipe. Check the pH to see if it drops between the suction and flow line indicating H_2S or a CO_2 reaction. Make a further check for H_2S with lead acetate paper.

13.3.2.1 Black Coating Iron Sulfide

Iron Sulfide (FeS_x) in one of its forms may show up as a hard black generally, *slightly* magnetic film. Look for pin point pits. Check for traces of H_2S with a lead acetate indicator paper or similar tests. In a high temperature well, check to see if this coating is only on the middle to upper sections of the pipe. Hydrogen sulfide is normally not corrosive above 135°F. The source of H_2S may be in a hotter section of the hole.

Scrape the pipe to get a good look at any pitting, check the hardness of the coating, and compare the top, bottom, and middle of the pipe. Check the pH of the system.

13.3.2.2 Black Coating Magnetite

Magnetite, (Fe_3O_4) takes a long time to form, in the range of months to years. Magnetite shavings will be hard, brittle and magnetic. They will stick to a knife or file blade. Magnetite is normally found on production tubing or old casing. Its presence in a drilling or work over operations does not mean corrosion is actively taking place. Flakes of magnetite stuck on the drillpipe or in mud log samples, or black water are usually the result of bit or pipe action against the old casing. Check for traces of H_2S with a lead acetate paper or similar tests.

13.3.2.3 Black Coating Calcium Carbonate

This is not actually corrosion of the steel. This is a scale. Generally it is very thin, but can be scraped off with a knife. The scale is not magnetic. It generally will appear in high pH drilling fluid.

13.3.2.3.1 Black Boils or Black Barnacles—Iron Hydroxide

Iron(II) hydroxide ($Fe(OH)_2$) on pipe is an indication of extensive corrosion (see Figure 13–5). It forms on casing or tubing in fluids *above* a pH of 7. Barnacles or scale is formed on the pipe and deep pits are hidden beneath the barnacles. The barnacles are soft but they are hard enough to protect the chemical reaction in the pits which grow large enough to cause pipe failure. This corrosion occurs slowly and is often the result of using formation water as the make up water. The corrosion is often misidentified as H_2S or CO_2 corrosion because it first appears as black stains on the pipe. Iron(II) hydroxide is most commonly found on production tubing and casing. It is less often found on drillpipe since the reaction is slow.

13.3.3 Erosion Corrosion

High velocity systems, particularly gaseated systems, in the upper part of the hole develop enough velocity to erode off the natural inhibitor coatings on steel (or aluminum) and expose the fresh metal to both further erosion and continuing corrosion.

13.3.4 Stray Electrical Currents

Stray currents are a big problem with pipe lines but an improperly grounded generator on the drill rig can produce corrosion of casing or pipe, or on any of the other components on the drill rig where there is water or fluids that can form an electric cell. Unexplained corrosion at the top of the pipe or on the well head should be a sign to check the grounding of the generators and the rig in general. Look for powdery or gummy reaction products in pits or crevices.

13.3.5 Hydrogen Sulfide (H_2S)

Hydrogen sulfide (H_2S) gas is a deadly gas and precautions need to be taken when it is present. However, this is a discussion of corrosion, and other sources should be consulted concerning the health effects of the gas.

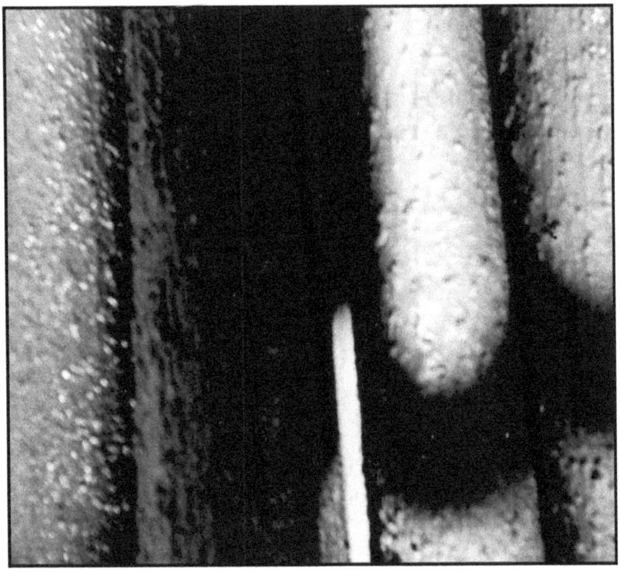

Figure 13–5 *Barnacles or black scale bubbles of iron hydroxide*

Sulfide can be presented in three different forms, H_2S (in acidic pH), HS^{-1} (middle range pH) and S^{2-} (alkaline pH) as shown in Figure 13–6.

Iron(II) sulfide has different forms. At high H_2S concentrations, a pitting type of corrosion results. At lower H_2S concentration, the FeS will be a scale and act as a protective layer to reduce the corrosion rate. A pH above 9 converts the H_2S to sulfide ion (S^{-2}) and/or bisulfide ion (HS^-). The moment H_2S is suspected, the pH should be maintained 10.5 to 11. However, H_2S will lower the pH, so a pH buffer should always be present.

Sulfides allow the entry of hydrogen atoms into the steel. This can cause embrittlement and cracking, shown in Figure 13–7.

Hydrogen sulfide, in trace amounts, causes a hard black film on the drillpipe. Corrosion tends to show up as deep long pits with a black coating, but the initial corrosion pit can be a deep pin point of a pit. The pit will shortly lead to stress cracking, shown in Figure 13–8. Corrosion and stress cracking from significant amounts of hydrogen sulfide can occur within minutes. More typically, trace amounts of H_2S appear in an area where it is not expected. Hydrogen sulfide sensors at the flow line may detect trace amounts. The pH will begin to drop with small amounts of hydrogen sulfide. An odor at the shaker is noticeable at higher quantities of H_2S.

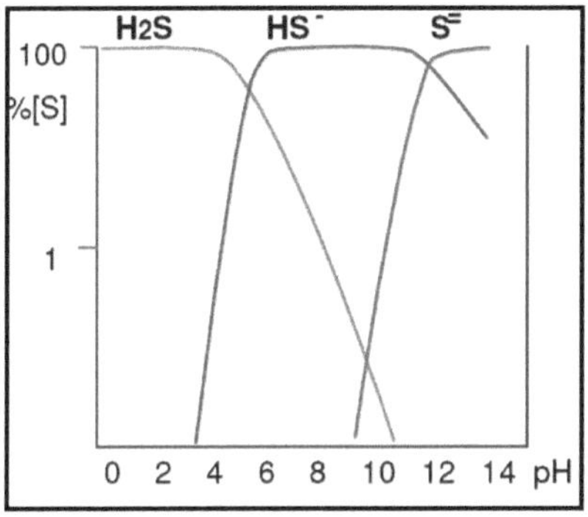

Figure 13–6 *Hydrogen sulfide in three forms (Clariant)*

Figure 13–7 *Embrittled and cracked drillpipe*

Figure 13–8 *Hydrogen sulfide pits and CO_2 pits*

Hydrogen sulfide has a moderate solubility in water (more soluble than oxygen). This solubility results in a weak acid which reduces the pH of the drilling fluid (see Table 13–2) and can also cause sour corrosion.

Hydrogen sulfide in a drilling fluid with no oxygen forms sodium hydrosulfide (NaHS) which tends to act as an inhibitor. However it is

Table 13–2 H_2S Reaction with Water and Iron

H_2S corrosion explained with H_2S gas in the drilling fluid:	
1) $H_2S + H_20 \leftrightarrow H^+ + HS^-$	
2) $2H^+ + Fe^{2-} \rightarrow Fe^{2+} + H_2\uparrow$	Rx2 Reduction Coupled with Rx3 Oxidation
3) $H^+ + HS^- \rightarrow 2H^+ + S^{2-}$	With H^+ being the basis for stress cracking
4) $Fe^{2+} + S^{2-} \rightarrow FeS\downarrow$	Basis for Iron Scavenger
5) $H_2S + CaO \rightarrow CaS\downarrow + H20$	
6) $H_2S + H_2O + Na(OH) \rightarrow NaHS + H_2O$	
$NaHS + Na(OH) \leftrightarrow Na2S\downarrow + H2O$	Sodium sulfide is inactive, but
$NaHS + H^+ \leftrightarrow H_2S$	the sodium sulfide will reverse to H_2S if the pH goes below 7

difficult to keep traces of oxygen out of the drilling fluid. In addition to air or membrane nitrogen injected into the mud stream, the mud guns, shale shaker, and desilter/desander all tend to beat air into the mud.

13.3.5.1 H_2S in Oil Mud and Invert Emulsions

Invert emulsions and oil muds offer significant protection against H_2S corrosion because the pipe is oil wet. However, H_2S is more soluble in oil than in water. When the pressure and temperature are reduced in the mud (when coming up hole or when reaching the surface) H_2S gas may be released or react with some free water. If an oil mud is treated with excess lime (CaO) the pH will remain high and the lime will also react with any available H_2S.

13.3.6 Carbon Dioxide

Carbon dioxide is found in various forms in the drilling fluid and can be a corrosive or problem agent since it is present in natural gas. The forms of carbon dioxide are mostly dependent upon the pH of the drilling fluid, and also affect the pH:

- carbon dioxide gas (CO_2)
- $CaCO_3$ (both as a sized additive in the drilling fluid and precipitated as scale on the drillpipe)
- carbonate ion (CO_3^{-2})
- bicarbonate ion (HCO_3^-)
- carbonic acid (H_2CO_3)

The relationship of the three phases is dependent on pH of the drilling fluid and its partial pressures.

- **pH < 7**—The CO_2 will be in equilibrium with carbonic acid (H_2CO_3) at a pH below 7. This is a strong acid that reduces the pH and will cause corrosion on steel.
- **pH 7 to pH 9**—The CO_2 will form the bicarbonate ion (HCO_3^{-1}). Bicarbonate ion is a natural buffer. In this case, the drilling mud will be buffered against pH change.
- **pH > 9**—HCO_3^{-1} starts to precipitate out as scale ($CaCO_3$) or a similar scale with other ions.

Corrosion resulting from the presence of CO_2 depends on the CO_2 partial pressure, temperature, water content, flow velocity, presence of oxygen, H_2S concentration, and chloride concentration (Stone et al., 1989).

Water content: dry CO_2 is not corrosive until the temperature is greater than 750°F. Therefore, the presence of CO_2 in a production well is of little consequence until it becomes wet. When CO_2 and water are present, a chemical reaction Eq. (13.10) that produces carbonic acid occurs:

$$CO_2 + H_20 \leftrightarrow H_2CO_3 \qquad (13.10)$$

A workover in a producing well that is oil wet will be protected from CO_2 corrosion when the oil remains in the external phase. An oil-wet condition occurs when low water concentrations are present, and the water is completely dissolved or surrounded by oil. When the water concentration in an oil well reaches 25 to 35%, the phase relationship is strained and can revert to being water-wet, i.e., the oil becomes dissolved or surrounded by water. This provides the H_2O for reaction with the CO_2 and metal surfaces, and the CO_2 becomes the corrosive agent.

Carbonic acid lowers the pH of the well fluids and increases the corrosiveness. It also reacts with iron to form an iron carbonate, Eq. (13.11) (scale), plus hydrogen (see Figure 13–9):

$$Fe + H_2CO_3 \rightarrow FeCO_3 + H_2\uparrow \qquad (13.11)$$

The partial pressure of CO_2 in general does not affect the corrosion of most stainless steels. The corrosion resistance of 9 Cr/1Mo and Martensite stainless steel is affected with increasing CO_2 partial pressure when chloride levels increase at temperatures greater than 200°F.

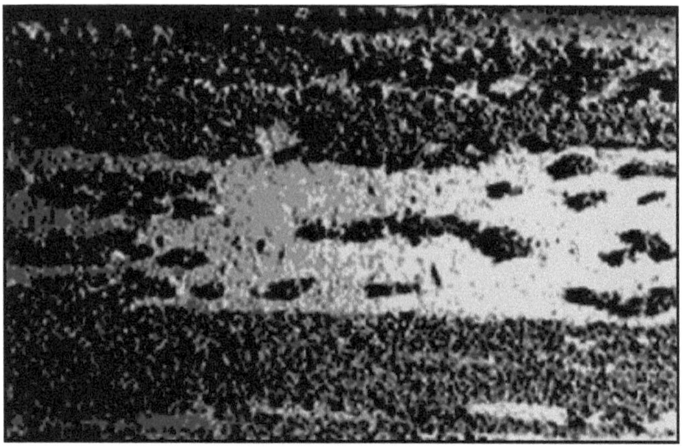

Figure 13–9 *Iron carbonate (scale)*

Laboratory testing and field data have shown that the pitting rate of carbon and low alloy steel increases with a CO_2 partial pressure of about 15 psia and above. This has led to the rule of thumb that states that when the partial pressure of CO_2 is less than 3 psia, the well is generally non-corrosive; when partial pressure is between 3 and 30 psia, corrosion is possible; once the CO_2 partial pressure is greater than 30 psi, the well is corrosive.

The most common form in a drilling fluid is the bicarbonate ion. The bicarbonate ion will combine with any calcium or magnesium in the system and precipitate out as calcium carbonate or magnesium carbonate (magnesium scale). This is not normally a problem in drilling operations since the scale disappears into the mud system. Upon occasion with high concentrations, it can plug fine screen shakers. In workover or production operations, every effort needs to be made to clean the makeup water and reduce the bicarbonate ion before proceeding down the hole.

Carbon dioxide can be introduced from the makeup water, from gas in the formation, from treatment of cement, or from decay or organic additives, generally in this order. Carbon dioxide induced corrosion shows up as long shallow pits that normally are black, but may also show a red rust color. Corrosion often shows up as shallow pits in the box and pin ends of pipes. The black film corrosion associated with carbon dioxide is not as hard or persistent as from H_2S.

13.3.7 Invisible or Long Term Corrosion

There is a tendency to use 3% to 4% KCl as a shale inhibitor in both drilling and fracturing. While this may be the most efficient concen-

Table 13–3 KCl Corrosion Potential

KCl at 3% to 4% is in the most corrosive range. As the water becomes more saturated with salt, the oxygen content of the water drops, and the system becomes less corrosive.	
At the cathode, $KCL + H_20 \rightarrow K0H + HCl$	Hydrochloric acid—very corrosive
At the anode, $2HCl + Fe^{2+} \rightarrow FeCl_2 + H_2$	Ferric Chloride—very corrosive keeps the anode clean and accelerates corrosion

tration for shale inhibition, it is at the peak of the corrosion curve and can increase the corrosion rate. In drilling operations, this normally shows up as some general rust on the pipe. If the well and near well bore formation is not washed clean after drilling or fracturing, long term corrosion will occur in the casing (see Table 13–3).

In general, salts in drilling fluids or fracturing fluids, (i.e. KCl, NaCl) have long term corrosive effects that are worst in the concentrations of 1–10%. The use of either fresh water or saturated systems avoids this problem. As the salt system becomes more saturated, it reduces the amount of oxygen in the fluid system.

At first, the added dissolved salt increases the corrosion rate, but then later the corrosion rate decreases as the dissolved salt increases further. Corrosion rate is at its highest value at 3 wt% in NaCl-based brines and is at the lowest value at 26 wt (see Figure 13–10). The increase in corrosion rate at 3 wt% NaCl is due to the increase conductivity preventing the formation of the protective layer of $Fe(OH)_2$ on the steel. The reason for the lower corrosion rate at NaCl saturation level is the reduced O_2 solubility in water at such high level of salinity (Chitty, 1998).

13.4 Corrosion Testing

Meters, chemicals, test kits and instructions of detailed test procedure can be obtained from most commercial water testing vendors. The normal tests used in drilling fluid are similar to boiler feed tests and most of the material is found under that category as well as general water testing. There are also some excellent web sites with explanations of the use of the test kits.

The best approach to corrosion control is to split measure the properties relative to corrosion at both the suction and flowline. This includes the pH, oxygen, total hardness, iron concentration, and

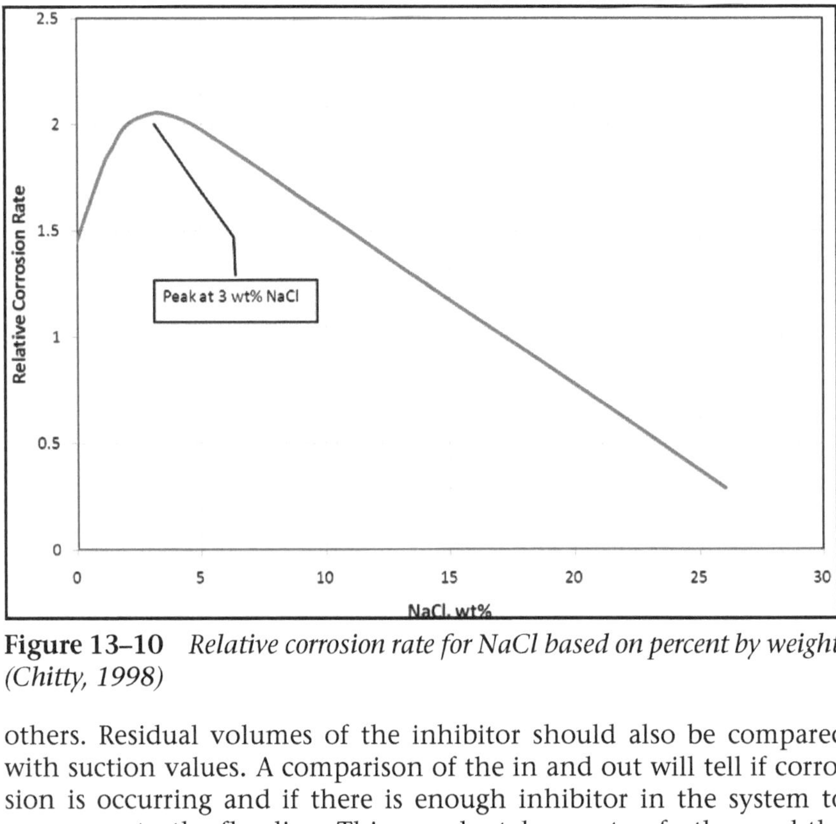

Figure 13–10 *Relative corrosion rate for NaCl based on percent by weight (Chitty, 1998)*

others. Residual volumes of the inhibitor should also be compared with suction values. A comparison of the in and out will tell if corrosion is occurring and if there is enough inhibitor in the system to carry over to the flowline. This may be taken a step further and the above tests run on the rig water, as this is a common source of problems. If using air drilling in areas where corrosion is a common production problem, or where it appears corrosion will be a problem, use natural gas or cryogenic nitrogen in place of air.

13.4.1 pH

pH is an essential measurement in corrosion control. Below a pH of 7, there may be significant problems with corrosion. The danger of corrosion decreases as the pH increases above 7.

There are pH meters as well as paper strip indicators. For most operations, the paper strip indicators are adequate for the degree of accuracy required in corrosion control. pH should be 10 or higher at the suction and at least 9 at the flow line. A large drop in pH is an indicator of potential problems

For use with aluminum drill-pipes, the pipe should not be exposed to pH below 7 or above 10 for long periods of time. The aluminum oxide coating degrades in the high and low pH range. With a loss of coating, the "raw" aluminum is exposed and is "chemically milled" off the pipe. With invert oil mud, the pH of the water should be kept in the range of 8–10. Additionally, chlorides above 180,000 ppm tend to destabilize the coating on an aluminum drillpipe. With high chlorides the pH should be kept in range of 7.5–9.

13.4.2 Oxygen

There are electric oxygen probes as well as paper strip tests. The paper strip tests appear to be most practical and inexpensive. Total oxygen content should be tested in and out. It is obviously not practical or effective with aerated drilling fluids, but reduction in oxygen at the flow line versus the suction may give a clue to potential problems in other systems. Oxygen content reduction between the suction and flowline indicates that something is reacting with the oxygen. It could be corrosion of the steel-check the iron content. Avoid beating air into the pits with the mud guns.

13.4.3 Total Hardness

The total hardness, or generally the magnesium and calcium content of the makeup water is not as critical to corrosion control as it is to the amount of chemical needed to control corrosion. Total hardness becomes more important in workovers since it a factor in scale formation.

Total hardness affects the efficiency of the treating chemicals. High total hardness needs to be treated out of the drill water before addition to the drilling fluid system. High total hardness at the flowline is probably the result of a low pH. Calcium and potassium ions, called "hardness" in the oilfield may be called "solids" by the water treatment personnel. Do not be confused by the terminology.

Total hardness as a field measurement is often titrated with a standard versenate solution and manver indicator.

Total Hardness (mg/L) = ml of versenate solution × 400/ml of sample.

13.4.4 Bicarbonates

Bicarbonates (HCO_3^-) above about 1,000 ppm can make it difficult to keep the pH high enough or to increase the pH. High bicarbonates are very often associated with drillpipe and casing corrosion. Keeping the pH above 9.5 or 10 will avoid most bicarbonate formation.

13.4.5 Iron

Iron is probably the most significant test for corrosion. There should be a measure of the iron at the shaker and iron in the suction water. An increase in iron at the flowline means the corrosion is occurring. The best policy is that the makeup water should be clear of iron to avoid masking the increase in iron at the flowline.

However, an increase in iron during workover or re-entry may only mean that corrosion products are being knocked off of the casing. In this case, the workover string needs to be examined within a day of operations. Magnetite on the shaker is a sure sign of casing corrosion, and not work string corrosion.

The typical iron test reduces the ferrous iron to ferric iron by heating or boiling the test solution and titrating the results. There are some iron test strips available that are much simpler to use and give adequate results.

For iron testing, an increase in soluble iron is an indication of corrosion.

$$Fe \rightarrow Fe^{2+} + 2e^- \qquad (13.12)$$

However, a decrease in iron concentration in solution might also indicate corrosion.

$$Fe^{2+} + H_2S \rightarrow FeS + H_2 \qquad (13.13)$$

Therefore, testing for iron alone is not enough to indicate corrosion.

13.4.6 Alkalinity

The alkalinity measures the ability of the water to neutralize acids. Carbonate and bicarbonate are two main components of alkalinity. Alkalinity is measured by titration as two values or in two steps. The "p" or phenolphthaline alkalinity is in ml (milliliters) or pH change relative to a standard sulfuric acid solution of 0.20 Normal, and the "T" or total alkalinity is in ml and measured with bromophenol blue indicator or pH change.

13.4.7 Phosphonates and Organo-Phosphate Esters

Residual phosphonates at the flowline (compared to the suction) are the best way to monitor the addition of phosphate inhibitors.

Residual inhibitor at the flowline is generally an indicator that enough of the material is being added to the system to treat the steel and still have a small excess. A comparison of the in and out will tell if corrosion is occurring and if there is enough inhibitor in the system to carry over to the flowline.

There are several test methods available. One involves the UV catalyzed oxidation of phosphonate to orthophosphate and another the chemical conversion of the phosphonate. The commercial test kits are the most practical method for the determination of the residual phosphonates at the flow line.

13.4.8 Hydrogen Sulfide (H_2S)

The most common measurement for hydrogen sulfide in the drilling mud is lead acetate paper.

Invert emulsions and oil mud offers significant protection against H_2S corrosion because the pipe is oil wet. However, H_2S is more soluble in oil than in water. When the pressure and temperature are reduced in the mud (when coming up hole or when reaching the surface) H_2S gas may be released or react with some free water. If an oil mud is treated with excess lime (CaO), the pH remains high and also reacts with any available H_2S.

13.4.9 Carbon Dioxide (CO_2)

Corrosion from CO_2 does not require oxygen in the drilling fluid or packer fluid. Corrosion is most commonly seen in production tubing and casing. It can be introduced from the makeup water, from gas in the formation, treatment of cement, or decay of organic additives, generally in the order listed.

Carbon dioxide induced corrosion shows up as long shallow pits that normally are black, but may also show a red rust color. Corrosion often shows up as shallow pits in the box and pin ends of pipes. The black film corrosion associated with carbon dioxide is not as hard or persistent as from H_2S. The pH of a water based mud system containing CO_2 products needs to be kept above 9.

13.5 Measuring Corrosion

Corrosion of the drillpipe and casing occurs as the steel seeks to return to a stable iron oxide form. More specifically, in terms of oilfield drilling, the concern is over the RATE of corrosion on the various grades of drillpipe and casing and how to measure and limit it.

Table 13–4 Simple CO_2 Reactions in Water

At a low or neutral pH:	
$CO_2 + H_2O \rightarrow H_2CO_3$	Carbonic acid, very corrosive
At the anode, $Fe^0 \rightarrow Fe^{2+} + 2e^-$	Fe^0 is metallic iron or steel
At the cathode, $H_2CO_3 + 2e^- \rightarrow CO_3^{-2} + H_2$	
The follow up reaction:	
$2Fe^{2+} + (CO_3)^{-2} \rightarrow Fe_2CO_3 + \uparrow H_2$	Black Iron Carbonate

The API standard on weight loss in pipe due to corrosion is less 200 mils per year (200 mpy). A mil is one thousandth of an inch, not a millimeter (one thousandth of a meter). The industry generally regards anything below 100 mpy as an acceptable corrosion rate (loss of metal) for tubular goods, but there is an effort to reduce the acceptable level to 50 mpy or less. One of the problems of measuring corrosion rate, is that the theoretical rate of corrosion doubles with every 54°F (30°C) increase in temperature (Chitty, 1998). Some other references claim corrosion doubles every 18°F or 10°C.

What is going on downhole may not be what is measured at the surface!!! (See Figure 13–11.)

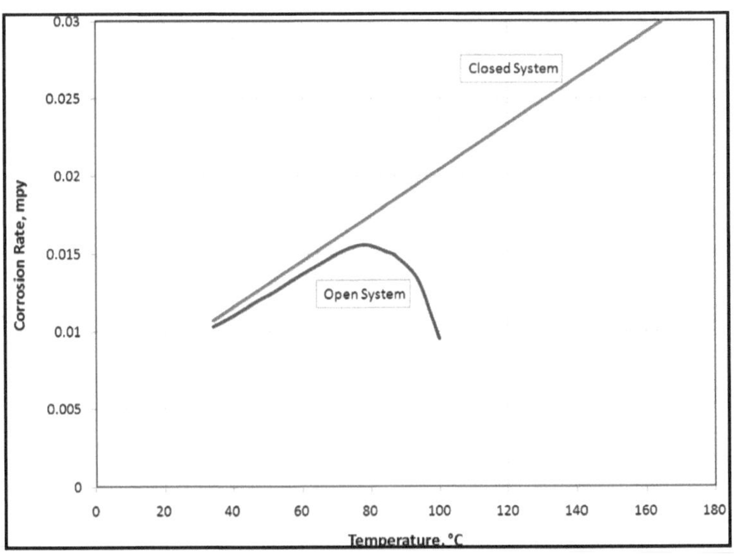

Figure 13–11 *Temperature (°C) versus corrosion rate (miles per year) (Chitty, 1998)*

13.5.1 Visual Inspection

Visual inspection of the pipe is one of the most important functions. Reports can be delayed or misunderstood. Physical drillpipe inspection can quickly point out a problem in the early phases. Identifying corrosion types by physical appearance was previously described in Section 13.3, page 582.

13.5.2 Measurement of Drilling Fluid Properties

As a complement to the visual, the earliest warning of corrosion or corrosion conditions is the measured fluid properties relative to corrosion at both the suction and flow line. This includes:

- pH
- oxygen content
- total hardness
- bicarbonates
- iron concentration
- residual volumes of the inhibitor

A comparison of the properties, in and out, will tell if corrosion is occurring (increased iron and/or significantly decreased pH), and if there is enough inhibitor in the system to carry over to the flow line. This may be taken a step further and pH and hardness tests run on the rig water to predict potential induced problems.

13.5.3 Drillpipe Rings

The corrosion rate for drillpipe is usually measured with a ring of steel similar to the tool joint. The ring is numbered and carefully weighed. The ring can be the same grade of steel as the drillpipe, or in some cases it can be a more reactive material to accelerate a warning of corrosion. Drillpipe rings are usually placed just above the drill collars and then again half way up the hole.

The drillpipe ring is left in the drillpipe for some period and retrieved on a trip and sent to the supplier to be weighed and the loss per year calculated. The obvious problem with the ring is that it reports corrosion after it has occurred at that point in the well. Changes in the corrosion treatment based on ring data are always lagging behind any problems that are occurring.

13.5.4 Corrosion Coupons

A corrosion coupon placed in the flowline can be weighed on a more immediate basis and so gives a quicker indication of material loss. It also can be observed for corrosion on a regular basis. The coupon can be a metal similar to the pipe or a more reactive metal, such as zinc or aluminum, that gives quicker responses to corrosive conditions. However, it is at surface conditions and the accelerating effects of downhole temperature and pressure are not measured which is why a more active metal is often used in place of a matching steel.

McNealy et al. (2009) showed how a special autoclave is used to test coupons for corrosion in advance of problems (see Figure 13–12). Testing conditions can be selected to simulate well conditions such as temperature, total pressure, gas mixture, brine, rotational speed, time, and types of coupons.

13.5.5 Electrical Resistance Probe

Electrical resistance probe measurement is an online method of measuring change due to corrosion or potential for corrosion. The electrical probe is excellent in operations like pipelines where pressures, temperatures, and volumes are constant over a long period, but it is difficult to interpret the drilling problems as they occur and thus to treat the fluid under drilling conditions. The technical problem with the probe is that it does not measure the downhole conditions which vary widely from the flow line condition and which also vary as drilling progresses.

13.6 General Corrosion Prevention and Treatment

Oxygen concentration in conventional drilling is low. The following can be used to control corrosion:

- Scavengers
- Corrosion resistant alloys
- Corrosion inhibitors for pipe coating

Corrosion prevention helps in minimizing corrosion products which might plug the formation or affect downhole tools. In underbalanced drilling, corrosion prevention is more difficult due to higher flow rate and higher oxygen concentration. Using oxygen scavenger is not practical due to the high concentration required (10 ppm scavenger per 1 ppm O_2).

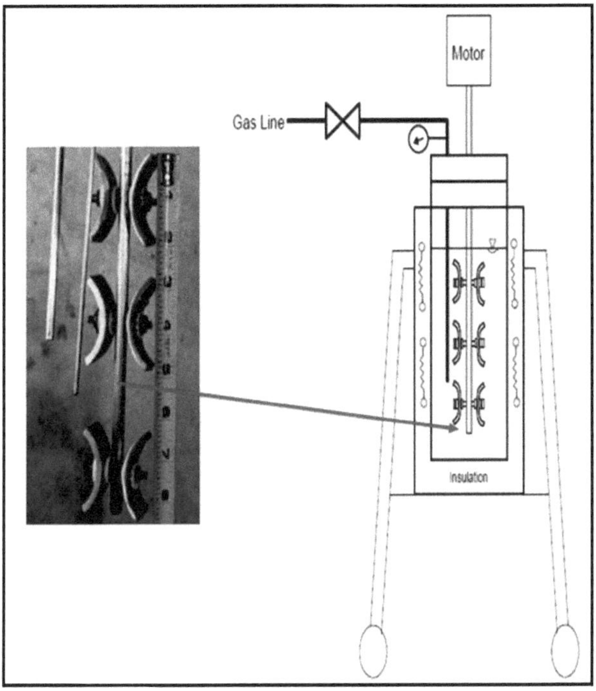

Figure 13–12 *Autoclave and coupons used for field condition tests (McNealy et al., 2009)*

The best methods to control corrosion rate in underbalanced drilling are:

- Use mechanical means to reduce O_2 concentration
- Use anodic inhibitors for pipe coating

There are two types of corrosion inhibitors:

- Cationic inhibitors such as amines that oil wet the pipe surface and form a protective layer. The cationic corrosion inhibitors are not effective in underbalanced drilling because they will be eroded by high annular velocities. In addition, high concentration of O_2 can penetrate the protective layer formed by cationic inhibitors and corrode the pipe.
- For underbalanced drilling, anodic corrosion inhibitors react with the surface to form a protective oxide layer. An anodic example is a phosphate ester inhibitor. The protective layer formed by this inhibitor is iron phosphate with hydrocarbon

coating. A concentration of at least 50 ppm is needed in drilling fluid for good inhibition.

Aerated mud has a high volume of air. Figure 13–13 shows corrosion rate using 50% GLR (Gas Liquid Ratio) in different drilling fluids, injection gases, and temperatures. Reducing the O_2 concentration greatly reduces the corrosion rate. Note the effect of increasing temperature and chloride.

Corrosion inhibitors are used to minimize the corrosion problem in coil tubing underbalanced operations. Without corrosion inhibitors coil tubing bottom-hole assemblies (BHA) can be plugged due to severe corrosion resulting in suspension of the operation. Figure 13–14 shows the BHA from coil tubing with and without corrosion inhibitors. The first picture shows how an underbalanced stimulation operation for a depleted reservoir in Brazil field using nitrogen (5% maximum oxygen) resulted in corroding the coil tubing. Acid was pumped after the rust has formed resulting in washing the corrosion product downhole and plugging the BHA. The inhibited coil tubing prevented corrosion of coil tubing and pumping acid did not plug the BHA.

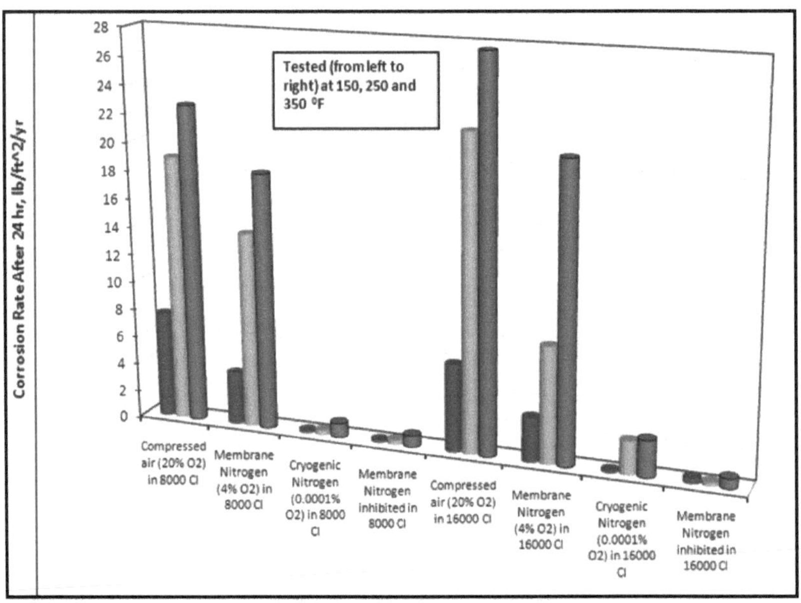

Figure 13–13 *Corrosion rate comparisons of 50% GLR at various conditions (Kippie et al., 2007)*

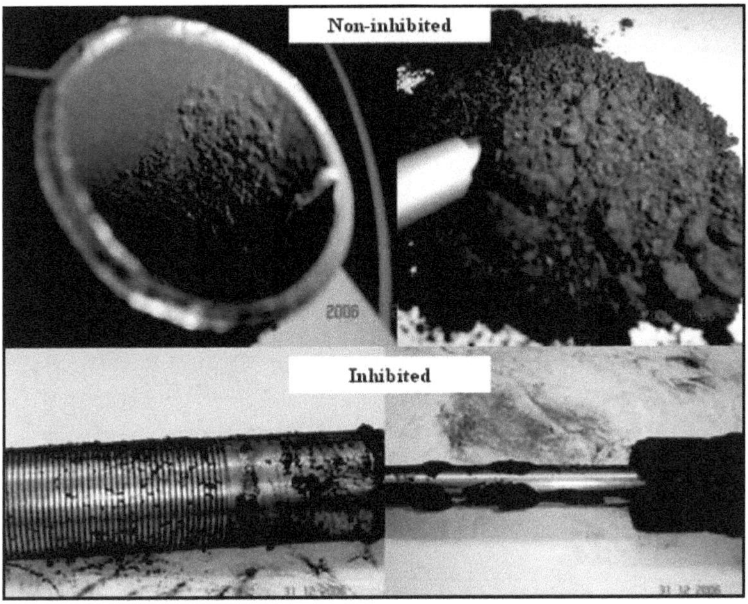

Figure 13–14 *Comparison between non-inhibited and inhibited BHA from the coil tubing used in UBD (Duque et al., 2008)*

13.6.1 Prevention

Prevention is more desirable with regards to corrosion, especially since testing only indicates after corrosion has occurred. Prevention takes not only the equipment form but also the chemical form in the profile of the drilling fluid.

13.6.1.1 Plastic Coating

Plastic coating is an option commonly used to prevent corrosion (see Figure 13–15). However any leak in the coating will result in severe local corrosion. Look for pitting as described above in any hole in the coating.

13.6.1.2 Corrosion Resistant Alloys

Corrosion resistant alloys (CRAs) such as Cr^{13+} can be used in high temperature within a corrosive environment. However, they are sensitive to low pH fluids such as hydrochloric acid (HCl). A strong acid, like HCl, will react with the protective layer of this type of alloy. Weak organic acids have minimum effect on such tubulars.

Figure 13–15 *Coated pipes (Clariant)*

13.6.1.3 pH Control

Corrosion is normally controlled with high pH. A pH of 7 is neutral but for corrosion control drilling fluids should be kept above pH 8 or 9. The common treating agents are:

- Caustic Soda (NaOH): Sodium hydroxide is the most common agent and generally the best choice from both a chemical and economical viewpoint.
- Potassium Hydroxide (KOH): Potassium hydroxide is normally used when it is desirable to keep the potassium concentration high to inhibit shale. From a corrosion viewpoint, it should not be used in saline or salt water.
- Soda Ash (Na_2CO_3): Sodium carbonate is often used as an indirect way of keeping the pH above 8 but below 10 while treating hard makeup water. It can cause some carbonate corrosion products if used in excess beyond treating hard water. It is an easy treatment for hard water especially when using foams.

In many areas the drilling fluid represses corrosion. When drilling with conventional mud systems, the pH is normally kept in the 8.5 to 10 range. This is a pH range where corrosion is limited due to the repression of the corrosive water born ions. To understand this, note first that in the pH range below about 8.5, the lower the pH, the more soluble Fe^{2+} becomes. The anode reaction, see Eq. (13.2), is more favored by the increased solubility of Fe^{2+} at pH values in the range below about 8.5; therefore, keeping the pH above about 8.5 tends to minimize the corrosion rate.

Furthermore, in the pH range above about 10, the higher the pH, the more soluble complex anions such as $Fe(OH)_3^-$ become and the anode reaction, Eq. (13.14), is more favored by the increased solubility of those complex anions at pH values in the range above about 10; therefore, keeping the pH below about 10 tends to minimize the corrosion rate.

$$Fe + 3OH^- \rightarrow Fe(OH)_3^- \tag{13.14}$$

With the higher hydroxide anion concentration, the Fe^{2+} tends to get complexed by the hydroxide anion to become a iron-containing complex anion which may be $Fe(OH)_3^-$, $Fe(OH)_4^{2-}$, $Fe(OH)_5^{3-}$, or $Fe(OH)_6^{4-}$, depending on pH.

A further technical note: In some cases, it has been noted that the downhole metallurgy is protected by a passivating layer of material such as $Fe(OH)_2$, as indicated in Figure 13–1. Other passivating materials could include Fe_3O_4, and mixed oxide-hydroxide compounds such as FeO(OH). Figure 13–1 shows only a partial coverage by the passivating material, but if the layer covers all exposed surfaces (i.e., is free of scratches, abrasions, etc., and the drilling fluid is not erosive of the passivating layer, then it may be observed that the pH does not affect the corrosion RATE at values between 4 and 10 for the drilling fluid because corrosion rate depends on speed of oxygen diffusion through the passivating layer (a very slow process) (Chitty, 1998) (see Figure 13–16).

13.6.1.4 Oxygen Control

Air contains about 21% oxygen, which is the reason for the old rule about submerging the mud guns so as not to beat air into the mud. The emulsifiers in the oil based mud systems oil-wet the pipe which forms a barrier against corrosion.

The net result is that for the most part very little corrosion occurs in most conventional drilling operations using oil or synthetic liquid as the external phase in the drilling fluids. The other extreme is that corrosion in conventional air/mist drilling is always a problem. As a general rule, drilling induced corrosion is most extreme in misting and gaseated systems.

Air drilling mist systems (Chapter 5) can normally be handled because the volume of water in the mist system is relatively low, about 6 bbl/hr (1m^3/hr). The pH can be kept above 9 and buffered with excess caustic soda (NaOH), potassium hydroxide (KOH), lime (CaO), or calcium hydroxide ($Ca[OH]_2$). A commercial phosphate inhibitor can also be used for a reasonable cost. Amines are not too effective because their coating is soft enough to rub or wash off.

When membrane nitrogen is used, there is still a corrosion problem because it normally contains about 5% oxygen (as opposed to 21% in air) and the same general precautions have to be taken.

Gaseated systems (IADC mud gas mixtures) using air have a significant problem with corrosion, especially in areas where corrosion is prevalent in the production process. Large quantities of air are used, and it is difficult to control oxygen and other types of corrosion because of high chemical costs. Extreme water and solids velocities in the upper part of the hole tend to wash off any passive coatings formed by the inhibitors. On the other hand, there are areas in the oil fields where corrosion is not a problem in the production process, and in those areas corrosion appears to be very limited even with air/water mixtures.

The best solution with gaseated systems using air or membrane nitrogen is to use a dual casing string, or parasite string, and only inject the air or membrane nitrogen into the fluid in the upper part of the hole. Chapter 2, Flow Drilling, and Chapter 3, Gaseated Fluids, contain extensive discussions about dual casing strings, and parasite strings.

Foam systems are a bit different. Foam uses a limited amount of air or membrane nitrogen and the oxygen can be treated out of the makeup water. A proper foam can form a strong enough chemical "skin" around the oxygen or air bubble that it cannot penetrate to the steel and start the corrosion process. In workover operations with foam, even very small amounts of oil in the system will react with the foaming agents to form an oil wet coating on the tubing which represses corrosion. Properly devised foam has proven very effective in controlling corrosion in workover operations in the Pinedale/Big Piney area of Wyoming where casing and tubing corrosion is a major problem.

However, oxygen can be effectively treated out of foam systems where a limited amount of water is used. In regular mud systems or gaseated systems, it is impractical from a material and cost standpoint to treat out the oxygen. The most common oxygen scavenger is ammonium bisulfate (NH_4HSO_3). In most cases, treatment is not economical because it takes 6:1 by weight of bisulfate /oxygen to neutralize oxygen, see Eq. (13.15).

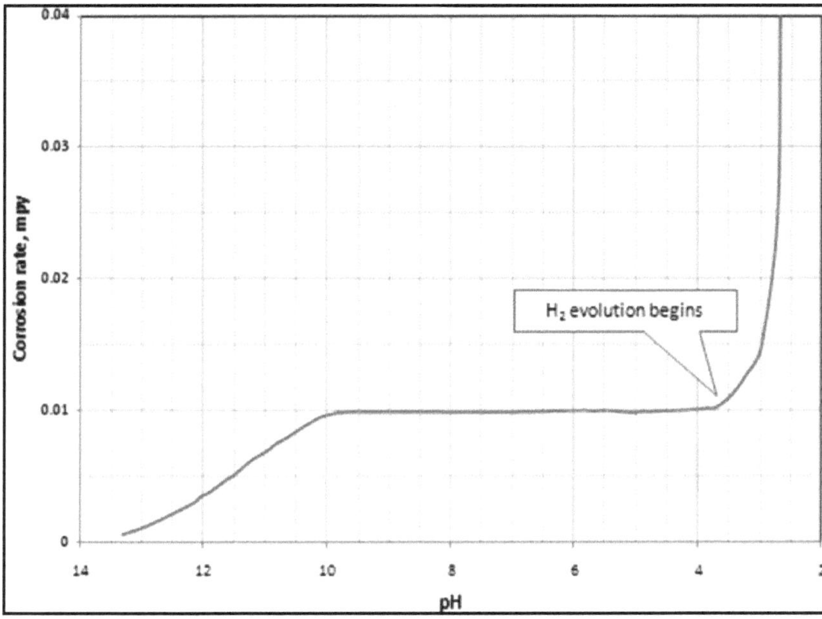

Figure 13–16 *pH versus corrosion rate mpy*

$$2(NH_4)(HSO_3) + O_2 \rightarrow (NH_4)_2SO_4 + H_2SO_4 \quad (13.15)$$

Please note that sulfuric acid (H_2SO_4) is a result of the oxygen reduction. Be sure that enough caustic soda is available to neutralize the acid and keep the pH above 8.

Lignite and lignosulfonate used as drilling mud additives are also effective oxygen scavengers. There are a number of other oxygen scavengers that are used in various water treatment and boiler treatment operations. In general, they do not lend themselves to temporary rig operations.

13.6.1.5 Emulsifiers and Oil

Many chemical emulsifiers are bi-polar and will oil-wet the steel of the pipe. The oil wetting of the steel pipe forms a coating and restricts the ability of the corrosive source to attack the steel. The addition of 1–2% by volume oil in work over or drilling operations will keep corrosion cells from forming. Invert emulsions and oil based fluids are an excellent protection against corrosion. The long term effect of preventing corrosion depends upon constant replenishment of the emulsifier and oil.

13.6.2 Treatments

The following materials limit corrosion by forming a film on the steel. The following chemical descriptions are an introduction to some of the more common commercial inhibitors. Acids or an acid environment will tend to weaken or destroy the various films. The films are also subject to erosion from high velocity fluids and particles, so a continuous flow of inhibitor is required to limit corrosion by isolating the steel of the drillpipe or casing from any corrosive elements. The corrosion inhibitor should be checked for concentration at both the suction and the flowline.

13.6.2.1 Organo-Phosphate Esters (Anionic)

Organo-phosphate esters are good general corrosion inhibitors, and their anionic nature aids in their reactive and inhibitive state. They are usually added in the range of 500 to 2,000 ppm. They are soluble, are easy to mix, will withstand temperatures up to about 150°F, are biodegradable and environmentally friendly. The organo-phosphate combines with oxygen to form a film on the steel drillpipe. As a result they help with oxygen control, and therefore are effective with chloride/oxygen embrittlement (McNealy et al., 2009).

They operate best in a pH of 8 to 12. When high calcium concentrations are present in the water, phosphates will produce a phosphate scale.

13.6.2.2 Phosphonates $(PO_3)^{-2}$

Phosphonates combine with a complex organic anion, are a primary scale inhibitor, and are not totally effective as a corrosion agent by themselves. They operate best in a pH of 8 to 12. They will also produce a phosphate scale with high calcium concentrations in the water. Phosphonates combine with oxygen to form a coating on the pipe and are temperature stable up to 250°F. They are easily soluble and work well with the organo-phosphate to form a good "1–2 punch" Polyphosphonates can be purchased and can be added directly to the system as a general broad inhibitor.

13.6.2.3 Amines (Cationic)

Amines have been used in drilling fluid for a long time. They are cationic and sometimes upset a mud or foam system. Amines form a thick black soft film on steel pipes. The film is porous to atomic oxygen and can be washed or rubbed off. The amines are good inhibitors when storing pipe but only partly effective in a flowing mud system.

13.6.2.4 Nitrates

Nitrates, as used in oil field operations, are a derivative of the amines. They form an effective hard resistant film but they require high concentrations and a high pH. With most drilling fluids, it is hard to get the combination of high concentration and the high pH that are required to form a film or coating using this chemical. In general, they are not in common use in the oil field

13.6.2.5 Silicates

Potassium silicate (K_2SiO_3) is the basis of silicate mud and available as a liquid or powder. The use of silicates to form a corrosion resistant film is common in the water treating industry. The powdered form of the material is the most practical for oil field use. A concentration of 400 ppm (about 8 lb/bbl) is an effective treatment level.

Silicates form a very hard resistant film over the pipe surface which further inhibits corrosion. The silicates will film over active corrosion pits, boils, and barnacles, and clear the water of iron oxides, which make it possible to measure if any further corrosion is taking place. On the other hand, it may not stop corrosion in active pits and so it is best to add the silicates to a clean pipe at the beginning of any operation when possible.

The ability to observe ongoing corrosion is particularly important in workovers and re-entries since old corrosion products tend to mask newer corrosion. Long-term corrosion probably continues in those silicate covered pits unless the down-hole environment is changed.

13.6.2.6 Lime (CaO)

Lime (CaO) is used for pH control with hydrogen sulfide. To control the H_2S in a drilling fluid, maintain a pH above 8. Caustic soda (NaOH) with H_2S forms sodium sulfide (Na_2S). While it will bind the sulfide, Na_2S is unstable and reverts at a low pH to H_2S gas (see Figure 13–6).

13.6.2.7 Zinc Carbonate ($ZnCO_3$)

Pre-treating with zinc carbonate ($ZnCO_3$) is one of the conventional solutions to neutralizing hydrogen sulfide gas that enters the mud stream. However, zinc compounds at high temperature affect rheology of drilling fluids.

The normal pre-treatment for traces (up to 100 ppm) of H_2S is 0.2 lb/bbl (0.6 Kg/m^3). When treating with zinc carbonate, add the sack of $ZnCO_3$ slowly (about 15 minutes to a sack) since it tends to flocculate or foam the drilling fluid. The simplified reaction is:

$$ZnCO_3 + H_2S \rightarrow ZnS\downarrow + HCO^{3-} \qquad (13.16)$$

13.6.2.8 H_2S Scavengers

Hydrogen sulfide scavengers are used to absorb the sulfides and remove them from the system. A finely divided iron oxide (hematite) is commonly used as the H_2S scavenger, see Eq. (13.17). Mixtures of different iron oxides such as iron (II) oxalate and iron sulfate are commonly used to improve performance of the H_2S scavengers.

$$Fe_2O_3\ H_2S \rightarrow FeS\downarrow\ H_2O \qquad (13.17)$$

For drilling fluids, a pretreatment of 20 lbs/bbl (50 kg/m^3) can be used. The material is also used to clean H_2S from gas streams. The reaction is slower in a very high pH environment. In high pH environments, ferrous gluconate (an organic iron chelating agent) is used. Ferrous gluconate is stable at high pH level up to 11.5, Eq. (13.18) (Fink, 2003).

$$2Fe(C_6H_{12}O_7)_2 + S^{2-} \rightarrow 2FeS\downarrow + 4C_6H_{12}O_7 \qquad (13.18)$$

Water soluble H_2S scavengers are based on amino alcohol solutions. Alkylamine-formaldehyde condensates can be used as H_2S scavenger in solvents such as diesel, kerosene or low molecular weight alcohols.

13.6.2.9 Other H_2S Treatments

With almost all acid gas problems, it is generally more practical and effective to use a commercial agent and an experienced technician. Generally, it takes up to five times as much agent as there is H_2S.

$$Mg\ (treatment)/1H_2S = (682 \times P_{mud})/SpG_{(mud)} \qquad (13.19)$$

There are a number of commercial treating agents that bind H_2S into a harmless insoluble material. These agents include, but are not limited to: oxidants, chlorine dioxide and aldehydes.

Oxidants

$$O_2 + H_2S \rightarrow 2S + 2H_2O \tag{13.20}$$

Chlorine dioxide

$$5H_2S + 8ClO_2 + 4H_2O \rightarrow 5SO_4^{2-} + 8Cl^- + 18H^+ \tag{13.21}$$

Aldehydes

Aldehydes can be also used as sulfide scavengers. They are cost effective but are limited by temperature and pH.

13.7 Make Up Water Problems and Solutions

Make up water that causes or accelerates corrosion is an induced problem. Local surface water needs to be chemically analyzed for excessive hardness, bicarbonates, or organic activity. Very hard water generally will not cause corrosion by itself, but it requires excessive corrosion treatment because the hardness ions (calcium and magnesium) attach to and neutralize the corrosion treatment.

13.7.1 Red Production Water

Red production water is commonly called gun barrel water. This is usually acidic and the red indicates it contains corroded iron. Do not use it if possible. If it must be used, keep the pH above 9 and see the treatment section.

13.7.2 Black Production Water

Black production water is also called gun barrel water and has the same problems as with red production water. The black is a sign of iron reduction (corrosion). It is an acidic solution. If it must be used, keep the pH above 9 and see the treatment section.

13.7.3 Potassium Chloride Solution

Used fracwater containing potassium chloride (KCl) is not an acceptable solution. Just don't use it. Three to four percent KCl is the most corrosive concentration (see Figure 13–10). Keep the pH above 9 if you must use it, but there is no short quick treatment.

13.7.4 Saltwater

Saline or salt water requires high treating concentrations for corrosion control. Keep the pH above 9 with caustic soda and be prepared to add excess treating chemicals.

13.7.5 Hard Water

Hard water or gypsum water (containing excessive Mg^{+2} and/or Ca^{+2}) reduces the corrosion treating chemicals effectiveness. The hardness needs to be treated out with caustic soda or soda ash. Limit the use of soda ash to the minimum amount required to precipitate the magnesium and calcium in the rig make up water.

13.7.6 Solution with High Bicarbonate Content

High bicarbonate ion, (HCO^-) content above 1,000 ppm, is often found in semiarid areas where there are extensive gypsum or anhydrite bearing formations. Treat with caustic soda (NaOH) to raise the pH above 8.5. With high bicarbonate ion content, it may take an excessive amount of caustic to get the pH to 9. High bicarbonate ion values are often a forerunner for bad corrosion problems with a low pH.

13.7.7 Wastewater

Wastewater from a sewage treatment plant may cause biological corrosion or it could give any number of other problems. The best clue may be that it smells bad. The smell of H_2S may indicate the water has been inoculated with sulfate-reducing bacteria. If these are present in the formation being drilled, the formation may get inoculated and turn sour. If possible, do not use it. If you must use it, have the water tested and get expert advice on how to treat it. In general you may have to use biocides or aerate the water.

13.8 Formation Water Quick Solutions

Most corrosion is a result of the local drill water and/or the formation water. Corrosion that is the result of formation water is localized to areas or fields with specific water conditions. Most of the following conditions can be treated before use. The simplest treatment is the use of caustic soda (NaOH) to raise the pH above 9. Soda ash (Na_2CO_3) can be used in place of caustic soda, but it runs the possible risk of scaling in workover or production operations, and it is probably not a good choice for hot wells.

13.8.1 Acidic Formation Water

Acidic formation water with a pH below 6 is usually black. This should be treated with caustic soda (NaOH) to raise the pH above 9.

13.8.2 Bicarbonate

High bicarbonate ion (HCO^-) levels above 1,000 ppm should be treated with caustic soda (NaOH) to raise the pH above 8.3. With high bicarbonates it may take an excessive amount of caustic to get the pH to 9.

13.8.3 Carbon Dioxide

Carbon dioxide (CO_2) should be treated with caustic soda or soda ash to raise the pH above 9.

13.8.4 Sulfate/Sulfide Ions

For various sulfate or sulfide ions, raising the pH above 10 with potassium hydroxide (KOH) or caustic soda (NaOH), will limit corrosion.

13.8.5 Hydrogen Sulfide Gas—Trace amounts

For hydrogen sulfide gas (H_2S) in trace amounts, keep the pH above 10 with caustic soda and lime and use one of the commercial iron sequestering agents. Pre treat the drill water with zinc carbonate at about 0.3 lb/bbl (0.6 kG.m^3), a scavenging agent at 10lb/bbl (30 kG/m^3), or another commercial agent. Hydrogen sulfide in more than trace amounts requires longer treatments.

13.8.6 Saline Water

It is best if saline water is not used with air injection fluids. Keep the pH above 7 when using saline water in other fluids. If possible a saturated solution would be less corrosive.

13.8.7 Potassium Chloride Water

The use of potassium chloride (KCl) water is not a bad short term corrosion problem if the pH is kept above 9. If it is not completely displaced out of the well there will be significant long term corrosion problems with tubing and casing.

13.9 General Recommendations to Minimize Corrosions in UBD Water Based Systems

13.9.1 Flow Drilling

- Do not use 3–10 wt% KCl or NaCl and try to use higher concentration to slow down the corrosion rate. For shale inhibition, use synthetic shale inhibitor.
- Avoid mixing air into the mud system.
- Maintain pH above 9.
- Add inhibitors if corrosion signs appeared.
- Keep mud guns submerged to avoid beating air into the drilling fluid.

13.9.2 Aerated or Gaseated Mud

- Do not use 3–10 wt% KCl or NaCl and try to use higher concentration to slow down the corrosion rate. For shale inhibition, use synthetic shale inhibitor.
- Maintain pH above 9.
- Use phosphate ester corrosion inhibitors.
- If pH is difficult to change, change drilling fluid type. Diesel oil and nitrogen system are the best to be used in this case.

13.9.3 Foam

- Use good foaming agents to form tight air emulsion.
- Do not use 3–10 wt% KCl or NaCl and try to use higher concentration to slow down the corrosion rate. For shale inhibition, use synthetic shale inhibitor.
- Use oxygen scavenger if you are not using a drilling motor.
- Maintain pH above 9.
- Use phosphate ester corrosion inhibitors.

13.9.4 Mist

- Use good quality fresh water.
- Do not use 3–10 wt% KCl or NaCl and try to use higher concentration to slow down the corrosion rate. For shale inhibition, use synthetic shale inhibitor.
- Maintain pH of 10–11.
- Use phosphate ester corrosion inhibitors.

13.10 Questions

1. Standing on the rig floor during a trip, what would you expect to see as the first sign of potential corrosion?

2. What is the most important mud property that you can measure to help limit corrosion?

3. In number 2 above, what number should that measurement show or what is the best range for the measurements?

4. The water truck brings make up water for the drilling fluid and it has just a little red tint and has 1,500 ppm chloride (It's just a little salty). What problems can you expect to have with the drilling fluid?

5. On a trip, if some of the drillpipe in the derrick looks black—as a black film on it—what would be the first thing you might check?

13.11 Harder Questions

6. On a re-entry of a well that is several years old, you are seeing black in the drilling fluid or black tones in the filtrate. On the shale shaker are black hard flakes.

   (a) What is it?

   (b) How can you tell from the shaker samples?

   (c) What is your problem and how do you solve it?

7. In a field known for corrosion you need 9.9ppg drilling fluid. You as the operator have the choice of buying made up saturated salt water at $7.00/bbl (including trucking) or 1,500 ppm brackish water from the frac and production pit for trucking at $1.50/bbl and building the mud weight with salt at the rig. What is your choice and why?

8. If you tested the iron, calcium, pH, and oxygen at the suction, is there anything else you should do in your general corrosion testing procedure?

9. (NH_4HSO_3) is a common chemical treating agent.

   (a) What is it?

   (b) What is it used for?

   (c) What are the limitations?

10. What are the signs of trace amounts of hydrogen sulfide entering the well bore?

13.12 Answers

1. Red rust on the pipe.

2. pH.

3. Keep the pH above 7 and preferably about 9.

4. It will take more chemicals to treat the mud and the red tint may mean that the water was in a well where corrosion was taking place.

5. The black film could come from carbon dioxide, hydrogen sulfide, excessive oxygen in the drilling fluid. The first thing to check would be the pH of the drilling fluid at the flow line (it should be above 7 or better still 8.5) to be sure to neutralize the corrosive conditions and especially neutralize any hydrogen sulfide.

   It could also be oil-wet pipe from an emulsifier and oil.

13.13 Answers to Harder Questions

6. (a) You are probably looking at magnetite.

   (b) The shaker samples are magnetic.

   (c) The magnetite is coming from the old casing and does not necessarily mean you have a problem. This is long term corrosion. Keep the pH and corrosion inhibitor concentration up.

7. The operator should be concerned about the long term effect of corrosion on the casing and possibly scale in the tubing. Chemical treating costs might also be significant. The choice based on unknown corrosion problems from production water and frac water and rig time would tend towards paying the higher price for clean saturated salt water.

8. It is important to test at the flow line to see the change on the trip through the hole. There is no way to directly measure the down-hole problems directly. The only opportunity is to see what the reaction has been.

9. (a) It is ammonium bisulfate.

   (b) It is used to treat oxygen out of water.

   (c) It takes a large amount of bisulfide to treat oxygen on a 6/1 ratio.

10. Trace amounts of hydrogen sulfide might:

    (a) lower the pH

    (b) make a black stain or coating on the drillpipe

13.14 References

Alcoa Aluminum Alloy Drillpipe, Publication AO1049, 2010.

Asrar, N. "Corrosion Control of Drilling Tools Through Chemical Treatments-Effectiveness and Challenges," SPE 130515 presented at the SPE International Conference on Oilfield Corrosion, Aberdeen, UK, May 24–25, 2010.

Chitty, G.H. "Corrosion Issues with Underbalanced Drilling in H_2S Reservoirs," SPE 46039 presented at the IADC/SPE Coiled Tubing Conference, Houston, TX, USA, March 15–16, 1998.

Duque, L.H., Guimarães, Z., Berry, S.L. and Gouveia, M. "Coiled Tubing and Nitrogen Generation Unit Operations: Corrosion Challenges and Solutions Found in Brazil Offshore Operations," SPE 113719 presented at the IADC/SPE Coiled Tubing and Well Intervention Conference and Exhibition, The Woodlands, TX, USA, April 1–2, 2008.

Fink, J.K. *Oil Field Chemicals*, Gulf Professional Publishing, Boston, MA, USA, 2003.

Kippie, D.P., Bellinger, C.E. and Scott, P.D. "Combining a Mechanical and Chemical Solution to Mitigate Corrosion in MPD/UBD/Air Drilling Operations," SPE 108338 presented at the IADC/SPE Managed Pressure Drilling and Underbalanced Operations Conference and Exhibition, Galveston, TX, USA, March 28–29, 2007.

McNealy, R., Hausler, R. and Tabinor, M. "Corrosion Inhibition of Low-Alloy Steels in Brine With Highly Oxygenated Nitrogen Membranes Gas for Underbalanced Drilling Applications," SPE 124044 presented at the SPE Annual Technical Conference and Exhibition, New Orleans, LA, USA, October 4–7, 2009.

Oilfield Production Chemicals and Microbiology, Clariant.

Rehm, B. *Practical Underbalanced Drilling and Workover,* Petroleum Extension Service at the University of Texas, Austin, TX, USA, 2003.

Roberge, P. R. *Handbook of Corrosion Engineering,* McGraw-Hill, New York, NY, 2000.

Stone, P.C., Steinberg, B.G. and Goodson, J.E. "Completion Design for Water floods and CO_2 Floods," *SPE Production Engineering,* 4, No. 4, November 1989, pp. 365–370.

Index

Numerics

60° zone 111

A

absolute pressure (psia) 298
acceleration pressure 181
access ladder 358
accumulator effect 146, 147, 152, 154
acoustic fluid level measuring device, *see* floating mud cap drilling (FMCD)
aerated drilling, *see* aerated system
aerated system 110, 201, 245, 270, 318
  aerated mud 110, 600
air system, *see* gas system
air volume, *see* gas volume
annular back pressure 124, 134, 146, 151, 285
  control valve 483
annular blow out preventer (BOP) 71, 364, 372, 395, 423, 477, 498, 522
  packoff 367
  rams 350
annular circulating friction 379
annular flow 175, 180, 339
  model 338, 344
annular friction pressure 136
annular liquid 122
annular pressure 114, 146, 201, 227, 299, 352, 379
  control 152
annular pressure loss (APL) 15, 48, 50, 52, 122, 124, 126, 221, 227, 415
  bubble expansion 15
annular preventer 129, 137, 350
annular sealing element 23
annular surface pressure 217
annular velocity 204, 206, 215, 325, 402
annulus 27, 48, 49, 53, 74, 96, 103, 131, 174, 299, 309
  pressure 117

B

balanced pressure drilling, *See* managed pressure drilling (MPD)
basic gas laws 11
bentonite drilling mud 110
Bernoulli equation 89, 149
Bingham plastic model 83, 88, 106, 261, 264, 284
Bingham-plastic fluid 198, 256, 259
bit float 317
Blasius equation 342

blooie line, *see* flow line
boosters 320, 323
bore-hole
  friction 212
  pressure 308
  stability 46, 468
  walls 298
bottom-hole assembly (BHA) 48, 70, 136, 160, 171, 301, 313, 344, 359, 415, 425
  commercial deployment systems 364
bottom-hole pressure 12, 15, 21, 48, 50, 52, 73, 98, 111, 114, 115, 124, 127, 131, 132, 181, 215, 272, 278, 285, 298, 299, 336, 498, 522
  bubble expansion 15
  gage 227
  measurements 168
  reduction 15, 112, 119, 198, 204, 211, 305
  requirement 118
Brzustowski method 549
Buckingham-Reiner equations 271

C

cable tool drilling 2
casing drilling 2, 8
  smear effect 8, 404, 406
casing pressure 55
choke pressure 15, 133
  gauge 134
choke system 15, 47, 52, 152
  choke manifold 479, 498, 499, 511
  choking 299, 325, 464, 498
  Power Choke drilling choke units 479
  pressure 213, 372, 522
Churchill equation 269
circulating pressure loss 1
Coanda effect, *see* flare tips
Colbrook-White equation 268
commercial deployment systems, *see* bottom-hole assembly (BHA)
compressors 198, 320, 323
concentric annular viscometer 261
concentric casing 33, 163, 164
  injection 74
  nitrogen injection 33, 169
  technique 130
  see dual casing string
concentric string 75, 145, 530
  see dual-casing string
connection fatigue life, *see* underbalanced liner drilling (UBLD)
constant bottom-hole pressure (CBHP) 394
constant circulating SUBS 5, 111, 127
  non stop driller 128, 220
constant circulating system 5
continuous annular injection 389, 394
  see also floating mud cap drilling (FMCD)
  see also pressured mud cap drilling (PMCD)
continuous core slim exploration holes 304
conventional mud pulse (MWD) 129, 131
corrosion 204, 577, 582, 595
  alloy corrosion 581
  black coating 584
  carbon dioxide 588–590
  chemicals 198, 241
  control 591, 598
  corrosion coupon 598
  corrosion curve 591
  corrosion rate 591, 595, 597
  electrical corrosion cell 579
  hematite 580, 582
  hydrogen sulfide 585
  identification 582
  iron corrosion 579
  magnetite 584
  measurement 597
    electrical resistance probe measurement 598
  OIL RIG 580

pitting 583
reactions 578
anode reaction 580
cathode reaction 580
stray currents 585
test procedure 591, 595
alkalinity 594
electric oxygen probes 593
iron testing 594
pH 592
total hardness 593
see also corrosion prevention
see also corrosion treatment
see also make up water
corrosion prevention 598, 601
emulsifiers 603, 605
inhibitors 211, 599
plastic coating 601
treating agents 602
corrosion resistant alloys (CRA) 601
corrosion treatment 606
amines 606
commercial treating agents 609
hydrogen sulfide scavengers 608
lime 607
nitrates 607
organo-phosphate esters 606
phosphonates 606
potassium silicate 607
zinc carbonate 607
critical flow 147, 152
critical pressure ratio 149
cutting-fluid mixture 291
cuttings transport 286, 299, 335
cyclic bending stress, *see* underbalanced liner drilling (UBLD)

D

data acquisition (DAQ) 497, 498, 509
packages 534
see also pseudo productivity index (PI)
deduster 320
see also muffler
see also separator
deepwater wells 394
see also mud cap drilling (MCD)
degassers 501
centrifugal degassers 501
vacuum degassers 501
differential pressure sticking 114
down-hole casing 55
down-hole deployment valve (DDV) 167, 532
down-hole fire 313–316, 318, 326, 330, 447
classic fire triangle 316
ignition source 316
potential 316
down-hole isolation valve (DIV) 32, 69, 164, 392, 395, 474
down-hole positive displacement motor (PDM) 477
down-hole pressure 42, 206
gauge 395
down-hole safety valve (DHSV) 131, 132, 363
down-hole tools 382
drag type bits 207
drift-flux 192
drill cuttings 91, 299, 300
slip velocity 105
drill pipe injection 21, 126, 127, 131
values 133
drill rate 43, 214, 298, 301, 302, 305
distance 207
instantaneous penetration rate 207
driller's method of well control 15, 18, 49
casing pressure constant 18
drilling annulus 148
drilling fluid 52, 74, 85, 105, 245, 271, 327, 330, 406
density 2
non-aqueous 90, 92
system 134
drilling the well on paper (DWOP) 496
drillpipe hole 24
drill-pipe pressure 213

drillstring 69, 74, 98, 103, 130, 174, 243, 317, 337, 338, 344, 348, 382, 399, 406, 419, 476, 530
  drill collars 337, 345
  liquid 122
dry gas 298, 306, 317, 324, 325
  injection rate 335
dual casing system 109
  dual casing string 7, 21, 55, 110, 131, 132
dual drill-pipe 304, 309
dual flapper float valve 160
dual float sub 476
dual gradient drilling 7
dune effect 111
Duns and Ros correlation 187
dusting 312

E

electric submersible pump 165
electromagnetic measurement while drilling (EMMWD) 78
  tool 33, 160
electromagnetic telemetry 167
emergency escape system 358
emergency shutdown systems (ESD) 475
emergency shut-down valve (ESD) 478
empirical correlations 174
empirical models 181
equipment layout drawings (ELD) 469
equivalent circulating density (ECD) 12, 37, 42, 44, 47, 48, 55, 73, 113, 117, 118, 122, 145, 148, 205, 235, 324, 372, 381, 385, 406
  management 402
expandable liners 2

F

Fanning friction factor 258, 343
final circulating pressure (FCP) 19
finite-difference method (FDM) 274, 284
fire float 317
fire stop float, *see* fire float
fishes, *see* bottom-hole assembly (BHA)
flare stack
  derrick supported stack 551
  guyed supported stack 550
  height 549, 569, 570
  self supporting stack 551
flare system 352, 464, 537
  combustion 543
  continuous non-emergency flaring 548
  dispersion analysis 538
  elevated flares 540, 550
  flame out conditions 538
  flammability limits 543
  flare header design 549
  flared gases 540
  flaring conditions 538
  fuel properties 542
  gas seals 557
  heat radiation 538
  ignition temperature 543, 546
  liquid seal drum 559
  non-smokeless flares 548
  pilot flame failure panel monitors 554
  primary function 538
  pulsation 564
  purge rates 558
  screen flame arresters 559
  smoke suppression systems 541
  smokeless flares 541
  testing 548
  toxic limits 538
  transmissivity 569
  turndown 540
  see also Brzustowski method
  see also ground flares
  see also stoichiometric combustion
flare tips 549, 556
  air assisted 553
  Coanda effect 551–553
  flame retention device 551
  flare pilots 554

thermocouple system 554
flare stabilization 551
non-smokeless 551
flashbacks 311
water leg 311
floating bed 300, 308, 309
floating mud cap drilling (FMCD) 381, 386, 390, 394
acoustic fluid level measuring device 387
oil based annular mud 390
floating rig 394
flow drilling 13, 16, 39, 46, 49
flow line 298
blooie line 298
flow pattern 308, 328, 338
laminar flow 79, 100
turbulent flow 79, 99, 104
flow rate 46, 340
flow simulation 517
fluid injection rates 60
foam density 272
foam drilling 39, 234, 241, 243, 267, 270, 271, 272, 285, 300
power units 243
surface choke 285
see also foam system
foam flow 273, 275
boundary 256
foam fluid 475
foam height (FH) 246
foam hydraulics 273
foam modeling 15
foam plastic viscosity 259
foam quality 264, 275, 278, 287
foam rheology 255, 274, 284
model 256
foam system 16, 42, 197, 199, 201, 205, 209, 211, 215, 222, 245, 253, 306, 313, 442, 448, 527, 529, 604
advantages 204
bubble film 237
bubble structure 234
carrying capacity 236
challenges 208–209
characteristics 198, 204
cleanout fluid 235
defoamers 210, 221
dry foam 202, 264
emulsion 197, 199, 201, 209
foam column 218, 225
foam enhancers 218
foam stability 234, 235
foam stiffeners 198, 221
foam volume 198
foaming agent 198, 202, 209, 215, 220, 222, 234, 241, 330, 604
gray area 236
guiding rule 234
incremental costs 198
limits 217
lost circulation material 219
mechanical equipment 208
milling fluid 235
modeling 237, 239
nuclear testing 201
oilfield practice 239
operations 12, 241
properties 234, 239
puff 236
recyclable foam 200, 245
robust foam 237
shear 234
single pass foam 200
stabile foam 202
stability 206, 236, 270
stiff foam 202, 219
texture 234, 236
two-phase foam 238
wet foam 202, 264
see also misting system
foam velocity 267, 271, 283
foam viscosity 256
formation flow 298
formation fluids 11, 24, 115
influx 274, 335
formation instability 309
formation pressure, *See* underbalanced drilling (UBD)
formation waters 209, 611
fracture gradient (FG) 496
fractured formation 377, 381, 383
friction controlled drilling 73

friction dominated regime 112, 121, 124, 125, 133, 214, 215
friction factor 79, 269, 281, 290, 342
  interfacial 342
  superficial 342
  see also Fanning friction factor
  see also Moody friction factor
friction pressure 206, 271, 522
frictional pressure loss 181, 192, 281, 299
front-end engineering and design (FEED) 469

G

gage pressure (psi) 298
gas bypass line 319
gas cap 30
gas compression 215, 299
gas core 341
  properties 342
gas cutting 11, 13
  strong-white equation 12, 13, 15
gas drilling 39, 297, 298, 305, 527, 529
  advantages 298, 305
  gas zones 305
  see also gas system
gas expansion 212
gas flow 133, 217, 302
gas holes 324
gas injection 35, 109, 112, 126, 130, 131, 148, 152, 153, 154, 172, 215, 217, 320, 427, 490
  rate 121, 335, 337, 346
  techniques 145
gas measurements 513
gas migration 374, 394
gas percolation 50
gas rates 513
gas ratio 214
gas slug 318
gas system 297, 312, 330, 442
  dampness 307
  depth limits 308
  gas influx 326
  gas velocity 330
  gas volume 323
  hole enlargement 308
  hole instability 308
  mist pump 321
  models 336
  nitrogen drilling 327
  oil influx 326
  operations 323, 328, 335, 338
  rig equipment 317
  shale formations 308
  unconventional gas 307
  see also misting system
gas tight, *see* underbalanced liner drilling (UBLD)
gas velocity 344
gas volume 13
  measurement 298, 300
gas volume fraction (GVF) 181, 234, 237
gaseated drilling 5, 39
  see also gaseated system
gaseated fluid 48, 109, 110, 119, 127, 145, 215, 235, 529
gaseated system 16, 21, 42, 109, 112, 113, 126, 134, 154, 174, 205, 206, 215, 313, 442, 448, 527, 585, 604
  advantages 114–115
  challenges 115–118
  characteristics 109
  compressor 122, 135
  critical issue 109
  cycle time 116
  flow pattern maps 179
  flow patterns 177, 182
  gas choice 109
  gaseated mud 15, 110
  homogenous fluid system 119
  incremental cost 114
  instability 116
  limits 123
  operations 12
  pressure surges 134
  see also multiphase flow
gas-flow meter 502
gasified fluid 475
gas-liquid injection ratio (GLR) 182, 266, 274, 278, 600

gas-liquid mixtures, *see* gaseated fluid
gas-oil ratio (GOR) 435
general gas law 11, 112, 442
general planning procedure 124
geological sample catching 514
geo-pressured shales 4
geothermal drilling 110
  see also aerated systems
Griffith correlation 187, 192
gross heating value, *see* high heating value (HHV)
ground flares 540, 561
  air assisted ground flare system 564
  multi-flare burners 562

H

Hagedorn and Brown correlation 175, 183–187
Hagen-Poiseuille equation 272
half-life (HL) 203, 220, 234, 239, 242, 246
  test 220, 222, 235
hammer drilling 323
hazard and operability (HAZOP) 71, 469, 471, 496
hazard identification (HAZID) 71, 469, 471, 496
hazardous area drawing (HAD) 469
hazards and effects management process (HEMP) 470
heat flux 562, 566
heavy weight drillpipe (HWDP) 476
Herschel-Bulkley model 87, 101, 260
high friction pressure 112
high heating value (HHV) 543
high pressure high temperature well (HPHT) 75
high-pressure well 507
hole cleaning 402, 415
homogeneous flow model 341
horizontal drilling 303
horizontal wells 46, 47, 48, 113, 158, 159, 381
HSE Management System 470
  see also hazard and operability (HAZOP)
  see also hazard identification (HAZID)
hydraulic modeling 181, 468, 495
hydraulic pressures 405
hydraulic work over (HWO) 349
  unit 69, 70
  see also snubbing
hydraulically-operated isolation tool 70
hydrocarbon bearing zone 317
hydrocarbon recovery 59
hydrocarbon sales, 45
hydrostatic pressure 98, 119, 160, 176, 181, 206, 271, 281, 285, 374, 387, 522
hydrostatic reduction 111
hydrostatic regime 111, 215

I

ideal gas law 265, 266, 274
ignition 555
  ignition system safety 556
  ignition transformer 556
  maintenance 557
  mixing igniter tube 556
  remote front flame generator 556
  see also flare system
inflatable plug (PIP) 161
Infra-Red Fourier Transform spectra 252
initial circulating pressure (ICP) 18, 19
injection annulus pressure 151
in-situ slurry density 288

J

jack work basket 358
Jain equation 270
jet subs 110, 127, 145, 146, 243
  string float 129
Joule-Thomson effect 274

K

kelly hose 247
Kemtron 516
kick size 18, 19
kinetic energy criterion 335
knock out drums (K.O.) 549, 560, 574
  horizontal drums 561
  liquid drop-out velocity 572
  vertical drums 561

L

liner drilling 2
liquid drilling fluid 305
liquid flow
  rate 341
liquid fraction 134
liquid injection 124
  limit 124
  rate 275, 335, 346
  rates 337
liquid measurements 513
liquid slug, *see* gas slug
liquid system 122, 245
liquid volume 214
Lisburne field 434
logging while drilling (LWD) 35, 45, 488
lost circulation material (LCM) 383
low heat value (LHV) 567
low heating value (LHV) 545, 548
low pressure AUTOCHOKE console (LPAC) 499
low-pressure well 506
lubrication 359
  down-hole barrier lubrication 364
  in-line lubrication 361
  pressure deployment lubricator 425
  requirements 364
  standard lubrication 360
  wireline assist lubrication 362
lubricator isolation valve 363

M

make up water 609
  black production water 609
  fracwater 610
  hard water 610
  high bicarbonate ion 610
  red production water 609
  saline 610
  waste water 610
managed pressure drilling (MPD) 3, 5, 7, 42, 115, 176, 297, 379, 498
  balanced pressure drilling 3
  concepts 1
  operations 52, 127, 152
maximum annular surface pressure (MASP) 417
measurement while drilling (MWD) 118, 476, 488, 530
mechanical energy balance 191
mechanically operated isolation valve 70
mechanistic models 174, 175, 193, 338
  development 175
membrane nitrogen
  generation units 504
membrane nitrogen system 482
  air separation 482
  gas separation 483
  pushbutton controls 483
mist drilling 198, 201, 297, 300
  anti-corrosion agents 300
  see also misting system
mist flow 346
misting system 205, 245, 306, 307, 323, 330, 335, 442
  gas quality 328
  mist limit 335
  misting agent 300, 324
  operation 328, 338
  operations 317
  see also foam system
  see also gas system
Moody friction factor 190, 268, 272, 275
motor drilling 206

mud cap drilling (MCD) 41, 46, 371, 376, 381, 390, 393
- drilling problems 377
- floating mud cap 373
- hole cleaning 391
- kill mud 372, 376
- operations 374
- pressure regime 393
- pressured mud cap 374
- static cap 373
- zonal isolation 395

mud caps 11, 25–27, 36, 38, 53, 55
- pure oil mud caps 27, 28
- water-based mud caps 27, 28
- *see also* mud cap drilling (MCD)

mud density 12, 15, 25, 37, 49, 372, 373

mud gas separator (MGS) 500

mud logging information 45

mud pulse 33
- tools 42, 169

mud pump 131, 247
- output 131

mud ring 300, 307, 335, 346, 444
- formation 337

mud system 205, 206
- contamination 114

mud velocity 7

mud weight 385, 386

muffler 320
- *see also* deduster
- *see also* separator

multiphase flow 174, 188
- equations 188

N

Navier-Stokes equation 174

net heating value, *see* low heating value (LHV)

net present value (NPV) 522, 524

Newtonian behavior 255

Newtonian fluid model 87

Newtonian shear rate 263

Newtonian viscosity 260, 272

Nikuradse equation 269

nitrogen generation units 532

nitrogen membrane system 484
- nitrogen injection 485
- process description 484
- standpipe bleed 485
- Sullair two-stage compressors 485

non-productive time (NPT) 59, 114, 168, 208, 316, 376, 404, 488, 524
- reduction 208

non-return valves (NRV) 129, 357, 464, 474, 532

nozzle 98, 103, 292

O

oil field hammer drill 301–303
- bit flounder 302
- carbide nodes 303
- hammer bit 303

oil flows 133

oil mud 3

oil soluble polymers 249
- organophilic clay 249

oil-based foam drilling fluid (OBFDF) 246

on-site gas compression equipment 432

operator specific safety factor, *see* underbalanced liner drilling (UBLD)

Ostwald-de Waele power law fluid 272

overbalanced drilling (OBD) 3, 488

P

parasite tubing string 110, 112, 129, 131, 145, 530

pay zones 115

PDC bits 43, 115, 207, 401

pipe handling system 357, 368
- operations 357

pipe heavy 27, 352, 366
- heavy mode 352

pipe light 11, 27, 137, 352

plug flow behavior 255

pore pressure (PP) 496
- *see also* formation pressure

potassium hydroxide (KOH) 324
power law fluids 261
power law model 85, 274, 284
power water injection wells (PWI) 59, 60, 65, 70
pressure adjustment 21
pressure fluids management system (PFMS) 502
  pressure control 502
  solids control 503
pressure profiles 299, 380
pressure propagation 21
pressure pulse, *see* pressure transient lag time
pressure transient lag time 21
  pressure pulse 21
pressure while drilling (PWD) 476, 496
pressured mud cap drilling (PMCD) 381, 382–386, 390, 392
  injectivity test 383
  static leakoff pressure 385
pressurized mud cap (PMC) 49
process and instrumentation diagram (P&ID) 495
process flow diagram (PFD) 469, 507
programmable logic controller system (PLC) 478, 499
pseudo productivity index (PI) 488
psi, *see* gage pressure (psi)
psia, *see* absolute pressure (psia)
pull out of the hole (POOH) 359
pulling out of the hole (POH) 24
pump pressure 125

R

Rabinowitch-Mooney equation 262
Rabinowitsch's theory 256
rate of penetration (ROP) 59, 66, 71, 115, 401, 402, 467
red beds 40
reservoir analysis 56
Reynolds Law 74
Reynolds number 79, 97, 99, 102, 192, 258, 268, 275, 280, 287, 342, 343, 344
  critical 80, 97, 99, 102
rheology 79, 92
  rheological behavior 82
  rheological models 82, 94
  rheological properties 90
rotary air drilling 301
rotary drilling tools 2
  oil booms 2
  rotary drill bit 2
rotating control device (RCD) 15, 22, 27, 37, 47, 48, 52, 53, 71, 136, 198, 218, 247, 311, 317, 328, 366, 372, 374, 382, 387, 394, 395, 464, 475, 477, 498, 532
  active 477
  packer 25
  passive 477
run in the hole (RIH) 359

S

Sayala crude 167
Scandpower 16
semi-static annular fluid column 371
separator 199, 200, 320, 352, 464, 477
  first-stage separator 480
  flush liquids 482
  four-phase separator skid 516
  geologic sampler skid 480
  make-up gas 482
  secondary flow line 479
  second-stage separator 481
  surface separation package 478
  system protection 482
  three-phase separation system 482
  see also deduster
  see also muffler
shallow wells 21
shear rate 81, 86, 102
shear stress 81, 86, 343
  equation 342
shear stress-shear rate equation 256
shut-in casing pressure (SICP) 18
shut-in drill pipe pressure (SIDP) 18

simple step-wise transition method 53
single-phase fluids 16, 39, 42, 44, 48, 53, 82, 450, 528
  Newtonian fluids 83
  Non-Newtonian fluid 83
single-phase gas hydraulic simulator 148
single-phase systems 527
slip velocity 206
slippage 217
slow rate circulating pressure (SRCP) 19
slug flow 300, 323, 328, 338, 346, 348
smoke suppression systems, *see* flare system
snubbing 22, 27, 55, 69, 117, 137, 218, 349, 352, 355, 365, 367, 474
  balance point 352
  basic unit 353–354
  components 353
  hydraulic fluid 350
  hydraulic system 350
  operation 364
  pressure control equipment 349
  rig assist 349
  wireline operations 364
  see also pipe handling
solid expandable tubing (SET) 32
solids 203
  control equipment 126, 211
spare tanks 352
standpipe gauge 48
  reaction 21
standpipe pressure 21, 55, 67, 95, 125, 323, 325
static leakoff pressure, *see* pressured mud cap drilling (PMCD)
static loss rate 386
steam injection 541
sticking 405
stoichiometric air required (SAR) 545
stoichiometric combustion 543
  complete combustion 546
  stoichiometric air 543
  stoichiometric condition 546
Stoke's law 256
stray 306
stripper rubber 366
stripping 11, 22, 27, 117, 136, 227, 352, 366, 367
  arrangement 355
  operations 22
  technique 367
  see also annular blow out preventer
strong-white equation, *See* gas cutting
subbing unit 170
Sullair two-stage compressors, *see* nitrogen membrane system
surface backpressure, *see* drill cuttings
surface data logging (SDL) 488
surface equipment package 474
  equipment setups 475
surface read out gauges (SRO) 75
swabbing effect 52
synthetic oil 123
synthetic oil based mud (SBM) 378

T

thixiotropic drilling fluids 122
through-tubing whipstock systems (TTRD) 432
torsional limit, *see* underbalanced liner drilling (UBLD)
total flow area (TFA) 151, 152
total gas containment system (TOGA) 505
transport velocity 288
tripping 53, 170, 218, 373, 375, 386, 389, 392
tubing-conveyed perforating guns (TCP) 349, 363
tubulars 352
two-phase drilling fluid 42
two-phase flow 12, 15, 16
two-phase systems 42

U

underbalanced coiled tubing drilling (UBCTD) 427, 434
  API 16ST 417
  bulkhead 420
  case histories 432–437
  challenges 428
  coiled tubing tractors 432
  CT reel 420
  CT rigs 424
  CT strings 417
  e-line 420
  engineering studies 436
  gooseneck 420
  limits 415, 432
  operations 424, 427
  pipe management 427
  preplanning 415
  service units 421, 424
  techniques 417
  tubing capacity 420
  yield strength 419
underbalanced drilling (UBD) 1, 4, 5, 7, 19, 29, 39, 59, 110, 114, 145, 153, 160, 166, 168, 176, 245, 297, 390, 399, 423, 522
  applications 145
  benefits 245
  completion technique 32, 48, 163, 164, 349
  conditions 44, 207
  decision trees 493
  drawdown tests 115
  engineering process 495
  equipment 6, 493
  fluids 44
  formation damage 30
  formation pressure 3, 5, 6, 36, 37, 41, 43, 44, 45
  gas-assist UBD 390
  injectivity 66
  invert oil emulsion muds 42
  key goal 495
  operations 1, 3, 11, 15, 19, 45, 64, 70, 117, 152, 179, 425, 464, 466, 470, 488
  performance improvement 62
  productivity tests 44
  pump rate 15, 48
  reservoir flow 46
  residual stress 45
  solid-separation equipment 30
  solution 5
  techniques 2, 11
  technology 64, 71, 155, 468, 508
  testing 496
  see also cable tool drilling
  see also foam system
  see also gas system
  see also gaseated system
  see also managed pressure drilling (MPD)
  see also misting system
  see also mud cap drilling (MCD)
underbalanced drilling gases 441
  air 443
    availability 443
    cost 443
    presence of oxygen 444–445
  carbon dioxide 456
    carbonic acid 458
    solubility 458
    special properties 456
    toxic 458
  cryogenic nitrogen 453
    boil off 455
    corrosion 454
    cost 454
    flare system 455
    injection pressures 454
  membrane nitrogen 451
    corrosion problem 453
    cost 453
    flare system 453
    oil continuous-phase fluids 453
  natural gas 446
    corrosion 446
    cost 447
    gas detectors 447
    methane 446, 448
    pressured lease line 451
underbalanced liner drilling (UBLD) 399, 404, 406, 407
  application 400

casing bit 401
connection fatigue life 403
cyclic bending stress 403
fluid systems 408
gas tight 404
hydraulic design 402
limitations 407
limiting factor 407
liner drilling bit assembly 400
liner tools 408
operator specific safety factor 403
special equipment 409
stabilization 407
torsional limit 402
well control considerations 408
wellbore ballooning 405
underbalanced string floats 476

V

valve numbering diagram (VND) 469
variable frequency drive (VFD) 554, 564
velocity criterion 336
viscosified seawater (VSW)
see also pressured mud cap drilling (PMCD)
viscosity 81
hysteresis 94
poise 81
volume equalized power law model 272

W

Wallis correlation 341
washed down 308
washout 308
water flows 133
weight on bit (WOB) 115, 432
well control 11
equipment 422
operations 11, 19, 213
principles 17
purpose 356
safety 355
tertiary system 357
wait and weight method 19
well kick 2, 47, 50, 51, 325
flow check 325
solution 326
well productivity index 161
wellbore 18, 21, 22, 25, 53, 59, 109, 174, 206, 349
ballooning, *see* underbalanced liner drilling (UBLD)
cavings 5
cuttings 5
fluids 4, 24, 43
skin damage 44
gas bubble 11
geometry 274
horizontal wellbores 50
hydrostatic pressure 65, 111
instability 4–5, 50, 116, 309
pressure 3, 4, 5, 6, 15, 16, 36, 49, 50, 90, 111, 112, 132, 136, 183, 207, 297, 298, 305, 352, 522
regions 338
stability 2, 5, 37, 45, 160
strengthening 1, 8
wellhead 11, 13, 130
wellpath 64
wetted perimeter 112, 118
wide area network system (WAN) 488
wired wrapped screen (WWS) 30, 160
wireline operations, *see* snubbing
work-string 359, 364
tool joints 367

Y

yield stress 203, 259

Z

zonal isolation, *see* mud cap drilling (MCD)

FSC
www.fsc.org
MIX
Papier | Fördert
gute Waldnutzung
FSC® C083411

Kolibri 360 GmbH
Ferdinand-Jühlke-Straße 7
99095 Erfurt, Deutschland
produktsicherheit@kolibri360.de